WITHDRAWN

Results and Problems in Cell Differentiation

48

Series Editors
Dietmar Richter, Henri Tiedge

Edward Koenig (ed.)

Cell Biology
of the Axon

Editor
Edward Koenig
70 Summer Hill Lane
Williamsville NY 14221
USA
ekoenig@buffalo.edu

Series Editors
Dietmar Richter
Center for Molecular Neurobiology
University Medical Center Hamburg-
Eppendorf (UKE)
University of Hamburg
Martinistrasse 52
20246 Hamburg
Germany
richter@uke.uni-hamburg.de

Henri Tiedge
The Robert F. Furchgott Center for Neural and
Behavioral Science
Department of Physiology and Pharmacology
Department of Neurology
SUNY Health Science Center at Brooklyn
Brooklyn, New York 11203
USA
htiedge@downstate.edu

ISBN 978-3-642-03018-5   e-ISBN 978-3-642-03019-2
DOI 10.1007/978-3-642-03019-2
Springer Heidelberg Dordrecht London New York

Results and Problems in Cell Differentiation ISSN 0080-1844

Library of Congress Control Number: 2009932721

© Springer-Verlag Berlin Heidelberg 2009
This work is subject to copyright. All rights are reserved, whether the whole or part of the material is concerned, specifically the rights of translation, reprinting, reuse of illustrations, recitation, broadcasting, reproduction on microfilm or in any other way, and storage in data banks. Duplication of this publication or parts thereof is permitted only under the provisions of the German Copyright Law of September 9, 1965, in its current version, and permission for use must always be obtained from Springer. Violations are liable for prosecution under the German Copyright Law.
The use of registered names, trademarks, etc. in this publication does not imply, even in the absence of a specific statement, that such names are exempt from the relevant protective laws and regulations and therefore free for general use.

Cover design: WMXDesign GmbH, Heidelberg

Printed on acid-free paper

Springer is part of Springer Science+Business Media (www.springer.com)

Contents

Myelination and Regional Domain Differentiation of the Axon 1
Courtney Thaxton and Manzoor A. Bhat

**Organizational Dynamics, Functions, and
Pathobiological Dysfunctions of Neurofilaments** 29
Thomas B. Shea, Walter K.-H. Chan, Jacob Kushkuley,
and Sangmook Lee

**Critical Roles for Microtubules in Axonal
Development and Disease** 47
Aditi Falnikar and Peter W. Baas

Actin in Axons: Stable Scaffolds and Dynamic Filaments 65
Paul C. Letourneau

**Myosin Motor Proteins in the Cell Biology
of Axons and Other Neuronal Compartments** 91
Paul C. Bridgman

Mitochondrial Transport Dynamics in Axons and Dendrites 107
Konrad E. Zinsmaier, Milos Babic, and Gary J. Russo

**NGF Uptake and Retrograde Signaling Mechanisms in
Sympathetic Neurons in Compartmented Cultures** 141
Robert B. Campenot

The Paradoxical Cell Biology of α-Synuclein 159
Subhojit Roy

Organized Ribosome-Containing Structural Domains in Axons 173
Edward Koenig

Regulation of mRNA Transport and Translation in Axons................ 193
Deepika Vuppalanchi, Dianna E. Willis, and Jeffery L. Twiss

**Axonal Protein Synthesis and the Regulation
of Local Mitochondrial Function** 225
Barry B. Kaplan, Anthony E. Gioio, Mi Hillefors,
and Armaz Aschrafi

**Protein Synthesis in Nerve Terminals
and the Glia–Neuron Unit**... 243
Marianna Crispino, Carolina Cefaliello, Barry Kaplan,
and Antonio Giuditta

Local Translation and mRNA Trafficking in Axon Pathfinding......... 269
Byung C. Yoon, Krishna H. Zivraj, and Christine E. Holt

**Spinal Muscular Atrophy and a Model for Survival
of Motor Neuron Protein Function in Axonal
Ribonucleoprotein Complexes** 289
Wilfried Rossoll and Gary J. Bassell

Retrograde Injury Signaling in Lesioned Axons 327
Keren Ben-Yaakov and Mike Fainzilber

**Axon Regeneration in the Peripheral
and Central Nervous Systems**.. 339
Eric A. Huebner and Stephen M. Strittmatter

Index... 353

Introduction

Prospective and Retrospective on Cell Biology of the Axon

Axons from projection macroneurons are elaborated early during neurogenesis and comprise the "hard wired" neuroanatomic pathways of the nervous system. They have been the subjects of countless studies from the time that systematic research of the nervous system had its beginnings in the 19th century. Microneurons (i.e., interneurons), which are generated in greater numbers later during neurogenesis, and form local neuronal circuits within functional centers, produce short axons that have not been studied directly, notwithstanding the fact that their sheared-off terminals probably contribute substantially to the heterogeneity of brain synaptosome fractions. Strictly speaking, therefore, for purposes of this volume, axons from projection neurons serve as the principal frame of reference.

In many instances, the mass of a projection neuron's axon can dwarf the mass of the cell of origin. This consideration, among others, has historically posed questions about the biology of the axon, not the least vexing of which have centered on the basis of axonal growth and steady state maintenance. A simple view has long prevailed until recently, in which the axon was regarded as to have essentially no intrinsic capacity to synthesize proteins. By default, structural and metabolic needs were assumed to be effectively satisfied by constant bidirectional trafficking between the cell body and the axon of organelles, cytoskeletal polymers, and requisite proteins. From this general premise, it was assumed that directed growth of axons in response to guidance cues during development was also governed solely by the cell body. Such a restricted view has been discredited in recent years by a significant body of research that has revealed a considerable complexity governing the local expression *within* axons, which has rendered the traditional conceptual model anachronistic. Many distinctive features and recent research developments that characterize the newfound complexity of the cell biology of axons – a complexity that has clear implications for pathobiology – are reviewed and discussed in the present volume, briefly highlighted as follows.

The first chapter by Thaxton and Bhat reviews the current understanding of signaling interactions and mechanisms that underlie myelination, while also governing differentiation of regional axonal domains, and further discusses domain disorganization in the context of demyelinating diseases.

The following three chapters focus on endogenous cytoskeletal systems that structurally organize the axon, confer tensile strength, and mediate intracellular transport and growth cone motility. Specifically, Shea et al. address issues of how organizational dynamics of neurofilaments are regulated, including mechanisms of transport, and how dysregulation of transport can contribute to motor neuron disease. Fainikar and Baas focus on organizational and functional roles of the microtubule array in axons and further consider mechanisms that regulate microtubule assembly and disassembly, which, when impaired, predispose axons to degenerate. Letourneau then reviews the characteristics of the actin cytoskeleton, including its organization and functions in mature and growing axons, regulated by actin-binding proteins, and the roles the latter play in transport processes and growth dynamics.

The next set of four chapters deals with selected aspects of intracellular transport systems in axons. Thus, Bridgman identifies several classes of myosin motor proteins intrinsic to the axon compartment and discusses their principal roles in the transport of specific types of cargoes, and in potential dynamic and static tethering functions related to vesicular and translational machinery components, respectively. Zinsmaier et al. review mitochondrial transport and relevant motor proteins, discussing functional imperatives and mechanisms that govern mitochondrial transport dynamics and directional delivery to specifically targeted sites. The following chapter about NGF transport by Campenot provides a critical discussion of mechanisms that mediate retrograde signaling associated with NGF's role in trophic-dependent neuronal survival. In the last chapter of this series, Roy discusses potential impairment of transport and/or subcellular targeting of α-synuclein that may account for accumulations of Lewy body inclusions in a number of neurodegenerative diseases characterized as synucleinopathies.

The succeeding series of five chapters center on historically controversial areas related to axonal protein synthesizing machinery and various aspects of how local expression of proteins are regulated in axons. The lead-off chapter by Koenig describes the occurrence and organizational attributes of discrete ribosome-containing domains that are identified in the cortex as intermittently spaced plaque-like structures in myelinated axons, and, while absent as such in the unmyelinated squid giant axon, appear as occasional discrete ribosomal structural aggregates within axoplasm. Next, Vuppalanchi et al. present an in-depth review of endogenous mRNAs, classes of proteins translated locally, and discussion of the intriguing and rapidly expanding area of ribonucleoprotein (RNP) trafficking in axons. This is followed with a chapter by Kaplan et al. which provides insight into the importance that local synthesis of nuclear encoded mitochondrial proteins plays in mitochondrial function and maintenance, as well as axon survival. In the following chapter by Crispino et al., evidence is reviewed that supports the occurrence of transcellular trafficking of RNA from glial cells to axons and further discusses the significance that glial RNA transcripts may play in contributing to local expression of proteins in the axon and axon terminals. A chapter by Yoon et al. examines RNA trafficking and localization of transcripts in growth cones and reviews the evidence that extracellular cues modulate directional elongation associated with axonal pathfinding through signaling pathways that regulate local synthesis of proteins. The final chapter of this set by Rossoll and Bassell addresses key genetic and molecular defects that underlie spinal muscular atrophy, a degenerative condition that especially affects α-motoneurons, and the roles the unaffected SMN gene product plays as a molecular chaperone involved in mRNA transport and translation in axons.

The final two chapters deal with neural responses to axon injury. Ben-Yaakov and Fainzilber review and discuss current understanding about how a local reaction to injury in axons triggers local protein synthesis of a protein that forms a signaling complex, which is then conveyed from the lesion site to the cell body to initiate regeneration. Lastly, Huebner and Strittmatter provide a review and discussion of recent developments in the current understanding of endogenous and exogenous factors that condition axonal regenerative capacity in the peripheral and central nervous systems and identify injury-induced activation of specific genes that govern regenerative activities.

Along with a cursory prospective of the current volume, it seems only fitting to highlight some of the early key antecedents that have led to recent developments in the field. The retrospective begins with selected neurohistologists of yesteryear, who initially established a cellular orientation in the context of nervous system organization and also framed significant issues related to axonal biology in the idiosyncratic language of the late 19th century. Although eclipsed after the turn of the century, the same issues reemerged many years later, when they were reframed in terms of contemporary cell biology. Also given some deserved consideration is the role large-sized axon models played to help advance early investigative efforts at a cellular level.

In his exhaustively documented to me, entitled ***The Nervous System and its Constituent Neurones*** (1899), Lewellys Barker credits Otto Deiters' descriptions of carefully hand-dissected nerve cells from animal and human brains and spinal cords, published posthumously in 1865, with identifying the distinctive characteristics of the "axone" among multiple neuronal processes. He observed that the *"...axis-cylinder ... consist(s) of a rigid hyaline, more resistant substance, which at short distance from its origin in the nerve cell passed directly over into a medullated nerve fibre."* Illustrations based on Deiters' deft manual isolation of nerve cells were informative and insightful, but there were fundamentally different concepts competing for acceptance at the time about the underlying functional organization of the nervous system, one of which centered on the notion of a continuous reticular fibrillar network. Conclusive evidence that firmly established the "neurone doctrine" as the basis was ultimately achieved in the last decade of the century, in which the Golgi silver impregnation method to stain neural tissue so aptly employed by Ramón y Cajal in his classical studies, played a key role. Deiters, nonetheless, focused attention on two important axonal features of a major class of projection neurons; namely, mechanical tensile strength, and the myelin sheath investment.

The contemporaneous classical degeneration studies performed on peripheral sensory and motor spinal nerve root fibers by Waller in 1850, and on CNS pyramidal track fibers by Türck in 1852 set the stage for research developments in cellular neurobiology for many years to come. The results made it clear that axons were dependent on cell bodies for structural integrity and viability, which gave rise to the concept of cell bodies as indispensible "trophic" centers. The overriding issue thereby became: How does the cell body actually perform its trophic function?

In attempting to address the conundrum of trophic influence at the turn of the 19th century, Barker (1899) posed the following rhetorical question: "**Does the axon actually receive all its nutrient material from the ganglion cell, or does it depend**, as would seem a priori much more likely, for the most part **upon autochthonous metabolism** needing only the influence of the cell to which it is connected to govern assimilation?" Barker then takes note of "... a very ingenious hypothesis" advanced by Goldscheider; namely, "...that it is most probable that there is an actual transport of a material perhaps a *fermentlike substance* [i.e., enzyme] from the cell along the whole course of the axone to its extremity, and that first through the influence of the chemical body the axone is enabled to make use for its nutrition of the material placed at its disposal in its anatomical course."

With these two explanations (see bold print above), Barker, in effect, articulated two potential modes of supplying proteins to the axon compartment well before the two corresponding lines of basic research on "local synthesis" and "axoplasmic transport" took root in the mid-20th century. These research foci and their offshoots over the years have yielded a large body of information about the biology of the axon, although, not without controversies along the way.

The era of axoplasmic transport research was ushered in by Paul Weiss' "nerve damming" experiments in the mid-1940s. Placement of an arterial cuff around a peripheral nerve, whether crushed, uncrushed, or regenerating, produced axoplasmic damming, which resulted in various forms of ballooning, telescoping, coiling, and beading of axons proximal to the compression site. Subsequent release of compression yielded a distal redistribution characterized as a continuous proximo-distal movement of axoplasm at a rate estimated to be about 2 mm/day (Weiss and Hiscoe, 1948). Actually, it was a few years earlier at a Marine Biological Laboratory meeting in Woods Hole that Weiss (1944) first invoked the concept of axoplasmic transport, not only to explain the experimental damming results, but also to suggest it as a general mechanism to account for natural growth and renewal of the axon, which was stated as follows. *"The neuron, as a living cell, is in a state of constant reconstitution. The synthesis of its protoplasm would be confined to the territory near the nucleus (perikaryon). New substance would constantly be added to the nerve processes from their base. The normal fiber caliber permits unimpeded advance of this mass, with central synthesis and peripheral destruction in balance. Any reduction of caliber impedes proximo-distal progress of the column and thus leads to its damming up, coiling, etc."*

Several reports appeared in the literature during the ensuing decade that supported the idea of axoplasmic transport. While studying the systemic uptake of [^{32}P] into cellular constituents of neurons, Samuels et al. (1951) observed movement of radiolabeled phosphoproteins along nerves at a rate of about 3 mm/day. Lubinska (1954) noted two asymmetrical bulbous enlargements juxtaposed to nodes of Ranvier on each side when examining dissected isolated axon segments, in which the larger of the two was invariably located on the central side of a node. Extrapolating from the cuff compression experiments of Weiss, Lubinska inferred that such perinodal asymmetry was probably caused by the natural constriction of the node that would presumably impede proximo-distal movement of axoplasm.

Studies of neurosecretory neurons in vegetative nervous centers also strongly suggested the transport of neurosecretory material from sites of synthesis in cell bodies to sites of secretion in axon terminals (Scharrer and Scharrer, 1954), while microscopic observations of neurons in culture directly revealed bidirectional movements of axonal granules and vesicular structures (Hughes, 1953; Hild, 1954). Later, the first preliminary report of axoplasmic transport of radiolabeled proteins in cats appeared, based on intrathecal injections of [^{14}C]methionine and [^{14}C]glycine, in which 1–3 cm radiolabeled protein "peaks" "moved" along peripheral nerves at rates of 4–5 mm/day, and 7–11 mm/day (Koenig, 1958).

In the next two decades, more than thousand papers on axoplasmic transport appeared (see Grafstein and Forman, 1980). In advance of the vast growth in the transport literature, Goldscheider's hypothesis, positing transport of a "fermentlike substance" from the cell body into the axon, was tested in the case of acetylcholinesterase (AChE), a peripheral membrane enzyme in cholinergic neurons anchored to plasma and cytomembranes. Most AChE in neural and muscle tissues was inhibited irreversibly by alkyl phosphorylation of the active center, using diisopropylflurophosphate (DFP), and recovery of enzymic activity, regarded as an indirect measure of resynthesis, was evaluated along several peripheral nerves and cognate nerve cell centers over time (Koenig and Koelle, 1960). AChE activity reappeared along peripheral nerves and in cell bodies analyzed in manners that were temporally and spatially independent. The findings suggested the likelihood of *local synthesis* in axons as a possible mechanism for enzymic recovery, but did not rule out axoplasmic transport as an alternate, or ancillary mechanism.

In the late 19th century, a basophilic "stainable substance" in nerve cell bodies was revealed by the so-called "method of Nissl" that employed a basic aniline dye to stain nerve cells in neural tissue. The significance of cytoplasmic Nissl substance/Nissl bodies was eventually elucidated with the advent of electron microscopy (EM), when basophilic aggregates were identified as ribosome-studded, rough endoplasmic reticulum (Palay and Palade, 1955). Palay and Palade also noted in their EM survey of the nervous system that while ribosomes were apparently absent from mature axons, they were present in dendrites, and that nerve cell bodies were richly endowed with ribosomes much like gland cells. The long recognized lack of Nissl staining in the axon hillock region (the funnel-like protuberance arising from the perikaryon) and initial segment became recognized as a characteristic hallmark of axons. Moreover, nerve cell bodies were thought to have more than sufficient capacity to synthesize and supply requisite proteins via axoplasmic transport to support growth and maintain mass of an extended axonal process. Nonetheless, negative results based on randomly selected thin sections viewed at an ultrastructural level could not be considered necessarily conclusive. The uncertainty issue made the corollary question of whether axons contained RNA compelling to answer.

During the mid-1950s, RNA distribution was investigated in immature neurons during development of the chick spinal cord (Hughes, 1955), and the guinea pig fetal cerebral cortex (Hughes and Flexner, 1956), using a microscope equipped with quartz optics, and a UV light source in a spectral region selective for RNA absorption. Ultraviolet microscopy revealed that RNA was diffusely distributed within immature axons, but then disappeared just before Nissl bodies formed in cell bodies. Efforts to investigate RNA, or endogenous protein synthesizing activity directly in the *mature* axon, however, was not possible until sensitive quantitative methods of detection and analysis at a cellular level were developed, including the availability of a suitable axon model.

Axon models played important roles in early studies at a cellular level, not only initially to investigate axonal electrophysiology, but also later to analyze axonal RNA, and rheological properties of axoplasm. The squid giant axon was the first experimental model to be employed for the purpose of intracellular electrophysiological recording. Its use by Hodgkin and Huxley (1939) made it possible to document the reversal of membrane polarity during the overshoot of an action potential. The findings refuted the Bernstein theory that prevailed since the turn of the 20th century, which predicted that an action potential would simply cause the negatively polarized membrane to collapse to 0 mV. The landmark experiment entailed inserting an electrode axially into a squid axon that was 500 μm in diameter, dissected from *Loligo pealeii*. Also, importantly, the experiments established for the first time the existence of a functional plasma membrane.

It is especially noteworthy that Hodgkin and Huxley acknowledged the English zoologist, anatomist, and neurobiologist, J.Z. Young, who had discovered the giant axon a few years earlier, for having recommended its use. Later, Young (1945) conjectured that axoplasm is a viscous fluid that is likely to exhibit non-Newtonian flow behavior, a property in which stress force produces nonlinear flow (e.g., initially resisting flow, but then finally yielding to flow at a critical force, with flow accelerating thereafter). Young drew his inference about non-Newtonian flow behavior on the basis of weak form birefringence of axoplasm shown by inspection in polarization microscopy. The form birefringence was attributed to the "neurofibrillar" organization of axoplasm.

The conjecture was later confirmed in Robert Allen's laboratory, at which time a rheological model was formulated as a result of experiments in living squid axon preparations. Specifically, in addition to showing that axoplasm was firmly attached to the plasma membrane and that it could be easily sheared, axoplasm was characterized as a "complex viscoelastic fluid", having an elastic modulus greater in the longitudinal direction than in the radial direction (Sato et al., 1984). Such rheological behavior is consistent with our present understanding of how the three major cytoskeletal systems are organized and interact in the axon; i.e., linearly oriented structural elements, comprising microtubules (e.g., see Fainikar and Baas, this volume) and neurofilaments that exhibit lateral crossbridging (e.g., see chapter by Shea et al., this volume), in addition to a diffuse actin filament network, which in part also forms a dense cortical layer, consisting of a submembraneous F-actin network, essential for membrane stability and anchoring many integral membrane proteins (e.g., see Letourneau, this volume).

The visco-elastic properties of axoplasm made it possible to extrude axoplasm out of a cut end of squid giant fibers, much like expressing toothpaste from a tube. On the other hand, its quasi solid-like properties also made it feasible to translate axoplasm out of its myelin sheath with microtweezers as an "axoplasmic whole-mount" (e.g., see Koenig, this volume). Indeed, the tensile strength and ease with which axoplasm can be translated as a whole-mount are enhanced with an increase in axon diameter due to the corresponding greater neurofilament content.

Allen also greatly advanced in research in the field of axoplasmic transport with the development of *video enhanced differential interference contrast microscopy*, which made it possible to visualize organelles transported along microtubules at a submicroscopic level in the squid giant axon (Allen et al., 1982), and in extruded axoplasm (Brady et al., 1982). The methodological approach was key to the subsequent discovery of kinesin (Vale et al., 1985), the first of a number of microtubule, and F-actin dependent molecular motor proteins later characterized.

The vertebrate's equivalent to the squid giant axon model is the large, heavily myelinated Mauthner axon in goldfish, and in other teleost fishes. While it is not at par with the squid axon with respect to size, it is exceptionally large for a vertebrate axon, in which the axoplasmic core can range from 20 to 90 μm in diameter. The paired axons originate from very large, electrophysiologically identifiable Mauthner nerve cells located in the hindbrain. The rapidly conducting Mauthner axons project the length of the spinal cord, giving off very short collaterals through the myelin sheath (e.g., see Koenig, this volume), which synapse with a neuronal network that triggers a "C-bend" reflex of the trunk musculature to initiate an escape response.

In the late 1950s, Jan-Erik Edström developed ultramicro analytic methods for RNA in the picogram range designed for isolated microscopic samples. The possibility that axons may have an intrinsic capacity to synthesize proteins (Koenig and Koelle, 1960), notwithstanding an apparent lack of ribosomes (Palay and Palade, 1955), prompted Edström et al. (1962) to analyze RNA extracted by ribonuclease digestion of isolated axoplasm micro-dissected from fixed goldfish Mauthner cell fibers. The landmark study, and subsequent analysis of axoplasm from Mauthner (Edström, 1964a; 1964b), and spinal accessory fibers of the cat (Koenig, 1965), documented the occurrence of RNA in adult vertebrate axons, and provided the first quantitative data about RNA content and nucleotide base composition. More than a decade elapsed before ribosomal RNA (rRNA) was demonstrated in axoplasm isolated from Mauthner fibers (Koenig, 1979), and unmyelinated squid giant fibers (Giuditta et al., 1980). Eventually, a systematic cortical distribution of novel ribosome-containing structural domains was revealed in isolated vertebrate axoplasmic whole-mounts (Koenig and Martin, 1996; Koenig et al., 2000), while in the squid giant axon, ribosomes were observed to be clustered in randomly distributed structural aggregates within the core of axoplasm (Martin et al., 1998; Bleher and Martin, 2001) (e.g., see Koenig, this volume).

Evidence of local protein synthesis and translational machinery in axons has long been held captive by the sway of the deeply ingrained view in neurobiology that axoplasmic transport is the sole source of all axonal proteins. Such a view was promulgated in early literature, and, later, reinforced periodically by dogmatic assertions in reviews of axoplasmic transport, illustrated, for example, by the following: *"Remarkably, synthesis of all axonal proteins is restricted to a cell body tens of micrometres in diameter. Every protein has to be transported from where it is made to where it is needed"* (Brady, 2000).

To the contrary, a newfound complexity, which is recognized and documented in the present volume, controverts long held shibboleths regarding the single mode of expression in axons. There are now clearly strong reasons for adopting a balanced, broadly based view about gene expression vis-à-vis axons of projection neurons. Intracellular transport systems not only deliver proteins to axons directly, but also deliver and localize mRNA transcripts for translation as an integral part of RNP-dependent RNA trafficking from the soma. In addition, a third source of gene products potentially reaches the axon via a local transcellular route from adjacent ensheathing cells. Differences in contributions of gene products to the axonal compartment from each of these potential sources will likely vary, depending on the specific neuronal phenotype. Differences among the three potential sources are also likely to depend upon the state of the neuron; i.e., during the growth of immature axons, during steady state maintenance and functional activity of mature axons, and during the reaction of axons to injury. Sorting out the relative importance of each source in the various exigency states of the neuron, as well as analyzing the roles that transport systems play on the supply side would not only deepen our understanding of the normal biology of the axon, but should also offer insight into the potential for pathobiological dysfunctions. At this juncture, these quests must be left for future "axonologists" to pursue.

Williamsville, NY, Edward Koenig
May 2009

References

Allen RD, Metuzals J, Tasaki I, Brady ST, Gilbert SP (1982) Fast axonal transport in the squid giant axon. Science 218:1127–1128
Barker L (1899) The Nervous System. Appelton, New York, pp. 1122
Bleher R, Martin R (2001) Ribosomes in the squid giant axon. Neurosci 107:527–E534
Brady ST, Lasek RJ, Allen RD (1982) Fast axoplasmic transport in extruded axoplasm from squid giant axon. Science 218:1129–1131
Brady ST (2000) Neurofilaments run sprints not marathons. Nat Cell Biol 2:E43–E45
Edström A (1964a) The ribonucleic acid in the Mauthner neuron of the goldfish. J Neurochem 11:309–314
Edström A (1964b) Effect of spinal cord transection on the base composition and content of RNA in the Mauthner nerve fibre of the goldfish. J Neurochem 11:557–559
Edström J-E, Eichner D, Edström A (1962) The ribonucleic acid of axons and myelin sheaths from Mauthner neurons. Biochim Biophys Acta 61:178–184
Hild H (1954) Das morphologische, kinetische und endokrinologische Vehalten von hypothalamischen und neurohypophyseren Gewebe in vitro. Z Zellforsch 40:257–312
Giuditta A, Cupello A, Lazzarini, G (1980) Ribosomal RNA in the axoplasm of the squid giant axon. J Neurochem 34:1757–1760
Graftstein B, Forman D (1980) Intracellular transport in neurons. Physiol Rev 60:1167–1283
Hodgkin AL, Huxley AF (1939) Action potentials recorded inside a nerve fibre. Nature 144:710–711
Hughes A (1953) The growth of embryonic neurites. A study on cultures of chick neural tissues. J Anat 87:15–162

Hughes A (1955) Ultraviolet studies on the developing nervous system of the chick. In: Waelsch H (ed) Biochemistry of the developing nervous system. Academic, New York, pp 166–169

Hughes A, Flexner LB (1956) A study of the development of cerebral cortex of foetal guinea pig by means of the ultraviolet microscope. J Anat 90:386–394

Koenig E, Koelle GB (1960) Acetylcholinesterase regeneration in peripheral nerve after irreversible inactivation. Science 132:1249–1250

Koenig E (1965) Synthetic mechanisms in the axon. Part II: RNA in myelin-free axons of the cat. J Neurochem 12:357–361

Koenig E (1979) Ribosomal RNA in Mauthner axon: Implications for a protein synthesizing machinery in the myelinated axon. Brain Res 174:95–107

Koenig E, Martin R (1996) Cortical plaque-like structures identify ribosome-containing domains in the Mauthner cell axon. J Neurosci 16:1400–1411

Koenig E, Martin R, Titmus M, Sotelo-Silveira JR (2000) Cryptic peripheral ribosomal domains distributed intermittently along mammalian myelinated axons. J Neurosci 20:8390–8400

Koenig H (1958) The synthesis and peripheral flow of axoplasm. Trans Amer Neurol Assoc 83:162–164

Lubinska L (1954) Form of myelinated nerve fibers. Nature 173:867–869

Martin R, Vaida B, Bleher R, Crispino M, Giuditta A (1998) Protein synthesizing units in presynaptic and postsynaptic domains of squid neurons. J Cell Sci 111:3157–3166

Palay SL, Palade, GE (1955) The fine structure of neurons. J Biophys Biochem Cytol 1:69–88

Samuels AJ, Boyarsky LI, Gerard RW, Libet B, Brust M (1951) Distribution, exchange and migration of phosphate compounds in the neurons. Am J Physiol 164:1–15

Sato M, Wong TZ, Brown DT, Allen RD (1984) Rheological properties of living cytoplasm: A preliminary investigation of squid axoplasm (Loligo pealei). Cell Motil 4:7–23

Scharrer E, Scharrer B (1954) Neuroseketion. In: von Moellendorff W, Bargmann W (eds) Handbuch der Mikroskopischen Anatomie des Menschen. Bd VI/5, Springer, Berlin, pp 953–1066

Young JZ (1945) Structure, degeneration and repair of nerve fibres. Nature 156:152–156

Vale RD, Reese TS, Sheetz MP (1985) Identification of a novel force-generating protein, kinesin, involved in microtubule-based motility. Cell 42:39–50

Weiss P (1944) Damming of axoplasm in constricted nerve: a sign of perpetual growth in nerve fibers? Biol Rec 87:160

Weiss P, Hiscoe HB (1948) Experiments on the mechanism of nerve growth. J Exp Zool 107:315–393

Myelination and Regional Domain Differentiation of the Axon

Courtney Thaxton and Manzoor A. Bhat

Abstract During evolution, as organisms increased in complexity and function, the need for the ensheathment and insulation of axons by glia became vital for faster conductance of action potentials in nerves. Myelination, as the process is termed, facilitates the formation of discrete domains within the axolemma that are enriched in ion channels, and macromolecular complexes consisting of cell adhesion molecules and cytoskeletal regulators. While it is known that glia play a substantial role in the coordination and organization of these domains, the mechanisms involved and signals transduced between the axon and glia, as well as the proteins regulating axo–glial junction formation remain elusive. Emerging evidence has shed light on the processes regulating myelination and domain differentiation, and key molecules have been identified that are required for their assembly and maintenance. This review highlights these recent findings, and relates their significance to domain disorganization as seen in several demyelinating disorders and other neuropathies.

1 Introduction

One of the most critical processes of both the central and peripheral nervous systems is myelination, involving the ensheathment and insulation of axons by glial cell membranes. As the glial cells, comprising Schwann cells in the peripheral nervous system (PNS) and oligodendrocytes in the central nervous system (CNS), contact and continually wrap their membranes around axons, they create polarized domains (Bhat 2003; Salzer 2003). These domains include the node, paranode, juxtaparanode, and internode.

C. Thaxton and M.A. Bhat (✉)
Department of Cell and Molecular Physiology, Curriculum in Neurobiology, UNC-Neuroscience Center and Neurodevelopmental Disorders Research Center, University of North Carolina School of Medicine, Chapel Hill, NC, 27599-7545, USA
e-mail: Manzoor_Bhat@med.unc.edu

During development, axo–glial interactions mediate the restriction of ion channels into specific membrane domains; that is, potassium channels in the juxtaparanode and sodium channels in the node of Ranvier, which, in turn, allow for the rapid and efficient conduction of the nerve impulse. The exact mechanisms governing the segmentation of ion channels into specific domains is elusive, but evidence has shown that disruption of paranodal axo–glial junctions leads to severe impairments of saltatory conduction, motor coordination, and myelination (Bhat 2003; Salzer 2003; Salzer et al. 2008). These phenotypes are often seen in demyelinating disorders and other neuropathies, which exemplify the importance of axo–glial junctions to the steady state kinetics of the action potential and proper nervous system functioning. Recent findings have identified several critical molecules and signaling pathways mediating the formation of axo–glial junctions and the regional organization of ion channel domains in the axonal membrane. In this review, new advancements in our knowledge of myelination and the differentiation of four domains in myelinated fibers will be highlighted, in addition to discussing the mechanisms regulating their formation, maintenance during normal functioning, and disease onset and progression.

2 Myelination of Axons

Myelination is a process whereby specialized cells of the nervous system, termed glia, elaborate double membrane wrappings around axons, creating an insulating layer that promotes the fast conduction of nerve impulses. The many wrappings effectively increase total membrane resistance and decrease total membrane capacitance between nodes of Ranvier, which greatly reduces "leakage" of current across the internodal membrane. The "sparing" of axoplasmic current, in combination with very fast internodal electrotonic conduction, rapidly depolarize the downstream nodal membrane to threshold.

While myelination is required in both the PNS and CNS, there are distinct differences between the myelin forming cells with respect to the proteins and the signals required for myelination in these two systems. In the PNS, as Schwann cells differentiate, they will assume one of the two fates: they will either (1) form a 1:1 relationship with an axon and myelinate it or (2) extend multiple processes that will ensheath several axons (Jessen and Mirsky 2005). Oligodendrocytes, on the other hand, extend multiple processes that will contact and myelinate several axons, up to forty separate axons at a time (Simons and Trotter 2007). While there are several factors that affect glial cell differentiation, such as growth factors and the extracellular matrix (ECM), the most notable is the axon phenotype, which determines its diameter. Those axons greater than 1 μm in diameter will be myelinated; whereas those smaller than 1 μm will be ensheathed. Interestingly, axonal diameter also determines the length of the internode, the segment of myelin between two nodes, as well as the thickness of the myelin layer(s), but the exact mechanisms governing the detection of axonal thickness by Schwann cells and oligodendrocytes remains elusive.

Premyelinating Schwann cells are distinctly bipolar, with processes extending longitudinally along the length of an axon. This extension will ultimately determine the location of the nodes of Ranvier and the internodal length. Once the internode is defined, signals from the axon and the ECM induce Schwann cells to extend their membrane laterally and spiral inwardly around the axon. The continuous wrapping of the Schwann cell membrane facilitates the development of the adaxonal (i.e., adjacent to the axon) and abaxonal (i.e., abutting the ECM) membrane layers. On the abaxonal side, Schwann cells are surrounded by a specialized ECM known as the basal lamina. The basal lamina is unique to the PNS and is formed by the Schwann cells to assist with their maturation and differentiation into a myelinating phenotype (Chernousov and Carey 2000; Court et al. 2006). Another unique feature of peripheral Schwann cells is the formation of nodal microvilli. These structures are small protrusions that extend beyond the distal-most paranodal loop and contact the underlying node. These structures are believed to participate in the formation of the node and mediate communication between the axonal node and the adjacent Schwann cell (Gatto et al. 2003; Ichimura and Ellisman 1991; Melendez-Vasquez et al. 2001).

Oligodendrocytes, unlike Schwann cells, are multipolar cells that have numerous processes extending from their cell bodies. These processes mediate the defasciculation and seperation of axons, to which, eventually, the majority of the processes will attach to, and also myelinate several different axons. As mentioned earlier, one oligodendrocyte has the ability to myelinate as many as forty axons (Simons and Trotter 2007). The ensheathment of multiple axons by oligodendrocytes suggest that different signaling mechanisms govern their ability to identify neighboring cells and to distinguish the node and other axonal domains compared to Schwann cells. At this time, little is known about these mechanisms or the molecules involved. Additional distinctions between oligodendrocytes and Schwann cells are the absence of microvilli overlying the node and a basal lamina, which is absent in the parenchyma of the central nervous system (Hildebrand et al. 1993; Melendez-Vasquez et al. 2001).

Because of the absence of the nodal protrusions, it is unclear how oligodendrocytes mediate intercellular signaling during the formation of the node. There is, however, evidence that oligodendrocytes secrete specific factors that coordinate the clustering of nodal components preceding myelination (Kaplan et al. 2001, 1997). Additionally, the existence of perinodal astrocytes are hypothesized to interact with nodal components, and thus may provide signaling cues to adjacent myelinating oligodendrocytes (Black and Waxman 1988; Hildebrand et al. 1993). The absence of basal lamina from oligodendrocytes suggests that other ECM components or environmental factors may provide the binding sites for anchorage requisite for myelination, but at this time this remains an unresolved issue. Although substantial differences exist between the mechanisms of myelination between PNS Schwann cells and CNS oligodendrocytes, one common feature is the ability of both types of glia to potentiate the development of polarized axonal domains during myelination. The formation of these domains (viz. the node, paranode, juxtaparanode, and internode) is crucial for proper saltatory conduction of the action potential. Key molecules are

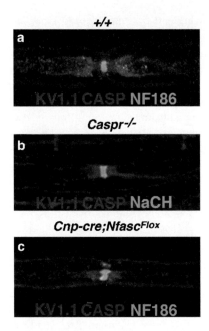

Fig. 1 Domain organization in myelinated PNS nerve fibers. Teased sciatic nerve fibers from wild-type (+/+; **a**), *Caspr* null (*Caspr*−/−; **b**), and *Neurofascin*^{NF155} (*NF155*) specific null mice (*Cnp-cre;Nfasc*^{Flox}; **c**) mice immunostained with antibodies against Kv1.1 (*red*), Caspr (*blue*), and Neurofascin 186 (NF186; *green*). In wild-type nerve fibers, localization of Kv1.1 fluorescence is restricted to the juxtaparanode. Caspr staining marks the paranode and NF186 is a marker of the nodal region. In *Caspr* null fibers, the lack of paranodal axo–glial junctions results in the diffusion of potassium channels into the paranode, as evident by the presence of Kv1.1 fluorescence adjacent to NF186 staining at the node (**b**). Loss of NF155 expression results in the lack of *Caspr* fluorescence at the paranode and the redistribution of potassium channels into the paranodal region, similar to *Caspr* mutants (**c**). In both mutants, the node remains unaltered as indicated by NF186 fluorescence

involved with the formation of these domains and their absence results in grave consequences as discussed below (Fig. 1).

3 Axonal Domains of Myelinated Axons

3.1 The Node of Ranvier

The nodes of Ranvier are short, myelin-free segments of axonal membrane that are distributed at regular intervals along myelinated nerve fibers, in which the action potential is regenerated in a saltatory manner. These regions are enriched in voltage-gated sodium (Nav) ion channels, which occupy a density of approximately 1,500 μm^{-2} (Waxman and Ritchie 1993). Nav channels are heterotrimeric complexes comprised of a pore-forming α-subunit that regulates ion flow and one or more transmembrane-spanning β-subunits that mediate both extracellular and intracellular interactions (Isom 2002; Yu and Catterall 2003). During development, a transition between the α-subunits occurs, in which Nav1.2, present in immature nodes, is replaced by Nav1.6 in adult nodes (Boiko et al. 2001). While all mature nodes in the PNS express Nav1.6 exclusively, subsets of adult CNS nodes express Nav1.2 and Nav1.8 (Arroyo et al. 2002). The significance of this exchange in subunits is currently unknown, but may pertain to varying activity of the subunits. What is known is that the Nav channels are essential for the proper conduction of the nerve impulse, because loss of Nav1.6 causes a dramatic decrease in conduction velocities, accompanied with abnormal nodal and paranodal structure (Kearney et al. 2002).

Several proteins expressed in the node are known either to interact with and/or mediate Nav channel function, including the cytoskeletal proteins ankyrin G (AnkG), βIV spectrin, αII spectrin, and the cell adhesion molecules (CAMs) *neurofascin* (NF186), and *NrCAM* (Salzer 2003). AnkG belongs to a family of scaffolding proteins that function to stabilize membrane-associated proteins by linking them to the actin–spectrin cytoskeleton within specialized domains (Bennett and Lambert 1999). It is expressed in both the axon initial segment (AIS) and the nodes of neurons, where it interacts with Nav channels through either their α-, or β-subunit(s) (Bouzidi et al. 2002; Kordeli et al. 1995; Lemaillet et al. 2003; Malhotra et al. 2002). This interaction is essential for the targeting of Nav channels to the AIS, as mice deficient in a cerebellar-specific AnkG show a loss of Nav channel clustering at the AIS of Purkinje neurons and the inability to fire action potentials (Zhou et al. 1998a). Presumably, the loss of AnkG within the nodes may result in a similar loss of Nav channels; however, while the AIS and nodes have many similarities in molecular composition, their functions may be differentially regulated.

AnkG associates with the cytoskeleton through its interaction with b-spectrins; specifically, b*IV spectrin*, which is also localized to the nodes of Ranvier and AISs (Berghs et al. 2000; Komada and Soriano 2002). This interaction is critical for the clustering of AnkG, and in turn, Nav channels to the node inasmuch as loss of bIV spectrin in mice results in reduced levels of these proteins in nodes and AISs, increased nodal axonal diameter, and severe tremors and impaired nerve conduction (Komada and Soriano 2002; Lacas- Gervais et al. 2004). Concomitantly, AnkG null Purkinje neurons show loss of βIV spectrin in the AIS, revealing a codependent relationship between AnkG and βIV spectrin and their localization to these critical areas of action potential propagation (Jenkins and Bennett 2001).

Recently, another spectrin, α*II spectrin*, was identified at the nodes, and proposed to have physiological significance to the nodal architecture. Normally, αII spectrin is associated with the paranode, but new findings have indicated the presence of aII spectrin in immature nodes (Garcia-Fresco et al. 2006; Ogawa et al. 2006). Initially, αII spectrin is expressed in both the nodes and the paranodes in developing nerves, but gradually becomes restricted to the paranode as myelination progresses. Although its final site of expression resides at the paranode, αII spectrin was proposed to play a role in the assembly of the nodes and the clustering of Nav channels, since loss of its expression in the neurons of zebra fish resulted in abnormal nodal dimensions (Voas et al. 2007). Although electrophysiological data from these mutants could not be assessed, it is probable that the increase in nodal length observed may perturb the propagation of the action potential and slow down nerve conduction. Interestingly, the progressive restriction of αII spectrin during myelination may suggest that βIV spectrin replaces it in mature nodes. Further analysis of the significance of αII spectrin to the node or the paranode may prove to be important to our understanding of how these domains are initially constructed.

Neurofascin (NF186) and *NrCAM* are members of the L1 subfamily of immunoglobulin (Ig) cell adhesion molecules (CAMs) that mediate cell–cell, and cell–matrix interactions (Grumet 1997; Volkmer et al. 1992). Both proteins are expressed in the nodes and AISs and interact with AnkG through a conserved region present in the cytoplasmic domain of each protein (Davis et al. 1996; Lustig et al. 2001; Zhang et al. 1998). The association of AnkG with NF186 is mediated by tyrosine phosphorylation. The unphosphorylated form of NF186 is able to associate with AnkG at the nodes, while phosphorylation perturbs the interaction of NF186 with AnkG (Garver et al. 1997; Zhang et al. 1998). Through their interaction with AnkG, both NrCAM and NF186 are thought to coordinate Nav channel clustering and node formation, because accumulation of these CAMs occurs prior to the presence of both AnkG and Nav channels in PNS nodes (Custer et al. 2003; Lambert et al. 1997; Lustig et al. 2001). Additionally, experiments utilizing function-blocking antibodies against the CAMs revealed that both Nav channels and AnkG failed to accumulate at the nodes in *in vitro* cocultures (Lustig et al. 2001). Furthermore, mice deficient in NrCAM expression resulted in delayed aggregation of Nav channels and AnkG to the nodes, although they did eventually cluster and the nodes functioned normally (Custer et al. 2003). Conversely, other findings suggest that AnkG is responsible for the initial assembling of Nav channels and CAMs to CNS nodes (Jenkins and Bennett 2002).

At this time, no mutational analysis of NF186 alone has been conducted, but a conventional null mutant lacking both isoforms of neurofascin, NF186, and the glial neurofascin (NF155) expressed in the paranode exhibited complete loss of nodal and paranodal formation (Sherman et al. 2005). These mice died at postnatal day 6 (P6), which prevented further characterization of their functions in axo–glial domain formation and maintenance. However, recent findings by the same group revealed that reexpression of NF186 in the mutant axons resulted in the relocalization of AnkG and Nav channels to the node, suggesting that NF186 coordinates the formation of the node and the clustering of the Nav channels (Zonta et al. 2008). These results are very compelling and further analysis of a true NF186 knockout would greatly contribute to our future understanding of its role in nodal development and organization.

A unique set of nodal proteins exists that is expressed specifically in the PNS. These proteins, which reside within the Schwann cell microvilli that extend from the outermost paranodal loop of myelin, contact the node and are proposed to function in nodal development and formation. An array of proteins are expressed within these small protrusions, including *gliomedin*, *ERM* (ezrin/radixin/moesin) proteins, *EBP-50* (ezrin binding protein 50), *dystroglycan*, *RhoA-GTPase*, and *syndecans* (Eshed et al. 2005; Gatto et al. 2003; Goutebroze et al. 2003; Melendez-Vasquez et al. 2004, 2001; Saito et al. 2003). Of particular interest is Gliomedin, which was shown to interact with NF186 and NrCAM in the nodal axolemma. This interaction was proposed to coordinate the clustering of these proteins into the PNS node (Eshed et al. 2005). Similarly, ablation of dystroglycan, a laminin receptor, in myelinating Schwann cells resulted in reduced Nav channel clustering at the nodes and disrupted nodal microvilli formation (Saito et al. 2003). These findings indicate that Schwann cells may function to coordinate the initial formation and clustering of nodal components, most likely through their interactions with NF186 and NrCAM, and further exemplify the importance of glial signals to nodal development, particularly in the PNS. While we have yet to discover the exact mechanisms regulating nodal development, it is evident that all the proteins discussed above play significant roles in nodal domain formation, maintenance, and function. Further studies to elucidate their mode of action may provide insight into the mechanisms regulating these processes in disease.

3.2 The Paranode

3.2.1 The Function of the Vertebrate Paranodal Region

The paranode is a region in myelinated nerve fibers where the terminal myelin loops form specialized septate-like junctions with the axolemma. These axo–glial junctions are directly contiguous to the nodes of Ranvier and are thought to act as a barrier or molecular sieve, which impedes free diffusion between the nodal space and juxtaparanodal periaxonal space (Pedraza et al. 2001). As myelination progresses, the internodal myelin layers are compacted. This compaction forces cytoplasm to redistribute outwardly towards the paranode, and results in the formation of the characteristic paranodal loops. These paranodal loops, representing the initial wraps of myelin, also function as an anchorage point to stabilize the glial cell as myelination proceeds. Accordingly, the appearance of transverse bands, or the septate-like junctions, is first observed at the distal-most loop of the preforming paranode. Formation of the bands then progresses inwardly towards the juxtaparanode. The continual wrapping of myelin and the formation of these "septae" serve to cluster the juxtaparanodal potassium (Kv) channels and separate them from the nodal Nav channels. Thus, formation of the paranodal axo–glial junctions is crucial to the demarcation and segmentation of axonal domains in nerve fibers that allow for proper conduction of the nerve impulse (see below).

3.2.2 Functional Relevance of Invertebrate Septate Junctions to Vertebrate Paranodal Axo–Glial Junctions

Septate junctions (SJs) are one of the most widely and diversely expressed junctions in invertebrates. These junctions play critical roles in governing cell polarity, cell adhesion, and the diffusion of molecules between cells (Banerjee et al. 2008, 2006a, b; Baumgartner et al. 1996). During epithelial cell development, SJs form in a region known as the apico-lateral domain, an area just basal to the apical hemiadherens junctions, and zonula adherens. These junctions form in a circumferential pattern and function to maintain epithelial cell polarization and integrity by sustaining a constant distance of 15 nm between adjoining cells. Additionally, SJs were found to act as a diffusion or paracellular barrier to restrict the movement of molecules between the apical and basolateral surfaces of epithelia (Banerjee et al. 2008, 2006a; Carlson et al. 2000; Tepass et al. 2001). Similarly, vertebrate paranodal axo–glial junctions function by providing a periaxonal barrier to ionic diffusion between regional longitudinal axonal domains of myelinated fibers and thus are considered orthologous to invertebrate SJs (Banerjee and Bhat, 2007; Bhat 2003; Salzer 2003). Perhaps the most relevant example of invertebrate SJs to vertebrate paranodal axo–glial junctions is found in the nervous system of *Drosophila* (Fig. 2). Similar to oligodendrocytes in the mammalian CNS, *Drosophila* glial cells encompass several axons with their membrane, but

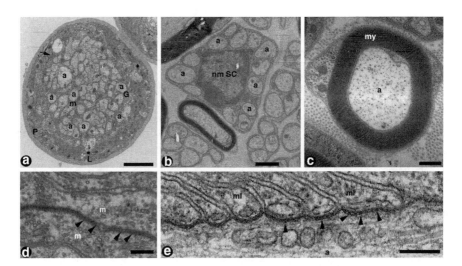

Fig. 2 Comparative ultrastructure of Drosophila and mouse unmyelinated and myelinated nerve fibers. (**a**) Cross-sections of peripheral nerve fibers from *Drosophila* show the inner glia (G) ensheathing axons (a). Electron dense, ladder-like structures (*arrow*) known as septate junctions form between the outer perineurial (P) and the inner ensheathing glial cell (G) membranes (m) of *Drosophila* (*arrowheads*, D). Electron micrograph cross-section of a Remak bundle in the mouse peripheral nerve (**b**). Remak bundles consist of several small diameter axons (a) that are ensheathed by a single nonmyelinating Schwann cell. These fibers do not acquire myelination, and are similar in structure to *Drosophila* nerve fibers. Ultrastructure of a single myelinated mouse peripheral nerve fiber in cross-section (**c**). The continual wrapping of the Schwann cell myelin membrane (my) forms a multilamellar layer that is electron dense. A longitudinal section of the paranodal region of a myelinated axon (a) shows the septate-like junctions that form between the myelin loops (ml) and the underlying axolemma. *Scale bars:* (**a**) 2 μm; (**b**) 1 μm; (**c**) 0.5 μm; (**d, e**) 0.2 μm

instead of myelinating the axons, they simply ensheath them (Banerjee and Bhat 2008). A separate layer of perineurial cells develops around the glia, similar to the layer of fibroblasts that form the perineurium in vertebrates, which surrounds myelinated nerve fasciculi in the PNS (Hildebrand et al. 1993; Jessen and Mirsky 2005). To limit the flow of ions from the axons, septate junctions are formed between glial cells in a homotypic fashion and heterotypically between the glial and perineurial cells. As is the case with epithelial cells, these SJs function as a paracellular barrier to regulate or prevent the diffusion of ions and molecules from the hemolymph (Banerjee and Bhat 2007; Tepass et al. 2001). Similarly, vertebrate paranodal axo–glial junctions behave as diffusion barriers between Nav channels in the node and Kv channels in the juxtaparanode. By maintaining the segregation of ion channels into their respective domains, the paranode facilitates proper saltatory conduction, while also ensuring repolarization of the action potential.

The identification of *Drosophila* SJs and the proteins involved in their formation and stabilization has lead to the elucidation of several homologues expressed in vertebrate paranodes (Fig. 3). Of note are the *Drosophila* cell adhesion molecules (CAMs) *neurexin IV*, *contactin*, and *neuroglian*, and the cytoskeletal protein, *coracle* (Banerjee et al. 2006b; Faivre-Sarrailh et al. 2004). Loss of expression of these genes has devastating effects on septate junction formation, and the stabilization of the paracellular barrier (Banerjee et al. 2006a; Baumgartner et al. 1996). The vertebrate counterparts of these molecules are *contactin-associated protein* (Caspr), contactin, the 155 kDa isoform of neurofascin (NF155), and *protein 4.1B*, respectively (Bhat et al. 2001; Boyle et al. 2001; Peles et al. 1997; Tait et al. 2000). All these proteins localize to the paranode, and through genetic ablation and biochemical analysis we have begun to understand their importance to the formation of the paranode and the maintenance and segregation of axonal domains.

3.2.3 Key Regulators of Paranodal Formation and Stability

Initial studies aimed towards elucidating the proteins involved in the formation of the paranodal axo–glial junctions were focused around the Neurexin/Caspr/ Paranodin (NCP) family of cell recognition molecules (Bellen et al. 1998). This superfamily is composed of five vertebrate homologues, that is, Caspr–Caspr5 (Spiegel et al. 2002). Of these isoforms, only Caspr is expressed at the paranodes, where it becomes enriched in the axolemma when myelination arrests (Arroyo et al. 1999; Bhat et al. 2001; Einheber et al. 1997; Menegoz et al. 1997). Caspr (aka Paranodin) is a Type 1 transmembrane protein that is comprised of a large expansive extracellular domain and a short intracellular domain. The extracellular domain of Caspr contains an array of subdomains implicated in cell–cell and cell–matrix interactions, including discoidin, EGF (epidermal growth factor), laminin G, and fibrinogen-like domains (Bellen et al. 1998; Bhat 2003; Denisenko-Nehrbass et al. 2002). While the specific function of each individual domain remains elusive, it is

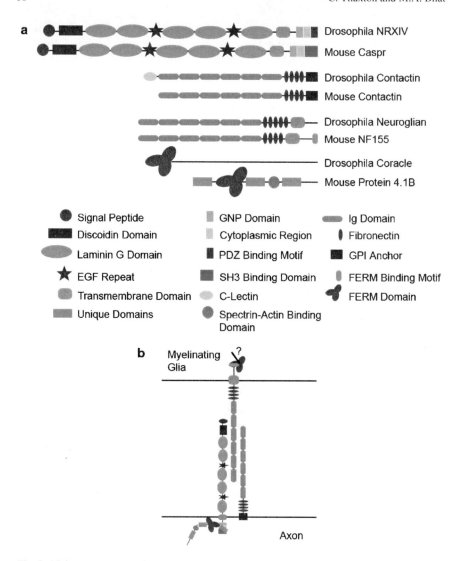

Fig. 3 Major components of septate junctions in Drosophila and mouse. The domain structure of *Drosophila* Nrx IV, Contactin, Neuroglian, and Coracle, and their vertebrate counterparts in mouse, Caspr, Contactin, NF155, and Protein 4.1B reveals significant homology between these proteins (**a**). Schematic representation of the proteins involved in the formation of the paranodal axo–glial septate junctions in mouse (**b**). NF155 is expressed strictly in the myelinating glial within the paranodal loops. The presence of a FERM binding domain within NF155 predicts an interaction with a FERM protein, which may mediate signaling to the glial cytoskeleton. NF155 is hypothesized to bind to either Caspr or Contactin, but the exact mechanisms are yet unknown

known that the extracellular domain of Caspr is vital for the formation of transverse septae, as evidenced by the absence of these junctions in Caspr-deficient mice (Bhat et al. 2001). The periaxonal space of the Caspr mutants was often invaded by astrocytic process in the CNS, Schwann cell microvilli in the PNS, and extracellular matrix components. The lack of stability resulted in the eversion of the paranodal loops, in which severe ataxia and reduced conduction velocity were also noted in these mice. Additionally, the juxtaparanodal rectifying *Shaker*-like potassium channels, Kv1.1 and 1.2, were frequently mislocalized to the paranodal region, while the nodal Nav channels remained unchanged. This alteration is likely to be mediated by the short intracellular domain of Caspr, which includes proline rich and glycophorin C domains that are known to interact with the actin cytoskeleton and coordinate it (Denisenko-Nehrbass et al. 2003a, b; Gollan et al. 2002; Menegoz et al. 1997). Thus, these findings exemplified the importance of Caspr to the formation of the axo–glial junctions and to the distribution, segregation, and organization of ion channels within axonal domains.

Support for the role of Caspr in the organization of axonal domains and the formation and stabilization of paranodes came with the discovery of its protein binding partners and regulators. On the axonal side, Caspr associates with contactin, a glycosylphosphatidylinositol (GPI)-anchored protein belonging to the Ig superfamily (Brummendorf and Rathjen 1996; Falk et al. 2002). Contactin is also enriched at the paranodes of myelinated fibers and forms a *cis* interaction with Caspr in the axolemma (Peles et al. 1997; Reid et al. 1994). This interaction is mediated by the extracellular domain of Caspr and the fibronectin III domains of contactin (Bonnon et al. 2003; Faivre-Sarrailh et al. 2000). Genetic ablation of contactin in mice results in an analogous phenotype as Caspr mutants, with the loss of transverse bands and attendant paranodal disorganization (Boyle et al. 2001). In addition, Caspr fails to localize to the plasma membrane, suggesting that contactin may function to transport and/or stabilize Caspr to the paranodal axonal membrane. Indeed, further characterization of the interaction between contactin and Caspr reveals that a mutually exclusive relationship exists between the two proteins. Without contactin, Caspr is retained in the endoplasmic reticulum and fails to traffic to the paranodal axolemma (Bonnon et al. 2003; Faivre-Sarrailh et al. 2000). Concomitantly, contactin cannot stably localize to the paranode in the absence of Caspr, but rather is found in the nodes in the CNS (Bhat et al. 2001; Rios et al. 2000).

Several scaffolding and cytoskeletal components reside at the paranodes, and emerging evidence suggests a critical role for these proteins in the maintenance of axo–glial junctions. Protein 4.1B belongs to the Band 4.1 superfamily of membrane cytoskeletal linking proteins (Hoover and Bryant 2000; Parra et al. 2000; Sun et al. 2002). It is present in the paranodes, and distributed diffusely in the juxtaparanodes of axons; to date, it is the only known protein 4.1 isoform localized to the axo–glial junctions (Ohara et al. 2000). Protein 4.1B contains a conserved FERM (four point one/ezrin/radixin/moesin) domain that mediates its binding with several transmembrane receptors, including Caspr at the paranodes and Caspr2 at the juxtaparanodes (Denisenko-Nehrbass et al. 2003b; Garcia-Fresco et al. 2006; Gollan et al. 2002). The conserved cytoplasmic GNP (glycophorin C/Neurexin IV/paranodin) domain of Caspr mediates its association with protein 4.1B (Gollan et al. 2002; Sousa and Bhat, 2007). Loss of the GNP domain resulted in the internalization of the Caspr–contactin complex from the axonal plasma membrane, suggesting an important role for Protein 4.1B in the stabilization of this complex and axo–glial junctions at the paranode (Gollan et al. 2002). It remains to be seen what effect the loss of protein 4.1B might have on the formation and maintenance of the paranode; however, studies of the *Drosophila* orthologue, coracle, suggest a key role for the FERM domain-containing proteins in the organization and stabilization of septate junctions (Baumgartner et al. 1996; Laval et al. 2008; Ward et al. 1998). Recent work has identified the presence of a macromolecular complex consisting of protein 4.1B, Ankyrin B (AnkB), and the αII and βII spectrins that associates with Caspr and contactin at the paranodal junctions (Garcia-Fresco et al. 2006; Ogawa et al. 2006). Loss of Caspr resulted in the absence of AnkB, or its diffusion out of the paranodes, which reflects the importance of Caspr in orchestrating the assembly of the underlying paranodal cytoskeleton (Garcia-Fresco et al. 2006; Pillai et al. 2007; Sousa and Bhat, 2007). Given that the junctional specialization of the paranode functions as a "fence" in regulating the diffusion of ions associated with functional activity in the axon, it will be interesting to see how future studies may elucidate the roles of these cytoskeletal proteins in segregating the axonal domains.

The interaction(s) between glia and axons is important in regulating myelination and the formation of axonal domains. As oligodendrocytes and Schwann cells associate with the axon, they initiate the clustering of the paranodal components in the axolemma. While several axonal proteins have been identified at the paranode, only one glial protein is known to localize to the paranodal axo–glial junctions, the 155 kDa isoform of neurofascin, NF155 (Collinson et al. 1998; Moscoso and Sanes 1995; Tait et al. 2000). NF155 is a CAM that belongs to the L1 subgroup of the Ig superfamily (Davis and Bennett 1993; Holm et al. 1996; Volkmer et al. 1992). It differs from the axonally expressed NF186, in that it lacks the mucin-like domain and contains an extra fibronectin type III domain (Davis et al. 1996). The exact mechanisms governing their varied expression pattern is not clear, but over 50 alternatively spliced isoforms exist that appear to be developmentally regulated (Hassel et al. 1997). Initial mutational analysis in mice resulted in the loss of both isoforms of neurofascin, NF186 and NF155 (Sherman et al. 2005). These mice were unable to form either the nodal complex or the paranodal axo–glial junctions and died at postnatal day 6. Caspr and contactin were diffused throughout the axon in these mutants. Interestingly, upon reexpression of NF155 in glia, the paranodes reorganized and Caspr and contactin relocalized to the axolemma. This coincided with previous reports that expression of NF155 colocalized with Caspr during the clustering of the paranodal loops during myelination (Tait et al. 2000). Further analysis of the proposed interaction(s) revealed that the extracellular domain of NF155 could associate with the Caspr–contactin complex in vitro (Charles et al. 2002). These results are somewhat controversial, however, as other studies employing similar in vitro techniques find that NF155 binds contactin and that the presence of Caspr perturbs this interaction (Gollan et al. 2003). Although it seems unclear as to how these proteins associate, these results do suggest a role for NF155 in paranodal organization, possibly through interactions with Caspr and/or contactin.

Recent findings utilizing Cre-loxP conditional knockout strategies has provided new insights into the specific role of NF155 in axo–glial formation and maintenance. Using a Cre recombinase driven by the CNPase promoter, which is specific for glia, it was shown that loss of NF155 expression alone results in the loss of septate-like junctions, paranodal disorganization, and the failure to segregate axonal domains (Pillai et al. 2009). Because of the strong correlation of phenotype with those of Caspr and contactin mutants, it was shown that both of these axonally expressed proteins were absent from the paranodal axolemma in NF155 mutant nerves. This suggested that NF155 and Caspr and/or contactin interact with the glial NF155 protein, and that this interaction mediates the formation and stability of the Caspr–contactin complex to the paranodal-domain of the axonal plasma membrane. The ability of NF155 to direct the formation of the paranode and stabilize the complex formation suggests that it may interact with cytoskeletal proteins. Although no protein-binding partners have been identified, it has been shown that the intracellular domain of NF155 contains a FERM domain-binding motif (Gunn-Moore et al. 2006). Given that FERM domain-containing proteins are cytoskeletal linkers, NF155 may coordinate signals originating from the axon that direct changes in the glial cytoskeleton prior to and during myelination; whereby, axonal domains are partitioned.

3.2.4 The Importance of Lipids to Paranodal Formation

As noted previously, myelin is the elaborated glial membrane sheath that circumferentially wraps axons, which becomes compacted to form a multilamellar layer of insulation. The major constituents of myelin are lipids, which make up approximately 70–85% of the dry weight (Morell et al. 1994). Nearly one-third of the lipid mass is comprised of gylcosphingolipids, galactosylceramide (GalC), and its sulfated derivative, sulfatide (Coetzee et al. 1996; Norton and Cammer 1984). These lipids were shown to have a significant role in oligodendrocyte development, the initiation of myelination, and the stabilization of the compacted myelin layers (Marcus and Popko 2002). Studies aimed towards elucidating the function of these lipids during myelination resulted in the development of knockout mice deficient in CGT (UDP-galactose:ceramide galactosyltransferase), the enzyme that synthesizes galactosylceramide (Coetzee et al. 1996). Surprisingly, these mice formed myelin, but displayed severe ataxia, tremors, and slowed nerve conduction. Further examination of these mice revealed the appearance of disorganized and everted paranodal loops and the absence of axo–glial junctions (Dupree et al. 1998).

Similar phenotypes were also observed in CST (cerebroside sulfotransferase) null mice (Honke et al. 2002; Ishibashi et al. 2002). These mice differ from CGT, in that they are unable to synthesize only sulfatide, whereas galactosylceramide is still produced. These findings implied that glial sulphosphingolipids were essential to the formation and/or maintenance of the paranode. The exact mechanisms in which these lipids function to promote axo–glial junction formation are unknown, but it is hypothesized that they are responsible for creating lipid microdomains, where paranodal components may reside and function. In this context, it is noteworthy that studies have indicated that NF155, Caspr, and contactin are present in lipid-rich microdomains (Bonnon et al. 2003; Schafer et al. 2004). Furthermore, NF155, Caspr, and contactin are mislocalized outside of the paranode in CGT mice (Dupree et al. 1999; Hoshi et al. 2007; Poliak et al. 2001). Alternatively, it may be that an unidentified protein(s) may be localized in the domain to interact with these lipids, either in *cis* or *trans* relation, and may thereby direct the localization of Caspr, contactin, and NF155 to the paranode.

3.3 The Juxtaparanodal Region

The juxtaparanode is a potassium channel-rich region that lies underneath the compact myelin sheath, just proximal to the paranodes. It has been proposed that the significance of localizing potassium channels in this domain is concerned with repolarization of the action potential, as well as counteracting instability of excitability, especially in the transition zone between myelinated and unmyelinated portions of distal motor nerve fibers (Chiu et al. 1999; Rasband et al. 1998; Zhou et al. 1998b). Specifically, two rectifying *Shaker*-like potassium channels, Kv1.1 and Kv1.2, occupy a majority of the domain, along with the CAMs, Caspr2, and TAG-1 (Arroyo et al. 1999; Mi and Berkowitz 1995; Rasband et al. 1998; Wang et al. 1993). Caspr2 is expressed strictly in the axon and is functionally related to Caspr, except that it contains a PDZ binding motif in its intracellular domain that Caspr lacks (Poliak et al. 1999). Caspr2 was shown to be critical to the maintenance of Kv channels in the juxtaparanode, as its absence in mice resulted in the diffusion of these channels throughout the internode (Poliak et al. 1999). Subsequently, it was found that the PDZ binding domain links Caspr2 to Kv1.1 and Kv1.2 (Poliak et al. 2003). Initial reports predicted that PSD-95, a PDZ-containing protein found within the juxtaparanode, might be a potential candidate for this interaction, as it has been shown to associate with Kvβ2; however, later studies reported that PSD-95 and Caspr2 do not interact (Baba et al. 1999; Poliak et al. 1999). Furthermore, Kv1.1 and Kv1.2 remained clustered at the juxtaparanodes of PSD-95 deficient mice; therefore, it remains to be elucidated what PDZ protein(s) links Caspr2 and Kv channels in the juxtaparanode (Rasband et al. 2002).

Another critical component of the juxtaparanode is the GPI-anchored adhesion molecule, TAG-1 (Furley et al. 1990; Karagogeos et al. 1991; Traka et al. 2002). TAG-1 belongs to the Ig superfamily and shares 50% sequence homology with contactin; it is expressed in both neurons and myelinating glia (Traka et al. 2002). Like Caspr and contactin in the paranode, TAG-1 and Caspr2 associate and form a *cis* complex within the juxtaparanodal axolemma. This interaction is mediated by the Ig-domains of TAG-1 (Traka et al. 2003; Tzimourakas et al. 2007). Interestingly, the axonal *cis* complex of TAG-1/Caspr2 binds with TAG-1 expressed in glia by homophilic interactions between the TAG-1 molecules (Traka et al. 2003). Mutational analysis in mice revealed that a codependent relationship between TAG-1 and Caspr2 exists, similar to that of Caspr and contactin, where the accumulation of either protein to the juxtaparanode relies on the expression of the other

(Poliak et al. 2003; Traka et al. 2003). Deletion of *TAG-1* in mice resulted in the exclusion of Kv1.1 and Kv1.2 channels from the underlying juxtaparanodal axonal membrane concomitant to Caspr2 null mice (Traka et al. 2003). Although the Kv channels were absent from the membrane, no change in conduction velocity was observed in TAG-1 mutant nerves (Traka et al. 2003). Accordingly, Caspr2 mutant mice displayed normal conduction (Poliak et al. 2003). While no electrophysiological abnormalities were observed in TAG-1 or Caspr2 mutant mice, the effects of extended loss of these proteins to action potential propagation and resting potential are unknown. Examining these effects may be of particular interest to disease pathology because Kv1.1 mutant mice display backfiring of the action potential and extended hyperexcitability (Zhou et al. 1999). Additionally, TAG-1 function has been recently implicated in learning and cognition and suggests an important role for these proteins in long-term plasticity (Savvaki et al. 2008).

3.4 The Internodal Region

The internode comprises the area between the juxtaparanodes, and accounts for 99% of the total length of a myelinated nerve segment (Salzer 2003). As previously mentioned, the length of this region is determined by the axonal diameter and represents the most extended region involving axo–glial interactions. Very little is known about the mechanisms regulating the ability of glia to detect the axonal diameter to determine the final internodal length. Additionally, as animals grow, this region lengthens to compensate for the extension of limbs (Abe et al. 2004). A handful of proteins have been found to localize to the adaxonal membrane at the inner-most lip of the myelin membrane. These proteins include the nectin-like proteins, *Necl1* and *Necl4*, the polarity protein *Par-3*, and the *myelin associated glycoprotein* (MAG). The nectin-like proteins (Necl) are cell adhesion molecules that belong to the Ig superfamily (Takai et al. 2003). They are related to the nectin cell adhesion proteins, but differ by their inability to bind to afadin. Necls are often associated with tight junctions and contain PDZ binding motifs that may facilitate their interaction with cytoskeletal scaffolding proteins. Recent findings have implicated Necl1 and Necl4 in myelination. These proteins were expressed in a polarized fashion at the inner mesaxon or at the initial contact of the glial myelin membrane with axons (Maurel et al. 2007). Heterophilic interactions between Necl1 on axons and Necl4 on glia mediate the initial attachment and wrapping of the glial membrane (Maurel et al. 2007). Perturbation of Necl4 by use of RNAi in Schwann cell-DRG neuron cocultures inhibited myelination (Maurel et al. 2007). Similar results were observed in mutant mice deficient in the axonally expressed Necl1 (Park et al. 2008). These results indicate the importance of these proteins to myelination and axo–glial recognition and adhesion.

Par-3 is a PDZ-containing adaptor protein, which regulates polarity in many cell types (Ohno 2001). The *Drosophila* homolog of Par-3, *Bazooka*, functions to establish cell polarity in epithelial cells through its restricted expression at the apical surface

(Pinheiro and Montell 2004). Like Bazooka, the vertebrate Par-3 is a polarized protein that is expressed asymmetrically in glia at the junction of the inner mesaxon or the adaxonal layer. It is proposed to facilitate the adhesion of glia to axons (Chan et al. 2006). Disruption of Par-3 function in Schwann cells resulted in the inability of Schwann cells to adhere to and myelinate axons (Chan et al. 2006). Further analysis determined that Par-3 is associated with *p75NTR*, a BDNF receptor that promotes myelination through its PDZ1 domain. While other binding partners of Par-3 are unknown, it is likely that it interacts with many cell adhesion receptors that contain PDZ binding motifs. Of particular interest is the possible association of Necl4 and Par-3. Given their colocalization to the inner mesaxon, the presence of a PDZ binding motif in Necl4, and the observation in mice that an association of Par-3 and nectins occurs, these proteins may interact to coordinate axo–glial adhesion prior to myelination (Takekuni et al. 2003).

4 Mechanisms Regulating Axonal Domain Formation and Maintenance

A key question regarding the assembly of axonal domains remains elusive; namely, what comes first, the node or the paranode? Is formation of the nodes dependent on or independent of paranodal formation? Furthermore, does the long-term maintenance of the node require intact paranodes? A central theme has emerged from the numerous studies on axo–glial domains, which is that distinct differences exist between the mechanisms regulating the partitioning of axonal domains in the PNS vs. the CNS. This is, to a degree, to be expected as myelination proceeds in different fashions between oligodendrocytes and Schwann cells. Later we attempt to interpret recent findings to shed light on the organization of axonal domains and their dependence on adjacent domains for long-term stability and maintenance.

4.1 Nodal Formation in the PNS and CNS: Extrinsic vs. Intrinsic; Dependent vs. Independent Mechanisms

In the PNS, Schwann cell development and myelination are intimately linked to axonal and ECM stimuli and adhesion. The maturation of Schwann cells into a myelinating phenotype and the segmentation of axonal domains rely on forming this intimate relationship. Studies have shown that PNS domain organization begins with the formation of the node and proceeds inwardly towards the internode (Melendez-Vasquez et al. 2001; Poliak et al. 2001). An extrinsic pattern of nodal assembly is suggested for the PNS as the presence of NF186 and NrCAM precedes that of Nav channels and the cytoskeletal components, AnkG and βIV spectrin (Peles and Salzer 2000; Salzer 2003). Additionally, the appearance of these nodal components is observed before the paranodal proteins, Caspr and contactin, which suggests that PNS nodes form prior to and independently of the paranodes (Melendez-Vasquez et al. 2001).

Schwann cell distal microvilli are believed to coordinate the formation of nodes through their association with NF186 and/or NrCAM. This association may occur through gliomedin, as it is expressed in the nodal microvilli and interacts with both NF186 and NrCAM (Eshed et al. 2005). Disruption of gliomedin by RNAi resulted in the absence of Nav channels at the node and suggests that glia facilitate the initial assembly of the node by mediating NF186 localization. This follows an extrinsic pattern of node formation, in which extracellular signals initiate node formation and also transmit signaling to the axonal cytoskeleton to stabilize the complex once formed. Interestingly, gliomedin can be secreted, resulting in soluble forms that are accumulated and incorporated into the surrounding basal lamina in the perinodal space (Eshed et al. 2007, 2005). In the absence of Schwann cells, axons exposed to the soluble form of gliomedin formed nodes, further supporting a role for glia in the formation of the node extrinsically. Dystroglycan, a laminin receptor expressed in Schwann cell nodal microvilli, may also mediate the extrinsic formation of PNS nodes, as genetic ablation of its gene in mice results in severe conduction blockade, dysmyelination, and the loss of nodal Nav channel clustering (Occhi et al. 2005; Saito et al. 2003). Additionally, these mice form normal axo–glial junctions at the paranodes, suggesting that paranodal formation does not require the formation of the node. It may also indicate that the formation of paranodes alone is not sufficient to cluster the nodal complex, although further studies in these mice revealed that AnkG and NF186 are still clustered at the node in the absence of Na_v channels. This suggests that dystroglycan may simply act to stabilize these ion channels to the node instead of coordinating the organization of the entire node. However, the presence of NF186 at these nodes still suggests that an extrinsic mechanism of assembly may occur albeit through alternative mechanisms.

Nodal formation in the CNS is considered to behave quite differently than that of the PNS. It is hypothesized that instead of an extrinsic glial mediated assembly, a more intrinsic form of coordination occurs, beginning with the cortical axonal cytoskeleton and assembling in association with the axolemma. One of the major factors contributing to this model of an intrinsic assembly is that oligodendrocytes do not extend nodal microvilli (Melendez-Vasquez et al. 2001). Additionally, NF186, NrCAM, and Nav channels were recruited to the node subsequent to the appearance of AnkG in intermediate nodes (Jenkins and Bennett 2002). An attempt to disrupt AnkG resulted in the generation of a cerebellum specific knockout (Zhou et al. 1998a). Aberrant Nav channel clustering was observed in the AISs of Purkinje neurons, yet AnkG was still present in the nodes of these mutants; therefore, the effects of loss of AnkG on CNS node formation could not be evaluated (Jenkins and Bennett 2001; Zhou et al. 1998a). While it is evident from these findings that AnkG expression is important for Nav channel clustering at AIS, they also present evidence that several isoforms of AnkG may exist that differentially localize to separate areas of Nav channel clustering. Although oligodendrocytes lack nodal microvilli, they still retain the ability to cluster Nav channels through the secretion of soluble factors (Kaplan et al. 1997). This mechanism of clustering is similar to the aggregation of Nav channels to PNS nodes by secreted gliomedin, and suggests that oligodendrocytes may, in fact, assemble nodes without direct contact (Eshed et al. 2007, 2005). Further examination revealed that oligodendrocyte conditioned medium clustered Nav1.2, the immature Nav channel isoform; whereas, Nav1.6 present in mature nodes required myelination (Kaplan et al. 2001). Additionally, the early assembly required an intact axonal cytoskeleton, and suggests that certain intrinsic signals may still be required for CNS nodal formation and maturation. The identification of the soluble factor(s) released by oligodendrocytes will certainly be critical to the future analysis and elucidation of the mechanisms governing CNS node formation.

An alternative method of CNS assembly may involve signaling from perinodal astrocytes that contact the nodal domain. The relationship of these astrocytes to nodal function is not well characterized, but interestingly, these cells were shown to express the cytoskeletal mediating ERM protein *ezrin* (Melendez-Vasquez et al. 2001). As mentioned earlier, ERM proteins are also present in the nodal microvilli of Schwann cells, and their expression at these domains is coordinated with the extrinsic formation of the node in the PNS. Taken together, it is possible that perinodal astrocytes may compensate for the lack of nodal microvilli in oligodendrocytes and provide nodal assembly cues in a similar manner as Schwann cell microvilli. It would be interesting to see if these perinodal astrocytes express gliomedin, or a derivative, as well as dystroglycan.

Yet another possible mechanism of CNS node development is the idea that nodes are formed by the sequestration of nodal components from the axolemma by the formation of the paranodes. In support of this idea is the finding that Caspr expression in the forming paranode preceded the expression of Nav channels in the nodes of optic nerves (Rasband et al. 1999). Since oligodendrocyte myelination does not depend on the demarcation of the internode prior to myelination, as it does with Schwann cells, it is possible that oligodendrocytes begin to wrap their membrane while progressively extending processes along the axon. In this manner, the oligodendrocytes create a barrier that will induce the aggregation of the nodal proteins. In support of this mechanism, it was found in mice that reexpression of NF155 alone, in a complete neurofascin null background, resulted in the clustering of Nav channels (Zonta et al. 2008). These findings implicate a paranodal dependent formation of nodes in the CNS, but these mice also did not live past the complete knockout mice, and no assessment of conduction velocities was performed in these animals. Furthermore, mice deficient in Caspr expression form nodes in the CNS and PNS in the absence of intact axo–glial junctions, suggesting that paranodal formation is not a prerequisite for CNS node formation (Bhat et al. 2001). Concomitantly, NF155-specific knockout mice also form nodes independently of paranodal formation, further supporting the hypothesis that nodes form through separate autonomous mechanisms than paranodes (Pillai et al. 2009). Taken together, it appears that nodal stability may rely on paranodal domains, but the initial assembly of the node forms independently of other axonal domains.

New evidence has emerged regarding the requirement of paranodal axo–glial junctions in the maintenance and long-term stability of the node. Recently, Pillai et al. (2009) tested this hypothesis by knocking out NF155 in adult mice through the use of tamoxifen inducible Cre recombinase, driven by the proteolipid protein (PLP) promoter (Doerflinger et al. 2003). They allowed the mice to reach adult stage (P23) and subsequently treated the mice with tamoxifen to induce glial-specific genetic ablation of neurofascin (NF155). They found that axo–glial junctions dissembled, and essentially unraveled, from the juxtaparanodal side distally outward towards the node. The progressive unraveling of the paranodes resulted in the invasion of the paranodal space with Kv channels. While the node remained unchanged initially, it was found that with extended time (up to 1 year posttreatment), NF186, Nav channels, and AnkG diffused out of the node (CT and MB, unpublished observation). Upon further examination, these same proteins were diffused out of the AIS, further revealing a progressive loss from the external axonal nodes sequentially upwards towards the AIS at the neuron cell body. This suggests that through development and maturation, the paranodes not only function to sequester ion channels to specific domains, but that they maintain segregation by stabilizing these components. Therefore, even though nodes may form independently of paranodes during development, their long-term integrity may rely upon paranodal axo–glial junctions. As intriguing as these results may be, further experimentation will need to be conducted to elucidate the signaling involved. Additionally, the use of this inducible system may help with future studies to elucidate the effects of progressive myelin loss on axonal domains, much like that seen in multiple sclerosis.

4.2 Disease Manifestation in the Absence of Segmented Axonal Domains

Several human neuropathies and disorders, such as multiple sclerosis and Charcot–Marie Tooth disease, result in the progressive demyelination of nerves, leading to axonal degeneration. There are varied causes to the development of these diseases, be it autoimmune attack or genetic predisposition, but the dissolution of axonal domain organization appears to be a central indicator of the development and progression of these disorders (Berger et al. 2006; Lubetzki et al. 2005; Nave et al. 2007; Oguievetskaia et al. 2005; Shy 2006; Trapp and Nave 2008). While NF155, NF186, Caspr, and contactin have not been directly associated with disease predisposition and onset, they have been shown to be disrupted in disease pathology (Coman et al. 2006; Howell et al. 2006; Mathey et al. 2007; Wolswijk and Balesar 2003). Emerging evidence has revealed that NF155 is targeted in multiple sclerosis (MS) (Howell et al. 2006; Maier et al. 2007, 2005; Mathey et al. 2007). Initially, it was found that NF155 expression was reduced in the paranodes of MS patients, and was considered an early marker for the ensuing demyelination of the nerve tracts (Howell et al. 2006; Maier et al. 2005). Further analysis revealed that NF155 had decreased association with lipid rafts in MS, and therefore, axo–glial junction stability was compromised (Maier et al. 2007).

The mislocalization of Kv channels to the paranodes confirmed the dissolution of axo–glial junctions in the diseased lesions (Howell et al. 2006). A recent report revealed the presence of autoantibodies to NF155 and NF186 in patients with MS (Mathey et al. 2007). Interestingly, increasing amounts of autoantibodies were found in patients with chronic progressive MS compared to those with relapsing MS, suggesting that as the disease becomes more severe it targets these critical proteins, preventing the reformation of the paranodes and nodes. The inability of new oligodendrocytes to myelinate the affected lesions and organize axonal domains, due to the lack of both Neurofascins, would promote the axonal degeneration observed in chronic MS individuals. In support of the degeneration of paranodes in MS lesions, it was also found that loss of Caspr expression precedes demyelination in MS patients (Wolswijk and Balesar 2003). Other compelling evidence to support the role of intact axo–glial junctions in the prevention of axonal degeneration is found in Caspr mutants, in which loss of Caspr expression in mice resulted in the presence of axonal swellings and cytoskeletal abnormalities in Purkinje neurons of the cerebellum (Garcia-Fresco et al. 2006). These abnormalities precede the degeneration of axons, as seen in multiple neuropathies (Fabrizi et al. 2007; Lappe-Siefke et al. 2003; Rodriguez and Scheithauer 1994). In addition to their importance to axonal domain organization, the juxtaparanodal proteins, Caspr2 and Tag-1, have been implicated in autism, language impairment, as well as learning and cognition disorders, respectively, suggesting other roles for these CAMs in nervous system function (Alarcon et al. 2008; Bakkaloglu et al. 2008; Savvaki et al. 2008; Vernes et al. 2008). While genetic ablation of axonal domain proteins has not been identified as the causative agent for disease manifestation, it is clear that these proteins are critical in preventing the progression of demyelinating neuropathies, and serve as possible therapeutic targets for the future treatment of these devastating disorders.

5 Concluding Remarks

To date, significant advancements through the development of mouse models have contributed greatly to our knowledge of the proteins involved in myelination and the segregation of axonal domains. While many of the proteins involved have been identified and characterized, future studies may provide more insight into the signaling mechanisms involved in the formation and stabilization of each domain. The future efforts are likely to be centered on cytoskeletal scaffolding proteins and the signaling mechanisms governing their function, expression, and localization to specific axonal as well as glial domains. Additionally, emerging advances in animal model systems are sure to facilitate research directed towards studying the effects of extended loss of axo–glial junctions. These studies will certainly shed light on the processes, proteins, and functions misregulated during disease, and may reveal new roles for many of the axo–glial junctional proteins in demyelinating disorders, axonopathies, and other neuropathies.

Acknowledgments The work in our laboratory is supported by the grants from NIGMS and NINDS of the National Institutes of Health, National Multiple Sclerosis Society, and funds from the State of North Carolina.

References

Abe I, Ochiai N, Ichimura H, Tsujino A, Sun J, Hara Y (2004) Internodes can nearly double in length with gradual elongation of the adult rat sciatic nerve. J Orthop Res 22:571–577

Alarcon M, Abrahams BS, Stone JL, Duvall JA, Perederiy JV, Bomar JM, Sebat J, Wigler M, Martin CL, Ledbetter DH (2008) Linkage, association, and gene-expression analyses identify CNTNAP2 as an autism-susceptibility gene. Am J Hum Genet 82:150–159

Arroyo EJ, Xu YT, Zhou L, Messing A, Peles E, Chiu SY, Scherer SS (1999) Myelinating Schwann cells determine the internodal localization of Kv1.1, Kv1.2, Kvbeta2, and Caspr. J Neurocytol 28:333–347

Arroyo EJ, Xu T, Grinspan J, Lambert S, Levinson SR, Brophy PJ, Peles E, Scherer SS (2002) Genetic dysmyelination alters the molecular architecture of the nodal region. J Neurosci 22:1726–1737

Baba H, Akita H, Ishibashi T, Inoue, Y, Nakahira K, Ikenaka K (1999) Completion of myelin compaction, but not the attachment of oligodendroglial processes triggers K+ channel clustering. J Neurosci Res 58:752–764

Bakkaloglu B, O'Roak BJ, Louvi A, Gupta AR, Abelson JF, Morgan TM, Chawarska K, Klin A, Ercan-Sencicek AG, Stillman AA (2008) Molecular cytogenetic analysis and resequencing of contactin associated protein-like 2 in autism spectrum disorders. Am J Hum Genet 82:165–173

Banerjee S, Bhat MA (2007) Neuron-glial interactions in blood-brain barrier formation. Annu Rev Neurosci 30:235–258

Banerjee S, Bhat MA (2008) Glial ensheathment of peripheral axons in Drosophila. J Neurosci Res 86:1189–1198

Banerjee S, Pillai AM, Paik R, Li J, Bhat MA (2006a) Axonal ensheathment and septate junction formation in the peripheral nervous system of Drosophila. J Neurosci 26:3319–3329

Banerjee S, Sousa AD, Bhat MA (2006b) Organization and function of septate junctions: an evolutionary perspective. Cell Biochem Biophys 46:65–77

Banerjee S, Bainton RJ, Mayer N, Beckstead R, Bhat MA (2008) Septate junctions are required for ommatidial integrity and blood-eye barrier function in Drosophila. Dev Biol 317:585–599

Baumgartner S, Littleton JT, Broadie K, Bhat MA, Harbecke R, Lengyel JA Chiquet-Ehrismann R, Prokop A, Bellen HJ (1996) A Drosophila neurexin is required for septate junction and blood-nerve barrier formation and function. Cell 87:1059–1068

Bellen HJ, Lu Y, Beckstead R, Bhat MA (1998) Neurexin IV, caspr and paranodin–novel members of the neurexin family: encounters of axons and glia. Trends Neurosci 21:444–449

Bennett V, Lambert S (1999) Physiological roles of axonal ankyrins in survival of premyelinated axons and localization of voltage-gated sodium channels. J Neurocytol 28:303–318

Berger P, Niemann A, Suter U (2006) Schwann cells and the pathogenesis of inherited motor and sensory neuropathies (Charcot-Marie-Tooth disease). Glia 54:243–257

Berghs S, Aggujaro D, Dirkx R Jr, Maksimova E, Stabach P, Hermel JM Zhang JP Philbrick W, Slepnev V, Ort T (2000) βIV spectrin, a new spectrin localized at axon initial segments and nodes of ranvier in the central and peripheral nervous system. J Cell Biol 151:985–1002

Bhat MA (2003) Molecular organization of axo-glial junctions. Curr Opin Neurobiol 13:552–559

Bhat MA, Rios JC, Lu Y, Garcia-Fresco GP, Ching W, St Martin M, Li J, Einheber, S, Chesler M, Rosenbluth J (2001) Axon-glia interactions and the domain organization of myelinated axons requires neurexin IV/Caspr/Paranodin. Neuron 30:369–383

Black JA, Waxman SG (1988) The perinodal astrocyte. Glia 1:169–183

Boiko T, Rasband MN, Levinson SR, Caldwell JH, Mandel G, Trimmer JS, Matthews G (2001) Compact myelin dictates the differential targeting of two sodium channel isoforms in the same axon. Neuron 30:91–104

Bonnon C, Goutebroze L, Denisenko-Nehrbass N, Girault JA, Faivre-Sarrailh C (2003) The paranodal complex of F3/contactin and caspr/paranodin traffics to the cell surface via a nonconventional pathway. J Biol Chem 278:48339–48347

Bouzidi M, Tricaud N, Giraud P, Kordeli E, Caillol G, Deleuze C, Couraud F and Alcaraz, G. (2002) Interaction of the Nav1.2a subunit of the voltage-dependent sodium channel with nodal ankyrinG. In vitro mapping of the interacting domains and association in synaptosomes. J Biol Chem 277:28996–29004

Boyle ME, Berglund EO, Murai KK, Weber L, Peles E, Ranscht B (2001) Contactin orchestrates assembly of the septate-like junctions at the paranode in myelinated peripheral nerve. Neuron 30:385–397

Brummendorf T, Rathjen FG (1996) Structure/function relationships of axon-associated adhesion receptors of the immunoglobulin superfamily. Curr Opin Neurobiol 6:584–593

Carlson SD, Juang JL, Hilgers SL, Garment MB (2000) Blood barriers of the insect. Annu Rev Entomol 45:151–174

Chan JR, Jolicoeur C, Yamauchi J, Elliott J, Fawcett JP, Ng BK, Cayouette M (2006) The polarity protein Par-3 directly interacts with p75NTR to regulate myelination. Science 314:832–836

Charles P, Tait S, Faivre-Sarrailh C, Barbin G, Gunn-Moore F, Denisenko-Nehrbass N, Guennoc AM, Girault JA, Brophy PJ, Lubetzki, C (2002) Neurofascin is a glial receptor for the paranodin/Caspr-contactin axonal complex at the axoglial junction. Curr Biol 12:217–220

Chernousov MA, Carey DJ (2000) Schwann cell extracellular matrix molecules and their receptors. Histol Histopathol 15:593–601

Chiu SY, Zhou L, Zhang CL, Messing A (1999) Analysis of potassium channel functions In mammalian axons by gene knockouts. J Neurocytol 28:349–364

Coetzee T, Fujita N, Dupree J, Shi R, Blight A, Suzuki K, Popko B (1996) Myelination in the absence of galactocerebroside and sulfatide: normal structure with abnormal function and regional instability. Cell 86:209–219

Collinson JM, Marshall D, Gillespie CS, Brophy PJ (1998) Transient expression of neurofascin by oligodendrocytes at the onset of myelinogenesis: implications for mechanisms of axonglial interaction. Glia 23:11–23

Coman I, Aigrot MS, Seilhean D, Reynolds R, Girault JA, Zalc B, Lubetzki C (2006) Nodal, paranodal and juxtaparanodal axonal proteins during demyelination and remyelination in multiple sclerosis. Brain 129:3186–3195

Court FA, Wrabetz L., Feltri ML (2006) Basal lamina: Schwann cells wrap to the rhythm of spacetime. Curr Opin Neurobiol 16:501–507

Custer AW, Kazarinova-Noyes K, Sakurai T, Xu X, Simon W Grumet M, Shrager P (2003) The role of the ankyrin-binding protein NrCAM in node of Ranvier formation. J Neurosci 23:10032–10039

Davis JQ, Bennett V (1993) Ankyrin-binding activity of nervous system cell adhesion molecules expressed in adult brain. J Cell Sci Suppl 17:109–117

Davis JQ, Lambert S, Bennett V (1996) Molecular composition of the node of Ranvier: identification of ankyrin-binding cell adhesion molecules neurofascin (mucin+/third FNIII domain-) and NrCAM at nodal axon segments. J Cell Biol 135:1355–1367

Denisenko-Nehrbass N, Faivre-Sarrailh C, Goutebroze, L, Girault JA (2002) A molecular view on paranodal junctions of myelinated fibers. J Physiol 96:99–103

Denisenko-Nehrbass N, Goutebroze L, Galvez T, Bonnon C, Stankoff B, Ezan P Giovannini M, Faivre-Sarrailh C, Girault JA (2003a) Association of Caspr/paranodin with tumour suppressor schwannomin/merlin and 1 integrin in the central nervous system. J Neurochem 84:209–221

Denisenko-Nehrbass N, Oguievetskaia K, Goutebroze L, Galvez T, Yamakawa H, Ohara O, Carnaud M, Girault JA (2003b) Protein 4.1B associates with both Caspr/paranodin and Caspr2 at paranodes and juxtaparanodes of myelinated fibres. Eur J Neurosci 17:411–416

Doerflinger NH, Macklin WB, Popko B (2003) Inducible site-specific recombination in myelinating cells. Genesis 35:63–72

Dupree JL, Coetzee T, Suzuki K, Popko B (1998) Myelin abnormalities in mice deficient in galactocerebroside and sulfatide. J Neurocytol 27:649–659

Dupree JL, Girault JA, Popko B (1999) Axo-glial interactions regulate the localization of axonal paranodal proteins. J Cell Biol 147:1145–1152

Einheber S, Zanazzi G, Ching W, Scherer S, Milner TA, Peles E, Salzer JL (1997) The axonal membrane protein Caspr, a homologue of neurexin IV, is a component of the septate-like paranodal junctions that assemble during myelination. J Cell Biol 139:1495–1506

Eshed Y, Feinberg K, Poliak S, Sabanay H, Sarig-Nadir O, Spiegel I, Bermingham JR Jr, Peles E (2005) Gliomedin mediates Schwann cell-axon interaction and the molecular assembly of the nodes of Ranvier. Neuron 47:215–229

Eshed Y, Feinberg K, Carey DJ, Peles E (2007) Secreted gliomedin is a perinodal matrix component of peripheral nerves. J Cell Biol 177:551–562

Fabrizi GM, Ferrarini M, Cavallaro T, Cabrini I, Cerini R, Bertolasi L, Rizzuto N (2007) Two novel mutations in dynamin-2 cause axonal Charcot-Marie-Tooth disease. Neurol 69:291–295

Faivre-Sarrailh C, Gauthier F, Denisenko-Nehrbass N, Le Bivic A, Rougon G, Girault JA (2000) The glycosylphosphatidyl inositol-anchored adhesion molecule F3/contactin is required for surface transport of paranodin/contactin-associated protein (caspr). J Cell Biol 149:491–502

Faivre-Sarrailh C, Banerjee S, Li J, Hortsch M, Laval M, Bhat MA (2004) Drosophila contactin, a homolog of vertebrate contactin, is required for septate junction organization and paracellular barrier function. Development 131:4931–4942

Falk J, Bonnon C, Girault JA, Faivre-Sarrailh C (2002) F3/contactin, a neuronal cell adhesion molecule implicated in axogenesis and myelination. Biol Cell 94:327–334

Furley AJ, Morton SB, Manalo D, Karagogeos D, Dodd J, Jessell TM (1990) The axonal glycoprotein TAG-1 is an immunoglobulin superfamily member with neurite outgrowth-promoting activity. Cell 61:157–170

Garcia-Fresco GP, Sousa AD, Pillai AM, Moy SS, Crawley JN, Tessarollo L, Dupree JL, Bhat MA (2006) Disruption of axo-glial junctions causes cytoskeletal disorganization and degeneration of Purkinje neuron axons. Proc Natl Acad Sci U S A 103:5137–5142

Garver TD, Ren Q, Tuvia S, Bennett V (1997) Tyrosine phosphorylation at a site highly conserved in the L1 family of cell adhesion molecules abolishes ankyrin binding and increases lateral mobility of neurofascin. J Cell Biol 137:703–714

Gatto CL. Walker BJ, Lambert S (2003) Local ERM activation and dynamic growth cones at Schwann cell tips implicated in efficient formation of nodes of Ranvier. J Cell Biol 162:489–498

Gollan L, Sabanay H, Poliak S, Berglund EO, Ranscht B, Peles E (2002) Retention of a cell adhesion complex at the paranodal junction requires the cytoplasmic region of Caspr. J Cell Biol 157:1247–1256

Gollan L, Salomon D, Salzer JL, Peles E (2003) Caspr regulates the processing of contactin and inhibits its binding to neurofascin. J Cell Biol 163:1213–1218

Goutebroze L, Carnaud M, Denisenko N, Boutterin MC, Girault JA (2003) Syndecan-3 and syndecan-4 are enriched in Schwann cell perinodal processes. BMC Neurosci 4:29

Grumet M (1997) Nr-CAM: a cell adhesion molecule with ligand and receptor functions. Cell Tissue Res 290:423–428

Gunn-Moore FJ, Hill M, Davey F, Herron LR, Tait S, Sherman D, Brophy PJ (2006) A functional FERM domain binding motif in neurofascin. Mol Cell Neurosci 33:441–446

Hassel B, Rathjen FG, Volkmer H (1997) Organization of the neurofascin gene and analysis of developmentally regulated alternative splicing. J Biol Chem 272:28742–28749

Hildebrand C, Remahl S, Persson H, Bjartmar C (1993) Myelinated nerve fibres in the CNS. Prog Neurobiol 40:319–384

Holm J, Hillenbrand R, Steuber V, Bartsch U, Moos M, Lubbert H, Montag D, Schachner M (1996) Structural features of a close homologue of L1 (CHL1) in the mouse: a new member of the L1 family of neural recognition molecules. Eur J Neurosci 8:1613–1629

Honke K, Hirahara Y, Dupree J, Suzuki K, Popko B, Fukushima K, Fukushima J, Nagasawa T, Yoshida N, Wada Y (2002) Paranodal junction formation and spermatogenesis require sulfoglycolipids. Proc Natl Acad Sci U S A 99:4227–4232

Hoover KB, Bryant PJ (2000) The genetics of the protein 4.1 family: organizers of the membrane and cytoskeleton. Curr Opin Cell Biol 12:229–234

Hoshi T, Suzuki A, Hayashi S, Tohyama K, Hayashi A, Yamaguchi Y, Takeuchi K, Baba, H (2007) Nodal protrusions, increased Schmidt-Lanterman incisures, and paranodal disorganization are characteristic features of sulfatide-deficient peripheral nerves. Glia 55:584–594

Howell OW, Palser A, Polito A, Melrose S, Zonta B, Scheiermann C, Vora AJ, Brophy PJ, Reynolds R (2006) Disruption of neurofascin localization reveals early changes preceding demyelination and remyelination in multiple sclerosis. Brain 129:3173–3185

Ichimura T, Ellisman MH (1991) Three-dimensional fine structure of cytoskeletal- membrane interactions at nodes of Ranvier. J Neurocytol 20:667–681

Ishibashi T, Dupree JL Ikenaka K Hirahara Y, Honke K, Peles E, Popko B, Suzuki K, Nishino H, Baba H (2002) A myelin galactolipid, sulfatide, is essential for maintenance of ion channels on myelinated axon but not essential for initial cluster formation. J Neurosci 22:6507–6514

Isom LL (2002) The role of sodium channels in cell adhesion. Front Biosci 7:12–23

Jenkins SM, Bennett V (2001) Ankyrin-G coordinates assembly of the spectrin-based membrane skeleton, voltage-gated sodium channels, and L1 CAMs at Purkinje neuron initial segments. J Cell Biol 155:739–746

Jenkins SM, Bennett V (2002) Developing nodes of Ranvier are defined by ankyrin-G clustering and are independent of paranodal axoglial adhesion. Proc Natl Acad Sci U S A 99:2303–2308

Jessen KR, Mirsky R (2005) The origin and development of glial cells in peripheral nerves. Nat Rev Neurosci 6:671–682

Kaplan MR, Meyer-Franke A, Lambert S, Bennett V, Duncan ID, Levinson SR, Barres BA (1997) Induction of sodium channel clustering by oligodendrocytes. Nature 386:724–728

Kaplan MR, Cho MH, Ullian EM, Isom LL, Levinson SR, Barres BA (2001) Differential control of clustering of the sodium channels Na(v)1.2 and Na(v)1.6 at developing CNS nodes of Ranvier. Neuron 30:105–119

Karagogeos D, Morton SB, Casano F, Dodd J, Jessell TM (1991) Developmental expression of the axonal glycoprotein TAG-1: differential regulation by central and peripheral neurons in vitro. Development 112:51–67

Kearney JA, Buchner DA, De Haan G, Adamska M, Levin SI, Furay AR, Albin RL, Jones JM, Montal M, Stevens MJ (2002) Molecular and pathological effects of a modifier gene on deficiency of the sodium channel Scn8a (Na(v)1.6). Hum Mol Genet 11:2765–2775

Komada M, Soriano P (2002) BetaIV-spectrin regulates sodium channel clustering through ankyrin-G at axon initial segments and nodes of Ranvier. J Cell Biol 156:337-348

Kordeli E, Lambert S, Bennett V (1995) AnkyrinG. A new ankyrin gene with neural- specific isoforms localized at the axonal initial segment and node of Ranvier. J Biol Chem 270:2352–2359

Lacas-Gervais S, Guo J, Strenzke N, Scarfone E, Kolpe M, Jahkel M, De Camilli P, Moser T, Rasband MN, Solimena M (2004) BetaIVSigma1 spectrin stabilizes the nodes of Ranvier and axon initial segments. J Cell Biol 166:983–990

Lambert S, Davis JQ, Bennett V (1997) Morphogenesis of the node of Ranvier: co- clusters of ankyrin and ankyrin-binding integral proteins define early developmental intermediates. J Neurosci 17:7025–7036

Lappe-Siefke C, Goebbels S, Gravel M, Nicksch E, Lee J, Braun PE, Griffiths IR, Nave KA (2003) Disruption of Cnp1 uncouples oligodendroglial functions in axonal support and myelination. Nat Genet 33:366–374

Laval M, Bel C, Faivre-Sarrailh C (2008) The lateral mobility of cell adhesion molecules is highly restricted at septate junctions in Drosophila. BMC Cell Biol 9:38

Lemaillet G, Walker B, Lambert S (2003) Identification of a conserved ankyrin-binding motif in the family of sodium channel alpha subunits. J Biol Chem 278:27333–27339

Lubetzki C, Williams A, Stankoff B (2005) Promoting repair in multiple sclerosis: problems and prospects. Curr Opin Neurol 18:237–244

Lustig M, Zanazzi G, Sakurai T, Blanco C, Levinson SR, Lambert S, Grumet M, Salzer JL (2001). Nr-CAM and neurofascin interactions regulate ankyrin G and sodium channel clustering at the node of Ranvier. Curr Biol 11:1864–1869

Maier O, van der Heide T, van Dam AM, Baron W, de Vries H, Hoekstra D (2005) Alteration of the extracellular matrix interferes with raft association of neurofascin in oligodendrocytes. Potential significance for multiple sclerosis? Mol Cell Neurosci 28:390–401

Maier O, Baron W, Hoekstra D (2007) Reduced raft-association of NF155 in active MS- lesions is accompanied by the disruption of the paranodal junction. Glia 55:885–895

Malhotra JD, Koopmann MC, Kazen-Gillespie KA, Fettman N, Hortsch M, Isom LL (2002) Structural requirements for interaction of sodium channel $\beta 1$ subunits with ankyrin. J Biol Chem 277:26681–26688

Marcus J, Popko B (2002) Galactolipids are molecular determinants of myelin development and axo-glial organization. Biochim Biophys Acta 1573:406–413

Mathey EK, Derfuss T, Storch, MK, Williams KR, Hales K, Woolley DR, Al-Hayani A, Davies SN, Rasband MN, Olsson T (2007) Neurofascin as a novel target for autoantibody-mediated axonal injury. J Exp Med 204:2363–2372

Maurel P, Einheber S, Galinska J, Thaker P, Lam I, Rubin MB, Scherer SS, Murakami Y, Gutmann DH, Salzer JL (2007) Nectin-like proteins mediate axon Schwann cell interactions along the internode and are essential for myelination. J Cell Biol 178:861–874

Melendez-Vasquez CV, Rios JC, Zanazzi G, Lambert S, Bretscher A, Salzer JL (2001) Nodes of Ranvier form in association with ezrin-radixin-moesin (ERM)-positive Schwann cell processes. Proc Natl Acad Sci U S A 98:1235–1240

Melendez-Vasquez CV, Einheber S, Salzer JL (2004) Rho kinase regulates schwann cell myelination and formation of associated axonal domains. J Neurosci 24:3953–3963

Menegoz M, Gaspar P, Le Bert M, Galvez T, Burgaya F, Palfrey C, Ezan P, Arnos F, Girault JA (1997) Paranodin, a glycoprotein of neuronal paranodal membranes. Neuron 19:319–331

Mi F, Berkowitz GA (1995) Development of a K+-channel probe and its use for identification of an intracellular plant membrane K+ channel. Proc Natl Acad Sci U S A 92:3386–3390

Morell P, Toews AD, Wagner, M, Goodrum JF (1994) Gene expression during tellurium- induced primary demyelination. Neurotoxicology 15:171–180

Moscoso LM, Sanes JR (1995) Expression of four immunoglobulin superfamily adhesion molecules (L1, Nr-CAM/Bravo, neurofascin/ABGP, and N-CAM) in the developing mouse spinal cord. J Comp Neurol 352:321–334

Nave KA, Sereda MW, Ehrenreich H (2007) Mechanisms of disease: inherited demyelinating neuropathies–from basic to clinical research. Nat Clin Pract Neurol 3:453–464

Norton WT, Cammer W (1984) Isolation and characterization of myelin. In: Morell P (ed) Myelin. Plenum, New York, pp 147–195

Occhi S, Zambroni D, Del Carro U, Amadio S, Sirkowski EE, Scherer SS, Campbell KP, Moore SA, Chen ZL, Strickland S (2005) Both laminin and Schwann cell dystroglycan are necessary for proper clustering of sodium channels at nodes of Ranvier. J Neurosci 25:9418–9427

Ogawa Y, Schafer DP, Horresh I, Bar V, Hales K, Yang Y, Susuki K, Peles E, Stankewich MC, Rasband M (2006). Spectrins and ankyrinB constitute a specialized paranodal cytoskeleton. J Neurosci 26:5230–5239

Oguievetskaia K, Cifuentes-Diaz C, Girault JA, Goutebroze L (2005) Cellular contacts in myelinated fibers of the peripheral nervous system. Med Sci 21:162–169

Ohara R, Yamakawa H, Nakayama M, Ohara O (2000) Type II brain 4.1 (4.1B/KIAA0987), a member of the protein 4.1 family, is localized to neuronal paranodes. Brain Res Mol Brain Res 85:41–52

Ohno S (2001) Intercellular junctions and cellular polarity: the PAR-aPKC complex, a conserved core cassette playing fundamental roles in cell polarity. Curr Opin Cell Biol 13:641–648

Park J, Liu B, Chen T, Li H, Hu X, Gao J, Zhu Y, Zhu Q, Qiang B, Yuan J (2008) Disruption of Nectin-like 1 cell adhesion molecule leads to delayed axonal myelination in the CNS. J Neurosci 28:12815–12819

Parra M, Gascard P, Walensky LD, Gimm JA, Blackshaw S, Chan N, Takakuwa Y, Berger T, Lee G, Chasis JA (2000) Molecular and functional characterization of protein 4.1B, a novel member of the protein 4.1 family with high level, focal expression in brain. J Biol Chem 275:3247–3255

Pedraza L, Huang JK, Colman DR (2001) Organizing principles of the axoglial apparatus. Neuron 30:335–344

Peles E, Salzer JL (2000) Molecular domains of myelinated axons. Curr Opin Neurobiol 10:558–565

Peles E, Nativ, M, Lustig M, Grumet M, Schilling J, Martinez R, Plowman GD, Schlessinger J (1997) Identification of a novel contactin-associated transmembrane receptor with multiple domains implicated in protein-protein interactions. EMBO J 16:978–988

Pillai AM, Garcia-Fresco GP, Sousa AD, Dupree JL, Philpot BD, Bhat MA (2007) No effect of genetic deletion of contactin-associated protein (CASPR) on axonal orientation and synaptic plasticity. J Neurosci Res 85:2318–2331

Pillai AM, Thaxton C, Pribisko AL, Cheng J-G, Dupree JL, Bhat MA (2009) Spatio- temporal ablation of myelinating glia-specific neurofascin (nfascnf155) in mice reveals gradual loss of paranodal axo-glial junctions and concomitant disorganization of axonal domains. J Neurosci Res Jan 30. [Epub ahead of print]

Pinheiro EM, Montell DJ (2004) Requirement for Par-6 and Bazooka in Drosophila border cell migration. Development 131:5243–5251

Poliak S, Gollan L, Martinez R, Custer A, Einheber S, Salzer JL, Trimmer JS, Shrager P, Peles E (1999) Caspr2, a new member of the neurexin superfamily, is localized at the juxtaparanodes of myelinated axons and associates with K+ channels. Neuron 24:1037–1047

Poliak S, Gollan L, Salomon D, Berglund EO, Ohara R, Ranscht B, Peles E (2001). Localization of Caspr2 in myelinated nerves depends on axon-glia interactions and the generation of barriers along the axon. J Neurosci 21:7568–7575

Poliak S, Salomon D, Elhanany H, Sabanay H, Kiernan B, Pevny L, Stewart CL, Xu X, Chiu SY, Shrager P (2003) Juxtaparanodal clustering of Shaker-like K+ channels in myelinated axons depends on Caspr2 and TAG-1. J Cell Biol 162:1149–1160

Rasband MN, Trimmer JS, Schwarz TL, Levinson SR, Ellisman MH, Schachner M, Shrager P (1998) Potassium channel distribution, clustering, and function in remyelinating rat axons. J Neurosci 18:36–47

Rasband MN, Peles E, Trimmer JS, Levinson SR, Lux SE, Shrager P (1999) Dependence of nodal sodium channel clustering on paranodal axoglial contact in the developing CNS. J Neurosci 19:7516–7528

Rasband MN, Park EW, Zhen D, Arbuckle MI, Poliak S, Peles E, Grant SG, Trimmer JS (2002) Clustering of neuronal potassium channels is independent of their interaction with PSD-95. J Cell Biol 159:663–672

Reid RA, Bronson DD, Young KM, Hemperly JJ (1994) Identification and characterization of the human cell adhesion molecule contactin. Brain Res Mol Brain Res 21:1–8

Rios JC, Melendez-Vasquez CV, Einheber S, Lustig M, Grumet M, Hemperly J, Peles E, Salzer JL (2000) Contactin-associated protein (Caspr) and contactin form a complex that is targeted to the paranodal junctions during myelination. J Neurosci 20:8354–8364

Rodriguez M, Scheithauer B (1994) Ultrastructure of multiple sclerosis. Ultrastruct Pathol 18:3–13

Saito F, Moore SA, Barresi R, Henry MD, Messing A, Ross-Barta SE, Cohn RD, Williamson RA, Sluka KA, Sherman DL (2003) Unique role of dystroglycan in peripheral nerve myelination, nodal structure, and sodium channel stabilization. Neuron 38:747–758

Salzer JL (2003) Polarized domains of myelinated axons. Neuron 40:297–318

Salzer JL, Brophy PJ, Peles E (2008) Molecular domains of myelinated axons in the peripheral nervous system. Glia 56:1532–1540

Savvaki M, Panagiotaropoulos T, Stamatakis A, Sargiannidou I, Karatzioula P, Watanabe K, Stylianopoulou F, Karagogeos D, Kleopa, KA (2008) Impairment of learning and memory in TAG-1 deficient mice associated with shorter CNS internodes and disrupted juxtaparanodes. Mol Cell Neurosci 39:478–490

Schafer DP, Bansal R, Hedstrom KL, Pfeiffer SE, Rasband MN (2004) Does paranode formation and maintenance require partitioning of neurofascin 155 into lipid rafts? J Neurosci 24:3176–3185

Sherman DL, Tait S, Melrose S, Johnson R, Zonta B, Court FA, Macklin WB, Meek S, Smith AJ, Cottrell DF (2005) Neurofascins are required to establish axonal domains for saltatory conduction. Neuron 48:737–742

Shy ME (2006) Peripheral neuropathies caused by mutations in the myelin protein zero. J Neurol Sci 242:55–66

Simons M, Trotter J (2007) Wrapping it up: the cell biology of myelination. Curr Opin Neurobiol 17:533–540

Sousa AD, Bhat MA (2007) Cytoskeletal transition at the paranodes: the Achilles' heel of myelinated axons. Neuron Glia Biol 3:169–178

Spiegel I, Salomon D, Erne B, Schaeren-Wiemers N, Peles E (2002) Caspr3 and caspr4, two novel members of the caspr family are expressed in the nervous system and interact with PDZ domains. Mol Cell Neurosci 20:283–297

Sun CX, Robb VA, Gutmann DH (2002) Protein 4.1 tumor suppressors: getting a FERM grip on growth regulation. J Cell Sci 115:3991–4000

Tait S, Gunn-Moore F, Collinson JM, Huang J, Lubetzki C, Pedraza L, Sherman DL, Colman DR, Brophy PJ (2000) An oligodendrocyte cell adhesion molecule at the site of assembly of the paranodal axo-glial junction. J Cell Biol 150:657–666

Takai Y, Irie K, Shimizu K, Sakisaka T, Ikeda W (2003) Nectins and nectin-like molecules: roles in cell adhesion, migration, and polarization. Cancer Sci 94:655–667

Takekuni K, Ikeda W, Fujito T, Morimoto K, Takeuchi M, Monden M, Takai Y (2003) Direct binding of cell polarity protein PAR-3 to cell-cell adhesion molecule nectin at neuroepithelial cells of developing mouse. J Biol Chem 278:5497–5500

Tepass U, Tanentzapf G, Ward R, Fehon R (2001) Epithelial cell polarity and cell junctions in Drosophila. Annu Rev Genet 35:747–784

Traka M, Dupree JL, Popko B, Karagogeos D (2002) The neuronal adhesion protein TAG-1 is expressed by Schwann cells and oligodendrocytes and is localized to the juxtaparanodal region of myelinated fibers. J Neurosci 22:3016–3024

Traka M, Goutebroze, L, Denisenko N, Bessa M, Nifli A, Havaki S, Iwakura Y, Fukamauchi F, Watanabe K, Soliven B (2003) Association of TAG-1 with Caspr2 is essential for the molecular organization of juxtaparanodal regions of myelinated fibers. J Cell Biol 162:1161–1172

Trapp BD, Nave KA (2008) Multiple sclerosis: an immune or neurodegenerative disorder? Annu Rev Neurosci 31:247–269

Tzimourakas A, Giasemi S, Mouratidou M, Karagogeos D (2007) Structure-function analysis of protein complexes involved in the molecular architecture of juxtaparanodal regions of myelinated fibers. Biotechnol J 2:577–583

Vernes SC, Newbury DF, Abrahams BS, Winchester L, Nicod J, Groszer M, Alarcon M, Oliver PL, Davies KE, Geschwind DH (2008) A functional genetic link between distinct developmental language disorders. N Engl J Med 359:2337–2345

Voas MG, Lyons DA, Naylor SG, Arana N, Rasband MN, Talbot WS (2007) αII-spectrin is essential for assembly of the nodes of Ranvier in myelinated axons. Curr Biol 17:562–568

Volkmer H, Hassel B, Wolff JM, Frank R, Rathjen FG (1992) Structure of the axonal surface recognition molecule neurofascin and its relationship to a neural subgroup of the immunoglobulin superfamily. J Cell Biol 118:149–161

Wang H, Kunkel DD, Martin TM, Schwartzkroin PA, Tempel BL (1993) Heteromultimeric K^+ channels in terminal and juxtaparanodal regions of neurons. Nature 365:75–79

Ward RE, Lamb RS, Fehon, RG (1998) A conserved functional domain of Drosophila coracle is required for localization at the septate junction and has membrane- organizing activity. J Cell Biol 140:1463–1473

Waxman SG, Ritchie JM (1993) Molecular dissection of the myelinated axon. Ann Neurol 33:121–136

Wolswijk G, Balesar R (2003) Changes in the expression and localization of the paranodal protein Caspr on axons in chronic multiple sclerosis. Brain 126:1638–1649

Yu FH, Catterall WA (2003) Overview of the voltage-gated sodium channel family. Genome Biol 4:207

Zhang X, Davis JQ, Carpenter S, Bennett V (1998) Structural requirements for association of neurofascin with ankyrin. J Biol Chem 273:30785–30794

Zhou D, Lambert S, Malen PL, Carpenter S, Boland LM, Bennett V (1998a) AnkyrinG is required for clustering of voltage-gated Na channels at axon initial segments and for normal action potential firing. J Cell Biol 143:1295–1304

Zhou L, Zhang CL, Messing A, Chiu SY (1998b) Temperature-sensitive neuromuscular transmission in Kv1.1 null mice: role of potassium channels under the myelin sheath in young nerves. J Neurosci 1:7200–7215

Zhou L, Messing A, Chiu SY (1999) Determinants of excitability at transition zones in Kv1.1-deficient myelinated nerves. J Neurosci 19:5768–5781

Zonta B, Tait S, Melrose S, Anderson H, Harroch S, Higginson J, Sherman DL, Brophy PJ (2008) Glial and neuronal isoforms of Neurofascin have distinct roles in the assembly of nodes of Ranvier in the central nervous system. J Cell Biol 181:1169–1177

Organizational Dynamics, Functions, and Pathobiological Dysfunctions of Neurofilaments

Thomas B. Shea, Walter K.-H. Chan, Jacob Kushkuley, and Sangmook Lee

Abstract Neurofilament phosphorylation has long been considered to regulate their axonal transport rate, and in doing so it provides stability to mature axons. We evaluate the collective evidence to date regarding how neurofilament C-terminal phosphorylation may regulate axonal transport. We present a few suggestions for further experimentation in this area, and expand upon previous models for axonal NF dynamics. We present evidence that the NFs that display extended residence along axons are critically dependent upon the surrounding microtubules, and that simultaneous interaction with multiple microtubule motors provides the architectural force that regulates their distribution. Finally, we address how C-terminal phosphorylation is regionally and temporally regulated by a balance of kinase and phosphatase activities, and how misregulation of this balance might contribute to motor neuron disease.

1 Introduction

The orderly assembly of initially soluble subunits to form the complex network collectively referred to as the cytoskeleton presents a formidable challenge to any cell. Neurons, unlike other cells, must selectively transport cytoskeletal elements over distances that vastly exceed their perikarya into their axons. The cytoskeleton of the axon differs considerably from that of the perikaryon, and even from that of dendrites. Key differences include the large number of longitudinally oriented neurofilaments (NFs), which are enriched in C-terminal modification by

---

T.B. Shea (✉), W.K.-H. Chan, J. Kushkuley, and S. Lee
Departments of Biological Sciences and Biochemistry, Center for Cellular Neurobiology and Neurodegeneration Research, University of Massachusetts Lowell, One University Avenue, Lowell, MA, 01854
e-mail: Thomas_Shea@uml.edu

phosphorylation. For decades, the network of crosslinked, fibrous images observed via electron and conventional fluorescence microscopy, coupled with pioneering but simple radiolabeling techniques, led us to consider the axonal cytoskeleton as a support mechanism that, once assembled, was relatively inert. However, the advent of fluorescently conjugated cytoskeletal proteins and the ability to manipulate kinase activities in situ have revealed that (1) the axonal cytoskeleton is an incredibly dynamic structure, and that (2) the "crosslinks" are often more like handshakes along a moving reception line than permanent bolts between girders.

Development and maintenance of a functional nervous system is by definition dependent upon orderly elaboration and maintenance of the axonal array, which in turn is critically dependent upon organization of the axonal cytoskeleton. Although it is clear that compromise in axonal cytoskeletal dynamics can foster a range of mental and neuromuscular disorders throughout life, many fundamental processes that regulate axonal cytoskeletal organization remain the subject of controversy. One such controversial area is the role of NF phosphorylation in axonal transport.

The nature of this controversy, as discussed herein, is in part due to the complexity of NF phosphorylation, which involves the action of multiple kinases and hierarchical phosphorylation of multiple loci on NF subunits. In addition, however, a significant portion of the controversy arises from semantics and from different laboratories using approaches that highlight different aspects of NF dynamics.

Neurofilaments (NFs) are among the most abundant constituents of the axonal cytoskeleton. NFs have classically been considered to consist of three subunits, termed NF-H, NF-M, and NF-L, corresponding to heavy, medium, and light in reference to their molecular mass (Julien and Mushynski 1998). More recently, it has been demonstrated that an additional neuronal intermediate filament, alpha-internexin, is actually a fourth subunit (Yuan et al., 2006b). The C-terminal regions ("sidearms") of NF-H and NF-M contain multiple phosphorylation sites (Julien and Mushynski 1998) and protrude laterally from the filament backbone when phosphorylated (Sihag et al. 2007). Phosphorylation of these C-terminal sidearms regulates the interactions of NFs with each other and with other cytoskeletal structures, and, in doing so, it mediate the formation of a cytoskeletal lattice that supports the mature axon (Nixon 1998; Pant and Veeranna 1995; Sihag et al. 2007). A considerable body of evidence, spanning several decades, from a number of laboratories using diverse systems and approaches, supports the notion that phosphorylation of C-terminal sidearms, in particular those of NF-H, regulates NF axonal transport (Ackerley et al. 2003; Collard et al. 1995; DeWaegh et al. 1992; Hoffman et al. 1983; Jung and Shea 1999, Jung et al. 2000a, b; Lewis and Nixon 1988; Komiya et al. 1987; Marszalek et al. 1996; Nixon 1993; Shea et al. 1993; Tu et al. 1995; Watson et al. 1993; Yabe et al. 2001a, b; Zhang et al. 1997; Zhu et al. 1998). Given the wealth of supporting information for this role, it was perhaps to be expected that the appearance of a report suggesting that NF-H played no role in the regulation of NF transport would stimulate debate, among which included a challenge to the extent to which their data supported this unanticipated claim (Rao et al. 2002, 2003; Shea et al. 2003; Yuan et al. 2006a, b). Herein, we evaluate the collective evidence to date for and against a role for NF-H C-terminal phosphorylation in regulation of NF axonal transport, and present a few suggestions for further experimentation in this area.

2 C-Terminal Phosphorylation Regulates NF Axonal Transport

A considerable amount of the prior evidence presented to support a role for NF-H C-terminal phosphorylation in the regulation of NF axonal transport was correlative rather than experimental (Archer et al. 1994; Hoffman et al. 1983; Lewis and Nixon 1988; Nixon and Logvinenko 1986); that is, regional and/or developmental slowing of transport rates were regionally or temporally associated with the increases in NF-H C-terminal phosphorylation. Conversely, however, simultaneous analyses within the same optic axons of transport rates of differentially phosphorylated forms of NF-H revealed that the least phosphorylated NF-H variants (which migrated at 160 kDa on SDS-gels) transported twice as fast as NF-H subunits phosphorylated to the extent that they migrated on SDS-gels at 200 kDa. The subset of these 200 kDa NFs that displayed a unique phospho-epitope recognized by the developmentally delayed monoclonal antibody RT97, and enriched in those NFs undergoing the slowest transport (Lewis and Nixon 1983; Yabe et al. 2001a, b), underwent a further twofold slower transport (Jung et al., 2000a, b). This fourfold range of transport rates displayed simultaneously within the same axons by differentially phosphorylated NF-H subunits in the absence of any experimental manipulation supports a role for NF-H C-terminal phosphorylation in regulation of NF axonal transport. Moreover, direct manipulation of phosphorylation state within these optic axons in situ by regional application of a phosphatase inhibitor within optic axons in situ both increased RT97 immunoreactivity and decreased the rate of NF transport without decreasing the transport of other cargo (Jung and Shea 1999). These latter experimental studies provide more than correlative evidence for a role of C-terminal NF phosphorylation in regulation of transport rate. The direct association of RT97 with slower-transporting NFs in these latter studies also argues against dismissing prior regional and developmental correlations between NF phosphorylation due to their correlative nature. Finally, overexpression of NF-H dramatically slowed NF axonal transport in situ (Collard et al. 1995; Marszalek et al. 1996).

Additional attempts to address the role of C-terminal phosphorylation in NF dynamics included deletion of NF-H, which resulted in an increased rate of NF transport (Jung et al. 2006; Zhu et al. 1998). However, this was accompanied by a significant increase in axonal microtubules (Zhu et al. 1998). As NFs undergo anterograde axonal transport, at least in part via the microtubule motor kinesin (Xia et al. 2003; Yabe et al. 1999, 2000), it can therefore be effectively argued that the observed increased rate of transport is due to increased availability of transport machinery, and perhaps also due to an increased NF-M content, because NF-M is thought to contribute to the association of NFs with kinesin (Yabe et al. 2000). Selective deletion of the sidearm, which does not invoke a compensatory increase in axonal microtubules, or NF-M (Rao et al. 2002), is indeed a more refined approach towards the investigation of any role for NF C-terminal sidearms than complete subunit deletion. To accomplish this, Rao and colleagues inserted a gene expressing a truncated, tail-less NF-H into mice, in which the endogenous full-length NF-H had been deleted.

However, this approach is still encumbered by compensatory increases in phosphorylation of the NF-M C-terminal sidearm. Most strikingly, this compensatory increase includes the de novo appearance of RT97 immunoreactivity on NF-M, indicating a degree of reciprocity between NF-H and NF-M (Rao et al. 2002; Sanchez et al. 2000; Tu et al. 1995), which may affect NF transport. In this regard, those NFs from NF-H mice bearing RT97 immunoreactivity on NF-M were selectively excluded from a standard kinesin motor preparation (Shea and Chan 2008), which is analogous to the exclusion from this motor preparation of NFs from normal mice that bear RT97 on NF-H (Jung et al. 2006). As the above correlative and experimental evidence suggests a relationship between C-terminal phosphorylation events that generate RT97 immunoreactivity with regulation of NF transport rate, compensatory phosphorylation of the NF-M C-terminal sidearm to foster RT97 immunoreactivity confounds interpretation of the impact of NF-H C-terminal deletion on NF transport in the same manner as the compensatory increase in microtubules and in NF-M subunits that accompany deletion of the entire NF-H molecules. Studies involving deletion of the C-terminal region of NF-M (Rao et al. 2003) similarly cannot effectively address the impact of NF-M C-terminal phosphorylation on transport regulation due to the presence of intact NF-H (Jung et al. 2006).

Studies of the impact of C-terminal deletion on NF transport in situ have thus far been confined to relatively early periods following the administration of radiolabel (3–14 days; Rao et al. 2002, 2003; Shea et al. 2003; Yuan et al. 2006a, b). There are two major flaws in this approach. The first problem is that the effect of C-terminal phosphorylation is confined to the trailing aspect of the wave, *not* the leading edge. "Extensively phosphorylated" NFs are those that exhibited a progressive slowing of transport rate, while nonphosphorylated NFs, instead, maintained a constant rate (Jung et al. 2000a, b). In these studies, extensively phosphorylated NF-H were confined to the most proximal segments, while the leading edge, which is completely lacking phospho-dependent RT97 immunoreactivity, continued to progress along axons. The second problem, directly related to the first, is that it is far too early at 3–14 days to attempt any definitive examination of the impact of phosphorylation on NF retention along axons. The slowest-moving NFs, which include those that partition within the so-called "stationary phase" retained for extend periods (e.g., 1 year; Nixon and Logvinenko 1986), are those that are the most enriched in C-terminal phosphorylation (Lewis and Nixon 1988). While C-terminal phosphorylation can clearly foster retention of some NFs along axons, the lack of C-terminal phosphorylation sites should not be expected to increase transport of the leading edge of the moving wave, *because NFs in this region are not phosphorylated even in intact NF-H* (Jung et al. 2000a, b). The failure of "truncated" NFs to speed up should not be extrapolated to indicate that they would also exhibit a subsequent failure to slow down; this possibility can only be addressed instead by long-term pulse-chase analyses, of the sort previously utilized to demonstrate that phosphorylation indeed slows transport of NFs comprised of full-length subunits (Lewis and Nixon 1988; Nixon and Logvinenko 1986).

Additional compelling experimental evidence supporting a role for NF-H C-terminal phosphorylation in the regulation of NF axonal transport comes from

Miller and colleagues (Ackerley et al. 2003), who demonstrated the transfection of cultured cortical neurons with constructs expressing site-directed mutated forms of NF-H along with nonmutated "wild-type" NF-H, each of which was tagged with green fluorescent protein to allow real-time monitoring of axonal transport within living neurons. In these analyses, seven consensus sites, known to be phosphorylated within the C-terminal region of NF, were mutated to alanine (to prevent phosphorylation), or aspartate (to mimic phosphorylation), generating "constitutively nonphosphorylated" and "constitutively phosphorylated" NF-H, respectively. Analyses of transport rates demonstrated that NFs containing constitutively nonphosphorylated NF-H transported faster than those containing wild-type NF-H, while NFs containing constitutively phosphorylated NF-H transported slower than those containing wild-type NF-H (Ackerley et al. 2003).

Site-directed mutagenesis of the NF-H C-terminal sidearm also interfered with the effects of kinases known to regulate NF transport. p24/44 MAP kinase-mediated NF phosphorylation is essential for anterograde NF transport (Chan et al. 2004), while phosphorylation by Cdk5 inhibits it (Ackerley et al. 2003; Shea et al. 2004). Constitutive phosphorylation of the C-terminal NF-H sidearm prevented the inhibition of anterograde NF axonal transport that normally accompanies MAP kinase inhibition, and prevented the acceleration of NF transport, which normally accompanies cdk5 inhibition (Ackerley et al. 2003; Shea et al. 2004). Finally, site-directed mutagenesis altered the association of NFs bearing these mutations with kinesin: mutation of C-terminal cdk5-consensus sites to alanines increased the association of NFs with kinesin as compared to wild-type NFs, while mutation to aspartate decreased their association (Shea and Chan 2008). These data further support a role for C-terminal NF-H phosporylation in the regulation of NF transport, and can be considered to represent a somewhat more refined approach than truncation of a substantial length of the sidearm (Rao et al. 2003).

3 NF Transport and Residence Time Along Axons is Regulated by a Combination of Microtubule Motors and C-Terminal NF-H Phosphorylation

Two interrelated, well-described phenomena provide mechanisms by which C-terminal NF phosphorylation can impact NF axonal transport: phospho-mediated dissociation of NFs from their anterograde transport system and phospho-mediated NF–NF associations that compete with transport. Studies from several laboratories demonstrate that the microtubule motors kinesin and dynein mediate anterograde and retrograde NF axonal transport, respectively (He et al. 2005; Motil et al. 2006a, b; Prahlad et al. 2000; Shah et al. 2000; Theiss et al. 2005; Yabe et al. 1999, 2000; Xia et al. 2003). In this regard, C-terminal phosphorylation of NF-H progressively restricts association of NFs with kinesin to the point where RT97-reactive NFs do not associate with kinesin at all (Yabe et al. 1999), but instead demonstrates selective affinity for dynein (Motil et al. 2006a, b); restriction of binding to an anterograde motor provides a mechanism by which C-terminal phosphorylation can slow NF axonal transport, especially when coupled with promotion of binding to a retrograde motor. The ability of kinesin and dynein (motors which undergo so-called "fast" axonal transport) to translocate cargo such as NFs (the majority of which undergo so-called "slow" transport) has been validated by the demonstration that NFs undergo a series of rapid excursions at a fast rate, interspersed with prolonged pauses, which averages out to slow transport (Wang et al. 2000; Roy et al. 2000). This also accounts for the ability of some NF populations to undergo rapid transport.

In addition to regulation of motor protein association, C-terminal NF phosphorylation promotes NF–NF associations, leading to the generation of NF "bundles" (Yabe et al. 2001a, b). So-called "bundled" NFs underwent transport and/or exchange at a rate of at least 2× slower than did the surrounding "individual" NFs (Yabe et al. 2001a). RT97 immunoreactivity is selectively concentrated within the bundles, and phosphatase inhibition increases bundle size within axonal neurites. NFs containing the above constitutively phosphorylated C-terminal NF–H are also selectively concentrated within the bundles, while the constitutively nonphosphorylated forms are selectively excluded (Chan et al. 2005). Notably, this is the opposite of their respective affinity with kinesin, which underscores the reciprocal influence of NF phosphorylation on anterograde transport and NF–NF associations that can lead to bundle formation. These analyses in culture are consistent with the findings of Lewis and Nixon (1988) that RT97-reactive NFs were the slowest-moving NFs within axons in situ. Phospho-mediated NF–NF associations are likely to generate a "macro-structure" that precludes effective transport (Shea and Flanagan 2001; Shea and Yabe 2000). Whether the formation of NF–NF associations requires additional phosphorylation events beyond those that restrict motor binding is not clear as yet. Nevertheless, as phospho-NFs can bind to each other, this gives rise to an additional competing force, unique to the extensively phosphorylated NFs, and not present for nonphosphorylated/less phosphorylated NFs that can interfere with NF-motor associations.

The dynamics of moving and pausing NFs in terms of their association and dissociation with motor systems have been mathematically modeled, and this model agrees with the published data from several in vivo systems (Brown et al. 2005; Craciun et al. 2005). This model describes a pool of NFs associated with motors and a pool that is dissociated from motors but can readily reassociate with them. While this model did not present any definitive regulatory factors, the above studies would suggest that NF C-terminal phosphorylation is one factor that regulates the shift of NFs between these two pools. In further support of this notion, comparative analyses revealed that, when they were actually moving, constitutively phosphorylated NFs underwent transport at the same rate as the constitutively nonphosphorylated NFs; however, constitutively phosphorylated NFs paused more frequently than the constitutively nonphosphorylated NFs (Ackerley et al. 2003). The model presented for NF transport dynamics by Brown and colleagues (Brown et al. 2005; Craciun et al. 2005; Trivedi et al. 2007) also describes a third "pool" of NFs, which are dissociated from motor systems and are restricted in some capacity from entering the pool of NFs, which can reversibly associate with motors; the above studies would identify this latter pool as those NFs that have formed NF–NF associations to the extent that they are incapable of undergoing transport at least pending dissociation of these inter-NF linkages. Shea and Yabe (2000) presented a model of how continued formation of NF–NF associations could "trap" some NFs within the central region of a bundle. Any NF localized within a bundle would be restricted from reassociation with motors pending (1) dissociation and removal of sterically interfering NFs, (2) dissociation of any direct NF–NF associations of the given NF itself (which may require dephosphorylation), (3) availability of a cargo-free motor, and (4) association of the NF with that motor (which may also require an additional dephosphorylation event). This concept of steric hindrance of bundled NFs from reassociating with motors has been given experimental support by ongoing studies from our laboratory (Chan et al. 2007) in which some NFs dissociated from bundles by calcium chelation readily underwent kinesin-dependent association with MTs.

Brown and colleagues have recently expanded their modeling to include an additional "stationary" phase (Trivedi et al. 2007), which, like the earlier in vivo analyses of Lewis and Nixon (1988) and our analyses in differentiated neuroblastoma (Yabe et al. 2001a), is comprised of the slowest-moving "population" of NFs. Our earlier model of how phosphorylation can regulate NF transport (Shea and Flanagan 2001; Shea and Yabe 2000) can be expanded according to the more recent transport model of Brown and colleagues (Brown et al. 2005; Craciun et al. 2005; Trivedi et al, 2007), and integrated with it as follows.

Let us consider that nonphosphorylated NFs, which cannot form NF–NF bundles, can be characterized as either being on their motor or off their motor. By contrast, the distribution of NFs that have undergone key phosphorylation events not only encompasses these two states, but also includes a third condition: NFs that have undergone bundling, which restricts them from motor-dependent transport (Shea and Flanagan 2001; Shea and Yabe 2000). For simplicity, we will not further subdivide these populations into groups such as NFs that are associated with the motor, but with the motor "off track" as included in the model by Brown and colleagues (Brown et al. 2005; Craciun et al. 2005; Trivedi et al. 2007). Similarly, we will not distinguish between simple NF-NF associations and association of NFs with a bundle of NFs, but will instead group all NFs that have formed NF–NF associations as one population. Also for simplicity, we will assume that a given NF can move between/among available states with equivalent probability. Accordingly, nonphospho NFs would be expected to be associated with their motor 50% of the time, and dissociated from their motor the remaining 50% of the time. Phospho-NFs, by contrast, would be associated with their motor 33.3% of the time and dissociated from their motor 33.3% of the time (but not necessarily associated with other NFs), and associated with one or more other NFs for the remaining 33.3% of the time. As NF transport is contingent upon motor association and the rate of transport is dependent upon the frequency of motor association (Blum and Reed 1989), the selective decreased probability of phospho-NFs association with their anterograde transport system (33.3% vs. 50% for nonphosphorylated NFs) immediately suggests that they would undergo decreased overall anterograde transport.

Notably, the dynamics of NF bundling can contribute to dramatic additional slowing of net transport of phospho-NFs; once embedded within a bundle, a given NF must rely on elimination of all surrounding NFs before it can dissociate from the bundle and be available to resume motor association. Notably, if each such step is reversible, an "embedded" NF is equally likely to toggle back and forth between levels within a bundle, as opposed to consistently progressing towards the surface and dissociating from the bundle. Finally, even if a given NF dissociate from the bundle and is available for motor association, any passing motors may already be occupied with cargo (Shea and Yabe 2000). Both of these latter considerations would impose additional reductions in net NF transport rate.

These multiple, phospho-dependent dynamics could easily generate the wide range of transport rates observed for NFs in the various systems where studied, and strongly support a role for C-terminal NF phosphorylation in the slowing of NF axonal transport, and selective retention of phospho-NFs along axons for extended periods after less phosphorylated variants have translocated the full length of the axon (Jung et al. 2000a; Lewis and Nixon 1988; Nixon and Logvinenko 1986; Yabe et al 2001a).

It could be argued that phosphorylation could progressively slow NF axonal transport and generate the observed resident population of NFs by fostering longer periods of dissociation of NFs from their anterograde motor without invoking the need for bundling. However, simple dissociation, by phosphorylation either of NFs or of kinesin/linker proteins, is apparently inadequate in and of itself to generate a long-lasting resident NF population, as inhibition of p42/44 MAP kinase, known to be essential for anterograde NF axonal transport, did not simply curtail anterograde NF transport and leave NFs "in place" along axons. Rather, it resulted in bulk retrograde NF transport (Chan et al. 2004). Retrograde NF transport had been observed within axons under normal conditions, and could be highlighted by manipulations such as photobleaching regions of axons (which revealed that the bleached zone filled in from distal as well as proximal edges) or axonal transection (which, by blocking continuous anterograde flow of radiolabeled subunits, allowed retrograde NF transport to be observed distal to the transection); however, NFs undergoing retrograde transport were a minority as compared to those undergoing anterograde transport (Koehnle and Brown 1999; He et al. 2005; Roy et al. 2000; Wang et al. 2000; Motil et al. 2006a; Yabe et al. 1999, 2000). In studies where NF particle movement was quantified, 69–77% of moving NFs/NF particles underwent anterograde transport (Wang and Brown 2000; Roy et al. 2000; Chan et al. 2004). Individual particles were clearly capable of movement in either direction, as some were observed to change directionality of movement (Roy et al. 2001; Wang and Brown 2001; Chan et al. 2004), and inhibition of MAP kinase (required for anterograde NF transport) caused a "flip" of 76.7% ± 9% particles moving in an anterograde direction to 75.6% ± 12% moving in a retrograde direction (Chan et al. 2004). Bulk retrograde NF transport following inhibition of anterograde transport as seen by Chan et al. suggested that virtually all axonal NFs were subject to retrograde forces at all times, but net retrograde translocation was in some way counterbalanced and, in the case of most of the moving wave, outweighed by anterograde forces. Studies from several laboratories demonstrated that dynein mediated retrograde NF transport, confirming the existence of an opposing motor force (He et al. 2005; Motil et al. 2006a, b; Shah et al. 2000; Theiss et al. 2005). In addition to mediating retrograde transport, dynein also propels MTs in an anterograde direction via cargo-based interactions with actin filaments, and/or larger MTs, and, in doing so, delivers MTs into axons (Baas et al., 2006). NFs interacting with such MTs would therefore undergo dynein-based translocation into axons as MT cargo (Motil et al. 2006a).

Notably, however, the entire NF bundle retracted following inhibition of anterograde transport, while the axon itself did not retract, nor was overall axonal transport altered. This finding challenges the notion that NFs within the bundle underwent slower transport because the bundle itself was simply too large to be translocated (Lewis and Nixon 1988; Yabe et al. 2001a, b). Rather, it suggests that the maintenance of this macrostructure along the axons was also in some manner dependent upon the counterbalance of opposing motor forces. In this regard, our prior studies demonstrate that dynein preferentially associated with phospho-NFs (Motil et al. 2006a, b). Notably, this is the opposite of what had been observed for kinesin, which preferentially associated with nonphosphorylated NFs (Yabe et al. 2000). These findings suggested that a drag imparted by dynein could be responsible for the selective slowing in the transport rate of phospho-NFs, while allowing less/nonphosphorylated NFs to continue transporting at relatively faster rate. This was perhaps to be anticipated, as the most highly phosphorylated NFs are selectively incorporated into bundles (Yabe et al. 2001a, b). In addition, however, we observed that manipulation of dynein activity influenced the bundle. Dynein-mediated slowing of transport, or cessation of transport due to simultaneous binding of kinesin and dynein, may facilitate or even be required for the formation of the divalent cation-mediated associations of phospho-NFs that lead to bundling. We interpret these data to indicate that both motors are likely to be critical for NF bundle formation. This could occur simply by increasing residence time of NFs within axons, which could foster an increase in NF C-terminal phosphorylation, including those events critical for NF–NF associations. This line of reasoning is consistent with the hypotheses of Glass and Griffin (1994), who suggested that retrograde NF transport was one mechanism contributing to increased residence time of NFs along axons. As NFs contain contiguous phosphorylated and nonphosphorylated domains (Brown 1998; Chan et al. 2005), NFs in various overall phosphorylation states may retain the capacity to associate simultaneously with both motors.

As C-terminal phosphorylation promotes NF association with dynein (Motil et al. 2006a, b), the bulk of axonal NFs are highly phosphorylated (Pant and Veeranna 1995), and equivalent amounts of NFs/NF particles are capable of translocation in either direction following kinase manipulation (above), the question remains as to why only a minority of axonal NFs undergo retrograde transport under normal conditions. The answer may lie in phospho-mediated NF bundling itself. Those NFs that are capable of association with dynein (i.e., those that have undergone extensive C-terminal phosphorylation) are also the ones that form/associate with NF bundles. Thus, few NFs are freely available to undergo retrograde transport at any given time due to their tendency to form NF–NF associations leading to bundling. The major function of dynein with regard to NFs may not be to generate retrograde NF translocation, but rather to provide a counterbalance to anterograde transport. This line of reasoning is supported by the observation of robust retrograde transport within growth cones, which lack NF bundles (Chan et al. 2002). The potential importance of a balancing force provided by dynein activity in balancing the effects of kinesin was demonstrated by the generation of aberrant NF aggregates within axons following overexpression of dynamitin. When cells were transfected with a construct expressing dynamitin, those cells exhibiting modest expression of exogenous dynamitin displayed a net shift of NFs towards the distal end of neurites. However, those cells displaying robust expression of exogenous dynamitin displayed large NF aggregates within central and distal axonal regions. Finally, those cells displaying the most extreme levels of exogenous dynamitin displayed NF aggregates within the most proximal regions of axons (Motil et al. 2006a,b). These observations support the notion that the opposing force generated by dynein is an important regulator of orderly anterograde transport, and that an imbalance in motor activities may foster the development of NF spheroids characteristic of human motor neuron disease (discussed more fully below). Notably, shifting of the entire bundle following inhibition of motors is consistent with the notion that the cytoskeleton may be able to undergo some degree of transport "en bloc".

In efforts to understand the biomechanics underlying this situation, we considered whether hypotheses and experimental evidence already advanced for another major axonal cytoskeletal constituent, microtubules (MTs) could also be relevant to NFs. MTs of varying lengths are observed along axons, but they display different motilities. Relatively short MTs can be observed to translocate, while long MTs remain stationary for extended periods. Both kinesin and dynein mediate translocation of relatively short MTs, which translocate along the longer MTs. In this case, a given motor would interact with a short MT via its cargo domain, interact with the longer MT via its MT-binding domain, and "walk" the short MT along the longer MT, and possibly the actin cortex. It seemed unlikely that these motors would somehow selectively interact with shorter rather than longer MTs, yet the longer MTs did not exhibit movement. This prompted the suggestion that multiple motors exerting anterograde and retrograde forces canceled each other out, and, more importantly, exerted a crosslinking effect. Just like with short MTs, a given motor would interact with one MT via its cargo domain, and an adjacent MT via its MT-binding domain. However, if both MTs were long, then multiple motors of differing directionality would interact with both of these MTs, and essentially cancel each other out (Ahmad et al. 2000; Baas et al. 2006). In contrast to MTs, however, crosslinking of NFs by either kinesin or dynein is unlikely because, while NFs are known cargo for both motors, there is no evidence that they act as tracks. Accordingly, a single motor can interact with an NF via its cargo domain, but it could not interact with an adjacent NF via its MT-binding domain. However, retrograde movement of the bundle following inhibition of MAP kinase confirms that dynein has a functional association with the bundle. We, therefore, hypothesize that the MT-binding domains of bundle-associated motors are associated with adjacent MTs, and that the bundle maintains its normal distribution along axons via crosslinking to adjacent MTs. Notably, the MT array may also be tethered to the actin cortex, which may contribute to NF transport and distribution (Jung et al. 2004).

4 NF Phosphorylation State and Localization in Neuropathological Conditions: NFs May Contribute to Motor Neuron Disease by Sequestering Motor Proteins and/or Mitochondria

One hallmark of affected motor neurons in amyotrophic lateral sclerosis ALS/MND is the accumulation of filamentous "spheroids" within proximal axons (Julien and Mushynski 1998; Rao and Nixon 2003; Sihag et al. 2007). Spheroids are comprised of disorganized neurofilaments NFs displaying epitopes normally found exclusively on NFs within distal axonal regions. The pattern of NF-H phosphorylation in spinal tissue from ALS patients is identical to that of normal individuals; one interpretation is that perikaryal/proximal axonal phospho-NF spheroids are comprised of normally phosphorylated NFs that are simply mislocated (Bajaj et al. 1999). We hypothesized some time ago that precocious activation of one or more critical NF kinases could perturb association of NFs with their anterograde motor and promote aberrant accumulation by fostering precocious NF–NF associations (Shea and Flanagan 2001). Several experimental manipulations support this possibility. The NF kinases cdk5 and p38MAP kinase each phosphorylate NFs to generate epitopes in common with NF spheroids and are associated with ALS (Ackerley et al. 2004; Bajaj et al. 1999; Strong et al. 2001). Overexpression of MEKK-1, which in turn overactivated the stress-activated /c-jun terminal kinase, inhibited translocation of NFs into growing axonal neurites and fostered accumulation of axonal-specific phospho-NF epitopes within perikarya (DeFuria et al. 2006). Notably, deletion of the C-terminal region of NF-H delayed motor neuron pathology in a murine model of ALS (Lobsiger et al. 2005). Finally, as described earlier, injection of the phosphatase inhibitor okadaic acid invoked rapid de novo accumulation of RT97 immunoreactivity within perikarya and proximal axons of optic pathway (areas in which this degree of phospho-NF immunoreactivity is normally excluded (Jung and Shea 1999; Sanchez et al. 2000)), and simultaneously slowed NF transport within murine retinal ganglion cells in situ (Sanchez et al. 2000). This finding confirms that the kinase(s) that generate extensive C-terminal NF phosphorylation are present and active within perikarya and proximal axons. This not only suggests that the extent of C-terminal NF phosphorylation within these regions is dependent upon a balance of kinase and phosphatase activities, but also supports the notion that decreased compensatory phosphatase activity could contribute to aberrant NF phosphorylation and mislocation.

A phospho-dependent decrease in association of NFs with kinesin (Yabe et al. 2000), a phospho-dependent increase in association of NFs with dynein (Motil et al. 2006a, b), and/or a phospho-dependent increase in association of NFs with each other (Shea et al. 2004; Yabe et al. 2001a) - in short, all of the phospho-dependent dynamics discussed thus far – could contribute to perikaryal accumulation of phospho-NFs, either by inhibiting anterograde NF transport, enhancing a retrograde "pull," and/or by inducing precocious NF–NF associations within perikarya instead of axons. Which of these forces is the major factor leading to accumulation of NFs within perikarya/proximal axons remains to be determined, but it is likely that they each contribute to some degree. Notably, NF spheroids accumulate kinesin and dynein (Toyoshima et al. 1998a, b), which may represent an apparent futile attempt of these motors to translocate NFs that are "trapped" within the spheroids. Motor entrapment could ultimately impair overall axonal transport, including that of non-cytoskeletal elements (Collard et al. 1995). Resultant increased residence time of NFs within perikarya may contribute further to their aberrant phosphorylation (Black and Lee 1988) in a deleterious feedbackloop. Even a subtle shift in the balance of NF kinases and phosphatases may, over time, generate the extent of NF mislocalization that accompanies motor neuron disease (Motil et al. 2006b). Similarly, comparisons of axonal transport to models of traffic modeling suggests that a critical imbalance in association of NFs with their transport motors could generate a long-lasting "pile up" of NFs (Shea and Beaty 2007). Importantly, these analyses also suggest that restoration of appropriate motor balance could diminish continued delivery of NFs into these and perhaps deplete NF accumulations, as has been shown for experimentally induced NF aggregates in cultured cells (Shea et al. 1997).

As perikaryal NF accumulations contain NFs that are apparently normally phosphorylated, but mislocated (Bajaj et al. 1999; Strong et al. 2001), and as NF phosphorylation affects both transport and NF–NF associations, one speculation arising from these collective findings is that delaying extensive phosphorylation until NFs are longitudinally oriented within the lateral confines of the axon promotes orderly bundling, rather than generation of spheroids (Chan et al. 2007).

Notably, translocation of mitochondria into axons is dependent upon kinesin, and, as NFs bind mitochondria (Dubey et al. 2007; Leterrier et al. 1994; Wagner et al. 2003), the NF content of axons may provide a scaffold to maintain axonal mitochondrial distribution. These latter considerations provide a further mechanism by which impaired transport, including that of NFs, may contribute to motor neuron disease (De Vos et al. 2008).

5 Conclusions and Future Directions

In summary, a considerable body of literature amassed over the past few decades from a number of laboratories using a variety of experimental models and approaches provides insight into the mechanisms by which C-terminal NF phosphorylation may regulate NF axonal transport. The identification of responsible motors and the concurrent demonstration that phosphorylation both interferes with association of NFs with their anterograde motor, as well as promotes the NF–NF associations that correlate with the slowest-moving NFs, provide mechanisms that support these earlier hypotheses. While the approach of sidearm deletion (Rao et al. 2002, 2003; Yuan et al. 2006a, b) may indeed represent an improvement over deletion of the complete NF-H subunit, it unfortunately remains encumbered by some of the same compensatory mechanisms (e.g., NF-M hyperphosphorylation) that confounded interpretation following full-length NF-H deletion (Jung et al. 2006; Zhu et al. 1998). Of interest would be to examine the impact of NF-H C-terminal deletion on the association of NF isoforms with the slowest-moving NF population (i.e., the so-called stationary cytoskeleton) that is retained for extended periods along optic axons in long-term radiolabeling analyses (Nixon and Logvinenko 1986).

Beyond the simple consideration of C-terminal phosphorylation as regulating NF axonal transport lies the task of sorting out what is likely to be a series of hierarchical phosphorylation events, with distinct events regulating various aspects of NF dynamics. While phosphorylation events that induce the RT97 epitope are clearly associated with a number of characteristics of NF "maturation," there are no data to indicate that all NFs bearing the RT97 epitope have the identical phosphorylation state. One series of phosphorylation events, accompanied by generation of the RT97 epitope, may promote association with dynein, while additional phosphorylation events (for which we have no distinct antigenic marker as yet) may be required to promote NF–NF bundling. By contrast, some C-terminal phosphorylation events (e.g., those that induce the SMI-31 epitope), which occur prior to those that generate the RT97 epitope (Sanchez et al. 2000), are apparently compatible with kinesin association.

While we (Chan et al. 2007) and others (Gou et al. 1998) have demonstrated a role for divalent cations such as calcium in NF bundling, simultaneous association with dynein and kinesin – not uncommon for cargo of these motors (Martin et al. 1999) – may be an integral aspect of NF bundling in a manner analogous to the demonstration that competing forces generated by kinesin and dynein exert an architectural role in the maintenance of the axonal cytoskeletal lattice (Ahmad et al. 2000; Baas et al. 2006). As NFs contain contiguous phosphorylated and nonphosphorylated domains (Brown 1998; Chan et al. 2005), NFs in various overall phosphorylation states may retain the capacity to associate simultaneously with both motors. Differential phospho-dependent association with kinesin and dynein could contribute to the wide range of anterograde transport rates observed for axonal NFs. Elucidation of the order of phosphorylation by C-terminal kinases, and which events mediate which of the above aspects of NF dynamics, remain important areas of investigation.

Acknowledgments This review was supported by the National Science Foundation.

References

Ackerley S, Thornhill P, Grierson AJ, Brownlees J, Anderton BH, Leigh PN, Shaw CE, Miller CJ (2003) Neurofilament heavy chain side-arm phosphorylation regulates axonal transport of neurofilaments. J Cell Biol 161:489–495

Ackerley S, Grierson AJ, Banner S, Perkinton MS, Brownlees J, Byers HL, Ward M, Thornhill P, Hussain K, Waby JS, Anderton BH, Cooper JD, Dingwall C, Leigh PN, Shaw CE, Miller CC (2004) p38alpha stress-activated protein kinase phosphorylates neurofilaments and is associated with neurofilament pathology in amyotrophic lateral sclerosis. Mol Cell Neurosci 26:354–364

Ahmad FJ, Hughey J, Wittmann T, Hyman A, Greaser M, Baas PW (2000) Motor proteins regulate force interactions between microtubules and microfilaments in the axon. Nat Cell Biol 2:276–280

Archer DR, Watson DF, Griffin JW (1994) Phosphorylation-dependent immunoreactivity of neurofilaments and the rate of slow axonal transport in the central and peripheral axons of the rat dorsal root ganglion. J Neurochem 62:1119–1125

Chan W K-H, Yabe JT, Pimenta AF and Shea TB (2002) Growth cones contain a highly-dynamic population of neurofilament subunits. Cell Motil Cytoskel 54:195–207

Baas PW, Nada CV, Myers KA (2006) Axonal transport of microtubules: the long and short of it. Traffic 7:490–498

Bajaj NPS, Al-Sarraj ST, Leigh PN, Anderson V, Miller CCJ (1999) Cyclin dependent kinase 5 (cdk5) phosphorylates neurofilament heavy (NF-H) chain to generate epitopes for antibodies that label neurofilament accumulations in amyotrophic lateral sclerosis (ALS) and is present in affected motor neurons in ALS. Neuro Psychopharm Biol Psychiatr 23:833–850

Black MM, Lee VM-Y (1988) Phosphorylation of neurofilament proteins in intact neurons: demonstration of phosphorylation in cell bodies and axons. J Neurosci 8:3296–3305

Blum JJ, Reed MC (1989) A model for slow axonal transport and its application to neurofilamentous neuropathies. Cell Motil Cytoskel 12:53–65

Brown A (1998) Contiguous phosphorylated and non-phosphorylated domains along axonal neurofilaments. J Cell Sci 111:455–467

Brown A, Wang L, Jung P (2005) Stochastic simulation of neurofilament transport in axons: the "stop-and-go" hypothesis. Mol Biol Cell 16:4243–4255

Chan WK-C, Dickerson A, Otriz D, Pimenta A, Moran C, Malik K, Motil J, Snyder S, Pant HC, Shea TB (2004) Mitogen activated protein kinase regulates neurofilament axonal transport. J Cell Sci 117:4629–4642

Chan WK-H, Yabe JT, Pimenta AF, Ortiz D, Shea TB (2005) Neurofilaments can undergo transport and cytoskeletal incorporation in a discontinuous manner Cell Motil Cytoskel 62:166–179

Chan WK-H, Kushkuley J, Leterrrier J-F, Eyer J, Shea TB (2007) Calcium-mediated "bridging" of phosphorylated neurofilaments (NFs) promotes NF-NF association and inhibits NF-microtubule association: a mechanism for selective slowing of axonal transport of phospho-NFs. Mol Biol Cell 18 (Suppl) 2349

Collard JF, Côté F, Julien JP (1995) Defective axonal transport in a transgenic mouse model of amyotrophic lateral sclerosis. Nature 375:61–64

Craciun G, Brown A, Friedman A (2005) A dynamical system model of neurofilament transport in axons. J Theor Biol 237:316–322

DeFuria J, Chen P, Shea TB (2006) Divergent effects of the MEKK-1/JNK pathway on NB2a/d1 differentiation: some activity is required for neurite outgrowth and maturation but overactivation inhibits both phenomena. Brain Res 1123:20–26

De Vos KJ, Grierson AJ, Ackerley S, Miller CC (2008) Role of axonal transport in neurodegenerative diseases. Annu Rev Neurosci 31:151–173

DeWaegh SM, Lee VM, Brady ST (1992) Local modulation of neurofilament phosphorylation, axonal caliber, and slow axonal transport by myelinating Schwann cells. Cell 68:451–463

Dubey M, Chaudhury P, Kabiru H, Shea TB (2007) Tau inhibits anterograde axonal transport and perturbs stability in growing axonal neurites in part by displacing kinesin cargo: neurofilaments attenuate tau-mediated neurite stability. Cell Motil Cytoskel 65:89–99

Gou JP, Gotow T, Janmey PA, Leterrier JF (1998) Regulation of neurofilament interactions in vitro by natural and synthetic polypeptides sharing Lys-Ser-Pro sequences with the heavy neurofilament subunit NF-H: neurofilament crossbridging by antiparallel sidearm overlapping. Med Biol Eng Comput 36:371–387

He Y, Francis F, Myers KA, Yu W, Black MM, Baas PW (2005) Role of cytoplasmic dynein in the axonal transport of microtubules and neurofilaments. J Cell Biol 168:697–703

Hoffman PN, Lasek RJ, Griffin JW, Price DL (1983) Slowing of the axonal transport of neurofilament protein during development. J Neurosci 3:1694–1700

Jung C, Chylinski TM, Pimenta A, Ortiz D, Shea TB.(2004) Neurofilament transport is dependent on actin and myosin. J Neurosci. 24:9486-9496

Julien J-P, Mushynski WE (1998) Neurofilaments in health and disease. Prog Nuc Acid Res Mol Biol 61:1–20

Jung C, Shea TB (1999) Regulation of neurofilament axonal transport by phosphorylation in optic axons in situ. Cell Motil Cytoskel 43:230–240

Jung C, Lee S, Ortiz D, Zhu Q, Julien J-P, Shea TB (2006) Phosphorylation of the high molecular weight neurofilament subunit regulates the association of neurofilaments with kinesin in situ. Brain Res 141:151–155

Jung C, Yabe JT, Shea TB (2000a) Hypophosphorylated neurofilament subunits undergo axonal transport more rapidly than more extenisvely phosphorylated subunits in situ. Cell Motil Cytoskel 47:120–129

Jung C, Yabe JT, Shea TB (2000b) C-terminal phosphorylation of the heavy molecular weight neurofilament subunit is inversely correlated with neurofilament axonal transport velocity. Brain Res 856:12–19

Jung CJ, Shea TB (2004) Neurofilament subunits undergo more rapid translocation within retinas than in optic axons. Mol Brain Res 122:188–192

Jung C, Chylinski TM, Pimenta A, Ortiz D, Shea TB. (2004) Neurofilament transport is dependent on actin and myosin. J Neurosci. 24:9486–9496

Koehnle TJ, Brown A (1999) Slow axonal transport of neurofilament protein in cultured neurons. J Cell Biol 144:447–458

Komiya Y, Cooper NA, Kidman AD (1987) The recovery of slow axonal transport after a single intraperitoneal injection of beta, beta'-iminodipropionitrile in the rat. J Biochem (Tokyo) 102:869–873

Lasek RJ, Paggi P, Katz MJ (1992) Slow axonal transport mechanisms move neurofilaments relentlessly in mouse optic axons. J Cell Biol 117:607–616

Lasek RJ, Paggi P, Katz MJ (1993) The maximum rate of neurofilament transport in axons: a view of molecular transport mechanisms continuously engaged. Brain Res 616:58–64

Leterrier JF, Rusakov DA, Nelson BD, Linden M (1994) Interactions between brain mitochondria and cytoskeleton: evidence for specialized outer membrane domains involved in the association of cytoskeleton-associated proteins to mitochondria in situ and in vitro. Microsc Res Tech 27:233–261

Lewis SE, Nixon RA (1988) Multiple phosphorylated variants of the high molecular mass subunit of neurofilaments in axons of retinal cell neurons: characterization and evidence for their differential association with stationary and moving neurofilaments. J Cell Biol 107:2689–2701

Lobsiger CS, Garcia ML, Ward CM, Cleveland DW (2005) Altered axonal architecture by removal of the heavily phosphorylated neurofilament tail domains strongly slows superoxide dismutase 1 mutant-mediated ALS. Proc Natl Acad Sci USA 102:10351–10356

Martin M, Iyadurai SJ, Gassman A, Gindhart JG Jr, Hays TS, Saxton WM (1999) Cytoplasmic dynein, the dynactin complex, and kinesin are interdependent and essential for fast axonal transport. Mol Biol Cell 10:3717–3728

Marszalek JR, Williamson TL, Lee MK, Xu ZS, Hoffman PN, Becher MW, Crawford TO, Cleveland DW (1996) Neurofilament subunit NF-H modulates axonal diameter by selectively slowing neurofilament transport. J Cell Biol 135:711–724

Motil J, Chan W K-H, Dubey M, Chaudhury P, Pimenta A, Chylinski TM, Ortiz DT, Shea TB (2006a) Dynein mediates retrograde neurofilament transport within axonal neurites and anterograde delivery of NFs from perikarya into axons: regulation by multiple phosphorylation events. Cell Motil Cytoskel 63:266–286

Motil J, Dubey M, Chan W K-H, Shea TB (2006b) Inhibition of dynein but not kinesin induces focal accumulation of neurofilaments in axonal neurites. Brain Res 1164:125–131

Nixon RA (1993) The regulation of neurofilament protein dynamics by phosphorylation: clues to neurofibrillary pathology. Brain Pathol 3:29–38

Nixon RA (1998) The slow transport of cytoskeletal proteins. Curr Opin Cell Biol 10:87–92

Nixon RA, Logvinenko KB (1986) Multiple fates of newly synthesized neurofilament proteins: evidence for a stationary neurofilament network distributed non-uniformly along axons of retinal ganglion cell neurons. J Cell Biol 102:647–659

Pant HC, Veeranna (1995) Neurofilament phosphorylation. Biochem Cell Biol 73:575–592

Prahlad V, Helfand BT, Langford GM, Vale RD, Goldman RD (2000) Fast transport of neurofilament protein along microtubules in squid axoplasm. J Cell Sci 113:3939–3946

Rao M, Nixon RA (2003) Defective neurofilament transport in mouse models of amyotrophic lateral sclerosis: a review. Neurochem Res 28:1041–1047

Rao MV, Campbell J, Yuan A, Kumar A, Gotow T, Uchiyama Y, Nixon RA (2003) The neurofilament middle molecular mass subunit carboxyl-terminal tail domain is essential for the radial growth and cytoskeletal architecture of axons but not for regulating neurofilament transport rate. J Cell Biol 163:1021–1031

Rao MV, Garcia ML, Miyazaki Y, Gotow T, Yuan A, Mattina S, Ward CM, Calcutt NA, Uchiyama Y, Nixon RA, Cleveland DW (2002) Gene replacement in mice reveals that the heavily phosphorylated tail of neurofilament heavy subunit does not affect axonal caliber or the transit of cargoes in slow axonal transport. J Cell Biol 158:681–693

Roy S, Coffee P, Smith G, Liem RKH, Brady ST, Black MM (2000) Neurofilaments are transported rapidly but intermittently in axons: implications for slow axonal transport. J Neurosci 20:6849–6861

Sanchez I, Hassinger L, Sihag RK, Cleveland DW, Mohan P, Nixon RA (2000) Local control of neurofilament accumulation during radial growth of myelinating axons in vivo: selective role of site-specific phosphorylation. J Cell Biol 151:1013–1024

Shah JV, Flanagan LA, Janmey PA, Leterrier J-F (2000) Bidirectional translocation of neurofilaments along microtubules mediated in part by dynein/dynactin. Mol Biol Cell 11:3495–3508

Shea TB, Beaty WJ (2007) Traffic jams: models for neurofilament accumulation in motor neuron disease. Traffic 8:445–447

Shea TB, Chan W K-H (2008) Regulation of neurofilament dynamics by phosphorylation. Eur J Neurosci 27:1893–1901

Shea TB, Flanagan L (2001) Kinesin, dynein and neurofilament transport Trends Neurosci 24:644–648

Shea TB, Yabe JT (2000) Occam's razor slices through the mysteries of neurofilament axonal transport: can it really be so simple? Traffic 1:522–523

Shea TB, Fischer I, Paskevich PA, Beermann ML (1993) The protein phosphatase inhibitor okadaic acid increases axonal NFs and neurite caliber, and decreases axonal MTs in NB2a/d1 cells. J Neurosci Res 35:507–521

Shea TB, Wheeler E, Jung C (1997) Aluminum inhibits neurofilament assembly, cytoskeletal incoporation and axonal transport: dynamic nature of aluminum-induced perikaryal neurofilament accumulations as revealed by subunit turnover. Mol Chem Neuropathol 32:17–39

Shea TB, Jung CJ, Pant HC (2003) Does C-terminal phosphorylation regulate neurofilament transport? Re-evaluation of recent data suggests that it does. Trends Neurosci 26:397–400

Shea TB, Yabe JT, Ortiz D, Pimenta A, Loomis P, Goldman RD, Amin N, Pant HC (2004) CDK5 regulates axonal transport and phosphorylation of neurofilaments in cultured neurons. J Cell Sci 117:933–941

Sihag RK, Inagaki M, Yamaguchi T, Shea TB, Pant HC (2007) Role of phosphorylation on the structural dynamics and function of types III and IV intermediate filaments. Exp Cell Res 313:2098–2109

Strong MJ, Strong WL, Jaffe H, Traggert B, Sopper MM, Pant HC (2001) Phosphorylation state of the native high-molecular-weight neurofilament subunit protein from cervical spinal cord in sporadic amyotrophic lateral sclerosis. J Neurochem 76:1315–1325

Theiss C, Napirei M, Karl-Meller K (2005) Impairment of anterograde and retrograde neurofilament transport after anti-kinesin and anti-dynein antibody microinjection in chicken dorsal root ganglia. Eur J Cell Biol 84:29–43

Toyoshima I, Kato K, Sugawara M, Wada C, Masamune O (1998a) Kinesin accumulation in chick spinal axonal swellings with beta,beta'-iminodipropionitrile (IDPN) intoxication. Neurosci Lett 249:103–106

Toyoshima I, Sugawara M, Kato K, Wada C, Hirota K, Hasegawa K, Kowa H, Sheetz MP, Masamune O (1998b) Kinesin and cytoplasmic dynein in spinal spheroids with motor neuron disease. J Neurol Sci 159:38–44

Trivedi N, Jung P, Brown A (2007) Neurofilaments switch between distinct mobile and stationary states during their transport along axons. J Neurosci 27:507–516

Tu P-H, Elder G, Lazzarini RA, Nelson D, Trojanowski JQ, Lee VM-Y (1995) Overexpression of the human NFM subunit in transgenic mice modifies the level of endogenous NFL and the phosphorylation state of NFH subunits. J Cell Biol 129:1629–1640

Wagner OI, Lifshitz J, Janmey PA, Linden M, McIntosh TK, Leterrier JF (2003) Mechanisms of mitochondria–neurofilament interactions. J Neurosci 23:9046–9058

Wang L, Ho C-I, Sun D, Liem RKH, Brown A (2000) Rapid movements of axonal neurofilaments interrupted by prolonged pauses. Nature Cell Biol 2:137–141

Watson DF, Glass JD, Griffin JW (1993) Redistribution of cytoskeletal proteins in mammalian axons disconnected from their cell bodies. J Neurosci 13:4354–4360

Xia CH, Roberts EA, Her LS, Liu X, Williams DS, Cleveland DW, Goldstein LS (2003) Abnormal neurofilament transport caused by targeted disruption of neuronal kinesin heavy chain KIF5A. J Cell Biol 161:55–66

Yabe JT, Chan W, Shea TB (2000) Phospho-dependent association of neurofilament proteins with kinesin in situ. Cell Motil Cytoskel 42:230–240

Yabe JT, Chylinski T, Wang F-S, Pimenta A, Kattar SD, Linsley M-D, Chan W K-H, Shea TB (2001a) Neurofilaments consist of distinct populations that can be distinguished by C-terminal phosphorylation, bundling and axonal transport rate in growing axonal neurites. J Neurosci 21:2195–2205

Yabe JT, Pimenta A, Shea TB (1999) Kinesin-mediated transport of neurofilament protein oligomers in growing axons. J Cell Sci 112:3799–3814

Yabe JT, Wang F-S, Chylinski T, Katchmar T, Shea TB (2001b) Selective accumulation of the high molecular weight neurofilament subunit within the distal region of growing axonal neurites. Cell Motil Cytoskel 50:1–12

Yuan A, Nixon RA, Rao MV (2006a) Deleting the phosphorylated tail domain of the neurofilament heavy subunit does not alter neurofilament transport rate in vivo. Neurosci Lett 393:264–268

Yuan A, Rao MV, Sasaki T, Chen Y, Kumar A, Veeranna, Liem RK, Eyer J, Peterson AC, Julien JP, Nixon RA (2006b) Alpha-internexin is structurally and functionally associated with the neurofilament triplet proteins in the mature CNS. J Neurosci 26:10006–10019

Zhang B, Tu P, Abtahian F, Trojanowski JQ, Lee VM-Y (1997) Neurofilaments and orthograde transport are reduced in ventral root axons of transgenic mice that express human SOD1 with a G93A mutation. J Cell Biol 139:1307–1315

Zhu Q, Lindenbaum M, Levavasseur F, Jacomy H, Julien J-P (1998) Disruption of the NF-H gene increases axonal microtubule content and velocity of neurofilament transport: Relief of axonopathy resulting from the toxin b'b'-iminodiproprionitrile. J Cell Biol 143:183–193

Critical Roles for Microtubules in Axonal Development and Disease

Aditi Falnikar and Peter W. Baas

Abstract Axons are occupied by dense arrays of cytoskeletal elements called microtubules, which are critical for generating and maintaining the architecture of the axon, and for acting as railways for the transport of organelles in both directions within the axon. Microtubules are organized and regulated by molecules that affect their assembly and disassembly, their stabilization, their association with other cytoskeletal elements, and their alignment and bundling with one another. Recent studies have accentuated the role of molecular motor proteins and microtubule-severing proteins in the establishment and maintenance of the axonal microtubule array. The growing body of knowledge on the proteins and mechanisms that regulate axonal microtubules has fostered a better understanding of how many debilitating diseases cause axons to degenerate. The purpose of this chapter is to provide an update on current knowledge of axonal microtubules and the proteins that regulate them, and to reflect on cutting-edge findings linking these proteins and mechanisms to diseases that afflict the human population.

1 Microtubules in the Axon

Each typical neuron of the body generates a single elongated axon that has the potential to traverse complex and often long journeys to reach its target tissue. The axon is effectively unlimited in its growth potential, as evidenced by the fact that it continues to grow unabated when neurons are transplanted into a culture dish (He and Baas 2003). This is particularly surprising because axons contain relatively little protein synthetic machinery, and hence need axonal transport from the cell body for their growth and maintenance (Baas and Buster 2004). In addition, the axon is an engineering marvel that clearly requires sophisticated architectural struts to generate and maintain its structure. The transport and architectural needs of the

---

A. Falnikar and P.W. Baas (✉)
Department of Neurobiology and Anatomy, Drexel University College of Medicine,
2900 Queen Lane, Philadelphia, PA, 19129, USA
e-mail: pbaas@drexelmed.edu

axon are fulfilled by cytoskeletal elements, and in particular, by microtubules. Microtubules are polymers of tubulin subunits that provide cells with their shape by resisting compression, while simultaneously acting as the substrate for the transport of organelles and proteins in both directions (Baas and Buster 2004). Microtubules are certainly not unique to axons, but rather fulfill architectural and motile needs of cellular apparatuses as diverse as mitotic spindles and cilia.

Microtubules form a continuous array within the axon, extending from the cell body into the growth cone at its distal tip (see Fig. 1). Almost every microtubule within the array is oriented with its assembly-favored "plus" end directed away from the cell body (Heidemann et al. 1981; Baas et al. 1987). Although the microtubule array is continuous, the individual microtubules that comprise the array are staggered along the length of the axon and assume a variety of lengths (Bray and

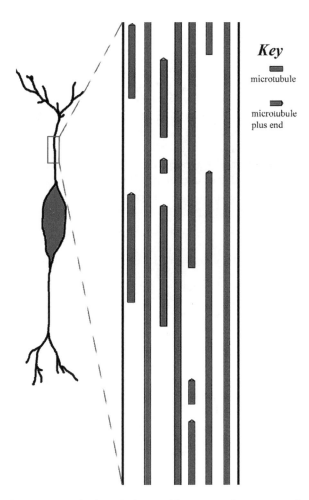

Fig. 1 A continuous array of microtubules provides structural support to the axonal shaft. Individual microtubules within the array vary in length. Almost all of the microtubules exhibit uniform polarity with plus ends oriented distally, away from the cell body

Bunge 1981; Yu and Baas 1994). Some microtubules are over a hundred microns long, while others are only a single micron in length, or even shorter. Early studies on the kinetics of tubulin transport suggested that radiolabeled tubulin moves slowly down the axon, and in a relatively coherent manner compared to diffusion (Black and Lasek 1980). It was posited that tubulin is transported in the form of the microtubules, and that this transport consists of a slow and synchronous "march" of the polymers. Live-cell imaging analyses over the past several years have refined this model substantially by demonstrating that the transport of microtubules down the axon is actually not slow and coherent at all, but it is fast, intermittent, asynchronous, and bidirectional (Wang and Brown 2002; Hasaka et al. 2004; He et al. 2005; Myers and Baas 2007). Moreover, it is only the very shortest microtubules, those less than about 7 µm in length, that are in transit.

Recent observations have focused a great deal of attention on the molecular motor proteins that transport the short microtubules, as well as a category of proteins called microtubule-severing proteins that break long microtubules into ones that are short enough to be transported (Baas et al. 2006). In addition, strong evidence suggests that the same molecular motor proteins that transport the short microtubules also impinge upon the long microtubules, and thereby play important functional roles in such matters as determining whether the axon grows or retracts, as well as the navigation of the tip of the axon, called the growth cone (Myers et al. 2006; Nadar et al. 2008). The relevant microtubule-severing proteins are enzymes called katanin and spastin. The relevant molecular motor proteins appear to be cytoplasmic dynein, as well as a small number of specialized kinesins, typically thought of as "mitotic" motors because they were originally identified as crucial for generating forces on microtubules in the mitotic spindle (Baas 1999; Baas et al. 2006).

In addition to being subjected to motor-driven forces and the potential for severing, microtubules are classically known to be dynamic polymers, which means that they have the capacity for rapid assembly and disassembly. In the absence of accessory and regulatory proteins and signals, microtubules display extremely rapid bouts of intermittent assembly and disassembly known as "dynamic instability." Such behavior is quite apparent in living cells as well; particularly, at the leading edge of motile cells (Wittmann et al. 2003), or in the case of the axon, within the growth cone (Suter et al. 2004). However, neurons are exceptionally rich in proteins that shift the dynamics toward assembly, and also stabilize the microtubules against disassembly. These proteins include classic fibrous microtubule-associated proteins (MAPs) such as tau, MAP1b, MAP1a, and MAP2, as well as other proteins such as STOP (stable tubule only protein), doublecortin, and crosslinking proteins such as plakins/plectins (Chapin and Bulinski 1992; Matus 1994; Bosc et al. 1996; Horesh et al. 1999; Leung et al. 2002). These proteins generally function by binding along the lattice of the microtubule, and thereby suppress the tendency of microtubules to disassemble. Other proteins, such as CRMP-2, may actually interact with the tubulin subunits to promote their assembly onto pre-existing microtubule polymers (Fukata et al. 2002). Notably, axons also contain proteins that promote microtubule disassembly, and these include stathmin and SCG10 (Curmi et al. 1999).

The relevant molecular motor proteins exert various complementary and antagonistic actions, so that desired effects can be achieved by regulating the balance of forces either globally or locally in the axon. The same is true of the relevant proteins that impact the assembly, disassembly, and stabilization of the microtubules. By having a number of different participants that impact the properties of the microtubules, the axon can judiciously regulate the microtubule array to participate appropriately in events such as axonal growth, retraction, branch formation, and navigation.

Figure 2 schematically shows molecular motor proteins transporting membranous cargo and short microtubules in the axon, while figure 3 schematically illustrates

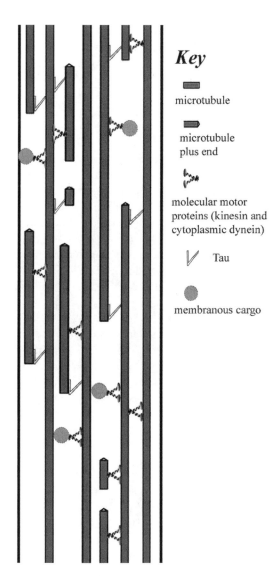

Fig. 2 Molecular motor proteins power microtubule-mediated transport of membranous cargo, as well as that of short microtubules in anterograde and retrograde directions within the axon

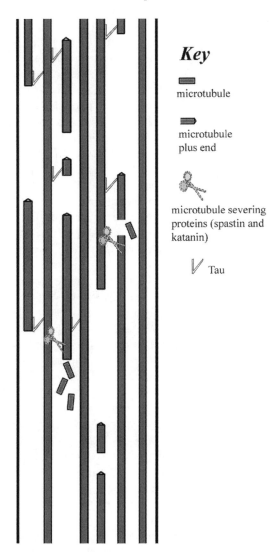

Fig. 3 Microtubule-severing proteins break long microtubules into shorter ones, thus regulating lengths, as well as numbers of microtubules within the axon. Tau limits the access of microtubule severing proteins, such as katanin to the microtubule lattice

the activity of microtubule-severing proteins. In both schematics, tau is also included, as it has been implicated in the potential regulation of molecular motors (Baas and Qiang 2005) and severing proteins (see Sect. 3.1).

Two other categories of proteins are worth mentioning with regard to the regulation of microtubules in the axon. The first is a relatively newly discovered family of proteins called +tips, which preferentially bind the plus ends of microtubules during bouts of assembly. Shortly after the relevant tubulin subunits become part of the polymer, the +tips tend to lose their association, resulting in high concentrations of these molecules at the growing plus-end of the microtubule relative to elsewhere along the microtubule's length. These +tips, which include "end binding" proteins such as EB1 and EB3, as well as CLIPs and CLASPs, impact the dynamic properties of the microtubules, and also their interactions with cortical structures, molecular motors, and other proteins (Vaughan 2005; Akhmanova and Hoogenraad 2005). These proteins may be important for several functions, including, for example, the interaction of the plus-ends of the microtubules with structures in the cell cortex during branch formation and growth cone navigation (Stepanova et al. 2003; Jimenez-Mateos et al. 2005; Kornack and Giger 2005).

The other category of proteins worth mentioning is actin, and any protein that impacts the organization, distribution, or properties of the actin cytoskeleton, given that microtubules and actin filaments are well known for interacting with one another. We have recently published a chapter elsewhere on this topic (Myers and Baas 2009), and will refrain from including a discussion of it here.

Much of what is known about the various microtubule-related molecules in the axon has been elucidated within the context of development. Adult axons are relatively "hard-wired," and do not undergo the degree of morphological plasticity as they do during development. Even so, it would appear that the vast majority of the microtubule-related molecules important for development are still present throughout the lifetime of the axon. Microtubules still need to be transported down the axon; albeit, somewhat less robustly compared to development. That means the molecular motor proteins and microtubule-severing proteins still have important roles to play. In addition, it is very important that a substantial portion of the microtubule mass is relatively stable to ensure that the axon has architectural support and railways for the ongoing transport and trafficking of organelles, RNA, and proteins needed for the vitality of the axon. Of course, damaged axons have some regenerative capacity, especially in the peripheral nervous system, and so preserving the necessary elements for axonal growth makes sense.

Over the past few decades, scientists interested in axonal microtubules have made the argument that microtubules are so important for the axon that undoubtedly studying how they are regulated would someday yield a huge payoff in terms of understanding the neurological disorders that afflict the human population. Recent years have proven especially fruitful in this regard. For example, we now know that mutations to microtubule-related proteins such as doublecortin and Lis1 can cause congenital flaws in brain lamination, resulting in mental retardation (Gleeson et al. 1998; Reiner et al. 1993), and it is clear that misregulation of tau is a critical component of Alzheimer's disease (see Sects. 3 and 3.1). It also appears that microtubule-based axonal transport is particularly sensitive to toxins and misfolded proteins, which can lead to nerve degeneration as a result of compromised axonal transport (De Vos et al. 2008). Therefore, in developing a thorough understanding of axonal microtubules, the microtubule research community is making rapid progress toward characterizing some of the most debilitating neurological diseases, as well as effective strategies for their treatment and prevention. The purpose of this chapter is not to be exhaustive, but rather to provide a few examples of how careful, in-depth study of axonal microtubules has led to breakthroughs in the understanding of diseases that impact the nervous system.

2 Microtubule-Severing and Hereditary Spastic Paraplegia

Hereditary spastic paraplegia (HSP) is a heterogeneous group of genetic diseases that mainly affect the corticospinal tracks, resulting in spasticity in the lower limbs (Bruyn 1992). Although there are variable forms of the disease, the age of onset is usually in early adulthood, with no developmental deficiencies. Mutations to several different genes can give rise to HSP, but roughly 40–60% of the cases result from autosomal dominant mutations to a gene called *SPG4*, which codes for a protein called spastin (Hazan et al. 1999). The discovery in 2002 that spastin is a microtubule-severing protein (Errico et al. 2002) prompted enormous interest, because it suggested that the axons of patients afflicted with this form of HSP may degenerate due to insufficient severing of microtubules. With a mechanism in hand, the potential existed for rapid progress on therapies.

Our interest in microtubule-severing activity first arose from our early work on the length-distribution of microtubules in the axon, which was conducted before katanin or spastin had been identified as severing proteins. We had determined that new microtubules do not spontaneously nucleate in the axon (Baas and Ahmad 1992), and hence, we speculated that microtubule number could only be increased in the axon by either release and transport of new microtubules from the centrosome (located in the cell body of the neuron) or by severing of pre-existing microtubules in the axon such that a single microtubule would be transformed into many short ones (Joshi and Baas 1993). The short microtubules could then either elongate into longer ones or disassemble to provide subunits used by their neighbors to elongate. We performed serial reconstructions from electron micrographs of cultured neurons and demonstrated the appearance of large numbers of short microtubules and the absence of long microtubules within the axon at sites where new branches were starting to form (Yu et al. 1994). These observations provided support for a model in which long microtubules are severed in a very localized and tightly regulated fashion during the formation of collateral branches. Using live-cell imaging, we directly observed the severing of short microtubules from looped bundles of microtubules within paused growth cones, and the subsequent movement of the short microtubules into filopodia (Dent et al. 1999). Producing more free ends of microtubules, via severing, may also be important because free ends of microtubules are known to interact with a variety of proteins and structures, such as those within the cell cortex (see discussion on +tips in Sect. 1). Microtubule-severing would certainly appear to be particularly important for the plasticity of the axon during development; however, as noted earlier, having a sufficient balance of long and short microtubules is crucial for proper axonal transport throughout the lifetime of the neuron (Yu et al. 2007).

Katanin was the first microtubule-severing enzyme to be well characterized. It was originally purified from sea urchin eggs, in which it was shown to sever microtubules by disrupting contacts within the polymer lattice using energy derived from ATP hydrolysis (McNally and Vale 1993). In studies on neurons, we showed that katanin is present at the centrosome, consistent with observations in other cell types, and that it is also widely distributed within the axon, and throughout all neuronal compartments (Ahmad et al. 1999; Yu et al. 2005). Inhibition of katanin by various experimental approaches prohibits microtubule release from the centrosome, and profoundly increases microtubule length throughout the neuronal cell body (Ahmad et al. 1999; Karabay et al. 2004). As a result, axonal outgrowth is severely compromised. In addition, we found that the levels of katanin are very high in axons that are actively growing toward their targets, but then decrease precipitously when the axon reaches its target and stops growing (Karabay et al. 2004).

We next expanded our studies to include spastin, starting with detailed quantitative and functional comparisons between katanin and spastin in the nervous system. We found that katanin has widespread expression in the various cells and tissues of the body; however, spastin is more enriched in the nervous system, with only low levels of expression elsewhere (Solowska et al. 2008). Nonetheless, during the development of the nervous system, the levels of katanin are many times higher than the levels of spastin. After development is complete, at which time katanin levels plunge, spastin levels are only slightly diminished. Despite this, the levels of katanin remain notably higher than the levels of spastin throughout the adult brain and spinal cord. These observations provided some support for a "loss of function" scenario for HSP, because a diminished level of functional spastin would presumably be more consequential in the adult, due to the lower levels of katanin. However, nothing in these studies suggested an obvious explanation of why the corticospinal tracks should be particularly sensitive to a diminution in functional spastin levels.

In terms of the functions of spastin, we found that over-expressing spastin in cultured neurons profoundly increased the frequency of axonal branch formation (Yu et al. 2008). This phenomenon did not occur when katanin was over-expressed (Yu et al. 2005), suggesting that spastin has some specialized properties to promote branch formation that are not shared by katanin. Depletion of spastin diminished, but did not eliminate branch formation, suggesting that katanin can provide sufficient severing of microtubules for branches to form, but that spastin is more optimal to perform in this capacity (Yu et al. 2008). In part, this may be due to the fact that spastin tends to accumulate at sites of branch formation far more than katanin. Patients with spastin mutations as well as spastin knockout mice do not show deficiencies in axonal branching during development (Fink and Rainier 2004), supporting the conclusion that katanin is sufficient to carry out this function in the absence of spastin.

Additional observations caused us to further doubt whether a "loss of function" scenario makes sense to explain the degeneration of the corticospinal tracks in HSP patients. The *spastin* gene is interesting in that it contains a second start-codon, not far downstream from the first one (Claudiani et al. 2005; Mancuso and Rugarli 2008). In rodents, this results in two spastin isoforms, which we refer to as M1 and

M85, with respective molecular weights of 68 and 60 kDa (note: they are named according to the relevant methionine; in humans, the shorter isoform would be M87 rather than M85; nonetheless, the term M85 will be used for simplicity). Notably, we found that M85 is expressed in all regions of the central nervous system at all times during development and in the adult (Solowska et al. 2008). By contrast, there is little or no detectable M1 at any time during development, or in the adult, with one exception. About 20% of the total spastin in the adult spinal cord is the M1 isoform. Given that this corresponds to the location and time where degeneration occurs in spastin-based HSP, we wondered if mutant M1 may pose a problem for the axon, while mutant M85 does not.

In this view, spastin-based HSP would be a "gain-of-function" disease in which axonal degeneration is not caused by insufficient microtubule-severing activity, but rather by a cytotoxic mutant protein. To test this hypothesis, we compared the effects of truncated versions of M1 and M85 spastin, lacking the AAA ATPase domain critical for severing function, on cultures of embryonic cortical neurons (Solowska et al. 2008). These studies demonstrated just how detrimental M1 mutants would be if they were robustly expressed during development. Neurons induced to express the truncated M1 are slower to develop, have shorter axons, and have generally less robust morphologies. Interestingly, the truncated M85-spastin, which would correspond to the mutant spastin expressed during development in HSP patients, did not cause any developmental problems in neuronal cell cultures.

To further investigate the mechanism of HSP, we tested the effects of truncated spastin M1 and M85 on fast axonal transport. In studies on squid axoplasm, perfusion of full-length M1 plus M85 spastins showed no effect on fast axonal transport, and the same was true for the truncated M85, which lacks the ability to sever microtubules (Solowska et al. 2008). However, the truncated M1 polypeptide strongly inhibited fast axonal transport, indicating that it is the 8 kDa amino terminal region of M1 that elicits these deleterious effects. One possibility is that pathogenic spastin mutations induce a conformational change that results in abnormal exposure of the 8 kDa amino terminal unique to M1. Consistent with this view, intragenic polymorphisms of spastin have been found within the 8 kDa amino terminus, which dramatically modify the HSP phenotype. In addition, spastin is known to interact with another HSP-related protein called atlastin via the 8 kDa amino-terminal region of M1 (Sanderson et al. 2006). Interestingly, recessive mutations in *atlastin* also lead to HSP (Zhao et al. 2001), suggesting that the binding of atlastin or other polypeptides to the amino terminal of M1 could help prevent M1-induced pathology (Zhao et al. 2001).

These observations not only provide insight into the underlying causes of the disease, but are also critically important for the development of effective treatment strategies. We do not believe the "loss/gain of function" issue is completely resolved; however, if our view on this problem is correct, then there is no utility in developing strategies to compensate for a hypothetical insufficiency in microtubule-severing activity in afflicted axons. Instead, a more productive approach would be to try to rid the afflicted neurons of the mutant M1 spastin. According to our results thus far, it would be preferable for the neurons to risk losing the viable spastin than to tolerate the pathogenic spastin molecules. Thus, a therapeutic RNAi approach may prove highly effective in depleting the problematic molecules from the neurons of patients with spastin-based HSP.

3 Tauopathies

Tauopathies are a class of neuro-degenerative disorders caused by abnormalities of the tau protein. As mentioned earlier, tau is one of the many microtubule-associated proteins that can influence microtubule dynamics in the axon. In addition, tau is thought to regulate the spacing between microtubules (Black 1987; Chen et al. 1992), as well as interactions between microtubules and other proteins and structures in the axon (Ebneth et al. 1998). The binding of tau to microtubules is regulated by its phosphorylation at specific sites (Fath et al. 2002). Tauopathies are generally caused by either mutations to tau that affect its capacity to bind to microtubules or by abnormalities in the pathways that phosphorylate tau. There are two proposed ways in which abnormal tau behavior is believed to lead to axonal degeneration. The first is that the microtubules are impaired because of the loss of normal levels of tau binding to their lattice. The second is that the abnormal tau may itself be toxic, as evidenced by its tendency to self-assemble into various types of aggregates, including straight filaments and paired helical filaments (Bunker et al. 2006; Arrasate et al. 1999), and by recent studies of the effects of the amino terminus of tau on axonal transport (Lapointe et al. 2009). At least 34 different pathology-associated mutations in various regions of the *tau* gene have been identified so far. Such pathologies include frontotemporal dementia with Parkinsonism, linked to chromosome 17 (FTDP-17). Other examples of tauopathies linked to specific tau mutations include progressive supranuclear palsy, corticobasal degeneration, and Pick's disease (Wang and Liu 2008). Interestingly, in Alzheimer's disease, there are no known mutations of tau, but only alterations in tau phosphorylation.

The *tau* gene, located on human chromosome 17, gives rise to six well-recognized isoforms of tau protein through the process of alternative splicing of mRNA. Missense mutations mentioned above act by causing misregulation of alternative splicing. Exon 2, 3, and 10 of *tau* RNA transcript are known to undergo alternative splicing. While the exons 2 and 3 encode amino terminal inserts, the exon 10 codes for a part of the microtubule-binding region of the tau protein. Exon 10 is also the exon in which the maximum numbers of disease-causing mutations have been identified. Alternative splicing is responsible for generating tau isoforms, containing either four or three microtubule binding repeats (4R tau, or 3R tau). Of the six known tau isoforms, three belong to each category. The number of microtubule binding repeats present dictates the microtubule binding ability of a given tau isoform. 4R tau isoforms have a stronger affinity for microtubules (Wang and Liu 2008). Interestingly, under normal physiological conditions, alternative splicing is regulated such that the ratio of 4R tau-to-3R tau is maintained at 1:1 (Donahue et al. 2007; D'Souza and Schellenberg 2005). It has been suggested that perturbation of this ratio lies at the root of tau-related neurodegeneration (Stanford et al. 2003). A mouse model of tauopathies generated by over-expression of the smallest tau isoform mimics characteristic features of human taupathies, such as progressive accumulation of tau inclusions and neurodegenration (Lee and Trojanowski 2001). However, it still remains controversial as to whether 3R or 4R tau isoforms are more toxic to the cell.

3.1 Tauopathies and Microtubule-Severing

One of the most common claims about tauopathies is that the loss of tau from the microtubules in the axon destabilizes the microtubules and leads to their depolymerization. This interpretation is hard to defend, however, in light of the fact that experimental depletion of tau from axons does not render the microtubules less stable, nor are microtubules less stable in tau knockout mice (Tint et al. 1998; Harada et al. 1994). These observations do not entirely invalidate the idea, however, because the mix of other microtubule-stabilizing molecules may largely be able to compensate for the loss of tau, with modest effects accumulating over time to give rise to microtubule loss.

Another scenario is suggested by our observations on microtubule-severing function. Quantitatively, neurons contain total levels of microtubule-severing proteins far higher than would be required to completely break down microtubules in the test tube (Solowska et al. 2008). Therefore, neurons must have regulatory mechanisms that either attenuate the activity of the severing proteins or protect the lattice of the microtubules from being fully accessed by the severing proteins. Whatever the case, it is not hard to understand how misregulation of these mechanisms could result in abnormal levels of microtubule-severing activity that could be profoundly deleterious to the axon.

The question arises as to how microtubule-severing function is regulated, such that severing of microtubules occurs when and where needed? We have proposed a model inspired by the observation that katanin-induced microtubule-severing becomes much more active in interphase extracts that are depleted of the frog homologue of MAP4, a fibrous MAP similar to tau (McNally et al. 2002). In addition, severing is more active in mitotic extracts compared to interphase extracts, and this difference is based on phosphorylation of proteins, but apparently not of katanin itself (Vale 1991; McNally et al. 2002). Interestingly, phosphorylation of MAP4 causes it to lose its association with the microtubules (McNally et al. 2002), consistent with a model that we call the "MAP protection model." In this model, fibrous MAPs protect the lattice of the microtubule from being accessed by katanin. Phosphorylation of the MAPs results in their detachment from the microtubule, thereby enabling katanin to gain access (Baas and Qiang 2005). In the axon, tau, rather than MAP4, would be the likely candidate to fulfill this role (see Fig. 3). The MAP protection model offers a potential means by which signaling cascades can regulate microtubule-severing activity in a spatially discrete manner, for example, at impending sites of axonal branch formation. The signaling cascades would cause tau (or other MAPs) to dissociate from the microtubules at the site where a branch is starting to form, thereby permitting katanin to break the microtubules into shorter, highly mobile pieces, precisely where needed.

To test the premise of this idea, we conducted studies in which we overexpressed katanin or spastin in fibroblasts in which we had first expressed a neuronal MAP, the katanin or spastin causes dramatic severing of microtubules and loss of microtubule mass. We found that robust levels of tau and MAP2 are able to protect the microtubules from being severed by katanin, but MAP1b was not able to do so (Qiang et al. 2006). Tau and MAP2 have a very similar microtubule-binding domain to MAP4, and so these results make sense. Tau, being the MAP enriched in axons, therefore, appeared to be the prime suspect for being the principal protector of microtubules in the axon. Indeed, when we depleted various MAPs from neurons with siRNA, it was only tau whose loss caused the microtubules in the axon to become notably more sensitive to the over-expression of katanin (Qiang et al. 2006). In subsequent studies, we found that tau is less effective at protecting microtubules from spastin (Yu et al. 2008), which is probably one reason why spastin is always expressed at much lower levels than katanin.

On the basis of these observations, we posited that one of the contributing factors to axonal degeneration in tauopathies (such as Alzheimer's disease) may be a gradually heightened sensitivity of the microtubules to abnormal microtubule-severing activity, mainly of katanin (Baas and Qiang 2005; Qiang et al. 2006). This hypothesis has not yet been tested, but if it is valid, it may suggest new avenues for therapies based on downregulating microtubule-severing activity in the afflicted axons.

4 Other Neurodegenerative Diseases

A number of review papers have been written on neurodegenerative orders over the past few years, and these include substantive sections on how flaws in microtubules and microtubule-related events such as axonal transport give rise to degeneration of axons (see, e.g., Roy et al. 2005; Chevalier-Larsen and Holzbaur 2006; Duncan and Goldstein 2006; Morfini et al. 2002). Although it is not possible to provide an in-depth discussion of the various relevant diseases within the limited space available here, three additional diseases will be discussed briefly below.

Amyotrophic lateral sclerosis (ALS) is a neurodegenerative disease that is believed to arise from accumulation of misfolded protein aggregates (Boillee et al. 2006). Motor neurons of the central nervous system are preferentially degenerated in the pathology of ALS. Degeneration of motor neurons leads to atrophy of muscles supplied by these neurons, which gradually leads to complete loss of voluntary muscle movement as the disease progresses. Although the exact cause of the disease is not known, mutations in the gene that encodes copper–zinc superoxide dismutase (SOD1) have been linked to cases of familial ALS (Rosen et al. 1993).

Under normal physiological conditions, SOD1 protects the cell from toxic superoxide radicals. A mouse model of ALS was created by over-expressing an ALS-linked SOD1 mutant; motor neurons in these mice exhibit defective axonal transport (Ligon et al. 2005). Similar to the situation discussed earlier with regard to spastin and HSP, loss of function of SOD1 seems unlikely to be the cause of the observed motor neuron defects in ALS (Boillee et al. 2006).

Motor neurons are large cells with long axons. Owing to their great lengths, these axons heavily depend on microtubule-driven transport of cargo for their maintenance and viability. For example, retrograde transport of trophic factors is thought to be essential for neuronal survival (Ye et al. 2003). The mechanism by which mutant SOD1 disrupts axonal transport is not entirely understood. Cytoplasmic dynein, the principal minus-end-directed microtubule-associated motor protein, is known to colocalize with aggregates of mutant SOD1 (Ligon et al. 2005). A recent report has shown interaction between mutant SOD1 and cytoplasmic dynein (Zhang et al. 2007). These observations suggest disruption of dynein-mediated transport of cargo along microtubules as an explanation for the degeneration and death of motor neurons caused by mutant SOD1. The hypothesis that perturbed axonal transport is the underlying cause of neuronal degeneration is further supported by a recent report linking mutations in the *spastin* gene to etiology of ALS (Munch et al. 2008). As mentioned earlier, mutations in the *spastin* gene can lead to impaired axonal transport. Finally, another gene whose mutations are linked with ALS, termed *ALS2*, generates a protein called alsin, which partially associates with the centrosome and appears to impact the microtubule system (Millecamps et al. 2005).

Potential links between impaired axonal transport, neuronal degeneration, and cell death are being actively explored in other neurological diseases as well. Huntington's disease (HD) is a progressive neurodegenerative disorder characterized by uncontrolled, uncoordinated jerky muscle movements along with decline in mental abilities. Mutations in the protein huntingtin, resulting in an abnormally large number of polyglutamine repeats in its sequence, is believed to be the cause of this disease. Despite widespread expression of huntingtin protein in non-nervous as well as nervous tissue, mutations of *huntingtin* gene lead to degeneration selectively in brain tissue. The mechanism of this degeneration is thus far unclear. However, it is known that mutant huntingtin exhibits a propensity to form aggregates within neurons (DiFiglia et al. 1997) and also has altered interactions with other proteins (Harjes and Wanker 2003). The hypothesis that neuronal death is caused by toxic gain of function by misfolded mutant proteins in the case of huntingtin is questioned by reports suggesting that neuronal dysfunction precedes aggregate formation (Gunawardena and Goldstein 2005). Recently, it has been demonstrated that huntingtin is a positive regulator of axonal transport and that it directly interacts with dynein (Caviston et al. 2007; Colin et al. 2008). Observations on mouse models of HD and also human HD patients provide evidence for disrupted axonal transport in the form of swollen neuronal projections, accumulation of vesicles and organelles, and mutant huntingtin (Gunawardena and Goldstein 2005). These pieces of evidence present a strong case for the role of disrupted axonal transport in the pathology of Huntington's disease. It has been hypothesized that, owing to the mutations, neuronal degeneration may be due to loss of function of huntingtin, thereby leading to the observed impairment in axonal transport (Her and Goldstein 2008; Szebenyi et al. 2003).

Perturbed axonal transport and also direct effects on neuronal microtubules are implicated in the pathology of Parkinson's disease (PD). PD is characterized by degeneration of dopaminergic neurons in the substantia nigra of the brain. Loss of these neurons leads to reduced dopaminergic input to the striatum, the part of the brain that controls voluntary muscle movement. The attendant functional disinhibition results in nonselective, excessive muscle tone, which produces muscle rigidity, and abnormalities in speech. Mutations in the *parkin* gene are associated with some of the familial cases of Parkinson's (Moore et al. 2005). *Parkin* codes for a protein–ubiquitin E3 ligase. Under normal physiological conditions, *parkin* independently binds to tubulin subunits and microtubules. Interaction of *parkin* with tubulin facilitates its degradation, whereas *parkin's* interaction with microtubules stabilizes them. PD-linked *parkin* mutants retain their ability to bind to microtubules, but are unable to bind to tubulin. This leads to intracellular accumulation of misfolded tubulin, leading to neurodegeneration (reviewed in Feng 2006). Furthermore, toxins implicated in PD, such as rotenone, cause depolymerization of microtubules (Marshall and Himes 1978). This disrupts microtubule-mediated transport of dopamine vesicles, leading to intracellular accumulation of leaky vesicles, which leads to neurodegeneration (Ren et al. 2005).

5 Concluding Remarks

In conclusion, as research investigators of microtubules, we are gratified that progress on the regulation of microtubules in neurons has led to a greater understanding of the normal physiology of the axon, as well as new mechanistic breakthroughs on neurodegenerative diseases that plague the human population. We look forward to further progress, and in particular to a new focus on potential therapies for treating, preventing, and perhaps even curing these microtubule-related diseases.

Acknowledgments The work in our laboratory is funded by the National Institutes of Health, the National Science Foundation, the State of Pennsylvania Tobacco Settlement Funds, the Alzheimer's Association, and the Hereditary Spastic Paraplegia Foundation. Aditi Falnikar is the recipient of the Doris Willig, M.D. Research Fellowship Award.

References

Ahmad FJ, Yu W, McNally FJ, Baas PW (1999) An essential role for katanin in severing microtubules in the neuron. J Cell Biol 145:305–315

Akhmanova A, Hoogenraad CC (2005) Microtubule plus-end-tracking proteins: mechanisms and functions. Curr Opin Cell Biol 17:47–54

Anonymous (1993) A novel gene containing a trinucleotide repeat that is expanded and unstable on Huntington's disease chromosomes. The Huntington's Disease Collaborative Research Group. Cell 72:971–983

Arrasate M, Perez M, Armas-Portela R, Avila J (1999) Polymerization of tau peptides into fibrillar structures. The effect of FTDP-17 mutations. FEBS Lett 446:199–202

Baas PW (1999) Microtubules and neuronal polarity: lessons from mitosis. Neuron 23–31

Baas PW, Ahmad FJ (1992) The plus ends of stable microtubules are the exclusive nucleating structures for microtubules in the axon. J Cell Biol 116:1231–1241

Baas PW, Buster DW (2004) Slow axonal transport and the genesis of neuronal morphology. J Neurobiol 58:3–17

Baas PW, Qiang L (2005) Neuronal microtubules: when the MAP is the roadblock. Trends Cell Biol 15:183–187

Baas PW, White LA, Heidemann SR (1987) Microtubule polarity reversal accompanies regrowth of amputated neurites. Proc Natl Acad Sci U S A 84:5272–5276

Baas PW, Vidya Nadar C, Myers KA (2006) Axonal transport of microtubules: the long and short of it. Traffic 7:490–498

Black MM (1987) Comparison of the effects of microtubule-associated protein 2 and tau on the packing density of in vitro assembled microtubules. Proc Natl Acad Sci U S A 84:7783–7787

Black MM, Lasek RJ (1980) Slow components of axonal transport: two cytoskeletal networks. J Cell Biol 86:616–623

Boillee S, Vande VC, Cleveland DW (2006) ALS: a disease of motor neurons and their nonneuronal neighbors. Neuron 52:39–59

Bosc C, Cronk JD, Pirollet F, Watterson DM, Haiech J, Job D, Margolis RL (1996) Cloning, expression, and properties of the microtubule-stabilizing protein STOP. Proc Natl Acad Sci U S A 93:2125–2130

Bray D, Bunge MB (1981) Serial analysis of microtubules in cultured rat sensory axons. J Neurocytol 10:589–605

Bruyn RP (1992) The neuropathology of hereditary spastic paraparesis. Clin Neurol Neurosurg 94 Suppl:S16–18

Bunker JM, Kamath K, Wilson L, Jordan MA, Feinstein SC (2006) FTDP-17 mutations compromise the ability of tau to regulate microtubule dynamics in cells. J Biol Chem 281:11856–11863

Caviston JP, Ross JL, Antony SM, Tokito M, Holzbaur EL (2007) Huntingtin facilitates dynein/dynactin-mediated vesicle transport. Proc Natl Acad Sci U S A 104:10045–10050

Chapin SJ, Bulinski JC (1992) Microtubule stabilization by assembly-promoting microtubule-associated proteins: a repeat performance. Cell Motil Cytoskel 23:236–243

Chen J, Kanai Y, Cowan NJ, Hirokawa N (1992) Projection domains of MAP2 and tau determine spacings between microtubules in dendrites and axons. Nature 360:674-677

Chevalier-Larsen E, Holzbaur EL (2006) Axonal transport and neurodegenerative disease. Biochim Biophys Acta 1762:1094–1108

Claudiani P, Riano E, Errico A, Andolfi G, Rugarli EI (2005) Spastin subcellular localization is regulated through usage of different translation start sites and active export from the nucleus. Exp Cell Res 309:358–369

Colin E, Zala D, Liot G, Rangone H, Borrell-Pages M, Li XJ, Saudou F, Humbert S (2008) Huntingtin phosphorylation acts as a molecular switch for anterograde/retrograde transport in neurons. EMBO J 27:2124–2134

Curmi PA, Gavet O, Charbaut E, Ozon S, Lachkar-Colmerauer S, Manceau V, Siavoshian S, Maucuer A, Sobel A (1999) Stathmin and its phosphoprotein family: general properties, biochemical and functional interaction with tubulin. Cell Struct Funct 24:345–357

D'Souza I, Schellenberg GD (2005) Regulation of tau isoform expression and dementia. Biochim Biophys Acta 1739:104–115

De Vos KJ, Grierson AJ, Ackerley S, Miller CC (2008) Role of axonal transport in neurodegenerative diseases. Annu Rev Neurosci 31:151–173

Dent EW, Callaway JL, Szebenyi G. Baas PW, Kalil K (1999) Reorganization and movement of microtubules in axonal growth cones and developing interstitial branches. J Neurosci 19:8894–8908

DiFiglia M, Sapp E, Chase KO, Davies SW, Bates GP, Vonsattel JP, Aronin N (1997) Aggregation of huntingtin in neuronal intranuclear inclusions and dystrophic neurites in brain. Science 277:1990–1993

Donahue CP, Ni J, Rozners E, Glicksman MA, Wolfe MS (2007) Identification of tau stem loop RNA stabilizers. J Biomol Screen 12:789–799

Duncan JE, Goldstein LS (2006) The genetics of axonal transport and axonal transport disorders. PLoS Genet 2:e124

Ebneth A, Godemann R, Stamer K, Illenberger S, Trinczek B, Mandelkow E (1998) Overexpression of tau protein inhibits kinesin-dependent trafficking of vesicles, mitochondria, and endoplasmic reticulum: implications for Alzheimer's disease. J Cell Biol 143:777–794

Errico A, Ballabio A, Rugarli EI (2002) Spastin, the protein mutated in autosomal dominant hereditary spastic paraplegia, is involved in microtubule dynamics. Hum Mol Genet 11:153–1 63

Fath T, Eidenmuller J, Brandt R (2002) Tau-mediated cytotoxicity in a pseudohyperphosphorylation model of Alzheimer's disease. J Neurosci 22:9733–9741

Feng J (2006) Microtubule: a common target for parkin and Parkinson's disease toxins. Neuroscientist 12:469–476.

Fink JK, Rainier S (2004) Hereditary spastic paraplegia: spastin phenotype and function. Arch Neurol 61:830–833

Fukata Y, Itoh TJ, Kimura T (2002) CRMP-2 binds to tubulin heterodimers to promote microtubule assembly. Nat Cell Biol 4:583–591

Gleeson JG, Allen KM, Fox JW (1998) Doublecortin, a brain-specific gene mutated in human X- linked lissencephaly and double cortex syndrome, encodes a putative signaling protein. Cell 92:63–72

Gunawardena S, Goldstein LS (2005) Polyglutamine diseases and transport problems: deadly traffic jams on neuronal highways. Arch Neurol 62:46–51

Harada A, Oguchi K, Okabe S (1994) Altered microtubule organization in small-calibre axons of mice lacking tau protein. Nature 369:488–491

Harjes P, Wanker EE (2003) The hunt for huntingtin function: interaction partners tell many different stories. Trends Biochem Sci 28:425–433.

Hasaka TP, Myers KA, Baas PW (2004) Role of actin filaments in the axonal transport of microtubules. J Neurosci 24:11291–11301

Hazan J, Fonknechten N, Mavel D (1999) Spastin, a new AAA protein, is altered in the most frequent form of autosomal dominant spastic paraplegia. Nat Genet 23:296–303

He Y, Baas PW (2003) Growing and working with peripheral neurons. Methods Cell Biol 71:17–35

He Y, Frances F, Myers KA, Black M, Baas PW (2005) Role of cytoplasmic dynein in the axonal transport of microtubules and neurofilaments. J Cell Biol 168:697–703

Heidemann SR, Landers JM, Hamborg MA (1981) Polarity orientation of axonal microtubules. J Cell Biol 91:661–665

Her LS, Goldstein LS. (2008) Enhanced sensitivity of striatal neurons to axonal transport defects induced by mutant huntingtin. J Neurosci 28:13662–13672

Horesh D, Sapir T, Francis F, Wolf SG, Caspi M, Elbaum M, Chelly J, Reiner O (1999) Doublecortin, a stabilizer of microtubules. Hum Mol Genet 8:1599–1610

Jimenez-Mateos EM, Paglini G, Gonzalez-Billault C, Caceres A, Avila J (2005) End binding protein-1 (EB1) complements microtubule-associated protein-1B during axonogenesis. J Neurosci Res 80:350–359

Joshi HC, Baas, PW (1993) A new perspective on microtubules and axon growth. J Cell Biol 121:1191–1196

Karabay A, Yu W, Solowska JM, Baird DH, Baas PW (2004) Axonal growth is sensitive to the levels of katanin, a protein that severs microtubules. J Neurosci 24:5778–5788

Kornack DR, Giger RJ (2005) Probing microtubule + TIPs: regulation of axon branching. Curr Opin Neurobiol 15:58–66

Lapointe NE, Morfini G, Pigino G, Gaisina IN, Kozikowski AP, Binder LI, Brady S T (2009) The amino terminus of tau inhibits kinesin-dependent axonal transport: implications for filament toxicity. J Neurosci Res 87:440–451

Lee VM, Trojanowski JQ (2001) Transgenic mouse models of tauopathies: prospects for animal models of Pick's disease. Neurology 56:S26–30

Leung CL, Green KJ, Liem RK (2002) Plakins: a family of versatile cytolinker proteins. Trends Cell Biol 12:37–45

Ligon LA, LaMonte BH, Wallace KE, Weber N, Kalb RG, Holzbaur EL (2005) Mutant superoxide dismutase disrupts cytoplasmic dynein in motor neurons. Neuroreport 16:533–536

Mancuso G, Rugarli EI (2008) A cryptic promoter in the first exon of the SPG4 gene directs the synthesis of the 60-kDa spastin isoform. BMC Biol 6:31

Marshall LE, Himes RH (1978) Rotenone inhibition of tubulin self-assembly. Biochim Biophys Acta 543:590–594

Matus A (1994) Stiff microtubules and neuronal morphology. Trends Neurosci 17:19–22

McNally FJ, Vale RD (1993) Identification of katanin, an ATPase that severs and disassembles stable microtubules. Cell 75:419–429

McNally KP, Buster D, McNally FJ (2002) Katanin-mediated microtubule severing can be regulated by multiple mechanisms. Cell Motil Cytoskel 53:337–349

Millecamps S, Gentil BJ, Gros-Louis F, Rouleau G, Julien JP (2005) Alsin is partially associated with centrosome in human cells. Biochim Biophys Acta 1745:84–100

Moore DJ, West AB, Dawson VL, Dawson TM (2005) Molecular pathophysiology of Parkinson's disease. Annu Rev Neurosci 28:57–87

Morfini G, Pigino G, Beffert U, Busciglio J, Brady ST (2002) Fast axonal transport misregulation and Alzheimer's disease. Neuromol Med 2:89–99

Munch C, Rolfs A, Meyer T (2008) Heterozygous S44L missense change of the spastin gene in amyotrophic lateral sclerosis. Amyotroph Lateral Scler 9:251–253

Myers KA, Baas PW (2007) Kinesin-5 regulates the growth of the axon by acting as a brake on its microtubule array. J Cell Biol 178:1081–1091

Myers KA, Baas PW (2009) Microtubule-actin interactions during neuronal development. In: Neurobiology of actin: from neurulation to synaptic function. Springer, Heidelberg (in press)

Myers KA, Tint I, Nadar CV, He Y, Black MM, Baas PW (2006) Antagonistic forces generated by cytoplasmic dynein and myosin-II during growth cone turning and axonal retraction. Traffic 7:1333–1351

Nadar VC, Ketschek A, Myers KA, Gallo G., Baas PW (2008) Kinesin-5 is essential for growth-cone turning. Curr Biol 18:1972–1977

Qiang L, Yu W, Andreadis A, Luo M, Baas PW (2006) Tau protects microtubules in the axon from severing by katanin. J Neurosci 26:3120–3129

Reiner O, Carrozzo R, Shen Y, Wehnert M, Faustinella F, Dobyns WB, Caskey CT, Ledbetter DH (1993) Isolation of a Miller-Dieker lissencephaly gene containing G protein beta-subunit-like repeats. Nature 364:717–721

Ren Y, Liu W, Jiang H, Jiang Q, Feng J (2005) Selective vulnerability of dopaminergic neurons to microtubule depolymerization. J Biol Chem 280:34105–34112

Rosen DR, Siddique T, Patterson D (1993) Mutations in Cu/Zn superoxide dismutase gene are associated with familial amyotrophic lateral sclerosis. Nature 362:59–62

Roy S, Zhang B, Lee VM, Trojanowski JQ. (2005) Axonal transport defects: a common theme in neurodegenerative diseases. Acta Neuropathol 109:5–13

Sanderson CM, Connell JW, Edwards TL, Bright NA, Duley S, Thompson A, Luzio JP, Reid E (2006) Spastin and atlastin, two proteins mutated in autosomal-dominant hereditary spastic paraplegia, are binding partners. Hum Mol Genet 15:307–318

Solowska JM, Morfini G, Falnikar A, Himes BT, Brady ST, Huang D, Baas PW (2008) Quantitative and functional analyses of Spastin in the nervous system: implications for hereditary spastic paraplegia. J Neurosci 28:2147–2157

Stanford PM, Shepherd CE, Halliday GM, Brooks WS, Schofield PW, Brodaty H, Martins RN, Kwok JB, Schofield PR (2003) Mutations in the tau gene that cause an increase in three repeat tau and frontotemporal dementia. Brain 126:814–826

Stepanova T, Slemmer J, Hoogenraad CC (2003) Visualization of microtubule growth in cultured neurons via the use of EB3-GFP (end-binding protein 3-green fluorescent protein). J Neurosci 23:2655–2664

Suter DM, Schaefer AW, Forscher P (2004) Microtubule dynamics are necessary for SRC family kinase-dependent growth cone steering. Curr Biol 14:1194–1199

Szebenyi G, Morfini GA, Babcock A (2003) Neuropathogenic forms of huntingtin and androgen receptor inhibit fast axonal transport. Neuron 40:41–52

Tint I, Slaughter T, Fischer I, Black MM (1998) Acute inactivation of tau has no effect on dynamics of microtubules in growing axons of cultured sympathetic neurons. J Neurosci 18:8660–8673

Vale RD (1991) Severing of stable microtubules by a mitotically activated protein in Xenopus egg extracts. Cell 64:827–839

Vaughan KT (2005) Microtubule plus ends, motors, and traffic of Golgi membranes. Biochim Biophys Acta 1744:316–324

Wang L, Brown A (2002) Rapid movement of microtubules in axons. Curr Biol 12:1496–1501

Wang JZ, Liu F (2008) Microtubule-associated protein tau in development, degeneration and protection of neurons. Prog Neurobiol 85:148–175

Wittmann T, Bokoch GM, Waterman-Storer CM (2003) Regulation of leading edge microtubule and actin dynamics downstream of Rac1. J Cell Biol 161:845–851

Ye H, Kuruvilla R, Zweifel LS, Ginty DD (2003) Evidence in support of signaling endosome-based retrograde survival of sympathetic neurons. Neuron 39:57–68

Yu W, Baas PW (1994) Changes in microtubule number and length during axon differentiation. J Neurosci 14:2818–2829

Yu W, Ahmad FJ, Baas, PW (1994) Microtubule fragmentation and partitioning in the axon during collateral branch formation. J Neurosci 14:5872–5884

Yu W, Solowska JM, Qiang L, Karabay A, Baird D, Baas PW (2005) Regulation of microtubule severing by Katanin subunits during neuronal development. J Neurosci 25:5573–5583

Yu W, Qiang L, Baas PW (2007) Microtubule-severing in the axon: implications for development, disease, and regeneration after injury. J Environ Neurosci Biomed 1:1–7

Yu W, Qiang L, Solowska JM, Karabay A, Korulu S, Baas PW (2008) The Microtubule- severing proteins Spastin and Katanin participate differently in the formation of axonal branches. Mol Biol Cell 19:1485–1498

Zhang F, Strom AL, Fukada K, Lee S, Hayward LJ, Zhu H (2007) Interaction between familial amyotrophic lateral sclerosis (ALS)-linked SOD1 mutants and the dynein complex. J Biol Chem 282:16691–16699

Zhao X, Alvarado D, Rainier S (2001) Mutations in a newly identified GTPase gene cause autosomal dominant hereditary spastic paraplegia. Nat Genet 29:326–331

Actin in Axons: Stable Scaffolds and Dynamic Filaments

Paul C. Letourneau

Abstract Actin filaments are thin polymers of the 42 kD protein actin. In mature axons a network of subaxolemmal actin filaments provide stability for membrane integrity and a substrate for short distance transport of cargos. In developing neurons dynamic regulation of actin polymerization and organization mediates axonal morphogenesis and axonal pathfinding to synaptic targets. Other changes in axonal shape, collateral branching, branch retraction, and axonal regeneration, also depend on actin filament dynamics. Actin filament organization is regulated by a diversity of actin-binding proteins (ABP). ABP are the focus of complex extrinsic and intrinsic signaling pathways, and many neurological pathologies and dysfunctions arise from defective regulation of ABP function.

1 Introduction

Polymerized filaments of the protein actin are a major cytoskeletal component of axons, along with microtubules and neurofilaments. The most significant function of actin filaments in mature axons is to create sub-plasmalemmal scaffolding that stabilizes the plasma membrane and provides docking for protein complexes of several membrane specializations. In addition, this cortical actin scaffold interacts with myosin motor proteins to transport organelles to short distances. These functions engage actin filaments as a stable structural component in axonal homeostasis. However, during axonal development, the dynamic organization of actin filaments and their interactions with myosins play critical roles in axonal elongation and branching, and in mediating the extrinsic influences that guide axons to their synaptic targets. This chapter will first briefly discuss the characteristics of actin and its regulation by actin-binding proteins (ABPs). Then, actin organization and functions

---

P.C. Letourneau
Department of Neuroscience, 6–145 Jackson Hall, University of Minnesota,
Minneapolis, MN, 55455
e-mail: letou001@umn.edu

in mature axons will be discussed, and finally, the dynamic roles of actin filaments in growing and regenerating axons will be discussed.

2 The Properties of Actin

2.1 Actin Protein

Actin is a globular 42 kD protein and one of the most highly conserved eukaryotic protein. Mammals have three actin isoforms, α, β and γ, of which α and mostly β are expressed in neurons. Under proper buffer conditions and in sufficient concentration, actin monomers (G-actin) spontaneously assemble into filaments (F-actin) with a diameter of 7 nm and lengths of a few μm. The bonds between actin monomers are specific but not of high strength, allowing actin solutions to be freely shifted from polymerized to unpolymerized states. Due to the intrinsic orientation of each monomer, an actin filament is polarized, in which one end, the barbed end, adds subunits at a lower monomer concentration than that occurs at the opposite end, the pointed end; while the latter end loses monomers at a lower concentration. If ATP is furnished to an actin solution, actin filaments form after nucleating, and reach a steady state, in which filaments undergo a dynamic "treadmilling," continually adding ATP-actin at the barbed ends, and losing ADP-actin at the pointed ends.

Actin is abundantly expressed in all tissues, which is critical for cell division, adhesion, and motility. In neurons actin comprises about 4–5% of total protein (Clark et al. 1983), though during brain development actin expression rises to 7–8% of cell protein (Santerre and Rich 1976). The critical concentration of actin in vitro is the G-actin concentration, in which monomers are in equilibrium with F-actin. For skeletal muscle actin-ATP, this is about 0.1 μM, and in the muscle essentially all actin is polymerized. The intracellular G-actin concentration of chick brain has been estimated as 30–37 μM, but only 50–60% of actin is estimated to be F-actin (Clark et al. 1983; Devineni et al. 1999), rather than >99%, as predicted by the actin content. This lower-than-expected degree of polymerization is due to neuronal G-actin binding proteins that sequester G-actin from being available for polymerization. G-actin sequestering proteins are just a few of the many ABPs that regulate the organization and functions of neuronal actin.

2.2 Actin Binding Proteins

Actin does not function in neurons in a "naked" state, as portrayed in models or cartoons. Rather, many intracellular proteins specifically bind F-actin or G-actin (see Table 1 in Dent and Gertler 2003; Pak et al. 2008). These ABPs regulate all aspects of actin organization and dynamics (Dos Remedios et al. 2003). In regulating actin functions, ABPs are the targets of extrinsic and intrinsic signaling pathways. ABPs are categorized by function: (1) sequester, or bind G-actin subunits; (2) nucleate actin polymerization; (3) cap F-actin barbed ends to inhibit polymerization; (4) cap pointed ends to inhibit depolymerization; (5) bind barbed ends to inhibit capping; (6) bind pointed ends to promote depolymerization; (7) bundle, crosslink or stabilize F-actin; (8) sever actin filaments; (9) move cargo along actin filaments, or move actin filaments; and (10) anchor F-actin to other cellular components.

Important examples of ABPs in each class described above include the following: (1) *β-thymosin*, which holds actin in a nonpolymerizable form, and *profilin*, which catalyzes exchange of actin-ADP to actin-ATP to promote polymerization; (2) *Arp2/3 complex*, which binds F-actin, and nucleates a new filament oriented at 70° to an existing actin filament, and *formins*, which promote polymerization of long actin filaments in filopodia; (3) *capZ*, which caps actin barbed ends in the Z-band of muscle cells; (4) *tropomodulin*, which binds pointed ends; (5) *ena (Drosophila)* (*mena*; murine homolog), which prevents barbed end capping to promote polymerization; (6) *actin depolymerizing factor* (ADF)/cofilin, which binds pointed ends and promotes depolymerization; (7) *filamin*, which crosslinks actin filaments into networks, in which *fascin* crosslinks actin filaments into close bundles, and *tropomyosin*, which binds along actin filaments to block other ABPs to stabilize them, and thereby regulates contraction/depolymerization, etc.; (8) *ADF/cofilin*, which binds and severs F-actin, and *gelsolin*, which depends on Ca^{2+} for actin filament severing; (9) multiple *myosin* motors that bind actin and move cargo toward the barbed or pointed ends; (10) *spectrin*, which mediates F-actin binding to the intracellular side of the plasma membrane, *vinculin*, which binds F-actin to integrin-mediated adhesion sites, and *ERM* proteins (ezrin, radixin, moesin), which bind actin to several membrane proteins. Some ABPs are expressed ubiquitously, such as β-thymosin, ADF/cofilin, and spectrin, while other ABPs are tissue-specific, such as muscle proteins Xin or myopodin, or the neuron-specific, drebrin A, which is only in the postsynaptic side of excitatory synapses.

3 Actin in Axons

3.1 The Sources of Axonal Actin

Actin is synthesized at high levels on polyribosomes in neuronal perikarya. However, actin is also synthesized in axons, and in vitro studies with developing neurons estimated that 1–5% of total neuronal actin is made in axons (Eng et al. 1999; Lee and Hollenbeck 2003). Though this is a small fraction of total neuronal actin, the temporal and spatial regulation of axonal actin mRNA translation is critical to functions of developing axons, mature terminals, and regenerating axons (Lin and Holt 2007). Within axons β-actin mRNA, complexed with a regulatory protein, zipcode-binding protein1, is localized in periaxoplasmic ribosomal plaques (PARPs), which are sites of protein synthesis in myelinated axons (Koenig 2009). Actin mRNA is also concentrated in developing axonal growth cones, where local β-actin synthesis in response to axonal guidance cues may be critical in wiring neural circuits (see Yoon et al. 2009). Local actin synthesis in presynaptic terminals may contribute to synaptic rearrangements underlying neural plasticity, and axonal injury triggers local actin synthesis that promotes axonal regeneration.

3.2 Axonal Transport of Actin and Actin mRNA

Both actin protein and actin mRNA are transported in axons. Actin mRNA is sorted for axonal transport by the recognition of 3′ untranslated regions (UTR), comprising β-actin mRNA sequences (i.e., zipcode) that are bound by a zipcode-binding protein (ZBP1), and assembled with other mRNA into ribonucleoprotein particles (RNPs). These complexes are bound by kinesin family motor proteins and transported along microtubules at fast rates up to >150 mm per day (Wang et al. 2007). Actin protein monomers or oligomers, synthesized in the perikaryon, are transported in the axons in macromolecular complexes associated with the slow component-b (Scb) transport group, which includes ≈200 proteins that are cotransported along microtubules at rates of 2–8 mm/day (Galbraith and Gallant 2000; Roy et al. 2008).

The difference is striking between the transport rates of β-actin mRNA in RNP and transport of actin monomer in Scb. In addition to the transport and translation of actin mRNA in axons, mRNAs for several ABP are also translated in developing and maturing axons (Lin and Holt 2007). There may be advantages for regulating actin structures by locally translating mRNA for actin and ABP after rapid mRNA delivery rather than by relying on slower transport of Scb proteins from the perikaryon. A greater immediacy and specificity of coordinating protein activities may result, which may be important in regulation of specialized axonal domains (see chapter by Yoon et al. 2009).

3.3 Actin Filament Organization in Axons: Ultrastructure of Axonal F-Actin

It has been a challenge to clarify the organization of axonal actin filaments. Hirokawa (1982) prepared frog spinal nerve axons for electron microscopy by quick-freeze and deep-etch methods. He observed two axoplasmic structural domains. A central region of microtubules, neurofilaments and membranous organelles, interconnected by cross links and a peripheral subaxolemmal space of about 100 nm wide that contained a dense network of thin filaments connected to the plasmalemma on one side, and connected to the central cytoskeletal networks on the other side (Fig. 1). Schnapp and Reese (1982) used similar methods with

Actin in Axons: Stable Scaffolds and Dynamic Filaments

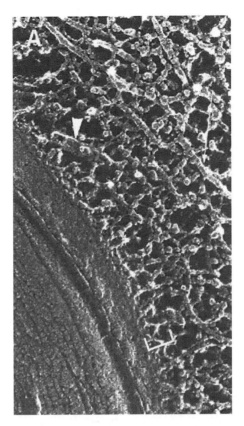

Fig. 1 This electron micrograph of a rapid-frozen, deep-etched myelinated frog axon shows the subaxolemmal space, which is filled with a scaffold of actin filaments (*brackets*). Actin filaments are obscured by other proteins aggregated onto the actin filaments. Microtubules (*arrowhead*) and neurofilaments (*arrows*) are visible near the subaxolemmal space. Myelin is present at the *lower left*. X130,000. Reprinted with permission from Hirokawa (1982)

turtle optic nerves, and observed a similar organization of central longitudinal bundles of neurofilaments and microtubules with organelles embedded in a granular matrix, and an 80–100 nm zone near the plasmalemma that was filled with a dense filament network. Hirokawa (1982) identified these subaxolemmal thin filaments as actin by labeling the axonal cortex with fluorescent phalloidin, which specifically binds F-actin.

The three dimensional arrangements of these cortical actin filaments are unclear. There is extensive overlapping of filaments interconnected by ABPs, such as filamin (Feng and Walsh 2004), and filament branching, mediated by Arp2/3 complex; moreover, actin filaments bind to the spectrin membrane skeleton and to other peripheral membrane proteins as well. Individual actin filaments in this axolemmal network are no longer than a few μm, and do not form filament bundles, such as those that exist in filopodia, microvilli or stress fibers of other cells.

3.4 Actin Filament Organization in Axons: F-Actin in the Central Axon Domain

The organization of actin filaments in the central domain, consisting of neurofilaments, microtubules, and organelles, is even less clear. Hirokawa (1982) observed no phalloidin-labeling of the central axonal domain, and though the crosslinks between the longitudinal cytoskeletal elements and organelles were abundant, few actin filaments were clearly observed. Schnapp and Reese (1982) described cross bridges between neurofilaments, and a granular matrix surrounding microtubules and associated organelles, but did not describe actin-like filaments in the central domain. In a later paper, Bearer and Reese (1999) examined extruded squid axoplasm and described actin filament networks oriented along longitudinal microtubule bundles. They suggested these actin filaments had roles in axonal transport. One documented function for F-actin in the central axoplasmic domain comes from the report of an axoplasmic filter that excludes the entry of BSA, or 70 kD dextran into axons, and is sensitive to the F-actin depolymerizing drug latrunculin A (Song et al. 2009). This filter begins at the initial segment of the axon, where a dense F-actin scaffold anchors a subaxolemmal protein complex of clustered sodium channels. Perhaps, this subaxolemmal F-actin scaffold is linked to the subcortical central network of actin and other cytoskeletal elements to comprise the molecular filter.

3.5 Actin Filament Organization in Axons: Actin Turnover in Axons

The striking electron micrographs (Fig. 1) convey an image of a complex cytoskeletal superstructure. However, movies of organelles rapidly moving through axoplasm reveal that this superstructure is flexible and dynamic. Cellular actin filaments undergo disassembly and polymerization, as regulated by local ABPs. Okabe and Hirokawa (1990) employed fluorescence recovery after photobleaching (FRAP) to examine the turnover of rhodamine-actin injected into dorsal root ganglion (DRG) neurons. They found that zones of bleached actin in DRG axons did not move, but the bleached zones recovered fluorescence with a half time of 15–30 min. They concluded that axonal actin filament networks were immobile, but turned over by regular cycles of depolymerization and F-actin assembly.

4 Actin Functions in Maintaining Axonal Structure

4.1 Maintenance of Axonal Integrity

Axons are subject to tensions and must resist strains, such as during vigorous limb movements. During neural development, growing neurons generate tensions on their substrate contacts that promote axon elongation. The actin scaffold beneath the axolemma and the associated peripheral membrane proteins comprise a membrane skeleton that resists strains, and protects axons from mechanical forces (Morris 2001). A primary component of the membrane skeleton is a network of α/β spectrin dimers that associate into tetramers, and bind actin filaments to form a flexible two-dimensional array that bind directly to membrane proteins of the axolemma, such as adhesion molecules L1, NCAM and protein 4.1N, and by binding to ankyrin adaptor proteins that bind other integral membrane proteins (Bennett and Baines 2001). Both spectrin and ankyrin are vital to axonal integrity. In mice that lack ankyrinB, which is broadly distributed along axons, axons are fragile, and the optic nerves degenerate (Susuki and Rasband 2008). Axons break in *C. elegans* mutants that lack β-spectrin (Hammarlund et al. 2007). The membrane-associated distribution of spectrin is sensitive to drugs that depolymerize F-actin, indicating an integral role of F-actin in spatially stabilizing this membrane skeleton.

4.2 Localization of Ion Channels at Nodes of Ranvier and Axon Initial Segment

The rapid propagation of action potentials along myelinated axons depends on the clustering of voltage-gated sodium (Nav) channels at the axon initial segment and at regularly spaced nodes of Ranvier. The primary components of these membrane domains at the nodes and axon initial segment are complexes of Nav channels, ankyrinG, L1-family adhesion molecules, βIV spectrin, protein 4.1N and F-actin (Susuki and Rasband 2008; Xu and Shrager 2005). Experimental results indicate that each of these molecules is required to form and/or maintain nodes of Ranvier and axon initial segments (see Thaxton and Bhat 2009).

4.3 Actin in PARPs

Associated with F-actin at the inner side of the axolemmal cortex are PARPs, which are periodically distributed as plaque-like protrusions in the cortex of myelinated axons (Sotelo-Silveira et al. 2004, 2006). As sites of protein synthesis of actin and other proteins, PARPs may have critical roles in axonal maintenance and in responses to both injury and physiological stimuli. Protein synthesis in PARPs is significantly inhibited by cytochalasin B, which causes actin depolymerization, indicating that cortical F-actin is critical to PARP localization and function (Sotelo-Silveira et al. 2008). PARPs contain the microtubule motor KIF3A and the actin motor myosin Va, which may mediate longitudinal and radial transport, respectively, of RNA/RNPs within axons (Sotelo-Silveira et al. 2004, 2006). Protein synthesis in PARPs is stimulated in vitro by cAMP, which suggests a role, that signaling pathways may play in upregulating local protein synthesis. Alternatively, or in addition, they may also regulate the transport and localization of RNPs, and/or modulate structural rearrangements within PARP domains (Sotelo-Silveira et al. 2008).

4.4 Actin in the Presynaptic Terminal

The actin-containing membrane skeleton continues into axonal terminals, to stabilize the presynaptic membrane and other components. In an additional role, an F-actin scaffold forms a corral around the reserve pool of synaptic vesicles, and is linked to vesicles via the ABP synapsin-1 (Dillon and Goda 2005). Whether F-actin has roles in vesicle release is unclear. Experiments with actin-depolymerizing drugs suggest that the F-actin corral is a barrier between the reserve vesicle pool and the readily releasable pool, and that F-actin impedes vesicle release (Halpain 2003). F-actin is also implicated in vesicle endocytosis and recycling, but, again, experimental results do not clearly reveal whether F-actin has more than a structural role (Dillon and Goda 2005). Actin is more concentrated in presynaptic terminals than it is in axons, and it is reported that a lengthy stimulus train increases the F-actin fraction at terminals from 25 to 50% F-actin (Halpain 2003). Perhaps, during short-term synaptic activity, F-actin provides a stable structural framework, but after prolonged activity, dynamic rearrangements of F-actin in the terminal may be critical in synaptic plasticity. There is increasing recognition that genetic aberrations in Rho GTPases, which are key ABP regulators (Luo 2002), and other ABP regulations are linked to defective learning and other mental functions (Bernstein et al. 2009). Although many of these defects involve actin function in dendrites, or other neuronal compartments, the dysregulation of presynaptic actin contributes to these malfunctions.

5 Actin Functions in Axonal Transport

As the cytoskeletal filament that binds myosin motors, F-actin has roles in axonal transport. In neurons several myosin isoforms have been identified, including myosins I, II, V, VI, IX, and X (Bridgman 2004). All these myosins have been implicated in neuronal migration and morphogenesis, and myosin II, especially, will be discussed in the section and on the role of actin in axonal growth and guidance. The strongest evidence for an actin role in axonal transport involves myosin Va (Bridgman 2004) (see also Bridgman 2009).

Long range, axial transport of organelles is mediated by microtubule-based motor proteins. When F-actin depolymerizing drugs (e.g., cytochalasins, latrunculin) were applied to axons, organelle transport continued unabated and mitochondrial transport even became faster (Morris and Hollenbeck 1995). This suggests that organelles are longitudinally transported along microtubule tracks, and perhaps, F-actin networks in the subaxolemmal cortex, and more centrally located; create physical impediments that can slow transport along microtubules. Other studies found that longitudinal rapid transport of mRNA-containing RNPs, and slow transport of Scb components are mediated by microtubule motors, and disrupting F-actin does not slow anterograde transport (Roy et al. 2008). However, when axonal microtubules were depolymerized in cultured neurons, robust anterograde and retrograde movements of mitochondria continue (Hollenbeck and Saxton 2005; Morris and Hollenbeck 1995). This is evidence that axonal actin filaments can be tracks for mitochondria movements (Bearer and Reese 1999). Other in vitro studies that used F-actin depolymerizing drugs found reduced rates of transport of fluorescently labeled microtubules and neurofilaments, supporting a hypothesis that these intact cytoskeletal elements might be moved for short distances along stable actin filaments (Hasaka et al. 2004; Jung et al. 2004). However, prolonged treatments with cytochalasins, or latrunculin to depolymerize F-actin might disrupt the axolemmal cortex and central axoplasmic meshworks, and indirectly interfere with microtubule-based transport.

Actomyosin-generated forces are involved in short distance transport of organelles (Bridgman 2004). The best candidate motor, myosin V, dimerizes to form two-headed proteins that move processively at rapid rates toward actin filament barbed ends. Complexes of myosin Va and kinesin motors are associated with many axonal cargoes, including ER vesicles, organelles that contain synaptic vesicle proteins, and RNPs (Langford 2002). By binding both microtubule and F-actin motors, cargos can be transported long distances along microtubules, and when the kinesin motors unbind from microtubules, myosin Va on the cargo can engage F-actin and transport the cargo radially to the axolemmal cortex, or into regions with few microtubules, such as the movement of synaptic vesicles into axon terminals, or movements of exocytotic or endocytotic organelles at the front of axonal growth cones (Evans and Bridgman 1995). BC1 RNA transport in Mauthner axon is an example supporting cooperative longitudinal-to-radial transport between microtubule and actin systems, respectively (Muslimov et al. 2002). Materials that are endocytosed at axon terminals, such as target-derived growth factors, are first moved retrogradely along F-actin until they engage microtubule motors for retrograde transport to the perikaryon (Reynolds et al. 1999). In mice with mutant myosin Va cargos accumulate in axonal terminals and regions that are sparsely populated with microtubules (Lalli et al. 2003), indicating the normal bidirectional myosin-mediated transport along actin filaments in these regions.

6 Actin Functions in Axonal Initiation, Elongation and Guidance

In mature axons, actin filaments maintain axonal integrity, localize components of specialized domains, and help transport cargo locally. In axon terminals, stable F-actin plays a scaffolding role, but in addition rapid reorganization of F-actin may reshape axonal structure during neural plasticity. However, unlike in mature axons, in which structural stability is a dominant actin role, the dynamic assembly and reorganization of actin filaments have major roles in axonal development. Dynamic actin structures contribute to the initiation, elongation, polarization and navigation of developing axons, and also axon regeneration. As important as recognizing the significance of actin functions in axonal morphogenesis is the understanding that this dynamic actin system is regulated through the action of ABPs.

6.1 Actin Function in Neurite Initiation

During neural development, neuronal precursors, or immature neurons migrate to various destinations before they settle, and sprout axons and dendrites. When put into tissue culture, immature hippocampal neurons attach to the substrate and are initially spherical before sprouting several neurites, which are undifferentiated processes. Eventually, one neurite accelerates its elongation and becomes the axon, while the other neurites become dendrites. This scenario provides a general model for neuronal morphogenesis (Craig and Banker 1994).

Upon settling on an in vitro substrate, immature neurons begin extending protrusions, as the plasma membrane is pushed out by the force of actin polymerization onto F-actin barbed ends beneath the plasmalemma (Fig. 2). Several ABPs promote actin polymerization at the leading edge of these protrusions. These include: (1) the Arp2/3 complex, which nucleates actin, (2) profilin, which supplies ATP-actin, and (3) ADF/cofilin, which releases G-actin from actin filament pointed ends and severs F-actin to generate new barbed ends for polymerization (Dent and Gertler 2003; Pak et al. 2008; Pollard and Borisy 2003). The activities of these ABPs are regulated by kinases, or phosphatases, by Ca^{2+} fluxes, and by membrane-derived PIP2 (phosphatidylinositol biphosphate). If Arp2/3 activity is high, a dendritic network of short, branched F-actin pushes out a broad lamellipodium. If ABPs, such as formins, or ena/VASP are present, long F-actin bundles push out filopodia. These protrusions are retracted as F-actin is severed by ADF/cofilin, and/or gelsolin, and depolymerized. Concurrently, myosin II filaments that are linked to adjacent structures pull actin filaments back from the leading margin in a retrograde flow that is coupled to actin depolymerization (Brown and Bridgman 2003). Myosin II filaments that bind to F-actin throughout the cortical meshwork generate tangential tensions that maintain the neuron as a sphere.

To sprout a neurite these protrusions must make adhesive contacts (Fig. 3). If neurons are plated on natural substrate ligands, such as laminin, surface receptors on lamellipodia and filopodia form adhesive interactions with substrate ligands that organize protein complexes at the cytoplasmic surfaces of these adhesion sites. The protein complexes serve to link newly polymerized F-actin to the adhesions. Integrin-mediated adhesions include ABPs vinculin, talin, and α-actinin (Shattuck and Letourneau 1989); adhesion complexes of Ig-family cell adhesion molecules (CAMs) include ABPs spectrin and ERM (ezrin, moesin, radixin; Ramesh 2004); and cadherin-mediated adhesions bind F-actin via the ABP α-catenin. These protein complexes also include signaling components that regulate actin polymerization and organization. These adhesions are critical to neurite initiation for two reasons. First, by connecting newly polymerized F-actin to adhesion bonds, the protrusions are stabilized against actomyosin tensions and retrograde actin flow. Secondly, by

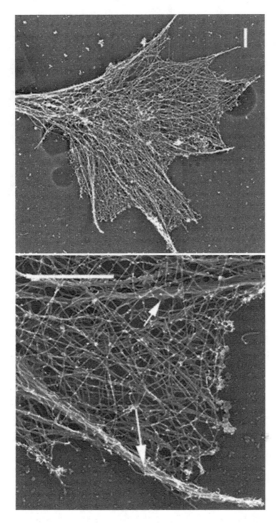

Fig. 2 Electron micrographs of the dense actin filament network at the leading edge of a hippocampal neuron axonal growth cone. Actin filament bundles (*arrows*) are enmeshed into the F-actin network. The lower panel is a high magnification view from the lower left of growth cone in the *upper panel*. *Scale bars* equal 0.5 µm. Reprinted with permission from Dr. Lorene Lanier

including signaling components, these adhesion-related complexes organize further actin polymerization.

This protrusion and adhesion of the neuronal perimeter is a prologue to neurite initiation. In a freshly plated neuron, microtubules encircle the cell periphery, contained by cortical tensions. As filopodial and lamellipodial protrusions expand the cell margin, microtubule plus ends jut into these protrusions, but they are swept back with retrograde F-actin flow. Where firm adhesions are made, retrograde actin flow slows, and microtubules can advance by polymerization and transport, and be directed

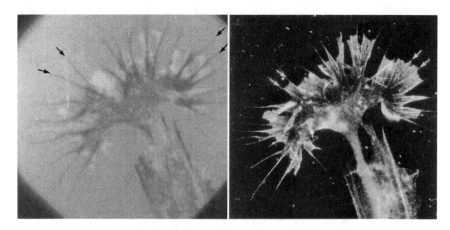

Fig. 3 Corresponding interference reflection micrograph (*left panel*) and fluorescence micrograph (*right panel*) images, showing adhesive contacts (*left*), and F-actin organization (*right*) at the front of a sensory neuron axonal growth cone. The growth cone is on the culture substrate after migrating off the upper surface of a Schwann cell visible at *lower right*. The *darkest areas* in the left panel are the closest adhesive contacts, *gray* areas are also areas of adhesion, while *bright* areas are farthest from the substrate. Many F-actin bundles (*arrows*) at the growth cone leading edge at associated with adhesive areas of the growth cone lower surface. From Letourneau (1981)

toward the adhesive sites (Fig. 4). Then, subsequent coupling of microtubule advance with actin-based protrusion and adhesion become the bud of a new neurite.

In this process actin has a dual role that both inhibits and promotes neurite initiation. Actomyosin tensions in the cell cortex, and in retrograde flow impede microtubule advance. On the other hand, actin polymerization drives the protrusion to expand the cell periphery, and F-actin links to adhesive sites to promote microtubule advance via actin-microtubule interactions and tensions that orient microtubule ends in the leading margin. The restrictive influence of F-actin in neurite initiation is indicated by experiments in which neurons are plated on a highly adhesive substrate in the media with the actin-depolymerizing drug cytochalasin B. The neurons still sprout neurites (Marsh and Letourneau 1984). In cytochalasin B, extensive surface adhesion promotes cell spreading despite the absence of protrusion; thus, without a cortical actin network, microtubule ends approach the cell margin. Eventually, a core array of microtubules advance and push a neurite outward. The idea that reduced cortical tensions allows microtubule advance to initiate a neuritic bud is supported by the result showing that under similar high adhesion, the myosin II inhibitor blebbistatin also increases neurite initiation and elongation (Ketschek et al. 2007; Kollins et al. 2009; Rosner et al. 2007).

However, not all actomyosin tensions inhibit neurite initiation. Two lines of evidence show that neurite initiation is promoted by pulling the cell cortex outward. Several labs used adhesive micropipettes, or microspheres to adhere and pull on neuronal surfaces to induce neurite formation (Fass and Odde 2003; Heidemann et al. 1995). With continued tension the neurites elongated further, and when they were released from their tether and attached to a substrate, the neurites elongated

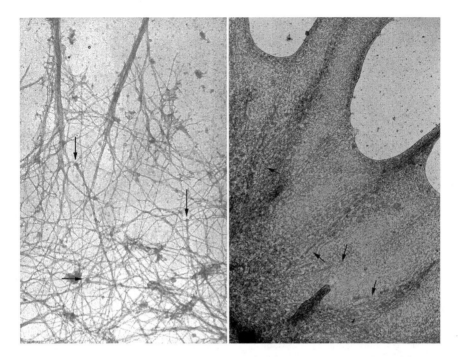

Fig. 4 Electron micrographs of the alignment of microtubules along actin filament bundles in the peripheral domains of whole mounted preparations of sensory neuron axonal growth cones. Microtubules are marked by *arrows*. The *left panel* was prepared by simultaneous fixation and extraction to better reveal F-actin network and microtubules. The *right panel* is an unextracted growth cone and reveals membrane organelles that track along the microtubules. *Left panel* is from Letourneau (1983). *Right panel* is from Letourneau (1979)

normally with a growth cone. When Dent et al. (2007) knocked out the three genes for all forms of the anticapping ABP ena/VASP, neurons from these mice could not initiate neurites on the ligand laminin. On laminin these neurons formed lamellipodial protrusions, but filopodia were not formed in the absence of ena/VASP. Neurite initiation by these mutant neurons was rescued by transfecting them to express any one of three ena/VASP genes, or one of two ABPs normally in filopodial, forming, or myosin X. These rescued neurons protruded filopodia, which became sites of neurite sprouting. These results illustrate how locally applied tension to pull out the cell cortex promotes organization of a microtubule core for a nascent neurite.

6.2 Actin Functions in Neurite Elongation

Neurite elongation is the distal movement of microtubules, organelles, and axoplasmic components with actin assembly, along a neurite and at its distal tip. The cytochalasin B experiments show that normal actin cytoskeletal activity is not required for neurite elongation; however, in the presence of cytochalasin B, the neurite elongation rate is reduced by 60% or more (Letourneau et al. 1987). Actin has two functions in neurite elongation that are localized at the enlarged tip of an elongating neurite, called the growth cone.

Growth cones are highly motile, and are enriched in actin (Fig. 2; Letourneau 1983, 1996). Actin polymerization and reorganization in growth cones is a major energy consumer, using up to 50% of neuronal ATP (Pak et al. 2008). Growth cones contain many ABPs (Cypher and Letourneau 1991; Shattuck and Letourneau 1989; see Table 1 in Dent and Gertler, 2003). An interesting analysis compared the ABPs expressed in developing neurons to a fibroblast cell line, and found that neurons are enriched in anticapping and F-actin bundling ABPs, such as ena and fascin, while expressing few filament capping proteins (Strasser et al. 2004). Such an ABP profile is consistent with the abundant F-actin and dynamic motility of growth cones. The growth cone motility is also promoted by GAP43, an ABP that is highly concentrated in growth cone plasma membranes (Benowitz and Routtenberg 1997). GAP43 is dramatically upregulated in regenerating axons and has roles in synaptic plasticity.

Actin organization and dynamics are regionally differentiated in growth cones (Dent and Gertler 2003; Pak et al. 2008). The axonal actin scaffold extends into the growth cone body, where the spectrin membrane skeleton ends. Further out in the peripheral domain, actin cytoskeleton and membrane structure are highly dynamic, characterized by abundant actin polymerization, rapid retrograde actin flow, a high rate of F-actin turnover (Okabe and Hirokawa 1991), and dynamic membrane exocytosis and endocytosis of myosin-transported vesicles. Actin filament polymerization pushes out the leading edge, from which F-actin bundles extend backward into the transition zone, where actin filaments are rearranged by myosin II and other ABPs, and are depolymerized. Behind this region, in the central domain, actin filaments are sparser, retrograde flow is slow, and actin turnover is slower (Okabe and Hirokawa 1991). In the base of the growth cone, F-actin shapes the cylindrical profile of the axon under tension generated by myosin II contractility (Bray and Chapman 1985; Dent and Gertler 2003; Loudon et al. 2006).

It is not clear how the dynamic motility of the peripheral domain is maintained. Certainly, regulation of ABP activity is critical, and local differences in activities of Rho GTPases in growth cones are particularly significant (Luo 2002). In addition, the paucity of microtubules in this distal-most region is a factor, as proteins or activities associated with microtubules have regulatory influences on actin dynamics (Bray et al. 1978; Dent and Gertler 2003).

Growth cones promote neurite elongation in two ways (Dent and Gertler 2003; Letourneau 1996; Suter and Forscher 2000). The leading margin extends, gains anchorage and expands to create cytoplasmic space into which microtubules and associated organelles advance (Figs. 3 and 4). The second activity is generation of traction on these adhesions beneath the peripheral and transition regions (Letourneau 1979, 1996). These traction forces are powered by myosin II, which is broadly distributed in growth cones, including in filopodia and lamellipodia (Brown and Bridgman 2003).

Growth cone traction promotes neurite elongation. Several laboratories have measured the traction exerted by growth cones and filopodia on their substrate attachments (Brown and Bridgman 2003; Fass and Odde 2003; Heidemann et al. 1995). When extrinsic tensile forces were applied to growth cones, the elongation rate increased in direct relation to the tension, up to a limit, after which the neurite detached or broke. Thus, two forces drive neurite elongation, the "push" of microtubule advance, and the "pull" of growth cone traction on adhesive contacts (Heidemann et al. 1995; Letourneau et al. 1987). Perhaps, the slow neurite elongation in the cytochalasin B and blebbistatin experiments was powered only by the "push" of microtubule advance. A biophysical model suggests that growth cone "pull" promotes neurite elongation by generating tensions that align, separate and stretch axoplasmic and axolemmal components in a manner that accelerates assembly and intercalation of components (Miller and Sheetz 2006). This model is consistent with continued axonal elongation after synaptogenesis, as an organism grows, and has relevance to promoting axonal regeneration. In addition, Van Essen (1997) incorporated the tensile forces in growing axons into a theory for brain morphogenesis, in which axonal tensions underlie the folding of the cerebral cortex, in essence "shrink wrapping" the large cortical surface into a manageably sized skull cavity.

To summarize, growth cone migration and neurite elongation involve three cytoskeletal activities; actin polymerization and protrusion of the leading edge, the advance of microtubules and transport of kinesin-powered organelles from the central domain into the peripheral domain, which has been called engorgement, and actomyosin II-powered sculpting, or consolidation of the posterior growth cone into a neuritic shaft (Dent and Gertler 2003; Letourneau 1996).

6.3 Actin Functions in Polarization of the Axon

As previously mentioned, an immature hippocampal neuron extends several neurites, which elongate slowly until one neurite accelerates, expands its growth cone, and becomes the axon. Neuronal polarization has been of long-term interest, and recently a variety of kinases and signaling components are implicated in neuronal polarization (Witte and Bradke 2008). Microtubules and actin filaments are the targets of these regulatory signals, as axonal specification ultimately involves determining which neurite acquires the most stable microtubules and most efficient transport system. Although any neurite can become the axon, the first neurite, which usually forms adjacent to the centrosome, is most likely to become the axon.

Actin filaments have facilitative and restrictive roles in neuronal polarization. The accelerated neurite growth that marks transition to an axon is accompanied by increased growth cone motility and F-actin content. ABPs that stimulate actin dynamics, such as GAP43 and ADF/cofilin, are enriched in the immature axon. Any neurite of an immature neuron can be manipulated to become the axon by pulling on the neurite (Lamoureux et al. 2002), or by presenting one neurite with a favorable substrate ligand, so it elongates faster than others (Esch et al. 1999). As predicted in the biophysical model above, the enhanced growth cone "pull" exerted by these manipulations promotes effective transport and cytoskeletal organization to tip the balance toward axonal polarization in that neurite. On the other hand, when immature neurons were treated with cytochalasin B, more than one neurite became an axon (Witte and Bradke 2008). Perhaps, cytochalasin-induced breakdown of neuritic actin networks accelerates microtubule advance and organelle transport in all neurites, eliminating any intrinsic advantage of ABPs, or transport properties that might be contained in any one neurite.

6.4 Actin Functions in Growth Cone Navigation

Correct wiring of neural circuits requires that axons navigate through embryos to their synaptic targets. The growth cone is the "navigator," a sensorimotor structure that detects and responds to features and molecules that define pathways of axonal elongation. In order to detect navigational cues, growth cones extend filopodia and lamellipodia that are much longer than needed for neurite elongation. The mean filopodial length is 10 μm, although long filopodia up to 100 μm in length are regularly observed in vitro, providing growth cones with a large search capacity in small embryos (Gomez and Letourneau 1994; Hammarback and Letourneau 1985). In vivo and in vitro experiments and genetic manipulations to limit growth cone protrusion do not stop axon elongation, but rather cause axon navigational errors (Chien et al. 1983; Marsh and Letourneau 1984; Zheng et al. 1996).

Growth cone ABPs are the focus of guidance and cue signaling. Many region-specific ABP functions have been revealed by chromophore-assisted laser inactivation of individual ABPs in growth cones (Buchstaller and Jay 2000). By regulating ABPs to spatially regulate actin assembly and organization, guidance cues induce growth cones to follow adhesive paths, avoid repulsive terrain, and turn toward targets. Positive or attractive cues promote actin polymerization and protrusion in growth cone regions closer to the guidance cue (Dent and Gertler 2003; Gallo and Letourneau 2000; Gallo et al. 1997), while negative or repulsive cues limit actin polymerization and protrusion in regions closer to the negative cue. Some ABPs are regulated in opposite directions by positive and negative cues. For example, in chick DRG and retinal growth cones, downstream signaling from attractive cues, NGF and netrin, respectively, activate ADF/cofilin to sever actin filaments, creating new barbed ends for actin polymerization (Marsick and Letourneau, in preparation). Conversely, Semaphorin3A, a repulsive cue, signals to deactivate ADF/cofilin, reducing actin dynamics and suppressing protrusive activity (Aizawa et al. 2001). ERM proteins are activated by phosphorylation and link actin filaments to the plasma membrane. NGF-mediated signals increase ERM phosphorylation in the DRG growth cone's leading margin, promoting protrusion (Marsick and Letourneau, in preparation). Conversely, Semaphorin3A signaling decreases ERM phosphorylation, which contributes to loss of adhesive contact and protrusions (Gallo 2008).

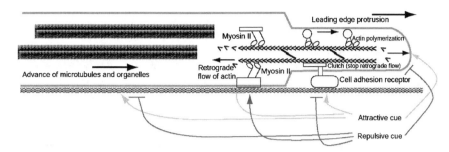

Fig. 5 A model of the mechanism of growth cone migration. Actin polymerization pushes the leading margin forward. Forces generated by myosin II pull actin filaments backwards, where filaments are disassembled. When growth cone receptors make adhesive contacts with a surface, a "clutch" containing ABPs links the adhesive contact to actin filaments of the leading edge, and the retrograde actin flow stops. This permits microtubule advance and promotes axonal elongation. Intracellular signaling generated by attractive and repulsive axonal guidance cues interacts with the molecular mechanisms of actin polymerization, myosin II force generation, adhesive contacts, and microtubule advance to regulate growth cone navigation

Growth cone guidance, however, is more complex than simply regulating one or two ABPs to promote or inhibit local actin assembly (Fig. 5). The activity of ABPs are integrated with other factors, such as the available pool of G-actin (Devineni et al. 1999), and posttranslational modifications of actin activity (Lin and Holt 2007). For example, experiments in vitro with Xenopus, spinal neurons growth cones turn toward the side in which ADF/cofilin activity is lower (Wen et al. 2007), while growth cones of chick DRG neurons turn toward the growth cone side in which ADF/cofilin activity is higher (Marsick and Letourneau, in preparation). In both cases F-actin levels are higher in the growth cone side towards the turn, but in Xenopus neurons a low G-actin pool, or high level of F-actin capping may favor F-actin accumulation where ADF/cofilin F-actin severing is less, while in chick neurons, G-actin is abundant and capping activity is low, so high ADF/cofilin activity creates many F-actin barbed ends to promote polymerization.

The roles of myosin II in growth cone turning provide another example that the outcome of ABP activity depends on additional factors. On the growth cone side toward an attractive cue, myosin II-generated tension on firm adhesive contacts promotes alignment and advance of microtubules, while on the side away from an attractive cue, actomyosin tensions on fewer adhesions retract the distal growth cone margin, as the growth cone turns (Loudon et al. 2006). Finally, growth cones contain mRNAs coding for β-actin and several ABPs, plus translational machinery (Lin and Holt 2007). Although actin is usually abundant in growth cones, conditions may exist when G-actin availability is limited, and local actin synthesis may add monomers to fuel local F-actin polymerization and a turning response. This notion is supported by evidence that attractive cues like neurotrophins and netrin stimulate β-actin synthesis in growth cones (Lin and Holt 2007). On the other hand, some repulsive responses may be mediated by proteolysis of actin and ABPs, leading to local suppression of protrusive motility and turning away from a negative cue.

After control of actin dynamics and protrusive activity by guidance cue signaling, the local advance of microtubules and organelles completes the navigational response. Microtubule ends probe into the peripheral domain, but most are swept back in retrograde actin flow. However, some microtubules enter adherent areas, and their advance is directed along actin filament bundles (Fig. 4; Letourneau 1979, 1983; Schaefer et al. 2008). These microtubules pioneer the advance of axonal structures, and their advance into protrusive areas is required for turning (Bentley and O'Connor 1994; Gallo and Letourneau 2000; Letourneau 1996; Suter and Forscher 2000; Zhou and Cohan 2000).

Ample evidence suggests that microtubule make specific connections with F-actin (Myers and Baas 2009). Microtubule plus ends bind proteins, called +TIPs, including CLIP-170, APC, EB proteins, and the dynein–dynactin complex (see Falnikar and Baas 2009). Interactions of these proteins and additional proteins, CLASPs and IQGAP, may mediate linkage to F-actin for transmission of forces, involving microtubule-associated proteins, or myosin motors. A classical MAP, MAP1B, is located in growth cones, and has been shown capable of linking microtubules and F-actin. Spectraplakins are large proteins with N-terminal F-actin binding and C-terminal microtubule binding domains. The spectraplakin MACF1 is highly expressed in mammalian brain, and genetic deletion of the Drosophila homolog shot results in defective axon initiation and navigation. A recent paper shows that the neuronal ABP drebrin is located in F-actin bundles of growth cone filopodia and mediates F-actin linkage to the +TIPs protein EB3 (Geraldo et al. 2008). Disruption of drebrin-EB3 interaction aborts neurite initiation, and impairs growth cone function. These potential microtubule-F-actin interactions may have multiple roles, such as mediating cytoskeletal interactions in central and cortical axonal domains, or in growth cones; changing these interactions may mediate the transition from axon shaft to growth cone body. However, linking microtubule plus ends to F-actin bundles in the growth cone leading margin is particularly important in growth cone navigation.

7 Actin Function in Axonal Branching

When axons reach their targets, they often form collateral branches to search for multiple potential synaptic partners. Axonal branching may also occur during postnatal refinement and strengthening of synaptic connections. Further, axonal branching often occurs when axon terminals or axons are injured. Axonal branching begins with actin polymerization and formation of F-actin bundles to protrude filopodia, or lamellipodia from a quiescent axonal shaft. When a bead bearing a positive guidance cue contacts a responsive axon, filopodia and lamellipodia appear, followed by a growth cone and then a neurite (Fig. 6; Dent et al. 2004; Gallo and Letourneau 1998; Kalil et al. 2000). Receptor-mediated signaling, in response to binding to the ligand-linked bead, activates ABPs to initiate actin polymerization. For example, ena/VASP is activated by netrin to induce filopodia along axonal shafts (Lebrand

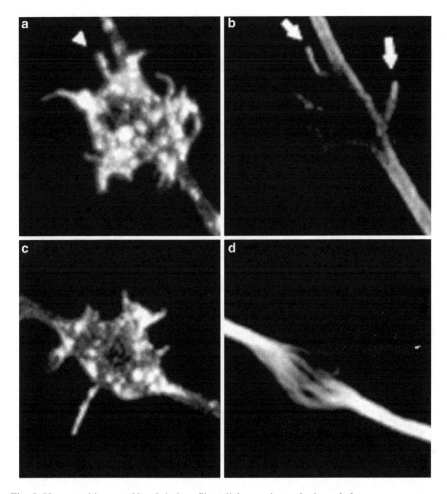

Fig. 6 Neurotrophin-coated beads induce filopodial sprouting and microtubule rearrangements to induce collateral branch along axonal shafts. Panels (**a**) and (**c**), show phalloidin-stained filopodial sprouts (*arrowheads*) at sites of NGF bead contact with a sensory neuron axon. Panels (**a**) and (**c**) demonstrate the axonal swelling and formation of F-actin bundles at sites of bead–axon contact. Panel (**b**) shows microtubule invasion of the filopodial sprouts shown in Panel (**a**) (stained with anti-β-tubulin; *arrows*). Panel (**d**) shows localized microtubule unbundling (stained with anti-β-tubulin) that occurred at sites of axon contact with NGF beads (shown in (**c**)). The translucent beads are not visible in the confocal images. From Gallo and Letourneau (1998)

et al. 2004). Receptor signaling may activate dynamic F-actin polymerization and protrusion from "hot spots" of F-actin along axonal shafts (Lau et al. 1999; Loudon et al. 2006), or localized signaling may activate local translation of β-actin mRNA along an axon (Willis et al. 2007). A key event in the transition from protrusion to branch formation is unbundling the axonal microtubules to unleash microtubule ends to interact with the F-actin bundles and advance into a nascent branch (Fig. 6; Gallo and Letourneau 1998; Kalil et al. 2000).

8 Actin Function in Axonal Retraction

As neural circuits are constructed, more axons and branches are formed than are eventually incorporated into circuits. Removal of these extra axons and branches occurs by an active process involving the axonal actin cytoskeleton (Luo 2002; Luo and O'Leary 2005). This "pruning" can involve short terminal branches, such as at the ends of motor axons during synapse elimination at the neuromuscular junction, or in the development of ocular dominance columns in the visual system. Pruning also eliminates long branches, as during sculpting of area-specific connections of layer 5 cortical neurons. Axonal pruning involves actomyosin contraction in the axolemmal cortex. RhoA GTPase is activated by Semaphorin3A and ephrin-A, which induce axon retraction, to increase myosin II contractility (Gallo 2006). Initially, actomyosin tensions rearrange the cortical scaffold into F-actin bundles, parallel to the axolemma, which allows myosin II contraction to more effectively shorten the axon into sinuous curves with microtubules and neurofilaments forced backwards (Gallo 2006). After this initial retraction, proteolytic and degenerative processes break down axonal components. Axonal retraction also occurs after injury, or severing of axons. The abrupt entry of Ca^{2+} ions into injured axons activates calpain proteinase, which cleaves the membrane-bound spectrin in the subaxolemmal cortex, and allows myosin II within the actin scaffolding to retract the axonal stump (Spira et al. 2001).

9 Actin Function in Axonal Regeneration

When axons are crushed or severed, the loss of membrane integrity induces the axon segment proximal to lesion site, but still connected to the perikaryon, to undergo proteolysis and cytoplasmic degradation. As described above, actomyosin-mediated retraction may also occur. In cases when axons recover from these events, protrusive motility reappears at the proximal stump, and a growth cone migrates forward. When axons of cultured developing neurons are severed, the appearance of a growth cone and axonal elongation can occur within minutes (Bray et al. 1978), indicating that developing axons contain sufficient actin and ABPs to sustain growth cone motility, when appropriate signals occur. Other studies with cultured adult neurons found that some adult axons could also form a growth cone, and initiate regeneration soon after severing (Verma et al. 2005). This regenerative response depended on the activity of proteases, and several kinases, TOR, and p38 MAPK, as well as on local protein synthesis. Signaling to renew F-actin dynamics and perhaps, β-actin mRNA translation, occurs within a short timeframe that excludes the possibility that these materials are transported from the cell body (Willis et al. 2007).

Although many adult axons contain sufficient components to initiate growth cone motility (Wang et al. 2007), sustained regeneration requires that an injured neuron return to an immature state, and express genes for tubulin subunits, actin and many ABPs (Raivich and Makwana 2007). This response may require retrograde signals from growth factors that a regenerating axon acquires from surrounding glia and tissue cells. However, other molecules and signals in the injured environment inhibit axon regeneration, especially in the CNS (Gervasi et al. 2008). These include re-expression of negative guidance cues like Semaphorin3A, and myelin proteins. Cytoplasmic signals from these negative factors activate RhoA, the upstream activator of myosin II, which lead to tensions and axon retraction. Cell-permeable inhibitors of RhoA have been developed for clinical trials to increase spinal cord regeneration (Kubo and Yamashita 2007).

10 Actin Dysfunction in Disease States of Axons

Abnormalities in F-actin dynamics and actin-containing structures contribute to axonal dysfunction in an increasingly recognized number of situations (Bernstein et al. 2009). Many of these defects arise from genetic or metabolic perturbations of ABP function. Genetic defects in ABP regulation of growth cone actin may lead to defective axonal connectivity that causes mental retardation and behavioral disorders. Because of the importance of the membrane-associated actin scaffold to axonal integrity and ion channel function, defective F-actin organization may perturb action potential propagation, and disrupt the supply, or organization of neurotransmitter vesicles in axon terminals and impair synaptic communication. Because of the high ATP demand to maintain dynamic F-actin organization in axons, ischemic events can disrupt actin organization with subsequent impaired physiological functions.

Some axonal dysfunction arises from abnormal formation of actin-containing inclusions that block axonal transport, or otherwise disorganize axonal components (Bernstein et al. 2009). Actin-containing structures described as aggregates, paracrystalline arrays, or rods are present in axons of individuals with several neurological diseases. Rods that contain actin complexed with cofilin are abundant in neurons of individuals with Alzheimers Disease. Actin and cofilin are concentrated in Hirano bodies, which are abundant in the neurons of individuals with amyotrophic lateral sclerosis and Parkinson's disease. These rods form rapidly in cultured hippocampal neurons that are subjected to oxygen deprivation, and other insults that cause ATP deprivation, Ca^{2+} spikes, or other forms of cellular stress.

11 Conclusion

Axons are the circuit elements that communicate neural information. The cortical actin filament network of mature axons provides axolemmal stability and a transport substrate that are necessary to maintain the electrical and chemical membrane properties that are vital for neural communication. In a developmental context, dynamic regulation of actin filament organization and actomyosin-mediated forces are responsible for initiating, stimulating and guiding axons, as neural circuits are formed. Dynamic actin filament organization mediates extrinsic regulation of axonal form and extent during development. Successful axonal regeneration requires re-expression of the dynamic actin organization of developing neurons. These actin functions depend on a diversity of actin-binding regulatory proteins, and future research to unravel the complex mechanisms by which ABPs determine actin organization which will help develop strategies for axonal repair, and will further our understanding of genetic and environmental causes of neuronal dysfunction and disorders.

Acknowledgments The preparation of this chapter and the author's research have been generously supported by the NIH (HD019950), NSF, and the Minnesota Medical Foundation. Dr. Gianluca Gallo provided valuable comments on the text, and Dr. Lorene Lanier generously provided images for Fig. 2.

References

Aizawa H, Wakatsuki S, Ishii A, Moriyama K, Sasaki Y, Ohasi K, Sekine-Aizawa Y, Sehara-Fujisawa A, Mizuno K, Goshima Y, Yahara I (2001) Phosphorylation of cofilin by LIM-kinase is necessary for semaphorin 3A-induced growth cone collapse. Nat Neurosci 4:367–373

Bearer EL, Reese TS (1999) Association of actin filaments with axonal microtubule tracts. J Neurocytol 28:85–98

Bennett V, Baines AJ (2001) Spectrin and ankyrin-based pathways: metazoan inventions for integrating cells into tissues. Physiol Rev 81:1353–1392

Benowitz LI, Routtenberg A (1997) GAP-43: an intrinsic determinant of neuronal development and plasticity. Trends Neurosci 20:84–91

Bentley D, O'Connor TP (1994) Cytoskeletal events in growth cone steering. Curr Opin Neurobiol 4:43–48

Bernstein BW, Maloney MT, Bamburg JR (2009) Actin and diseases of the nervous system. In:Gallo G, Lanier L (eds) Neurobiology of actin: from neuralation to synaptic function. Springer, Berlin (in press)

Bray D, Chapman K (1985) Analysis of microspike movements on the neuronal growth cone. J Neurosci 5:3204–3213

Bray D, Thomas C, Shaw G (1978) Growth cone formation in cultures of sensory neurons. Proc Natl Acad Sci U S A 75:5226–5269

Bridgman PC (2004) Myosin-dependent transport in neurons. J Neurobiol 58:164–174

Bridgman PC (2009) Myosin motor proteins in the cell biology of axons and other neuronal compartments. Results Probl Cell Differ. doi: 10.1007/400_2009_10

Brown J, Bridgman PC (2003) Role of myosin II in axon outgrowth. J Histochem Cytochem 51:421–428

Buchstaller A, Jay DG (2000) Micro-scale chromophore-assisted laser inactivation of nerve growth cone proteins. Microsc Res Tech 48:97–106

Chien CB, Rosenthal DE, Harris WA, Holt CE (1993) Navigational errors made by growth cones without filopodia. Neuron 11:237–251

Clark SE, Moss DJ, Bray D (1983) Actin polymerization and synthesis in cultured neurons. Exp Cell Res 147:303–314

Craig AM, Banker G (1994) Neuronal polarity. Annu Rev Neurosci 17:267–310

Cypher C, Letourneau PC (1991) Identification of cytoskeletal, focal adhesion and cell adhesion proteins in growth cone particles isolated from developing chick brain. J Neurosci Res 30:259–265

Dent EW, Gertler FB (2003) Cytoskeletal dynamics and transport in growth cone motility and axon guidance. Neuron 40:209–222

Dent EW, Barnes AM, Tang F, Kalil K (2004) Netrin-1 and Semaphorin 3A promote or inhibit cortical axon branching, respectively by reorganization of the cytoskeleton. J Neurosci 24:3002–3012

Dent EW, Kwiatkowski AV, Mebane LM, Philippar U, Barzik M, Rubinson DA, Gupton S, Van Veen JE, Furman C, Zhang J, Alberts AS, Mori S, Gertler FB (2007) Filopodia are required for cortical neurite initiation. Nat Cell Biol 9:1347–1359

Devineni N, Minamide LS, Niu M, Safer D, Verma R, Bamburg JR, Nachmias VT (1999) A quantitative analysis of G-actin binding proteins and the G-actin pool in developing chick brain. Brain Res 823:129–140

Dillon C, Goda Y (2005) The actin cytoskeleton: integrating form and function at the synapse. Annu Rev Neurosci 28:25–55

Dos Remedios CG, Chhabra D, Kekic M, Dedova IV, Tsubakihara M, Berry DA, Nosworthy NJ (2003) Actin-binding proteins: regulation of cytoskeletal microfilaments. Physiol Rev 83:433–473

Eng H, Lund K, Campenot RB (1999) Synthesis of β-tubulin, actin and other proteins in axons of sympathetic neurons in compartmented cultures. J Neursci 19:1–9

Esch T, Lemmon V, Banker G (1999) Local presentation of substrate molecules directs axon specification by cultured hippocampal neurons. J Neurosci 19:6417–6426

Evans LL, Bridgman PC (1995) Particles move along actin filament bundles in nerve growth cones. Proc Natl Acad Sci U S A 92:10954–10958

Falnikar A, Baas PW (2009) Critical roles for microtubules in axonal development and disease. Results Probl Cell Differ. doi: 10.1007/400_2009_2

Fass JN, Odde DJ (2003) Tensile force-dependent neurite elicitation via anti-β1 integrin antibody-coated magnetic beads. Biophys J 85:623–636

Feng Y, Walsh CA (2004) The many faces of filamin: a versatile molecular scaffold for cell motility and signaling. Nat Cell Biol 6:1034–1038

Galbraith JA, Gallant PE (2000) Axonal transport of tubulin and actin. J Neurocytol 29:889–9111

Gallo G (2006) Rho-kinase coordinates F-actin organization and myosin II activity during semaphorin3A-induced axon retraction. J Cell Sci 119:3413–3423

Gallo G (2008) Semaphorin 3Ainhibits ERM phosphorylation in growth cone filopodia through inactivation of PI3K. Dev Neurobiol 68:926–933

Gallo G, Letourneau PC (1998) Localized sources of neurotrophins initiate axon collateral sprouting. J Neurosci 18:5403–5414

Gallo G, Letourneau PC (2000) Neurotrophins and the dynamic regulation of the neuronal cytoskeleton. J Neurobiol 44:159–173

Gallo G, Lefcort FB, Letourneau PC (1997) The trkA receptor mediates growth cone turning toward a localized source of nerve growth factor. J Neurosci 17:5445–5454

Geraldo S, Khanzada UK, Parsons M, Chilton JK, Gordon-Weeks PR (2008) Targeting of the F-actin-binding protein drebrin by the microtubule plus-tip protein EB3 is required for neuritogenesis Nat Cell Biol 10:1181–1189

Gervasi NM, Kwok JC, Fawcett JW (2008) Role of extracellular factors in axon regeneration in the CNS: implications for therapy. Regen Med 3:907–923

Gomez TM, Letourneau PC (1994) Filopodia initiate choices made by sensory neuron growth cones at laminin/fibronectin borders in vitro. J Neurosci 14:5959–5972

Halpain S (2003) Actin in a supporting role. Nat Neurosci 6:101–102

Hammarback JA, Letourneau PC (1985) Neurite extension across regions of low cell-substratum adhesivity: implications for the guidepost hypothesis of axonal pathfinding. Dev Biol 117:655–662

Hammarlund M, Jorgensen EM, Bastiani MJ (2007) Axons break in animals lacking β-spectrin. J Cell Biol 176:269–275

Hasaka TP, Myers KA, Baas PW (2004) Role of actin filaments in the axonal transport of microtubules. J Neurosci 24:11291–11301

Heidemann SR, Lamoureux P, Buxbaum RE (1995) Cytomechanics of axonal development. Cell Biochem Biophys 27:135–155

Hirokawa N (1982) Cross-linker system between neurofilament, microtubules, and membranous organelles in frog axons revealed by the quick-freeze, deep-etching method. J Cell Biol 94:129–142

Hollenbeck PJ, Saxton WM (2005) The axonal transport of mitochondria. J Cell Sci 118:5411–5419

Jung C, Chylinski TM, Pimenta A, Ortiz D, Shea TB (2004) Neurofilament transport is dependent on actin and myosin. J Neurosci 24:9486–9496

Kalil K, Szebenyi G, Dent EW (2000) Common mechanisms underlying growth cone guidance and axon branching. J Neurobiol 44:145–158

Ketschek AR, Jones SL, Gallo G (2007) Axon extension in the fast and slow lanes: substrate-dependent engagement of myosin II functions. Dev Neurobiol 67:1305–1320

Koenig E (2009) Organized ribosome-containing structural domains in axons. Results Probl Cell Differ, doi: 10.1007/400_2008_29

Kollins KM, Hu J, Bridgman PC, Huang YQ, Gallo G (2009) Myosin-II negatively regulates minor process extension and the temporal development of neuronal polarity. Dev Neurobiol 69:279–298

Kubo T, Yamashita T (2007) Rho-ROCK inhibitors for the treatment of CNS injury. Recent Pat CNS Drug Discov 2:173–179

Lalli G, Gschmeissner S, Schiavo G (2003) Myosin Va and microtubule-based motors are required for fast axonal retrograde transport of tetanus toxin in motor neurons. J Cell Sci 116:4639–4650

Lamoureux P, Ruthel G, Buxbaum RE, Heidemann SR (2002) Mechanical tension can specify axonal fate in hippocampal neurons. J Cell Biol 159:499–508

Langford GM (2002) Myosin-V, a versatile motor for short range vesicle transport. Traffic 3:859–865

Lau P-m, Zucker RS, Bentley D (1999) Induction of filopodia by direct elevation of intracellular calcium ion concentration. J Cell Biol 145:1265–1276

Lebrand C, Dent EW, Strasser GA, Lanier LM, Krause M, Svitkina TM, Borisy GG, Gertler FB (2004) Critical role of Ena/VASP proteins for filopodia formation in neurons and in function downstream of netrin-1. Neuron 42:37–49

Lee SK, Hollenbeck PJ (2003) Organization and translation of mRNA in sympathetic axons. J Neurosci 23:8618–8624

Letourneau PC (1979) Cell-substratum adhesion of neurite growth cones and its role in neurite elongation. Exp Cell Res 124:127–138

Letourneau PC (1981) Immunocytochemical evidence for colocalization in neurite growth cones of actin and myosin and their relationship to cell-substratum adhesions. Dev Biol 85:113–122

Letourneau PC (1983) Differences in the distribution of actin filaments between growth cones and the neurites of cultured chick sensory neurons. J Cell Biol 97:963–973

Letourneau PC (1996) The cytoskeleton in nerve growth cone motility and axonal pathfinding. Perspect Dev Neurobiol 4:111–123

Letourneau PC, Shattuck TA, Ressler AH (1987) Push and pull in neurite elongation: observations on the effects of different concentrations of cytochalasin B and taxol. Cell Motil Cytoskeleton 8:193–209

Lin AC, Holt CE (2007) Local translation and directional steering in axons. EMBO J 26:3729–3736

Loudon RP, Silver LD, Yee HF, Gallo G (2006) RhoA-kinase and myosin II are required for the maintenance of growth cone polarity and guidance by nerve growth factor. J Neurobiol 66:847–867

Luo L (2002) Actin cytoskeleton regulation in neuronal morphogenesis and structural plasticity. Annu Rev Cell Dev Biol 18:601–635

Luo L, O'Leary DDM (2005) Axon retraction and degeneration in development and disease. Annu Rev Neurosci 28:127–156

Marsh LM, Letourneau PC (1984) Growth of neurites without filopodial or lamellipodial activity in the presence of cytochalasin B, J Cell Biol 99:2041–2047

Miller KE, Sheetz MP (2006) Direct evidence for coherent low velocity axonal transport of mitochondria. J Cell Biol 173:373–381

Morris CE (2001) Mechanoprotection of the plasma membrane in neurons and other non-erythroid cells by the spectrin-based membrane skeleton. Cell Mol Biol Lett 6:703–720

Morris RL, Hollenbeck PJ (1995) Axonal transport of mitochondria along microtubules and F-actin in living vertebrate neurons. J Cell Biol 131:1315–1326

Muslimov IA, Titmus M, Koenig E, Tiedge H (2002) Transport of neuronal BC1 RNA in Mauthner axons. J Neurosci 22:4293–4301

Myers KA, Baas PW (2009) Microtubule-actin interactions during neuronal development. In: Gallo G, Lanier L (eds) Neurobiology of actin: from neuralation to synaptic function. Springer, Berlin (in press)

Okabe S, Hirokawa N (1990) Turnover of fluorescently labeled tubulin and actin in the axon. Nature 343:479–482

Okabe S, Hirokawa N (1991) Actin dynamics in growth cones. J Neurosci 11:1918–1929

Pak CW, Flynn KC, Bamburg JR (2008) Actin-binding proteins take the reins in growth cones. Nat Rev Neurosci 9:136–147

Pollard TD, Borisy GG (2003) Cellular motility driven by assembly and disassembly of actin filaments. Cell 112:453–465

Raivich G, Makwana M (2007) The making of successful axonal regeneration: genes, molecules and signal transduction pathways. Brain Res Rev 53:287–311

Ramesh V (2004) Merlin and the ERM proteins in Schwann cells, neurons and growth cones. Nat Rev Neurosci 5:462–470

Reynolds AJ, Heydon K, Bartlett SE, Hendry IA (1999) Evidence for phosphatidylinositol 4-kinase and actin involvement in the regulation of 125I-β-nerve growth factor retrograde axonal transport. J Neurochem 73:87–95

Rosner H, Moller W, Wassermann T, Mihatsch, Blum M (2007) Attenuation of actomyosin II contractile activity in growth cones accelerates filopodia-guided and microtubule-based neurite elongation. Brain Res 1176:1–10

Roy S, Winton MJ, Black MM, Tronjanowski JQ, Lee VMY (2008) Cytoskeletal requirements in axonal transport of slow component-b. J Neurosci 28:5248–5256

Santerre RF, Rich A (1976) Actin accumulation in developing chick brain and other tissues. Dev Biol 54:1–12

Schaefer AW, Schoonderwoert VTG, Lin J, Mederios N, Danuser G, Forscher P (2008) Coordination of actin filament and microtubule dynamics during neurite outgrowth. Dev Cell 15:146–162

Schnapp BJ, Reese TS (1982) Cytoplasmic structure in rapid-frozen axons. J Cell Biol 94:667–679

Shattuck TA, Letourneau PC (1989) Distribution and possible interactions of actin-associated proteins and cell adhesion molecules of nerve growth cones. Development 105:505–519

Song A-h, Wang D, Chen G, Li Y, Luo J, Duan S, Poo M-m (2009) A selective filter for cytoplasmic transport at the axon initial segment. Cell 136 1148–1160. doi:10.1016/j. cell. 2009.01.016

Sotelo-Silveira JR, Calliari A, Cardenas M, Koenig E, Sotelo JR (2004) Myosin Va and kinesin II motor proteins are concentrated in ribosomal domains (periaxoplasmic ribosomal plaques) of myelinated axons. J Neurobiol 60:187–196

Sotelo-Silveira JR, Calliari A, Kun A, Koenig E, Sotelo JR (2006) RNA trafficking in axons. Traffic 7:508–515

Sotelo-Silveira J, Crispino M, Puppo A, Sotelo JR, Koenig E (2008) Myelinated axons contain β-actin mRNA and ZBP-1 in periaxoplasmic ribosomal plaques and depend on cyclic AMP and F-actin integrity for in vitro translation. J Neurochem 104:545–557

Spira ME, Oren R, Dormann A, Ilouz N, Lev S (2001) Calcium, protease activation, and cytoskeleton remodeling underlie growth cone formation and neuronal regeneration. Cell Mol Neurobiol 21:591–604

Strasser GA, Rahim NA, VanderWaal KE, Gertler FB, Lanier LM (2004) Arp2/3 is a negative regulator of growth cone translocation. Neuron 43:81–94

Susuki K, Rasband MN (2008) Spectrin and ankyrin-based cytoskeletons at polarized domains in myelinated axons. Exp Biol Med 233:394–400

Suter DM, Forscher P (2000) Substrate-cytoskeletal coupling as a mechanism for the regulation of growth cone motility and guidance. J Neurobiol 44:97–113

Thaxton C, Bhat MA (2009) Myelination and regional domain differentiation of the axon. Results Probl Cell Differ. doi: 10.1007/400_2009_3

Van Essen DC (1997) A tension-based theory of morphogenesis and compact wiring. Nature 385:313–318

Verma P, Chierzi S, Codd AM, Campbell DS, Meyer RL, Holt CE, Fawcett JW (2005) Axonal protein synthesis and degradation are necessary for efficient growth cone regeneration. J Neurosci 25:331–342

Wang W, van Nierkerk E, Willis DE, Twiss JL (2007) RNA transport and localized protein synthesis in neurological disorders and neural repair. Dev Neurobiol 67:1166–1182

Wen Z, Han L, Bamburg JR, Shim S, Ming GL, Zheng JQ (2007) BMP gradients steer nerve growth cones by a balancing act of LIM kinase and slingshot phosphatase on ADF/cofilin. J Cell Biol 178:107–119

Willis DE, van Niekerk EA, Sasaki Y, Mesngon M, Merianda TT, Williams GG, Kendall M, Smith DS, Bassell GJ, Twiss JL (2007) Extracellular stimuli specifically regulate localized levels of individual neuronal mRNAs. J Cell Biol 178:965–980

Witte H, Bradke F (2008) The role of the cytoskeleton during neuronal polarization. Curr Opin Neurobiol 18:479–487

Xu X, Shrager P (2005) Dependence of axon initial segment formation on Na+ channel expression. J Neurosi Res 79:428–441

Yoon BC, Zivraj KH, Holt CE (2009) Local translation and mRNA trafficking in axon pathfinding. Results Probl Cell Differ. doi: 10.1007/400_2009_5

Zheng JQ, Wan JJ, Poo M-m (1996) Essential role of filopodia in chemotropic turning of nerve growth cones induced by a glutamate gradient. J Neurosci 16:1140–1149

Zhou FQ, Cohan CS (2000) How actin filaments and microtubules steer growth cones to their targets. J Neurobiol 58:84–91

Myosin Motor Proteins in the Cell Biology of Axons and Other Neuronal Compartments

Paul C. Bridgman

Abstract Most neurons of both the central and peripheral nervous systems express multiple members of the myosin superfamily that include nonmuscle myosin II, and a number of classes of unconventional myosins. Several classes of unconventional myosins found in neurons have been shown to play important roles in transport processes. A general picture of the myosin-dependent transport processes in neurons is beginning to emerge, although much more work still needs to be done to fully define these roles and establish the importance of myosin for axonal transport. Myosins appear to contribute to three types of transport processes in neurons; recycling of receptors or other membrane components, dynamic tethering of vesicular components, and transport or tethering of protein translational machinery including mRNA. Defects in one or more of these functions have potential to contribute to disease processes.

1 Introduction

Studies of axonal transport have a long history that derives from the importance of transport to neural function. Interest in axonal transport peaked following the discovery of the kinesin and dynein families of microtubule motor proteins (Schnapp et al. 1986; Goldstein and Yang 2000; Hirokawa 1998). Microtubule-dependent motor proteins are recognized as the key components responsible for fast axonal transport and for some components of slow transport (Almenar-Queralt and Goldstein 2001). Soon after the discovery of microtubule associated motor proteins, there was an explosion in the discovery of myosin motors (Berg et al. 2001). This family now has over 18 classes and many of these (approximately eight) are expressed in neurons or sensory cells either during development or following maturation of the nervous

P.C. Bridgman
Department of Anatomy and Neurobiology, Box 8108, Washington University School of Medicine, 660 Euclid Avenue, St. Louis, MO, 63110, USA
e-mail: bridgmap@pcg.wustl.edu

system (Brown and Bridgman 2004). Although several different classes of myosin have been implicated in axonal transport, a definitive role has remained elusive (Bridgman 2004).

More recently, high interest in axonal transport has been revived by the recognition that transport plays an important role in neurodegenerative diseases (Chevalier-Larsen and Holzbaur 2006; De Vos et al. 2008). This has led to a closer examination of the components that regulated transport and the specific motor proteins that are involved. In this review, we focus on recent studies of the most abundant myosin motor protein classes found in neurons and their roles as transporters in axons and other neuronal compartments.

2 Axonal Transport and Neurodegenerative Disease

Axonal transport is essential for the normal function of neurons. For most neurons, the majority of the cytoplasm resides in the axon and its terminals. As the main protein synthesizing machinery is located in the cell soma, most structural and chemical components of the axon and the presynaptic terminals must be transported, often for long distances. The transport system has three main components: motor proteins, tracks composed of protein polymers, and cargo. All motor proteins also require ATP, and so a source of ATP at all points along the transport route is also essential. Most long distance transport in the axon occurs along the microtubule tracks using kinesin and dynein motor proteins to move materials in anterograde or retrograde directions, respectively. Classically, axonal transport has been divided into fast and slow components (Brown 2003). Microtubule-based motors contribute to both transport rate components. The only other identified transport system in axons involves actin filaments as tracks for myosin motors (Kuznetsov et al. 1992; Brown et al. 2004). Much less is known about myosin-based transport compared to microtubule-dependent transport.

Defects in axonal transport may contribute to a wide variety of neurodegenerative diseases. These include amyotrophic lateral sclerosis, distal hereditary motor neuropathy, spinal muscular atrophy, hereditary spastic paraplegia, Huntington's disease, Parkinson's disease, Charcot-Marie-Tooth disease, and Alzheimer's disease (De Vos et al. 2008). Although most of these appear to be primarily defective in microtubule-based transport, it is important to note that disruptions of microtubule tracks, cargo or mitochondrial damage, and misrouting will also affect myosin-based transport. Defects in microtubule organization can affect actin filament organization. Several types of cargo have been shown to interact with multiple motors (Bridgman 1999; Brown et al. 2004). Myosins may act as passive cargo during microtubule-based vesicular transport, and so disruption of this transport may eliminate the supply of myosin motors to distal sites. Defects in mitochondrial transport along microtubules could have profound affects on myosin transport because of the dependence of myosin motors on ATP. Similarly, defects in myosin motors may also have secondary affects on microtubule-based transport.

2.1 Myosin-Based Transport in Neurons

Some recent studies on axonal transport have excluded or provided conflicting data on the involvement of myosin in several forms of transport, including slow axonal transport. For instance, in one study it was shown that neurofilament transport in growing axons, which has classically been defined as a form of slow transport, may require actin and myosin (Jung et al. 2004). However, another study showed that neurofilament transport requires microtubules, but not actin (Francis et al. 2005). A similar requirement for microtubules, but not actin, was shown for axonal transport slow component-b (Roy et al. 2008). Despite such findings, a close look at various studies reveals that myosin motor proteins are likely to have both a distinct role in short-range transport and play a more subtle role in facilitating long-range axonal transport.

As already mentioned and described in detail in other chapters of this volume, most axonal transport occurs along microtubules, using kinesin or dynein motors. The advantages of a microtubule-based transport system over the one using actin filaments and myosins are clear. Microtubule tracks are relatively rigid, long, stable, nonbranched, and usually are not tightly bundled. In mature axons, microtubules are often segregated to mainly occupy the interior core of the axon (Schnapp and Reese 1982). Furthermore, in axons, the orientation of microtubules is uniform, with more than 95% of the plus or fast growing ends directed towards the distal end of the axon (Heidemann et al. 1981). Although microtubules do not extent the entire length of an axon, the ends of adjacent microtubules overlap, providing a continuous set of accessible transport tracks (Bray and Bunge 1981). Therefore, microtubules are the optimal transport tracks for rapid, long distance movement of cargo in either direction along the axon.

In contrast to microtubules, neuronal actin filaments are shorter, less stable, more flexible, and can be tightly bundled or highly branched. Their orientation in axons is likely to be mixed, although this has not been directly shown. In axons, actin filaments are concentrated in the region just under the membrane (Schnapp and Reese 1982), although a population of short filaments appears to interdigitate between microtubules more centrally located (Bearer and Reese 1999). Thus, in general, actin filaments are less than ideal tracks for efficient rapid, long-range transport of cargo along the axon. However, because actin filaments are highly dynamic, they can use their ability to rapidly polymerize to extend their tracks as the myosin motor moves with its cargo, effectively increasing the transport run length (Semenova et al. 2008).

These differences in the track systems as well as the properties of the motor proteins described in brief below have given rise to the proposal that microtubules provide the highway system for long-range transport in axons, while actin filaments provide the local street system for short-range transport within specific regions (cell body, or presynaptic terminals) or to transport cargo to the plasma membrane (Langford 2002). However, there is the additional possibility that myosins may contribute to the efficiency of microtubule-based long-range transport of cargo by acting as short-range transport intermediates or dynamic tethers, as microtubule motor protein carrying cargo transitions between different microtubules, or encounters obstacles.

Differences in the properties of the motor proteins associated with the two different transport track systems also reflect an advantage for the microtubule-based motors. The directional differences are clearly segregated between the two distinct microtubule motor families. The family of kinesins provides motors for most anterograde transport, while dynein provides retrograde transport. While kinesins represent a diverse family of proteins, the two-headed axonal kinesins that act as transport motors for the majority of cargo are highly processive (i.e., remain attached while moving along the microtubule for relatively long times), which is important for long run lengths along tracks (Goldstein and Yang 2000). They are also relatively fast. Dynein is also a processive motor, consistent with its role as a long-range transport motor.

Myosins generally move toward the barbed or fast growing end of the actin filament, although there is one important exception. Myosin VI moves toward the pointed or slow growing end of actin filaments in vitro (Sweeney and Houdusse 2007). This indicates that, similar to microtubule-based motor transport, actin-based motor transport is bidirectional. Furthermore, some two-headed myosins such as myosin V have some transport properties in common with the two-headed kinesins. Myosin V is a processive motor that walks "hand-over-hand" along actin filaments similar to kinesin-1's movement along microtubules (Mehta et al. 1999; Yildiz et al. 2003, 2004). It makes relatively large steps compared to kinesin-1, but moves more slowly because of a slower step frequency. Myosin V is also a highly flexible motor that easily switches tracks at actin filament branch points (Ali et al. 2007). This potentially allows myosin V to easily navigate through the more complex network of actin tracks found in specific compartments of the neuron. Myosin Va also has another unique property: the ability to associate through an electrostatic mechanism and move by diffusion along microtubules. This may facilitate switching between the two different systems of tracks. Thus, the classes of myosin that have been characterized as most important for transport have properties that facilitate efficient movements through complex actin-rich environments.

Table 1 summarizes the roles that have been identified for the different classes of myosin in neurons. The general characteristics and functions are then further described for each class in the following section.

Table 1 Axonal transport function and myosin class

Vesicular, organelle transport	Protein complex transport	Protein polymer transport	RNA/ribosome transport	Receptor transport	Endocytosis, exocytosis
Myosin class/isoform					
Ib ×					×
Ic					×
Ie ×					×
IIA		×			×
IIB ×	×	×			
Va ×	×		×	×	×
Vb ×	×			×	×
VI ×				×	×
X				×	

3 Myosin Transport Motors in neurons

3.1 Class I Myosins

This class of single headed myosins is relatively large, and functionally diverse containing eight members (Berg et al. 2001). At least four types of myosin I (Ib, Ic, Id, Ie) have been detected in nervous tissue (Wagner et al. 1992; Ruppert et al. 1993; Sherr et al. 1993; Bahler et al. 1994; Stoffler et al. 1995). There is no evidence that any of these isoforms directly participate in long-range axonal transport. Myosin Ib has wide expression during development (Ruppert et al. 1993; Sherr et al. 1993), and has been shown to associate with tubulovesicular structures in developing rat peripheral nerve axons (Lewis and Bridgman 1996). Although myosin Id also has wide expression in the nervous system, it appears to be excluded from the axonal compartment (Bahler et al. 1994). Its role in transport processes remains unknown. Myosin Ie associates with the two proteins, synaptojanin-1 and dynamin that play roles in presynaptic endocytosis (Krendel et al. 2007). Myosin Ic is also found in the axonal compartment, and may play a role in regulating growth cone filopodial extension (Diefenbach et al. 2002; Brown and Bridgman 2003). In other cell types, myosin Ic has been shown to play a role in restraining actin polymerization during endocytosis (Sokac et al. 2006). Thus, the main role of this class of myosins seems to be in endocytosis.

Defects in endocytosis could have profound affects on neuronal viability because of the requirement for the uptake of neurotrophins and other factors that influence or regulate cell survival. Currently, there is no information on the relationship between the defects in class I myosin motors, endocytosis, and neurodegenerative diseases.

3.2 Class II Myosins

Nonmuscle class II myosins have been designated "conventional" myosins because of their early discovery and close relationship to muscle myosin II. The heavy chains form dimers, and each dimer associates with two light chains (Shohet et al. 1989; Moussavi et al. 1993). The regulatory light chain plays an important role in regulating activity and the formation of bipolar filaments. Although the bipolar filaments resemble those found in muscle, they are much smaller and are often referred to as mini-bipolar filaments. Mini-bipolar filaments form in developing neurons (Bridgman 2002). Three isoforms (IIA, IIB, IIC) are present in many vertebrate cells (Kawamoto and Adelstein 1991; Golomb et al. 2004). While peripheral neurons appear to contain all three isoforms (Rochlin et al. 1995; Turney and Bridgman 2005; Brown et al. 2009), myosin IIC is absent from developing hippocampal neurons (Kollins et al. 2009).

Class II myosins play important roles in a variety of neuronal functions, including neural migration (Ma et al. 2004), axonal growth (Bridgman et al. 2001; Wylie and

Chantler 2001, 2003; Medeiros et al. 2006; Burnette et al. 2008; Schaefer et al. 2008), guidance (Turney and Bridgman 2005; Gallo 2006; Kubo et al. 2008; Brown et al. 2009), and possibly several aspects of synaptic function (Takagishi et al. 2005; Ryu et al. 2006; Tokuoka and Goda 2006; Srinivasan et al. 2008). The most interesting report of a synaptic-like function occurs in catecholamine and neuropeptide release in adrenal chromaffin cells (Doreian et al. 2008). Myosin II contributes to the control of exocytosis: stabilizing kiss-and-run fusion events.

A direct role for myosin II in axonal transport is controversial. Myosin II has been shown to associate with organelles in axons (DeGiorgis et al. 2002) and has been implicated in the transport of neurofilaments in neuroblastoma cells (Jung et al. 2004). However, more recently, a role for myosin II and actin filaments in neurofilament transport has been disputed. In rat sympathetic neurons, actin filaments were not required for transport, but microtubules were required (Francis et al. 2005). Developing hippocampal and sympathetic neurons from myosin IIB knockout mice have normal distributions of neurofilaments (Bridgman, unpublished observation). Similarly, slow component-b among the axonal transport rate groups, which contains a variety of proteins including synaptic proteins, does not appear to depend upon actin filaments (Roy et al. 2008). A possible role for myosin IIB in neurodegenerative disease has been suggested by an influence on the trafficking or processing of amyloid precursor protein (Massone et al. 2007).

3.3 Class V Myosins

Three isoforms of class V myosins have been identified in vertebrates, but only myosin Va appears to be widely expressed in the nervous system (Trybus 2008). It is also the only isoform that is clearly essential for normal neuronal function, and has an identified role in human disease (Mercer et al. 1991; Menasche et al. 2003). Myosin Vb has more limited expression in the nervous system, but appears to play important roles in synaptic receptor recycling (Hales et al. 2002; Lise et al. 2006; Wang et al. 2008). Myosin Vc is primarily associated with epithelial cells (Rodriguez and Cheney 2002), and appears to act as a nonprocessive transport motor (Watanabe et al. 2008).

Class V myosins form heavy chain dimers, but not bipolar filaments (Cheney et al. 1993). The tail region has a globular specialization that has been shown to interact with a variety of proteins including microtubules (Cao et al. 2004), and is considered to be the major cargo binding domain (Trybus 2008). The cargo domain can regulate activity (Thirumurugan et al. 2006), suggesting that binding of cargo can alter transport properties. However, because of alternative splicing of the myosin Va forms, the neuronal form lacks the portion of the domain found to be important for cargo binding in melanocytes (Wu et al. 2002b). Therefore, it remains to be determined whether the properties that have been characterized in the nonneuronal myosin Va apply to the neuronal form. Only the neuronal form binds an additional light chain that also interacts with dynein (Wagner et al. 2006). The neck region contains multiple binding sites for calmodulin that acts as light chains to regulate motility (Cheney et al. 1993). Movements of myosin Va along actin filaments have been shown to involve a hand-over-hand mechanism, and the vertebrate form is processive (Sellers and Veigel 2006; Yildiz et al. 2003). Mechanistically, there is an obvious similarity to kinesin-1.

Recent work has focused on the role of myosin V in dendrites and dendritic spines. Myosin Va has been shown to transport mRNA/protein complexes into hippocampal dendritic spines (Yoshimura et al. 2006), suggesting a role in regulating some aspects of postsynaptic function. It has also been shown that myosin Va interacts indirectly with PSD-95 through the light chain that it shares with dynein (Naisbitt et al. 2000). However, electrophysiological tests on myosin Va null mice did not reveal any functional defects in synaptic transmission of hippocampal neurons (Schnell and Nicoll 2001), and studies on glutamate receptor targeting to spines also were negative (Petralia et al. 2001). Despite these negative findings, more recently, it has been shown that myosin Va may play a role in the transport of AMPA receptors into spines from dendritic shafts in response to high levels of activity that usually lead to long-term potentiation (Correia et al. 2008).

The vertebrate form of myosin Vb is also a processive motor, although with different biochemical properties compared to myosin Va (Watanabe et al. 2006). Its interactions with Rab 11, and involvement in receptor recycling originally suggested that it may operate differently than myosin Va in neurons (Hales et al. 2002; Lise et al. 2006). It has also been shown to be involved in vesicle recycling in nonneuronal cells (Lapierre et al. 2001), but recently its precise role in this pathway has been questioned. It may act as a dynamic tether for endocytotic vesicles in nonneuronal cells (Provance et al. 2008). In hippocampal neurons, myosin Vb has been shown to contribute to AMPA receptor delivery to spines and recycling during synaptic plasticity (Lise et al. 2006; Wang et al. 2008).

The apparent similarities in the function of myosin Va and Vb for transporting AMPA receptors into dendrite spines of hippocampal neurons are hard to reconcile (Correia et al. 2008; Wang et al. 2008). Both studies conclude that myosin V transport of AMPA receptors is associated with increased activity and LTP, but they differ on the specific isoform responsible. At the center of the debate is the ability of these different isoforms to interact with cargo. While the nonneuronal form of myosin Va interacts with cargo via Rab 27 and melanophilin (Wu et al. 2002b), the neuronal form lacks the binding domain for this interaction. The study by Correia et al. (2008) is the first to show that the neuronal form of myosin Va, like myosin Vb, may interact with vesicular structures through Rab 11. If Rab 11 acts as the vesicle linker for both isoforms, then this would suggest that each isoform could easily substitute for one another. However, the data from these papers is not consistent with this possibility. Therefore, it remains to be determined which form of myosin V is essential for AMPA receptor transport into spines.

The studies described above indicate that class V myosin represents the main class for which there is substantial support for a vesicular transport role. This role is for short-range transport that is usually associated with recycling pathways (Langford 2002; Roder et al. 2008). In addition to a role in receptor recycling pathways, myosin Va also transports endoplasmic reticulum (ER) in axons and into actin rich regions such as spines (Takagishi et al. 1996; Langford 1999). The absence of spine ER has functional consequences on synaptic transmission and long term depression for Purkinje cells (Miyata et al. 2000) that likely contributes to the behavioral abnormalities observed in myosin Va null mice (Mercer et al. 1991).

In addition to the roles described earlier, myosin V contributes to tethering organelles to enhance processivity (Ali et al. 2008), and to halting, or delaying vesicle movements (Desnos et al. 2007; Bittins et al. 2009). Another interesting property is the ability of myosin Va to navigate through actin networks composed of randomly oriented or branched actin filaments (Ali et al. 2007). This suggests that myosin Va is capable of navigating through complex actin meshworks in vivo. It remains to be seen if myosin Vb also has this capability. This property could be important for understanding the function of myosin V, because the orientation of actin filaments in meshworks may determine whether net movement, tethering, or cross-linking occurs. Most surprising is the ability of myosin Va to diffuse, through electrostatic interactions, along microtubules for relatively long distances. It remains to be determined whether this property is physiologically relevant to transport in vivo.

In axons, myosin Va has been shown to associate with vesicles containing synaptic vesicle precursors and move as cargo along microtubules (Bridgman 1999). This suggests that myosin Va plays mainly a passive role during long-range, microtubule-based transport, although potential interactions with kinesin during transport (Huang et al. 1999), and recent observations on tethering interactions between kinesin and myosin Va (Ali et al. 2008) provide motivation for further investigation of a more active role. Dilute lethal (i.e., myosin Va null) mice show abnormally large presynaptic terminals in the cerebellum, suggesting that short-range transport of synaptic vesicle precursors may be defective (Bridgman 1999). Consistent with this possibility is that myosin Va associates with synaptic vesicle proteins (Prekeris and Terrian 1997) and may regulate some aspects of presynaptic vesicle exocytosis (Watanabe et al. 2005). Alternatively, myosin Va may restrain or help dock secretory granules at release sites (Desnos et al. 2007). Myosin Va also contributes to the regulation of neurofilament density in axons (Rao et al. 2002).

Perhaps related to the interactions between myosin Va and kinesin in vesicle transport is the observation that both myosin Va and kinesin are concentrated in ribosomal domains of myelinated axons (Sotelo-Silveira et al. 2004). Microtubule-based transport of mRNA is well documented (Hirokawa 2006). The peripheral location of the ribosomal domains (i.e., periaxoplasmic ribosomal plaques (PARPs)) may indicate an interactive role for these motor proteins, kinesin-based transport to the site of the plaques, and then myosin Va dependent tethering in actin rich peripheral regions of the axon (also, see Koenig 2009). Data that support this possibility, but is also relevant in the context of cooperative transport between the microtubule and actin-based systems, comes from the work of Muslimov et al. (2002). Injection of heterologous radiolabeled BC1 RNA, a small, noncoding (regulatory) RNA, into Mauthner cells is targeted by its 5' region to the axon (and dendrites) by microtubule dependent axial transport, but distribution to the cortical zone (and putative PARP domains) depends on F-actin.

3.4 Class VI Myosin

Although first discovered in *Drosophila* (Kellerman and Miller 1992), an important functional role for myosin VI was originally identified in hair cells (Avraham et al. 1995). Mutations in myosin VI cause deafness in mice and humans (Avraham et al. 1995; Melchionda et al. 2001; Ahmed et al. 2003).

Myosin VI is probably the most unusual class of myosin. It is the only type that has been shown to move towards the pointed end of actin filaments (Wells et al. 1999). In addition, the native form exists as monomer, but can be forced to undergo dimerization (Sweeney and Houdusse 2007). As a dimer, it acts as a processive motor. There is some evidence that, in cells, myosin VI can undergo cargo-mediated dimerization (Spudich et al. 2007). Thus, it may act as pointed end directed cargo carrying motor in cells. The main role for myosin VI in other cell types appears to be endocytosis (Buss et al. 2001). It has also been implicated in exocytosis, cell migration, and growth cone motility (Suter et al. 2000; Sweeney and Houdusse 2007). In some cases it may also contribute to actin filament cross-linking.

In neurons, myosin VI has been shown to play an important role in glutamate receptor endocytosis in presynaptic terminals (Osterweil et al. 2005), and in BDNF-mediated enhancement of neurotransmitter release (Yano et al. 2006). Postsynaptically, it also interacts with SAP97, which plays a role in anchoring AMPA receptors (Wu et al. 2002a). Thus, similar to myosin V, this class of myosin appears to function in short-range transport processes.

3.5 Class X Myosin

A single vertebrate form of myosin X is expressed in the nervous system (Sousa et al. 2006). It forms dimers through a short coiled-coil region, and contains multiple protein binding domains in its tail (Berg and Cheney 2002). It has been shown to be involved in a highly specific form of transport: the movement of receptors (integrins, or DCC – the receptor for netrin-1) towards the tip of filopodia or leading edge of lamellipodia (Berg and Cheney 2002; Zhu et al. 2007). This suggests that it has an important role in axonal guidance during development. Consistent with this role, it preferentially associates with regions of dynamic actin that is bundled by fascin (Nagy et al. 2008). However, it is also expressed postnatally in the central nervous system with especially high levels in Purkinje cells of the cerebellum (Sousa et al. 2006). This suggests that it may have other transport roles in mature neurons that have yet to be discovered.

4 Summary

The original proposal that axonal transport functions can be segregated into microtubule-based, long-range transport and actin-based, short-range transport continues to be supported by multiple recent studies. However, a new concept has been added to the transport model by several studies on the tethering function of myosins. The ability of myosin dependent tethering to increase the processivity of microtubule-based motors in vitro needs careful consideration in future studies of axonal transport in vivo. If myosins contribute to the efficiency of microtubule-based transport, then it is possible that defects in myosin function may also contribute to transport defects that underlie some neurodegenerative diseases. The recent focus on myosin dependent receptor recycling pathways indicates a bright future for further studies of myosin function in neuronal cell biology.

References

Ahmed ZM, Morell RJ, Riazuddin S, Gropman A, Shaukat S, Ahmad MM, Mohiddin SA, Fananapazir L, Caruso RC, Husnain T, Khan SN, Griffith AJ, Friedman TB, Wilcox ER (2003) Mutations of MYO6 are associated with recessive deafness, DFNB37. Am J Hum Genet 72:1315–1322

Ali MY, Krementsova EB, Kennedy GG, Mahaffy R, Pollard TD, Trybus KM, Warshaw DM (2007) Myosin Va maneuvers through actin intersections and diffuses along microtubules. Proc Natl Acad Sci USA 104:4332–4336

Ali MY, Lu H, Bookwalter CS, Warshaw DM, Trybus KM (2008) Myosin V and Kinesin act as tethers to enhance each others' processivity. Proc Natl Acad Sci USA 105:4691–4696

Almenar-Queralt A, Goldstein LS (2001) Linkers, packages and pathways: new concepts in axonal transport. Curr Opin Neurobiol 11:550–557

Avraham KB, Hasson T, Steel KP, Kingsley DM, Russell LB, Mooseker MS, Copeland NG, Jenkins NA (1995) The mouse Snell's waltzer deafness gene encodes an unconventional myosin required for structural integrity of inner ear hair cells. Nat Genet 11:369–375

Bahler M, Kroschewski R, Stoffler HE, Behrmann T (1994) Rat myr 4 defines a novel subclass of myosin I: identification, distribution, localization, and mapping of calmodulin-binding sites with differential calcium sensitivity. J Cell Biol 126:375–389

Bearer EL, Reese TS (1999) Association of actin filaments with axonal microtubule tracts. J Neurocytol 28:85–98

Berg JS, Cheney RE (2002) Myosin-X is an unconventional myosin that undergoes intrafilopodial motility. Nat Cell Biol 4:246–250

Berg JS, Powell BC, Cheney RE (2001) A millennial myosin census. Mol Biol Cell 12:780–794

Bittins CM, Eichler TW, Gerdes HH (2009) Expression of the dominant-negative tail of myosin Va enhances exocytosis of large dense core vesicles in neurons. Cell Mol Neurobiol doi: 10.1007/s10571-009-9352-2

Bray D, Bunge MB (1981) Serial analysis of microtubules in cultured rat sensory axons. J Neurocytol 10:589–605

Bridgman PC (1999) Myosin Va movements in normal and dilute-lethal axons provide support for a dual filament motor complex. J Cell Biol 146:1045–1060

Bridgman PC (2002) Growth cones contain myosin II bipolar filament arrays. Cell Motil Cytoskeleton 52:91–96

Bridgman PC (2004) Myosin-dependent transport in neurons. J Neurobiol 58:164–174

Bridgman PC, Dave S, Asnes CF, Tullio AN, Adelstein RS (2001) Myosin IIB is required for growth cone motility. J Neurosci 21:6159–6169

Brown A (2003) Axonal transport of membranous and nonmembranous cargoes: a unified perspective. J Cell Biol 160:817–821

Brown ME, Bridgman PC (2003) Retrograde flow rate is increased in growth cones from myosin IIB knockout mice. J Cell Sci 116:1087–1094

Brown ME, Bridgman PC (2004) Myosin function in nervous and sensory systems. J Neurobiol 58:118–130

Brown JA, Wysolmerski RB, Bridgman PC (2009) Dorsal root ganglion neurons react to Semaphorin 3A application through a biphasic response that requires multiple myosin II isoforms. Mol Biol Cell 20:1167–1179

Brown JR, Stafford P, Langford GM (2004) Short-range axonal/dendritic transport by myosin-V: a model for vesicle delivery to the synapse. J Neurobiol 58:175–188

Burnette DT, Ji L, Schaefer AW, Medeiros NA, Danuser G, Forscher P (2008) Myosin II activity facilitates microtubule bundling in the neuronal growth cone neck. Dev Cell 15:163–169

Buss F, Arden SD, Lindsay M, Luzio JP, Kendrick-Jones J (2001) Myosin VI isoform localized to clathrin-coated vesicles with a role in clathrin-mediated endocytosis. EMBO J 20:3676–3684

Cao TT, Chang W, Masters SE, Mooseker MS (2004) Myosin-Va binds to and mechanochemically couples microtubules to actin filaments. Mol Biol Cell 15:151–161

Cheney RE, O'Shea MK, Heuser JE, Coelho MV, Wolenski JS, Espreafico EM, Forscher P, Larson RE, Mooseker MS (1993) Brain myosin-V is a two-headed unconventional myosin with motor activity. Cell 75:13–23

Chevalier-Larsen E, Holzbaur EL (2006) Axonal transport and neurodegenerative disease. Biochim Biophys Acta 1762:1094–1108

Correia SS, Bassani S, Brown TC, Lise MF, Backos DS, El-Husseini A, Passafaro M, Esteban JA (2008) Motor protein-dependent transport of AMPA receptors into spines during long-term potentiation. Nat Neurosci 11:457–466

De Vos KJ, Grierson AJ, Ackerley S, Miller CC (2008) Role of axonal transport in neurodegenerative diseases. Annu Rev Neurosci 31:151–173

DeGiorgis JA, Reese TS, Bearer EL (2002) Association of a nonmuscle myosin II with axoplasmic organelles. Mol Biol Cell 13:1046–1057

Desnos C, Huet S, Fanget I, Chapuis C, Bottiger C, Racine V, Sibarita JB, Henry JP, Darchen F (2007) Myosin va mediates docking of secretory granules at the plasma membrane. J Neurosci 27:10636–10645

Diefenbach TJ, Latham VM, Yimlamai D, Liu CA, Herman IM, Jay DG (2002) Myosin 1c and myosin IIB serve opposing roles in lamellipodial dynamics of the neuronal growth cone. J Cell Biol 158:1207–1217

Doreian BW, Fulop TG, Smith CB (2008) Myosin II activation and actin reorganization regulate the mode of quantal exocytosis in mouse adrenal chromaffin cells. J Neurosci 28:4470–4478

Francis F, Roy S, Brady ST, Black MM (2005) Transport of neurofilaments in growing axons requires microtubules but not actin filaments. J Neurosci Res 79:442–450

Gallo G (2006) RhoA-kinase coordinates F-actin organization and myosin II activity during semaphorin-3A-induced axon retraction. J Cell Sci 119:3413–3423

Goldstein LS, Yang Z (2000) Microtubule-based transport systems in neurons: the roles of kinesins and dyneins. Annu Rev Neurosci 23:39–71

Golomb E, Ma X, Jana SS, Preston YA, Kawamoto S, Shoham NG, Goldin E, Conti MA, Sellers JR, Adelstein RS (2004) Identification and characterization of nonmuscle myosin II-C, a new member of the myosin II family. J Biol Chem 279:2800–2808

Hales CM, Vaerman JP, Goldenring JR (2002) Rab11 family interacting protein 2 associates with Myosin Vb and regulates plasma membrane recycling. J Biol Chem 277:50415–50421

Heidemann SR, Landers JM, Hamborg MA (1981) Polarity orientation of axonal microtubules. J Cell Biol 91:661–665

Hirokawa N (1998) Kinesin and dynein superfamily proteins and the mechanism of organelle transport. Science 279:519–526

Hirokawa N (2006) mRNA transport in dendrites: RNA granules, motors, and tracks. J Neurosci 26:7139–7142

Huang JD, Brady ST, Richards BW, Stenolen D, Resau JH, Copeland NG, Jenkins NA (1999) Direct interaction of microtubule- and actin-based transport motors. Nature 397:267–270

Jung C, Chylinski TM, Pimenta A, Ortiz D, Shea TB (2004) Neurofilament transport is dependent on actin and myosin. J Neurosci 24:9486–9496

Kawamoto S, Adelstein RS (1991) Chicken nonmuscle myosin heavy chains: differential expression of two mRNAs and evidence for two different polypeptides. J Cell Biol 112:915–924

Kellerman KA, Miller KG (1992) An unconventional myosin heavy chain gene from Drosophila melanogaster. J Cell Biol 119:823–834

Koenig E (2009) Results and problems in cell differentiation. Organized ribosome-containing structural domains in axons. doi: 10.1007/400_2008_29

Kollins KM, Hu J, Bridgman PC, Huang YQ, Gallo G (2009) Myosin-II negatively regulates minor process extension and the temporal development of neuronal polarity. Dev Neurobiol 69:279–298

Krendel M, Osterweil EK, Mooseker MS (2007) Myosin 1E interacts with synaptojanin-1 and dynamin and is involved in endocytosis. FEBS Lett 581:644–650

Kubo T, Endo M, Hata K, Taniguchi J, Kitajo K, Tomura S, Yamaguchi A, Mueller BK, Yamashita T (2008) Myosin IIA is required for neurite outgrowth inhibition produced by repulsive guidance molecule. J Neurochem 105:113–126

Kuznetsov SA, Langford GM, Weiss DG (1992) Actin-dependent organelle movement in squid axoplasm. Nature 356:722–725

Langford GM (1999) ER transport on actin filaments in squid giant axon: implications for signal transduction at synapse. FASEB J 13(Suppl 2):S248–S250

Langford GM (2002) Myosin-V, a versatile motor for short-range vesicle transport. Traffic 3:859–865

Lapierre LA, Kumar R, Hales CM, Navarre J, Bhartur SG, Burnette JO, Provance DW Jr, Mercer JA, Bahler M, Goldenring JR (2001) Myosin Vb is associated with plasma membrane recycling systems. Mol Biol Cell 12:1843–1857

Lewis AK, Bridgman PC (1996) Mammalian myosin 1 alpha is concentrated near the plasma membrane in nerve growth cones. Cell Motil Cytoskeleton 33:130–150

Lise MF, Wong TP, Trinh A, Hines RM, Liu L, Kang R, Hines DJ, Lu J, Goldenring JR, Wang YT, El-Husseini A (2006) Involvement of myosin Vb in glutamate receptor trafficking. J Biol Chem 281:3669–3678

Ma X, Kawamoto S, Hara Y, Adelstein RS (2004) A point mutation in the motor domain of nonmuscle myosin II-B impairs migration of distinct groups of neurons. Mol Biol Cell 15:2568–2579

Massone S, Argellati F, Passalacqua M, Armirotti A, Melone L, d'Abramo C, Marinari UM, Domenicotti C, Pronzato MA, Ricciarelli R (2007) Downregulation of myosin II-B by siRNA alters the subcellular localization of the amyloid precursor protein and increases amyloid-beta deposition in N2a cells. Biochem Biophys Res Commun 362:633–638

Medeiros NA, Burnette DT, Forscher P (2006) Myosin II functions in actin-bundle turnover in neuronal growth cones. Nat Cell Biol 8:215–226

Mehta AD, Rock RS, Rief M, Spudich JA, Mooseker MS, Cheney RE (1999) Myosin-V is a processive actin-based motor. Nature 400:590–593

Melchionda S, Ahituv N, Bisceglia L, Sobe T, Glaser F, Rabionet R, Arbones ML, Notarangelo A, Di Iorio E, Carella M, Zelante L, Estivill X, Avraham KB, Gasparini P (2001) MYO6, the human homologue of the gene responsible for deafness in Snell's waltzer mice, is mutated in autosomal dominant nonsyndromic hearing loss. Am J Hum Genet 69:635–640

Menasche G, Ho CH, Sanal O, Feldmann J, Tezcan I, Ersoy F, Houdusse A, Fischer A, de Saint Basile G (2003) Griscelli syndrome restricted to hypopigmentation results from a melanophilin defect (GS3) or a MYO5A F-exon deletion (GS1). J Clin Invest 112:450–456

Mercer JA, Seperack PK, Strobel MC, Copeland NG, Jenkins NA (1991) Novel myosin heavy chain encoded by murine dilute coat colour locus. Nature 349:709–713

Miyata M, Finch EA, Khiroug L, Hashimoto K, Hayasaka S, Oda SI, Inouye M, Takagishi Y, Augustine GJ, Kano M (2000) Local calcium release in dendritic spines required for long-term synaptic depression. Neuron 28:233–244

Moussavi RS, Kelley CA, Adelstein RS (1993) Phosphorylation of vertebrate nonmuscle and smooth muscle myosin heavy chains and light chains. Mol Cell Biochem 127–128:219–227

Muslimov IA, Titmus M, Koenig E, Tiedge H (2002) Transport of neuronal BC1 RNA in Mauthner axons. J Neurosci 22:4293–4301

Nagy S, Ricca BL, Norstrom MF, Courson DS, Brawley CM, Smithback PA, Rock RS (2008) A myosin motor that selects bundled actin for motility. Proc Natl Acad Sci USA 105:9616–9620

Naisbitt S, Valtschanoff J, Allison DW, Sala C, Kim E, Craig AM, Weinberg RJ, Sheng M (2000) Interaction of the postsynaptic density-95/guanylate kinase domain-associated protein complex with a light chain of myosin-V and dynein. J Neurosci 20:4524–4534

Osterweil E, Wells DG, Mooseker MS (2005) A role for myosin VI in postsynaptic structure and glutamate receptor endocytosis. J Cell Biol 168:329–338

Petralia RS, Wang YX, Sans N, Worley PF, Hammer JA III, Wenthold RJ (2001) Glutamate receptor targeting in the postsynaptic spine involves mechanisms that are independent of myosin Va. Eur J Neurosci 13:1722–1732

Prekeris R, Terrian DM (1997) Brain myosin V is a synaptic vesicle-associated motor protein: evidence for a Ca2+-dependent interaction with the synaptobrevin-synaptophysin complex. J Cell Biol 137:1589–1601

Provance DW Jr, Addison EJ, Wood PR, Chen DZ, Silan CM, Mercer JA (2008) Myosin-Vb functions as a dynamic tether for peripheral endocytic compartments during transferrin trafficking. BMC Cell Biol 9:44

Rao MV, Engle LJ, Mohan PS, Yuan A, Qiu D, Cataldo A, Hassinger L, Jacobsen S, Lee VM, Andreadis A, Julien JP, Bridgman PC, Nixon RA (2002) Myosin Va binding to neurofilaments is essential for correct myosin Va distribution and transport and neurofilament density. J Cell Biol 159:279–290

Rochlin MW, Itoh K, Adelstein RS, Bridgman PC (1995) Localization of myosin II A and B isoforms in cultured neurons. J Cell Sci 108(Pt 12):3661–3670

Roder IV, Petersen Y, Choi KR, Witzemann V, Hammer JA, III, Rudolf R (2008) Role of Myosin Va in the plasticity of the vertebrate neuromuscular junction in vivo. PLoS One 3:e3871

Rodriguez OC, Cheney RE (2002) Human myosin-Vc is a novel class V myosin expressed in epithelial cells. J Cell Sci 115:991–1004

Roy S, Winton MJ, Black MM, Trojanowski JQ, Lee VM (2008) Cytoskeletal requirements in axonal transport of slow component-b. J Neurosci 28:5248–5256

Ruppert C, Kroschewski R, Bahler M (1993) Identification, characterization and cloning of myr 1, a mammalian myosin-I. J Cell Biol 120:1393–1403

Ryu J, Liu L, Wong TP, Wu DC, Burette A, Weinberg R, Wang YT, Sheng M (2006) A critical role for myosin IIb in dendritic spine morphology and synaptic function. Neuron 49:175–182

Schaefer AW, Schoonderwoert VT, Ji L, Mederios N, Danuser G, Forscher P (2008) Coordination of actin filament and microtubule dynamics during neurite outgrowth. Dev Cell 15:146–162

Schnapp BJ, Reese TS (1982) Cytoplasmic structure in rapid-frozen axons. J Cell Biol 94:667–669

Schnapp BJ, Vale RD, Sheetz MP, Reese TS (1986) Microtubules and the mechanism of directed organelle movement. Ann N Y Acad Sci 466:909–918

Schnell E, Nicoll RA (2001) Hippocampal synaptic transmission and plasticity are preserved in myosin Va mutant mice. J Neurophysiol 85:1498–1501

Sellers JR, Veigel C (2006) Walking with myosin V. Curr Opin Cell Biol 18:68–73

Semenova I, Burakov A, Berardone N, Zaliapin I, Slepchenko B, Svitkina T, Kashina A, Rodionov V (2008) Actin dynamics is essential for myosin-based transport of membrane organelles. Curr Biol 18:1581–1586

Sherr EH, Joyce MP, Greene LA (1993) Mammalian myosin I alpha, I beta, and I gamma: new widely expressed genes of the myosin I family. J Cell Biol 120:1405–1416

Shohet RV, Conti MA, Kawamoto S, Preston YA, Brill DA, Adelstein RS (1989) Cloning of the cDNA encoding the myosin heavy chain of a vertebrate cellular myosin. Proc Natl Acad Sci USA 86:7726–7730

Sokac AM, Schietroma C, Gundersen CB, Bement WM (2006) Myosin-1c couples assembling actin to membranes to drive compensatory endocytosis. Dev Cell 11:629–640

Sotelo-Silveira JR, Calliari A, Cardenas M, Koenig E, Sotelo JR (2004) Myosin Va and kinesin II motor proteins are concentrated in ribosomal domains (periaxoplasmic ribosomal plaques) of myelinated axons. J Neurobiol 60:187–196

Sousa AD, Berg JS, Robertson BW, Meeker RB, Cheney RE (2006) Myo10 in brain: developmental regulation, identification of a headless isoform and dynamics in neurons. J Cell Sci 119: 184–194

Spudich G, Chibalina MV, Au JS, Arden SD, Buss F, Kendrick-Jones J (2007) Myosin VI targeting to clathrin-coated structures and dimerization is mediated by binding to Disabled-2 and PtdIns(4,5) P2. Nat Cell Biol 9:176–183

Srinivasan G, Kim JH, von Gersdorff H (2008) The pool of fast releasing vesicles is augmented by myosin light chain kinase inhibition at the calyx of Held synapse. J Neurophysiol 99:1810–1824

Stoffler HE, Ruppert C, Reinhard J, Bahler M (1995) A novel mammalian myosin I from rat with an SH3 domain localizes to Con A-inducible, F-actin-rich structures at cell-cell contacts. J Cell Biol 129:819–830

Suter DM, Espindola FS, Lin CH, Forscher P, Mooseker MS (2000) Localization of unconventional myosins V and VI in neuronal growth cones. J Neurobiol 42:370–382

Sweeney HL, Houdusse A (2007) What can myosin VI do in cells? Curr Opin Cell Biol 19:57–66

Takagishi Y, Oda S, Hayasaka S, Dekker-Ohno K, Shikata T, Inouye M, Yamamura H (1996) The dilute-lethal (dl) gene attacks a Ca2+ store in the dendritic spine of Purkinje cells in mice. Neurosci Lett 215:169–172

Takagishi Y, Futaki S, Itoh K, Espreafico EM, Murakami N, Murata Y, Mochida S (2005) Localization of myosin II and V isoforms in cultured rat sympathetic neurones and their potential involvement in presynaptic function. J Physiol 569:195–208

Thirumurugan K, Sakamoto T, Hammer JA, III, Sellers JR, Knight PJ (2006) The cargo-binding domain regulates structure and activity of myosin 5. Nature 442:212–215

Tokuoka H, Goda Y (2006) Myosin light chain kinase is not a regulator of synaptic vesicle trafficking during repetitive exocytosis in cultured hippocampal neurons. J Neurosci 26:11606–11614

Trybus KM (2008) Myosin V from head to tail. Cell Mol Life Sci 65:1378–1389

Turney SG, Bridgman PC (2005) Laminin stimulates and guides axonal outgrowth via growth cone myosin II activity. Nat Neurosci 8:717–719

Wagner MC, Barylko B, Albanesi JP (1992) Tissue distribution and subcellular localization of mammalian myosin I. J Cell Biol 119:163–170

Wagner W, Fodor E, Ginsburg A, Hammer JA III (2006) The binding of DYNLL2 to myosin Va requires alternatively spliced exon B and stabilizes a portion of the myosin's coiled-coil domain. Biochemistry 45:11564–11577

Wang Z, Edwards JG, Riley N, Provance DW, Jr, Karcher R, Li XD, Davison IG, Ikebe M, Mercer JA, Kauer JA, Ehlers MD (2008) Myosin Vb mobilizes recycling endosomes and AMPA receptors for postsynaptic plasticity. Cell 135:535–548

Watanabe M, Nomura K, Ohyama A, Ishikawa R, Komiya Y, Hosaka K, Yamauchi E, Taniguchi H, Sasakawa N, Kumakura K, Ushiki T, Sato O, Ikebe M, Igarashi M (2005) Myosin-Va regulates exocytosis through the submicromolar Ca2+-dependent binding of syntaxin-1A. Mol Biol Cell 16:4519–4530

Watanabe S, Mabuchi K, Ikebe R, Ikebe M (2006) Mechanoenzymatic characterization of human myosin Vb. Biochemistry 45:2729–2738

Watanabe S, Watanabe TM, Sato O, Awata J, Homma K, Umeki N, Higuchi H, Ikebe R, Ikebe M (2008) Human myosin Vc is a low duty ratio nonprocessive motor. J Biol Chem 283:10581–10592

Wells AL, Lin AW, Chen LQ, Safer D, Cain SM, Hasson T, Carragher BO, Milligan RA, Sweeney HL (1999) Myosin VI is an actin-based motor that moves backwards. Nature 401:505–508

Wu H, Nash JE, Zamorano P, Garner CC (2002a) Interaction of SAP97 with minus-end-directed actin motor myosin VI. Implications for AMPA receptor trafficking. J Biol Chem 277:30928–30934

Wu XS, Rao K, Zhang H, Wang F, Sellers JR, Matesic LE, Copeland NG, Jenkins NA, Hammer JAIII (2002b) Identification of an organelle receptor for myosin-Va. Nat Cell Biol 4:271–278

Wylie SR, Chantler PD (2001) Separate but linked functions of conventional myosins modulate adhesion and neurite outgrowth. Nat Cell Biol 3:88–92

Wylie SR, Chantler PD (2003) Myosin IIA drives neurite retraction. Mol Biol Cell 14:4654–4666

Yano H, Ninan I, Zhang H, Milner TA, Arancio O, Chao MV (2006) BDNF-mediated neurotransmission relies upon a myosin VI motor complex. Nat Neurosci 9:1009–1018

Yildiz A, Forkey JN, McKinney SA, Ha T, Goldman YE, Selvin PR (2003) Myosin V walks hand-over-hand: single fluorophore imaging with 1.5-nm localization. Science 300:2061–2065

Yildiz A, Tomishige M, Vale RD, Selvin PR (2004) Kinesin walks hand-over-hand. Science 303:676–678

Yoshimura A, Fujii R, Watanabe Y, Okabe S, Fukui K, Takumi T (2006) Myosin-Va facilitates the accumulation of mRNA/protein complex in dendritic spines. Curr Biol 16:2345–2351

Zhu XJ, Wang CZ, Dai PG, Xie Y, Song NN, Liu Y, Du QS, Mei L, Ding YQ, Xiong WC (2007) Myosin X regulates netrin receptors and functions in axonal path-finding. Nat Cell Biol 9:184–192

Mitochondrial Transport Dynamics in Axons and Dendrites

Konrad E. Zinsmaier, Milos Babic, and Gary J. Russo

Abstract Mitochondrial dynamics and transport have emerged as key factors in the regulation of neuronal differentiation and survival. Mitochondria are dynamically transported in and out of axons and dendrites to maintain neuronal and synaptic function. Transport proceeds through a controlled series of plus- and minus-end directed movements along microtubule tracks (MTs) that are often interrupted by short stops. This bidirectional motility of mitochondria is facilitated by plus end-directed kinesin and minus end-directed dynein motors, and may be coordinated and controlled by a number of mechanisms that integrate intracellular signals to ensure efficient transport and targeting of mitochondria. In this chapter, we discuss our understanding of mechanisms that facilitate mitochondrial transport and delivery to specific target sites in dendrites and axons.

1 Introduction

Mitochondria are functionally diverse organelles that produce about 15 times more ATP than glycolysis, critically affect intracellular Ca^{2+} homeostasis, and are central to the synthesis of steroids, terpenes (hemes), various lipids, the generation of free

---

K.E. Zinsmaier (✉)
Arizona Research Laboratories, Division of Neurobiology,
Department of Molecular and Cellular Biology,
University of Arizona, Tucson, AZ, 85721, USA
e-mail: kez@neurobio.arizona.edu

K.E. Zinsmaier and M. Babic
Arizona Research Laboratories, Division of Neurobiology,
Graduate Interdisciplinary Program in Neuroscience, University of Arizona,
Tucson, AZ, 85721, USA

K.E. Zinsmaier and G.J. Russo
Arizona Research Laboratories, Division of Neurobiology,
Graduate Program in Biochemistry and Molecular & Cellular Biology,
University of Arizona, Tucson, AZ, 85721, USA

radical species (ROS), and apoptosis (reviewed in Scheffler 2008). Moreover, mitochondria are not static but astonishingly dynamic organelles whose morphology, distribution, and activity are adaptable to physiological stresses and changes in metabolic demands of the cell. Mitochondrial shape ranges from cell-wide, interconnected tubular networks to individual tubules of various lengths. In neurons, the mitochondrial population consists mostly of short and long tubules that are distributed even into the farthest processes of axons and dendrites to maintain ATP levels and Ca^{2+} homeostasis. Consistently, mitochondria are enriched in regions of intense energy consumption like active growth cones, nodes of Ranvier and synaptic terminals (Fig. 1) (reviewed in Chan 2006; Chang and Reynolds 2006; Kann and Kovacs 2007; Mattson 2007).

More so than any other cell type, neurons critically depend on efficient long-distance transport of mitochondria for two major reasons: First, neurons exhibit extraordinarily high energy demands that almost exclusively (95%) depend on oxidative ATP production by mitochondria. Second, neurons are highly polarized cells with a complex morphology. Neuronal processes can extend over hundreds of centimeters, and exhibit unique cellular specializations with extraordinary energy demands. The great distance between the neuronal cell body and major sites of synaptic signaling in dendrites and axons creates a unique problem for neurons, and makes them particularly sensitive to impairments in the machinery distributing mitochondria (reviewed in Chang and Reynolds 2006; Kann and Kovacs 2007; Salinas et al. 2008).

A simple scaling operation illuminates the magnitude of the problem (Goldstein 2003). In humans, a cell body, 30–50 μm in diameter, of motor, or sensory neurons must typically support an axon that runs from the spinal cord to the toe, bridging about 1 m. Scaled to human proportions, this is equivalent to a circular room, 9–15 m in diameter, from which mitochondria are transported through a narrow tunnel to a distant outpost that is 320 km away. Assuming an average transport velocity of approximately $0.5\,\mu m\ s^{-1}$, as reported from mature axons of cultured hippocampal neurons (Ligon and Steward 2000a), a mitochondrion will reach its synaptic target in approximately 23 days. Even at known peak velocities of ~2 $\mu m\ s^{-1}$, it still takes ~6 days, although it is not clear whether such peak velocities can be sustained over long distances.

Rapidly accumulating evidence suggests that long-distance transport systems are an "Achilles heel" of neurons that can be easily disturbed by genetic or environmental insults (Goldstein 2001). Indeed, impairing mitochondrial transport causes a similar pathology as impairing mitochondrial function. Both lead to metabolic deficiencies, oxidative damage, excitotoxicity, and/or apoptosis that can lead to various forms of muscular dystrophy, neuropathy, paraplegia, and neurodegeneration. For example, mitochondrial transport in neurons is directly, or indirectly compromised in diseases such as Hereditary Spastic Paraplegia, Charcot-Marie-Tooth Type 2A Neuropathy, Chorea-Acanthocytosis, Niemann-Pick Type C1 disorder, Amyotrophic Lateral Sclerosis 2, Retinitis Pigmentosa, Alzheimer's, Huntington's, and other protein aggregation diseases (reviewed in Chang and Reynolds 2006; Finsterer 2006; Lin and Beal 2006; Mattson et al. 2008; Reeve et al. 2008; Salinas et al. 2008).

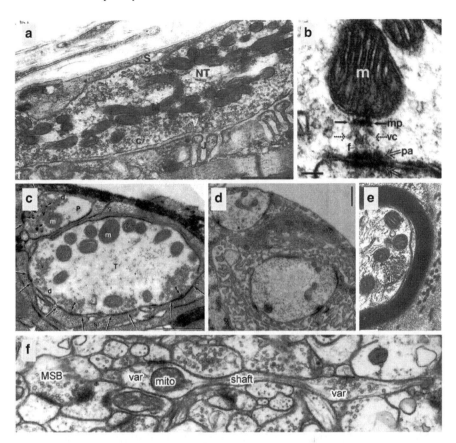

Fig. 1 Mitochondria in axons and dendrites. (**a**). Mitochondria in neuromuscular junction of a tonic soleus muscle in 7-month old mouse. (**b**). Mitochondria-associated adherens complex (MAC) of the Calyx of Held synapse. MAC elements: mitochondrion (m), mitochondrial plaque (mp, *solid arrows*), filaments (f), punctum adherens (pa, *open arrows*), and vesicular chain (vc, *dotted arrows*). (**c**). Mitochondria in phasic (P) and tonic (T) axon terminals of crayfish neuromuscular junction. Structures labeled: vesicles (v); mitochondrion (m); granular sarcoplasm (s); glial cell processes (g); dense-core vesicles (d). Scale bar, 1 μm. (**d**). Mitochondria in larval *Drosophila* neuromuscular junction of type 1b (large bouton) and 1s (small bouton) axon terminals (unpublished, H.L. Atwood, L. Marin, K.E. Zinsmaier). (**e**). Mitochondrial cluster close to ER-associated ribosomes in myelinated axons from sensory spinal nerves (*arrow*). (**f**). CA3 varicosities and axons in stratum radiatum of area CA1. A mitochondrion occupied the right, but not the left, bouton when it was examined three-dimensionally. *Scale bars*: panels **a, b**, 100 nm; panel **c**, 1 μm. Panel **a** was adapted with permission from (Fahim and Robbins 1982), panel **b** from (Rowland et al. 2000), panel **c** from (King et al. 1996), panel **e** from (Pannese and Ledda 1991), and panel **f** from (Shepherd and Harris 1998)

The combination of changing neuronal energy demands, complexity and length of neuronal processes, and activity-dependent plasticity of mitochondrial distribution coupled with the need of mitochondria to eventually return to the cell body requires a transport system that is tightly interweaved with two functionally different signaling pathways: one that communicates the "energetic state of the neuron" and one that communicates the "state of the mitochondrion" to the transport machinery. However, besides the advances in understanding the intricate signaling pathways that control mitochondrial transport to specific neuronal target sites discussed here, we know very little about mechanisms that return mitochondria from axons and dendrites to cell bodies after residing there for an unspecified period.

2 How Mitochondria Move in Axons and Dendrites

A steady stream of cargo moves from their major sites of biogenesis in the cell body to the distant reaches of axons and dendrites. A similar transport system runs in the opposite direction, moving cargo from synapses back to the cell body. The long-distance transport of mitochondria in axons and dendrites relies entirely on the tracks of microtubules (MTs) and their associated motors while actin-based transport is typically employed for short-distance movements (Morris and Hollenbeck 1995). Actin- and MT-based transport mechanisms of mitochondria differ in many aspects, such as the use of specific motors and their average net velocities. Actin-based transport velocities are typically slower, ranging from ~0.02 to 0.04 $\mu m\ s^{-1}$ (Krendel et al. 1998), while MT-based transport velocities range from ~0.1 to 0.7 $\mu m\ s^{-1}$ (Hollenbeck 1996; Ligon and Steward 2000a; Pilling et al. 2006; Misgeld et al. 2007; Louie et al. 2008).

2.1 Mitochondria "Cycle" Through Axons and Dendrites

Mitochondria are prominent members of axonally and dendritically transported cargo, but differ significantly from most cargos in the respect that they are not only transported into neuronal processes, but out as well. This "cycling" of mitochondria through neuronal processes is biased by the growth and activity of the neuronal process. In actively growing axons of cultured neurons, more than half of all motile mitochondria move anterogradely. This anterograde bias is apparently dependent on axonal growth, since it is reduced in nongrowing axons and is much less pronounced in mature neurons (Morris and Hollenbeck 1993; Ruthel and Hollenbeck 2003; Pilling et al. 2006; Misgeld et al. 2007; Louie et al. 2008).

The true biological significance for the cycling of mitochondria between neuronal processes and the cell body remains undetermined, but may be explained by the need of "aging" mitochondria to return to the cell body to get "refurbished," or degraded. Mitochondrial fusion and fission are required to maintain mitochondrial health. Repetitive rounds of selective fusion and fission govern mitochondrial turnover by segregating severely dysfunctional and depolarized mitochondria, and targeting them for degradation by autophagy (reviewed in Detmer and Chan 2007; Twig et al. 2008). JC-1 dye measurements of mitochondrial membrane potentials in axons of ganglion cells suggest that anterogradely transported mitochondria exhibit a higher membrane potential than retrogradely moving mitochondria (Miller and Sheetz 2004). However, ratiometric measurements using TMRM dye show no difference in the membrane potential among stationary, anterogradely and retrogradely moving mitochondria (Verburg and Hollenbeck 2008). Therefore, signals triggering transport of mitochondria back to the cell body remain enigmatic.

2.2 Directional Imprinting of Bidirectional Mitochondrial Transport

MT-based mitochondrial transport consists of a series of movements (Fig. 2a) that are frequently interrupted by brief stationary phases (stops), and by changes in the direction of movement, which results in a "saltatory" appearance (Morris and Hollenbeck 1993). The bidirectional motility of mitochondria is facilitated by the activities of two opposing MT-based motors: MT plus end-directed kinesin, and MT minus end-directed dynein motors. To simply relocate mitochondria, unidirectional, but not bidirectional transport superficially appears as the better economic choice in terms of both speed and energy demand. Yet, bidirectional transport is common and may provide distinct biological advantages, like the ability to circumvent obstacles, to correct errors, and to set up polarized distributions (reviewed in Hollenbeck 1996; Welte 2004).

Despite the superficially chaotic appearance of the saltatory and bidirectional movements of mitochondria, mitochondrial transport proceeds effectively in one direction, reaching peak net velocities of up to $3.4 \mu m \ s^{-1}$ for net anterograde transport, and $2.8 \mu m \ s^{-1}$ for net retrograde transport (Ligon and Steward 2000a; Pilling et al. 2006; Misgeld et al. 2007; Louie et al. 2008). The directional control underlying this effective transport is achieved by a poorly understood molecular mechanism that biases the time mitochondria employ one of the two opposing motor activities (Morris and Hollenbeck 1993; Pilling et al. 2006; Louie et al. 2008). Insights into this control mechanism are mainly derived from high-resolution analyses of mitochondrial transport in axons, where the uniform polarity of MTs (plus ends oriented outward) allows an unambiguous distinction of individual plus end- and minus end-directed movements (Baas et al. 1988) (also, see chapter by Falnikar and Baas 2009).

Motile mitochondria in axons fall into two distinct classes: mitochondria that are transported in a net anterograde direction toward synaptic terminals (AM), and mitochondria that are transported in a net retrograde direction (RM) toward the cell body. The two classes of mitochondria exhibit plus end-directed kinesin, and minus end-directed dynein movements, but to very different degrees. This differential use of motors by AM and RM mitochondria is best illustrated by the "duty cycle" of motile mitochondria (Fig. 2b, c), which describes the percentage of time that mitochondria allocate to stops, and plus- or minus-end directed movements (Morris and Hollenbeck 1993).

Net anterogradely transported (AM) mitochondria spend the majority of their time engaged in plus end-directed movements, but less than 10% of their time

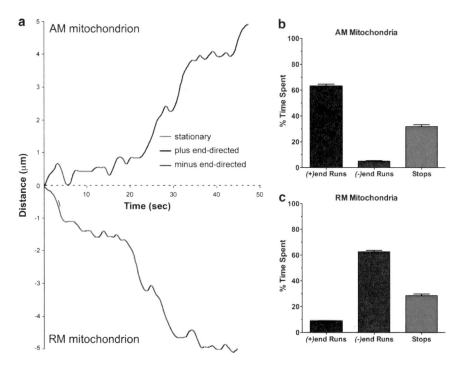

Fig. 2 Directional programming of mitochondrial transport in *Drosophila* motor axons. (**a**). Net anterogradely moving (AM; upward) and net retrogradely moving (RM; downward) mitochondria exhibit a mix of plus end-directed (*blue*) and minus end-directed movements (*red*), termed runs, that are separated by pauses (*green*) and reversals in direction. Plots of mitochondrial tracks were obtained from time-lapse images of GFP-labeled mitochondria (rate 1.006 s^{-1}, duration 200 s). For comparison, the start of individual tracks is set to zero. (**b**)., (**c**). Duty cycles of AM (**b**) and RM (**c**) mitochondria, in which the average percentage of time mitochondria spend in plus end-directed movements [(+)end Runs], minus end-directed movements [(−)end Runs] and short stationary phases (Stops)

engaged in minus end-directed movements (Fig. 2b, c). In contrast, net retrogradely transported mitochondria (RM) do the opposite, and spend the majority of their time undergoing minus end-directed movements, but less than 10% of their time undergoing minus end-directed movements. Both, AM and RM mitochondria spend a similar amount of time in brief stationary phases that are typically not longer than 1–4 s. As a consequence, plus end-directed movements of AM mitochondria are several fold longer in duration and distance than minus end-directed movements. In contrast, RM mitochondria exhibit the opposite: their minus end-directed movements are several-fold longer than plus end-directed movements. Accordingly, a program that defines how long a mitochondrion recruits a particular motor activity effectively determines the transport direction (Morris and Hollenbeck 1993; Hollenbeck 1996; Ligon and Steward, 2000a, b; Pilling et al. 2006; Louie et al. 2008). However, the nature of this directional programming is poorly understood. Firstly, it is not clear how the opposing forces by kinesin and dynein motors are controlled to favor movements in a given direction. Secondly, the signals and underlying molecular machinery governing the respective programs for net anterograde, or retrograde mitochondrial transport remain largely unknown.

3 Motors of Mitochondrial Transport

Transport of organelles requires three different types of proteins (Fig. 3): (1) cytoskeletal proteins forming a "track" like F-actin, microtubules, or neurofilaments (NFs), (2) molecular motors and (3) a host of linker, scaffolding and regulatory proteins that link and anchor motors in the mitochondrial membrane, and regulate the motors in response to cellular demands.

The complex movements of mitochondria in axons and dendrites are mediated by both MT and actin filaments. Since many regions contain one, or the other filament (e.g., spines, or growth cones), mitochondria must be able to switch smoothly between MT- and actin-based transport (Langford 2002). Both, actin filaments and MTs also facilitate long-term stationary phases of mitochondria (Chada and Hollenbeck 2004; Kang et al. 2008). Accordingly, we need to understand not only which motor proteins move mitochondria, but also how mitochondria are linked to motor proteins; how motor proteins are regulated to move either antero- or retrograde; how mitochondria are targeted and anchored to specific sites; and how the activities of all of these proteins are regulated to adjust mitochondrial distributions to cellular demand.

Motor proteins are grouped into three major families: MT-based kinesin and dynein motors, and actin-based myosin motors (Fig. 3). Most members of the kinesin family constitute plus end-directed MT-based motor proteins, while members of the cytoplasmic dynein family constitute minus end-directed motor proteins. Minus end-directed kinesins are typically slow, and except for a few examples, little is known about their role in neurons (reviewed in Hirokawa and Takemura 2005; Levy and Holzbaur 2006; Hirokawa and Noda 2008). Like MTs, actin filaments also mediate bidirectional transport (reviewed in Vale 2003).

3.1 Kinesin Motors

The kinesin superfamily (KIF) is a large gene family of MT-dependent motors with at least 45 members in mice and humans. Kinesins are ATPases that undergo a mechano-chemical cycle to move along a MT. Translocation is achieved by hydrolyzing ATP, and converting the released energy into mechanical work. Typically, kinesins contain a globular motor domain containing MT- and ATP-binding sequences, and

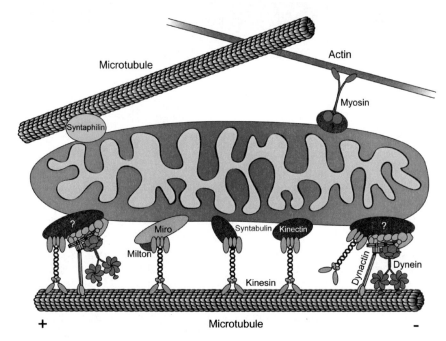

Fig. 3 Interactions of mitochondria with molecular motors and the cytoskeleton. Syntaphilin immobilizes and cross-links mitochondria with MTs. Myosin motors couple to mitochondria by an unknown mechanism, mediating short-distance transport. Far left and right depict a speculative motor control complex that couples both kinesin and dynein motors to mitochondria, and mediates bidirectional transport by either controlling motor activity and number, or the engagement of each motor with MTs. Kinesin motors can be coupled to mitochondria by interactions with Miro, a Miro–Milton/GRIF1/OIP106 complex, Syntabulin, or Kinectin

a cargo-binding domain that is apparently unique for each KIF. The diversity of these cargo-binding domains may explain how kinesins can specifically transport numerous different cargos. In vitro, KIFs exhibit MT-dependent movement velocities that range from 0.2 to 1.5 μm s^{-1}, consistent with speeds of fast axonal transport in vivo (reviewed in Hirokawa and Takemura 2005; Hirokawa and Noda 2008).

Two prevailing models describe how kinesin motors may facilitate movement. One model is deemed the "inch-worm" mechanism, in which the MT-binding domain slides along the MT after hydrolysis of ATP (Mather and Fox 2006). In this model, the MT-binding domain is always the leading domain in any movement, with the other globular domain catching up after the initial movement. The alternate model is the "hand-over-hand" description, in which kinesin performs a power stroke that actively moves the motor down the MT (Yildiz et al. 2004), and predicts alternating binding of each globular domain; thereby, generating movement in a sequential stepping fashion.

Two kinesins have been implicated in mitochondrial transport; namely, KIF1Bα and KIF5. Monomeric KIF1Bα is enriched in neurons, where it physically associates with mitochondria and supports their movement along MTs in vitro at rates that are comparable to in vivo conditions (Nangaku et al. 1994). Mutations in human KIF1Ba have been associated with Charcot-Marie-Tooth disease type 2A (Zhao et al. 2001). The fact that mitochondria are transported by both KIF5 and KIF1Bα might not be surprising since mitochondrial transport must reach targets in dendrites, and axons that differ in their MT organization, and because mitochondria resemble a large cargo that could have many potential motor protein binding sites.

KIF5 consists of three closely related subtypes: KIF5A, KIF5B, and KIF5C. KIF5B is expressed ubiquitously; KIF5A and KIF5C are neuron specific. KIF5 proteins contain an N-terminal globular motor domain as well as a neck, stalk and tail domain, and form homo- or heterodimers among themselves through a coiled-coil region in their stalk domains (Fig. 3). About half of KIF5 dimers form tetramers by recruiting two light chain molecules (KLCs) via their stalk and tail domains (reviewed in Hirokawa and Takemura 2005; Hirokawa and Noda 2008).

KIF5 proteins associate with mitochondria, but also with numerous other cargos (reviewed in Hirokawa and Noda 2008). In mice, deletion of KIF5B is embryonic lethal and causes abnormal perinuclear clusters of mitochondria (Tanaka et al. 1998). Since this phenotype can be rescued by exogenous expression of KIF5A, KIF5B, or KIF5C, any type of KIF5 can apparently transport mitochondria (Kanai et al. 2000). Knockout (KO) of KIF5A in mice does not impair viability and axonal organelle transport, but causes abnormal accumulations of NFs, and a reduction in brain size and the number of motor neurons (Kanai et al. 2000; Xia et al. 2003). Mutations in human KIF5A have been associated with Hereditary Spastic Paraplegia (Fichera et al. 2004).

In contrast to mammals, the *Drosophila* genome contains only one KIF5 gene (termed KHC). Lack of *Drosophila* KIF5 causes embryonic lethality. Partial loss of KIF5 activity severely impairs antero- and retrograde axonal transport of many types of cargos, including mitochondria (Saxton et al. 1991; Hurd and Saxton 1996; Martin et al. 1999). A similar defect is also caused by the lack of the only *Drosophila* kinesin light chain (KLC) gene (Gindhart et al. 1998). Upon partial loss of KIF5 activity, the rate of antero- and retrograde mitochondrial transport is severely reduced in *Drosophila* motor axons (Pilling et al. 2006).

3.2 Dynein Motors

Cytoplasmic dynein shares some basic structural similarities with kinesin, but is generally much more complex in organization. Like kinesin motors, dynein consists of two conserved heavy chains that produce motion via ATP hydrolysis. Unlike kinesin, dynein is capable of hydrolyzing ATP in each of its globular domains, which are both responsible for creating a walking-like movement along MTs (reviewed in Levy and Holzbaur 2006).

Cytoplasmic dynein is a multimeric super protein complex of ~2 MDa (Fig. 3), composed of 2-3 dynein heavy chains (Dhc), multiple intermediate, light intermediate, and light chains. Dhc consists of an N-terminal domain that forms the base of the molecule, to which most of the accessory subunits bind, and a motor domain. Typically, two heavy chains form two motor domains on relatively flexible stalks that dimerize at their ends. The motor domain consists of multiple AAA ATPase units that form a ring-shaped structure from which a stalk-like MT-binding domain projects out. ATP binding to the first AAA repeat motif induces dissociation from MTs. ATP hydrolysis then induces a conformational change that enables dynein to produce a power stroke by rebinding the MT (reviewed in Hook and Vallee 2006; Levy and Holzbaur 2006).

Cytoplasmic dynein activity typically requires a second multisubunit protein complex, dynactin. Dynactin binds dynein directly, and allows the motor to move along MTs over long distances in vitro. The dynactin complex (1 MDa) resembles a ~10×40 nm base with a sidearm containing two globular heads. The actin-related protein 1 (Arp1) forms a short octomeric filament at the base of the complex, which is capped at one side by actin capping protein (CapZ), while the opposite site is capped by ARP11 and the dynactin subunits p62, p25, and p27. A dimer of p150Glued forms the sidearm, which is connected to the Arp1 base by a tetramer of the p50 dynactin subunit dynamitin. p150Glued interacts directly with both cytoplasmic dynein and MTs, but only the dynein–MT interaction is nucleotide-sensitive. Both, dynein and dynactin are required for mitosis, vesicular trafficking, retrograde signaling, mRNA localization, and protein recycling and degradation. Accordingly, knockouts of cytoplasmic dynein and dynactin are lethal during an early stage in embryogenesis in both flies and mice (reviewed in Schroer 2004; Levy and Holzbaur 2006).

Two dynein (intermediate chain and heavy chain 1) and three dynactin subunits (Arp1, p62, and p150Glued) can not only interact with mitochondria, but also many other cargos (Habermann et al. 2001). Antibody injections or genetic manipulations of dynein change the distribution of mitochondria and other organelles (Brady et al. 1990; Bowman et al. 1999; Martin et al. 1999; LaMonte et al. 2002; Koushika et al. 2004). Targeted disruption of the dynein–dynactin complex in motor neurons of mice causes a progressive degeneration of motor neurons (LaMonte et al. 2002; Lai et al. 2007).

To date, live imaging of mitochondrial transport in dynein mutant motor axons of *Drosophila* provides the best evidence that dynein is required for minus end-directed movements of mitochondria. Loss of *Drosophila* Dhc64 reduces the rate of retrograde mitochondrial transport, and impairs the length and duration of minus end-directed mitochondrial movements. This, combined with the lack of evidence for an alternative fast minus-end motor, suggests that cytoplasmic dynein is likely the primary motor for retrograde mitochondrial transport in axons (Pilling et al. 2006).

3.3 Myosin Motors

Long-distance transport of mitochondria in axons critically requires MT-based motors. However, mitochondria also employ actin filaments for short-distance movements (Morris and Hollenbeck 1995; Ligon and Steward 2000b). There are a number of actin-based myosin motors in neurons that could mediate mitochondrial movements. For example, myosin V motors are widely used for organelle transport (reviewed in Vale 2003), but direct evidence associating axonal mitochondria with myosin motors is lacking (also, see chapter by Bridgman 2009).

Initially, it has been thought that the yeast myosin V ortholog, Myo2p, may facilitate mitochondrial transport. However, myosin motors do not serve as force generators for mitochondrial movements in yeast. Apparently, anterograde mitochondrial movements are driven by actin polymerization, and retrograde movements by the retrograde translocation of actin filaments (reviewed in Fagarasanu and Rachubinski 2007).

Interestingly, actin polymerization, and not myosin motor activity may also drive mitochondrial movements in axons and dendrites. WAVE1, a member of the Wiskott–Aldrich syndrome protein (WASP)-family 1, is associated with the outer mitochondrial membrane and activates the Arp2/3 complex, a key regulator of actin polymerization (Danial et al. 2003; Kim et al. 2006). Since WAVE1 is required for activity-dependent mitochondrial trafficking into dendritic spines and filopodia (Sung et al. 2008), it is possible that Arp2/3-dependent actin polymerization mediates mitochondrial translocation into dendritic protrusions. However, whether this is indeed the case has not yet been directly tested.

4 Coupling Mitochondria to Motors

4.1 KIF1Bα and KIF1 Binding Protein

KIF1 binding protein (KBP) may control the coupling of KIF1Bα to mitochondria. Overexpression of a dominant-negative mutation in KBP or RNA antisense expression decreases the activity of KIF1Bα, and leads to an aggregation of mitochondria (Wozniak et al. 2005). Loss of KBP activity in zebrafish reduces mitochondrial targeting to axons, and impairs axonal outgrowth and maintenance (Lyons et al. 2008). Mutations in human KBP have been associated with Goldberg–Shprintzen syndrome (Brooks et al. 2005). However, KIF1Bα may be directly coupled to other cargos since its seven residues in its C-terminal region selectively interact with PDZ domains from a number of scaffolding proteins, including PSD-95, PSD-97, and SSCAM (Mok et al. 2002).

4.2 KIF5 Versus Kinesin Light Chain

KIF5 has been implicated in the transport of numerous cargos, and thus may require diverse coupling mechanisms. KLC contains six tetratricopeptide repeat (TPR) motifs that can mediate a number of protein–protein interactions. Since KLC associates with the C-terminal tail of KIF5, there is some controversy whether KIF5 cargos couple directly, or indirectly through KLC to KIF5 (reviewed in Hirokawa and Noda 2008). Consistent with a critical role of KLC, tumor necrosis factor (TNF) induced hyperphosphorylation of KLC inhibits KIF5 activity and causes perinuclear clusters of mitochondria (De Vos et al. 1998, 2000). In addition, genetic manipulations of *Drosophila* KLC cause accumulations of organelles in axons (Gindhart et al. 1998). However, there is also evidence from *Drosophila*, suggesting that KLC may not be required for linking KIF5 to mitochondria (Glater et al. 2006).

4.3 Syntabulin

Syntabulin interacts with the C-terminal tail of KIF5, and may couple KIF5 to several different cargos, including mitochondria and syntaxin-containing vesicles (Su et al. 2004; Cai et al. 2005). Syntabulin is associated with the outer mitochondrial membrane through its C-terminal, which contains a predicted hydrophobic anchor with three conserved positively charged residues (Wattenberg and Lithgow 2001; Cai et al. 2005). Knockdown of syntabulin, and competitive blocking of the syntabulin-KIF5 interaction reduces mitochondrial density in neuronal processes, and impairs anterograde transport of mitochondria. Thus, syntabulin may couple mitochondria to KIF5 by associating with phospholipids, an unknown receptor in the outer mitochondrial membrane (Fig. 3). Alternatively, syntabulin could be a part of a scaffold that is required for the assembly of the mitochondrial transport machinery. However, it is unclear whether syntabulin has a universal role for KIF5-mediated transport of mitochondria since there is no obvious homolog in invertebrates.

4.4 The Miro–Milton Adaptor Complex

The *Drosophila* protein Milton, and its mammalian homologues γ-aminobutyric acid A receptor-interacting factor-1 (GRIF1; also called TRAK2) and O-linked *N*-acetylglucosamine–interacting protein 106 (OIP106; also called TRAK1), have been suggested to form a critical adaptor complex coupling mitochondria to KIF5 (Fransson et al. 2006; Glater et al. 2006; MacAskill et al. 2009a; Wang and Schwarz 2009). Deletion of Milton in *Drosophila* is lethal during larval development, and neurons lack mitochondria in axons and dendrites (Stowers et al. 2002; Gorska-Andrzejak et al. 2003; Glater et al. 2006).

The atypical mitochondrial GTPase Miro (Miro1 and 2) exhibits two GTPase domains, two Ca^{2+} binding domains, and a C-terminal membrane domain that tail-anchors the protein in the outer mitochondrial membrane (Fransson et al. 2003; Frederick et al. 2004; Guo et al. 2005). Overexpression and dominant mutations of Miro alter mitochondrial distributions, and morphology in cultured nonneuronal cells (Fransson et al. 2003, 2006). The yeast homolog of Miro, Gem1p, is required for maintaining a normal tubular mitochondrial network and mitochondrial inheritance (Frederick et al. 2004). Null mutations in *Drosophila* Miro (dMiro) disrupt axonal and dendritic transport of mitochondria, restricting mitochondria to the neuronal cell body (Guo et al. 2005).

Milton/GRIF1/OIP106 proteins contain coiled-coil domains, and a N-terminal Huntington Associated Protein 1 (HAP-1) domain (Brickley et al. 2005). Milton/GRIF1/OIP106 colocalizes with mitochondria and other cargos, and directly interacts with KIF5 (Fransson et al. 2006; Glater et al. 2006). In addition to KIF5, Milton/GRIF1/OIP106 also binds directly to the GTPase Miro (Fig. 3), forming a complex that, in principal, couples mitochondria to kinesin motors (Giot et al. 2003; Fransson et al. 2006; Glater et al. 2006). However, Miro is not the only mitochondrial membrane anchor for Milton, since Milton can also bind to an unknown mitochondrial protein through a Miro-independent domain that is located at its C-terminal (Glater et al. 2006).

The biochemical interactions among KIF5, Milton/GRIF1/OIP106, and Miro support the notion that this complex couples mitochondria to kinesin motors. However, further studies indicate that the role of this complex is much more complex than originally assumed. For example, Miro can also directly bind KIF5 in a Ca^{2+}-dependent manner (MacAskill et al. 2009b; Wang and Schwarz 2009). In addition, Miro, and possibly Milton/GRIF1/OIP106 are critical for executing directional programming of mitochondrial transport (Russo et al. 2009).

4.5 Kinectin

The concept of membrane anchors linking molecular motors to specific cargos was first suggested by the discovery that kinesin interacts with kinectin (Toyoshima et al. 1992; Ong et al. 2000). Kinectin antibodies block motor binding to microsomes, and reduce kinesin- and dynein-mediated organelle motility, suggesting that it may regulate, or coordinate motors rather than provide a mere membrane anchor for kinesin (Kumar et al. 1995; Sheetz and Dai 1996).

Kinectins comprise a large and complex protein family that includes 16 isoforms that is not present in *Caenorhabditis*, or *Drosophila* genomes. A 120 kDa kinectin isoform is specifically associated with mitochondria, and overexpression of its kinesin-interaction domain alters mitochondrial distributions in HeLa cells, while RNAi-induced knockdown of kinectin causes a collapse of the ER and mitochondrial network (Santama et al. 2004). However, kinectin KO mice show no defects in organelle motility (Plitz and Pfeffer 2001). This apparent contradiction may be explained by redundant systems, but further studies are required to resolve this controversial issue.

5 Regulation of Mitochondrial Motor Activities

Mitochondria exhibit bidirectional motility, and frequently reverse course (Fig. 2a). This raises the question: how does a mitochondrion control the opposing movements to reach its axonal destination, or return to the cell body? Apparently, the bidirectional motility of mitochondria is controlled by a program that defines how long a mitochondrion employs a particular motor activity (Morris and Hollenbeck 1993; Ligon and Steward, 2000a, b; Pilling et al. 2006; Louie et al. 2008). This control results in a strong bias of mitochondria toward movements in the primary transport direction (Fig. 2b, c) that ensures effective net transport in one direction. Accordingly, there must be two mechanisms controlling kinesin and dynein motors: one, achieving effective anterograde, and the other, achieving retrograde transport. However, the mechanisms of directional programming are poorly understood. Firstly, it is not clear how the opposing forces by kinesin and dynein motors are controlled to favor movements in a given direction. Secondly, the signals and underlying molecular machinery activating the respective programs remain poorly understood.

5.1 Motor Coordination Versus Tug-of-War

In principal, three different mechanisms could achieve effective net transport. In the simplest case, only one of the two opposing kinesin and dynein motors is bound to mitochondria at all times. However, this scenario has been principally ruled out. Therefore, gaining net distance must require tuning of the opposing motor activities, which can be achieved by either a "tug-of-war," or a "motor coordination" mechanism (reviewed in Gross 2003; Welte 2004; Gross et al. 2007).

The tug-or-war scenario assumes that both types of motors are simultaneously active, and engaged with MTs. As each type of motor tries to pull the cargo in its given direction, a tug-of-war ensues, in which the stronger or more abundant motor type will determine the direction of movement at any particular moment. Mechanistically, this scenario could be achieved by controlling either the force production of a given motor, or the average number of active motor molecules. However, the evidence for a tug-of-war mechanism is mostly theoretical (Muller et al. 2008). In contrast, a motor coordination mechanism ensures that both motors do not interfere with each other's function. In this case, each motor is turned "on" or "off" independently, such that plus end-directed kinesin motors are active when minus end-directed dynein motors are not, and vice versa (Gross 2003; Welte 2004). Both scenarios require the presence of higher-order regulatory mechanisms to either up-regulate one motor activity over the other, or coordinate the activities, but also to integrate signals from the cell to activate, terminate, or change the direction of net transport in response to cellular demands.

To ensure effective transport, motor coordination also requires a mechanism that controls the processivity of motors (Welte et al. 1998; Suomalainen et al. 1999;

Gross et al. 2000, 2002b; Smith et al. 2001; Rodionov et al. 2003). For example, the distance of minus end-directed movements during retrograde mitochondrial transport is several-fold longer than that of minus-end runs during anterograde transport, and vice versa (Fig. 2a). Consequently, movements opposite to the direction of net transport may reflect a basal or repressed state of motor processivity, which is selectively up-regulated by signals that switch motor activities, and convey the direction of net transport. It is unclear whether the mechanisms controlling motor coordination and processivity act hierarchically, or in parallel, but it is likely that their actions are coordinated.

Experimental evidence supporting motor coordination comes from several transport systems, including melanophore, and lipid-droplet transport (reviewed in Gross 2003; Welte 2004). Cellular changes alter the direction of lipid-droplet transport in one direction through a change in the length of plus end-directed movements; whereas, in the melanophore system the length of minus end-directed movements is changed. Importantly, in both cases, movements in the opposite direction are unaltered (Welte et al. 1998; Gross et al. 2002a). In addition, stall forces during lipid transport are similar during plus- and minus-end directed movements, and independent of the net direction of transport (Welte et al. 1998). Genetic manipulation of dynein primarily reduces motility of lipid droplets in both directions by decreasing the distance or velocity of movements (Gross et al. 2000, 2002b). In the melanophore system, manipulation of the dynein cofactor dynactin impairs transport in both directions (Deacon et al. 2003). Dynein and kinesin also do not compete during peroxisome transport. Instead, multiple kinesins, or dyneins cooperate in vivo and produce up to ten times the in vitro speed (Kural et al. 2005).

Supporting evidence for motor coordination of mitochondrial transport in axons comes from *Drosophila*. Upon genetic manipulation of kinesin, or dynein, there is no evidence that force production by the two opposing motor is competitive, which supports a motor coordination, but not a tug-of-war scenario (Pilling et al. 2006). In addition, lack of dMiro selectively impairs either kinesin- or dynein-mediated movements, depending on the direction of mitochondrial net transport in axons. Since loss of the primary motor activity for antero- or retrograde transport does not increase the activity of the opposing motor, these data also support a motor coordination, and not a tug-of-war scenario (Russo et al. 2009).

The lack of experimental evidence makes the tug-of-war scenario less likely. However, this conclusion could be premature since modeling of a tug-of-war scenario using load-dependent transport properties of individual motors revealed complex patterns of movement that could be erroneously interpreted as coordinated transport in experimental systems (Muller et al. 2008). In contrast to previous expectations, the modeled tug-of-war scenario is highly cooperative, and produces a number of different motility patterns. Modifications of motor properties mimicking the effect of either regulatory mechanisms, or mutations affected bidirectional motility in complex ways such that: (1) motility was affected only in one direction, (2) motility was impaired in one direction and enhanced in the other, or (3) motility was similarly altered in both directions. Interestingly, this variability largely agrees with observations upon experimental modifications in different systems, suggesting that the modeled tug-of-war scenario represents a realistic model for bidirectional transport in vivo (Muller et al. 2008).

5.2 Possible Molecular Mechanisms of Motor Control

The mechanisms mediating motor control are a likely target for the action of numerous *trans*-acting regulators. Consequently, it is critical to unravel the properties of the motor control machinery to gain a comprehensive understanding of when and where transport occurs. Various proteins that are necessary for the proper function or regulation of motor transport have been identified. Prominent examples are the dynein cofactor dynactin (Gross 2003), huntingtin in the BDNF transport system (Colin et al. 2008), and various proteins like halo, klar, and LDS2 in the lipid-droplet system (Welte et al. 2005). However, except for dynactin, none of these may play a significant role in mitochondrial transport.

5.2.1 Kinesin–Dynein Interactions

Bidirectional motility may be coordinated by a direct interaction between kinesin (KIF5) and dynein that is mediated through dynein intermediate chain 1 (DIC1), and KLCs 1 and 2 (Ligon et al. 2004). Consistently, a number of studies have shown that inhibition of dynein function, via function-blocking antibodies, or by genetic manipulations of dynein expression, inhibits bidirectional vesicular motility, and axonal transport (Waterman-Storer et al. 1997; Martin et al. 1999; He et al. 2005). In regard to mitochondria, loss of *Drosophila* KIF5 (Khc) profoundly reduces the rate of retrograde mitochondrial transport, while loss of dynein activity (Dhc64) reduces the velocity of kinesin movements (Pilling et al. 2006). In addition, genetic interactions between *Drosophila* Khc (KIF5) and dynein (Dhc64) mutations support interactions between both motors (Martin et al. 1999). However, the true significance of DIC for motor coordination has not yet been specifically tested. In addition to a possible DIC-mediated interaction, motor coordination may also depend on the dynein-associated dynactin complex (Martin et al. 1999; Gross et al. 2002b; Deacon et al. 2003; Gross 2003).

5.2.2 Dynactin–Dynein–Kinesin Interactions

The dynactin complex is well known for its physical association with cytoplasmic dynein, and for promoting the processivity of dynein motors in vitro (King and Schroer 2000; Schroer 2004). However, it's role for mitochondrial transport specifically, and bidirectional transported cargo in general, is complex and remains controversial.

Dominant-negative mutations in dynactin's subunit p150 Glued interfere with organelle transport (Martin et al. 1999). More importantly, p150Glued mutations reduce the force generated during lipid-droplet plus end-directed movements, possibly inducing a partial tug-of-war (Gross et al. 2002b). This raised the possibility that dynactin usually turns the minus end motor off when the plus end motor is active. Consistently, dynactin interacts with both plus, and minus end-directed motors, at least in melanophores (Deacon et al. 2003). Dynactin's p150Glued subunit binds DIC. In addition, p150Glued can bind directly to KAP, a nonmotor subunit of the kinesin II motor, driving transport of pigment granules (Deacon et al. 2003). One of several possible models proposes that contact with dynactin stabilizes a motor's interaction with the track; thus, turning it on (Gross 2003; Gross et al. 2003). Accordingly, only one motor can be in the "on" state at any given time, if dynactin alternately contacts the two motors. However, in regard to mitochondrial transport, the role of the dynactin complex is rather unclear. RNAi-mediated knockdown of P150Glued, or dominant-negative mutations enhanced, rather than reduced kinesin and dynein movements in *Drosophila* motor axons (Pilling et al. 2006). This finding confirms that P150Glued indeed influences both motors, but provides no evidence that its function is critical for motor coordination.

Further evidence for a role of dynactin for motor coordination comes from studies altering the dynactin subunit p50/dynamitin (Echeverri et al. 1996). Overexpression of dynamitin disrupts the dynactin complex, and abolishes both plus- and minus-end directed motion of several bidirectionally moving cargos (Echeverri et al. 1996; Valetti et al. 1999; Deacon et al. 2003). However, the significance of dynamitin for mitochondrial transport remains to be examined.

Similar to the effects of dynamitin overexpression, mutations and RNAi-induced knockdown of the dynactin subunit Arp1 disrupt the dynactin complex (Haghnia et al. 2007). Loss of Arp1 severely disrupts the rate of antero- and retrograde transport of APP-containing vesicles and mitochondria in *Drosophila* motor axons. Contrary to the expectation, normal amounts of dynein are still associated with its cargos in Arp1 mutants, despite the absence of a fully assembled dynactin complex. In addition, Arp1 mutant APP-YFP labeled cargo vesicles spend less time on antero- and retrograde movements, and more time stationary in Arp1 mutant axons (Haghnia et al. 2007). While the mutant effects on retrograde transport were predicted by the proposed function of dynactin as a regulator of dynein processivity, the additional effects on anterograde transport are consistent with a role of dynactin coordinating motors (King and Schroer 2000; Gross 2003; Schroer 2004).

Notably, the rate of antero- and retrograde mitochondrial transport is dramatically reduced in Arp1 mutant axons (Haghnia et al. 2007), which is in sharp contrast to the finding that RNAi-induced knockdown, or a dominant-negative mutation of p150Glued has no significant effect on the rate of mitochondrial transport (Pilling et al. 2006). Moreover, manipulations of p150Glued increased the velocity of plus and minus end-directed movements for both net antero- and retrogradely moving mitochondria, suggesting that p150Glued acts as drag for kinesin and dynein motors (Pilling et al. 2006). Superficially, the two studies appear incompatible; however, the underlying assumption that the dynactin complex acts as a functional unit may not be true. In particular, the subunits forming the base of the dynactin complex (including Arp1 and dynamitin) may have a significantly different role than the MT-interacting p150Glued arm. To resolve this issue, systematic in vivo analyzes examining the significance of individual subunits of the dynactin complex will be necessary.

An interesting twist to the discussion of motor coordination is the surprising finding that purified, fluorescently labeled dynein–dynactin complexes exhibit in vitro bidirectional motility towards both the plus and minus ends of MTs (Ross et al. 2006). Although previous studies raised this possibility (Wang and Sheetz 1999; Mallik et al. 2005), the employed single-molecule assay provides convincing evidence for a bidirectional nature of the dynein–dynactin complex. Hence, it is possible that dynein may contribute to plus and minus end-directed transport in the cell. Consistent with this finding is that quantum dots undergoing motor-driven transport in cells take similar large 16-nm steps in the plus and minus end directions, which is consistent with dynein, rather than kinesin movements (Nan et al. 2005). However, the ability of dynein to switch the direction of transport could be significantly affected by load, and therefore not relevant for the transport of large organelles like mitochondria.

5.2.3 Miro–Milton-Motor Interactions

As discussed in the linker section (see Sect. 4.3), the mitochondrial membrane protein Miro binds KIF5 either directly, or indirectly over Milton/GRIF1/OIP106 to couple and likely regulate kinesin motors (Fransson et al. 2006; Glater et al. 2006; MacAskill et al. 2009a; Wang and Schwarz 2009). Null mutations of Milton, or Miro essentially abolish axonal and dendritic transport of mitochondria in *Drosophila* neurons, restricting mitochondria to the neuronal cell body (Stowers et al. 2002; Gorska-Andrzejak et al. 2003; Guo et al. 2005; Glater et al. 2006; Russo et al. 2009).

Lack of dMiro reduces the rate (flux) of antero- and retrograde mitochondrial transport by more than 98% (Russo et al. 2009). The few motile mitochondria in dMiro null mutant motor axons exhibit a profound defect that mirrors the expected effects upon loss of control over motor processivity. Specifically, net anterogradely transported mitochondria exhibit reduced kinesin-mediated, but normal dynein-mediated movements. Conversely, net retrogradely transported mitochondria exhibit much shorter dynein-mediated movements, while kinesin-mediated movements were minimally affected. This finding suggests that Miro promotes effective antero- and retrograde mitochondrial transport by extending the processivity of kinesin and dynein motors according to a mitochondrion's programmed direction of transport (Russo et al. 2009).

How Miro achieves control over the processivity of kinesin and dynein motors is unclear. However, at least in part, this control is likely facilitated by its known interactions with KIF5 and Milton/GRIF1/OIP106 (Fransson et al. 2006; Glater et al. 2006; MacAskill et al. 2009a, b). Since at least mammalian Miro1 can also bind directly to KIF5 (MacAskill et al. 2009b), it is possible that Miro may control the processivity of kinesin motors through its direct interaction with KIF5, or its indirect interaction with Milton. In addition, Milton/GRIF1/OIP106 proteins could also serve as a scaffold for Miro up-regulating the processivity of dynein motors. Consistent with such a "bidirectional" role of Milton is the finding that Milton is required for dynein movements during the formation of the "mitochondrial cloud" in fly oocytes (Cox and Spradling 2006).

Mechanistically, Miro's control over both motors could be facilitated by the two slightly different isoforms of Milton in *Drosophila* (Cox and Spradling 2006; Glater et al. 2006), and the two homologous mammalian genes GRIF1/TRAK2 and OIP106/TRAK2 (Beck et al. 2002; Iyer et al. 2003; Brickley et al. 2005; Fransson et al. 2006; Gilbert et al. 2006). The two mammalian proteins, as well as the different Milton RNA splice forms differ in their N-terminally located HAP-1 domain, which has been predicted to mediate interactions with dynactin (Li et al. 1998). Accordingly, one isoform may interact with kinesin motors, while the other may interact with dynein motors via dynactin. The latter is consistent with dynactin's in vitro effects, increasing the processivity of dynein-driven movements (King and Schroer 2000). However, clear evidence that specific Milton/GRIF1/OIP106 isoforms bind to dynactin or dynein subunits is lacking.

6 Targeting of Mitochondria to Specific Sites in Axons and Dendrites

A unique feature of mitochondria in axons and dendrites is their ability to switch between motile and long-term stationary states. The stationary state can last longer than 15 min (Chang et al. 2006; Chang and Reynolds 2006; Louie et al. 2008). However, for essentially all cases, it is not known how long mitochondria reside at specific axonal or dendritic target sites. The pattern of stationary mitochondria is highly variable between axons and dendrites of an individual neuron, but also among different types of neurons, and different developmental and activity stages (reviewed in Chang and Reynolds 2006). Considering the limited diffusion of ATP and Ca^{2+} ions in an intracellular environment (Belles et al. 1987; Hubley et al. 1996), stationary mitochondria likely serve as local service stations, providing ATP and potentially Ca^{2+} signaling functions.

Once mitochondria are delivered to a specific region, they often become stationary for extended periods (often referred to as "docking"), but eventually they resume their motility, and move back to the cell body. Hence, docking is a dynamic process that likely includes reversible interactions with the cytoskeleton, morphological and functional changes of mitochondria. In addition, mitochondria are targeted to several different sites in neurons, which likely require different signals that "attract" mitochondria to specific sites, disrupt their transport, and "immobilize" them in a reversible manner. Importantly, targeting of mitochondria to specific sites is probably never permanent, and thus requires signals that remobilize mitochondria for transport. Although many of these mechanisms are still obscure, a number of advances have been made during the past few years, unraveling mechanisms that target mitochondria to pre- and postsynaptic sites.

A number of in vivo studies have implicated the interactions of mitochondria with F-actin, MTs, or NFs as potential docking mediators (Smith et al. 1977; Hirokawa 1982; Benshalom and Reese 1985; Hirokawa and Yorifuji 1986; Pannese et al. 1986; Price et al. 1991; Morris and Hollenbeck 1995). Consistently, purified mitochondria bind MT-associated proteins and NFs in vitro, and exhibit physical links to both cytoskeletal elements, which could be dynamic (i.e., motor proteins), or static (e.g., docking, or adaptor proteins) links between mitochondria and the cytoskeleton (Linden et al. 1989a, b; Jung et al. 1993; Leterrier et al. 1994a, b; Wagner et al. 2003). However, many of the sites at which mitochondria accumulate, such as presynaptic terminals and growth cones, are actin-rich and MT- and NF-poor (Peters et al. 1991). Consistently, disruption of actin filaments in cultured neurons affects mitochondrial movement (Morris and Hollenbeck 1995). In addition, the ability of nerve growth factor (NGF) and phosphoinositide 3-kinase (PI3K) signaling to halt mitochondrial movement critically depends on the presence of actin filaments (Chada and Hollenbeck 2004), suggesting F-actin as a critical docking partner. Hence, mitochondria may have multiple mechanisms that mediate anchoring to various structural components.

EM studies provide no obvious insights as to how pre- or postsynaptic mitochondria may become stationary long-term, with one notable exception; namely, the highly specialized Calyx of Held contains a physical substrate for keeping mitochondria in place; the mitochondrion-associated adherens complex (Rowland et al. 2000). However, the molecular composition of this complex remains unresolved.

6.1 Ca^{2+} Dependent Arrest of Mitochondrial Transport

In neurons, high elevations of intracellular Ca^{2+} occur especially at pre- and postsynaptic sites. Synapses are also sites of high metabolic demand since maintaining ion homeostasis is energetically expensive. Since cytosolic Ca^{2+} spikes activate Ca^{2+}-sensitive mitochondrial dehydrogenases, which in turn modulate ATP production (Hajnoczky et al. 1995; Babcock and Hille 1998; Denton 2009), Ca^{2+} signaling is an effective way to retain mitochondria temporarily at synaptic sites to prevent local ATP depletion.

Elevations of intracellular Ca^{2+} arrest MT-based mitochondrial transport in many cell types (Rintoul et al. 2003; Yi et al. 2004; Brough et al. 2005; Chang et al. 2006; Mironov 2006). In cultured myoblasts and HEK-293 cells, mitochondrial movements are suppressed after Ca^{2+} release from internal stores (Yi et al. 2004; Brough et al. 2005). In neurons, synaptic activity and Ca^{2+} entry through glutamate receptors, or Ca^{2+} channels reduces mitochondrial motility, while mitochondrial mobility is accelerated after inhibition of Na^+ or Ca^{2+} channels (Rintoul et al. 2003; Chang et al. 2006; Mironov 2006). However, mitochondrial movements are also inhibited after membrane depolarization in Ca^{2+}-free solutions, raising the possibility that Ca^{2+} is either not directly involved in this inhibition, or acts in parallel with a signal that reflects local ATP depletion during prolonged activity (Brough et al. 2005; Mironov 2006). In addition, Ca^{2+}-induced inhibition of mitochondrial movement may not be universal since mitochondrial motility in processes of cortical neurons is immune to intracellular Ca^{2+} fluctuations (Beltran-Parrazal et al. 2006).

The mitochondrial GTPase Miro is probably the major Ca^{2+} sensor mediating Ca^{2+}-induced inhibition of mitochondrial motility. Miro contains two EF-hand Ca^{2+} binding domains that are sandwiched between an N- and C-terminal GTPase domain (Fransson et al. 2003; Frederick et al. 2004; Guo et al. 2005). Coexpression of EF-hand mutant Miro protein in hippocampal neurons, or nonneuronal cells does not impair mitochondrial motility per se. However, Ca^{2+}-induced inhibition of mitochondrial motility is abolished upon loss of Miro's Ca^{2+} binding domains, suggesting that it serves as a Ca^{2+} sensor controlling mitochondrial mobility (Saotome et al. 2008; MacAskill et al. 2009b; Wang and Schwarz 2009). In neurons, Miro-mediated Ca^{2+}-induced inhibition of mitochondrial motility is likely important for localizing mitochondria to postsynaptic sites in dendrites of hippocampal cells, and for minimizing the effects of glutamate-induced excitotoxicity (MacAskill et al. 2009b; Wang and Schwarz 2009). Currently, it is not known whether mitochondria at axon terminals use a similar mechanism for their presynaptic localization.

The precise mechanism of how Miro mediates Ca^{2+}-induced inhibition of mitochondrial motility is still controversial. The first model (Fig. 4a) proposes that Miro mediates mitochondrial transport primarily by a direct interaction between Miro and KIF5, which is disrupted upon Ca^{2+} binding by Miro's EF hands (MacAskill et al. 2009b). Thus, Ca^{2+}-binding in this model triggers a dissociation of mitochondria from kinesin motors. In contrast, the second model builds on the previously suggested adaptor model, in which Miro binds to *Drosophila* Milton, or its mammalian homologs GRIF1/TRAK2 and OIP106/TRAK1, which in turn directly bind to the kinesin motor KIF5 (Fransson et al. 2006; Glater et al. 2006). The updated model (Fig. 4b) suggests that in the absence of Ca^{2+}, the C-terminal tail of KIF5 is bound to the mitochondrion through the Milton–Miro complex such that KIF5's motor domain can engage with MTs, facilitating plus end-directed movements. Upon Ca^{2+} binding to Miro's EF hands, a resulting conformational change in Miro then allows direct binding of KIF5 to Miro, preventing KIF5's engagement with MTs (Wang and Schwarz 2009). Accordingly, Ca^{2+}-binding triggers a switch mechanism that disrupts the motor-MT engagement, but not the coupling between kinesin and mitochondria.

It is important to note that Ca^{2+}-induced inhibition of mitochondrial motility may not be the only Ca^{2+}-dependent function of Miro. In yeast, where mitochondrial transport does not depend on MTs, the EF-hand domains of Miro (Gem1p) are still required for maintaining a normal tubular mitochondrial network, and mitochondrial inheritance (Frederick et al. 2004, 2008). Consistently, mammalian Miro may

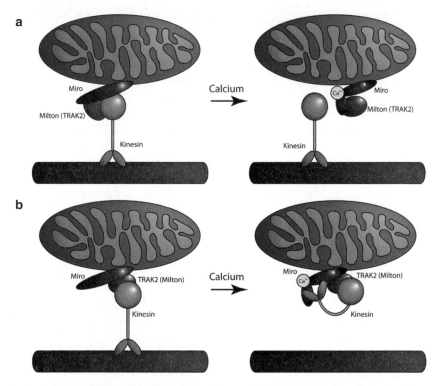

Fig. 4 Proposed models of Ca^{2+}-induced inhibition of mitochondrial transport. (**a**). Model 1: in the absence of Ca^{2+}, kinesin motors are coupled to mitochondria via a KIF5-Miro protein complex that facilitates transport. Upon Ca^{2+} binding Miro undergoes a conformational change that dissociates KIF5 from Miro, uncoupling kinesin motors from mitochondria. (**b**). Model 2: in the absence of Ca^{2+}, kinesin motors are coupled to mitochondria by binding of KIF5 to Milton, which in turn binds to Miro. Upon Ca^{2+} binding Miro undergoes a conformational change that exposes a KIF5 binding domain, which competes with KIF5's interaction with MTs such that KIF5 disengages from MTs and binds to Miro. Panel **a** adapted with permission from (MacAskill et al. 2009b); Panel **b** adapted with permission from (Wang and Schwarz 2009)

also serve as a Ca^{2+}-sensitive switch that indirectly controls mitochondrial fusion–fission dynamics (Saotome et al. 2008).

6.2 Syntaphilin-Mediated Docking of Mitochondria

Syntaphilin is the only protein that has been clearly implicated in physically docking mitochondria to the MT cytoskeleton (Kang et al. 2008). Syntaphilin is a neuron specific protein that was initially identified as an inhibitor of transmitter release (Lao et al. 2000). Injection of syntaphilin's coiled-coil domain into cultured neurons interfered with transmitter release; presumably, because the sequence of this domain is ~80% identical to the syntaxin-binding domain of ocsyn and syntabulin (Safieddine et al. 2002; Su et al. 2004). However, immunolabeling suggests that syntaphilin is predominantly a mitochondrial protein (Kang et al. 2008).

Syntaphilin's mitochondrial localization is mediated by two C-terminally located transmembrane domains (Kang et al. 2008). Typically, only stationary, but not motile mitochondria in axons contain significant amounts of syntaphilin (Das et al. 2003; Kang et al. 2008). Overexpression of GFP-tagged Syntaphilin increases the number of stationary mitochondria, an effect that is abolished upon deletion of syntaphilin's microtubule-binding domain. Overexpressed syntaphilin is primarily associated with stationary, but not motile mitochondria, suggesting that syntaphilin inhibits the motility of axonal mitochondria by binding to MTs. The syntaphilin-mediated stationary state of mitochondria is not dependent on actin filaments, or NFs. Deletion of the "axon-sorting domain" (ASP) in syntaphilin causes a uniform association with all mitochondria in axons and dendrites, suggesting that syntaphilin may use this domain for axonal localization independent of its mitochondrial targeting (Kang et al. 2008).

Syntaphilin KO mice are viable, fertile, and morphologically normal (Kang et al. 2008). Consistent with the effects of overexpression, loss of syntaphilin decreased the density of mitochondria in axons, but also increased the motility of mitochondria. The mechanism underlying the latter, likely to be an independent effect is not clear. Notably, deletion of syntaphilin reduces only slightly the number of presynaptic terminals containing mitochondria in the hippocampus (Kang et al. 2008). Hence, it is probably controlling long-term mitochondrial retention at nonsynaptic sites within axons. Given that syntaphilin is not homologous to any known MT-binding protein, it likely acts as a docking receptor through a unique interaction with MTs (Kang et al. 2008). Syntaphilin-mediated docking of mitochondria could be initiated by a number of signals including the Ca^{2+}-induced arrest of mitochondrial transport that is mediated by Miro proteins. However, there is currently no evidence that links these pathways. In addition, it is not clear whether syntaphilin-mediated docking of mitochondria is a specialization of vertebrates since there are no structurally or functionally related proteins in invertebrates.

6.3 Nerve Growth Factor-Mediated Docking of Mitochondria

A question that remains controversial is whether specific mitochondrial target sites use only short-range signals to terminate the transport of mitochondria as they pass by, or also employ long-range signals that attract mitochondria, perhaps even from the cell body.

The best evidence for a potentially long-range, chemo-attractive signal potentially attracting mitochondria to growth cones, and perhaps mature synapses is NGF (Fig. 5). In cultured chick sensory neurons, mitochondrial transport responds to growth cone activity that controls the transition to a stationary state, and also

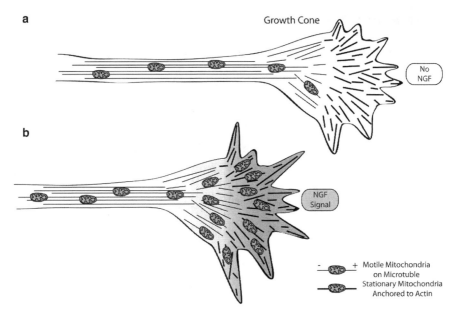

Fig. 5 Nerve growth factor (NGF)-mediated docking of mitochondria. Transport and docking of mitochondria in axons of cultured cells depend on the state of the growth cone. (**a**). In inactive axons, mitochondria travel does not significantly enrich at the axon terminal. (**b**). In the presence of NGF, mitochondrial transport increases towards the origin of NGF signaling, and mitochondrial transport away from the NGF source decreases, or is arrested in the presence, but not absence of actin filaments. Adapted from (Chada and Hollenbeck 2004)

modulates antero- and retrograde transport activity (Morris and Hollenbeck 1993). Positioning NGF-coated beads close to axons of cultured neurons increases mitochondrial movement into the region of focal NGF application, but decreases, and in many cases, arrests movement out of the region (Chada and Hollenbeck 2003, 2004). A similar phenomenon was observed in *Xenopus* spinal neurons (Lee and Peng 2006). This suggests that NGF may act over some distance to modulate mitochondrial transport, and essentially "attract" mitochondria.

The increase of mitochondrial movements towards NGF-coated beads, and subsequent arrest of movements requires the TrkA receptor and phosphoinositol 3 (PI3) kinase-mediated signaling; neither of which is necessary for mitochondrial movement per se (Chada and Hollenbeck 2003, 2004). Consistently, expression of pleckstrin homology (PH) domains, which bind phosphatidylinositols, increases anterograde mitochondrial movements and decreases retrograde movements without altering velocities, the amount of motors linked to cargos, and motor binding to tracks (De Vos et al. 2003). In addition, the ability of NGF/PI3 kinase signaling to arrest mitochondrial transport requires the presence of actin filaments (Chada and Hollenbeck 2004), suggesting that mitochondria are immobilized on actin filaments (Fig. 5). However, it remains unclear how mitochondria become dissociated from their MT motors, how mitochondria are immobilized on actin filaments, and whether growth factor signals also dock mitochondria at mature synapses.

6.4 NMDA-Induced Arrest of Mitochondrial Transport in Dendrites

Most dendritic mitochondria are localized at, or in the vicinity of postsynaptic sites (Fig. 1f) in an activity-dependent manner (Overly et al. 1996; Chang et al. 2006). Consistent with EM studies (Adams and Jones 1982; Kageyama and Wong-Riley 1984; Shepherd and Greer 1988; Cameron et al. 1991; Peters et al. 1991; Chicurel and Harris 1992), more than 90% of mitochondria are confined to the shaft of the dendrite in cultured neurons (Li et al. 2004; Sung et al. 2008). Less than 10% of dendritic protrusions (e.g., dendritic spines and filopodia) of mature cultured hippocampal neurons contain mitochondria (Li et al. 2004; Sung et al. 2008).

Extension or movement of mitochondria into dendritic protrusions correlates with the development and morphological maturation of spines. Molecular manipulations of mitochondrial fission/fusion proteins that reduce mitochondrial content in dendrites reduce the number of synapses and dendritic spines; whereas, increasing mitochondrial content or activity enhances the number and plasticity of spines and synapses. Hence, the dendritic distribution of mitochondria is critical and limiting for the support of synapses. Reciprocally, synaptic activity modulates the distribution, motility and fusion/fission balance of mitochondria in dendrites (Rintoul et al. 2003; Li et al. 2004). In addition, the number of mitochondria in dendritic protrusions increases as spines mature, either developmentally, or in response to local repetitive stimulation in an NMDA receptor-dependent manner (Li et al. 2004). Synaptic stimulation by excitotoxic doses of glutamate arrests mitochondrial movements and disrupts the typically elongated shape of most mitochondria in dendrites (Rintoul et al. 2003).

NMDA-induced inhibition of mitochondrial motility in dendrites is mediated in the mitochondrial Ca^{2+} sensor Miro (Fig. 4). Overexpression of mutant Miro protein that is unable to bind Ca^{2+} renders mitochondria "immune" to the usual arrest of mitochondrial motility upon NMDA receptor-dependent elevation of intracellular Ca^{2+} (MacAskill et al. 2009b; Wang and Schwarz 2009). Without the typically occurring NMDA-dependent arrest of mitochondrial motility, the survival rate of neurons upon NMDA-induced glutamate excitotoxicity is significantly reduced (Wang and Schwarz 2009).

The NMDA- and Ca^{2+}-dependent arrest of MT-based mitochondrial transport close to spines together with subsequent fission is likely required for translocating mitochondria into spines (Li et al. 2004; Sung et al. 2008). After the arrest of mitochondrial motility, the activity-dependent translocation of mitochondria into spines probably requires actin and WAVE1 (WASP-family verprolin homologous protein 1). WAVE1 is a key regulator of actin polymerization and dendritic spine morphology (Kim et al. 2006). WAVE1 is associated with the outer mitochondrial membrane (Danial et al. 2003). RNAi-induced knockdown of WAVE1 decreases the number of mitochondria in dendritic protrusions under basal conditions, and prevents the activity-dependent translocation of mitochondria into spines (Sung et al. 2008).

WAVE1's ability to regulate Arp2/3-dependent actin polymerization is inhibited by phosphorylation through cyclin-dependent kinase 5 (Cdk5) (Kim et al. 2006). In mature neurons, WAVE1 is highly phosphorylated under basal conditions, and, consequently, largely inactive (Sung et al. 2008). Repetitive stimulation and subsequent NMDA receptor activation causes dephosphorylation of WAVE1 at Cdk5 phosporylation sites, and is probably necessary for mitochondrial translocation into dendritic protrusions. Accordingly, WAVE1-dependent translocation of mitochondria into dendritic protrusions is likely to be mediated by Arp2/3-dependent actin polymerization (Sung et al. 2008). Since mitochondria are rarely found in dendritic spines, it is also likely that mitochondria move into spines only for a relatively short time.

7 Perspective

In the past years, major advances have been made in understanding the molecular machinery that dynamically drives and regulates the transport of mitochondria in neurons. Given the advances in imaging mitochondrial transport, one may expect the full array of the mitochondrial transport machinery to be identified in the near future. However, a comprehensive understanding of mitochondrial transport in neurons will also require a broader view: the life cycle of mitochondria in long-lived neurons. For example, how do mitochondria manage to remain functional during their prolonged excursions into axons and dendrites? It seems unlikely that mitochondria can survive in neuronal process without proper logistics; or, do they? Currently, we have only a limited appreciation of mitochondrial protein logistics in axons and dendrites. In addition, little is known about the value of mitochondrial fusion and fission dynamics in neuronal process, or how "aging" of mitochondria is managed in neuronal process. Recent advances suggest that mitochondrial transport may be tightly interwoven with a quality control mechanism that underlies mitochondrial fusion and fission governing mitochondrial turnover. Understanding the relationship between mitochondrial transport, dynamics and biogenesis in neurons will likely be critical to counteract the neuronal pathologies of the numerous disorders that are associated with mitochondria.

References

Adams I, Jones DG (1982) Quantitative ultrastructural changes in rat cortical synapses during early-, mid- and late-adulthood. Brain Res 239:349–363

Baas PW, Deitch JS, Black MM, Banker GA (1988) Polarity orientation of microtubules in hippocampal neurons: uniformity in the axon and nonuniformity in the dendrite. Proc Natl Acad Sci USA 85:8335–8339

Babcock DF, Hille B (1998) Mitochondrial oversight of cellular Ca2+ signaling. Curr Opin Neurobiol 8:398–404

Beck M, Brickley K, Wilkinson HL, Sharma S, Smith M, Chazot PL, Pollard S, Stephenson FA (2002) Identification, molecular cloning, and characterization of a novel GABAA receptor-associated protein, GRIF-1. J Biol Chem 277:30079–30090

Belles B, Hescheler J, Trube G (1987) Changes of membrane currents in cardiac cells induced by long whole-cell recordings and tolbutamide. Pflugers Arch 409:582–588

Beltran-Parrazal L, Lopez-Valdes HE, Brennan KC, Diaz-Munoz M, de Vellis J, Charles AC (2006) Mitochondrial transport in processes of cortical neurons is independent of intracellular calcium. Am J Physiol Cell Physiol 291:C1193–1197

Benshalom G, Reese TS (1985) Ultrastructural observations on the cytoarchitecture of axons processed by rapid-freezing and freeze-substitution. J Neurocytol 14:943–960

Bowman AB, Patel-King RS, Benashski SE, McCaffery JM, Goldstein LS, King SM (1999) *Drosophila* roadblock and Chlamydomonas LC7: a conserved family of dynein-associated proteins involved in axonal transport, flagellar motility, and mitosis. J Cell Biol 146:165–180

Brady ST, Pfister KK, Bloom GS (1990) A monoclonal antibody against kinesin inhibits both anterograde and retrograde fast axonal transport in squid axoplasm. Proc Natl Acad Sci USA 87:1061–1065

Brickley K, Smith MJ, Beck M, Stephenson FA (2005) GRIF-1 and OIP106, members of a novel gene family of coiled-coil domain proteins: association in vivo and in vitro with kinesin. J Biol Chem 280:14723–14732

Bridgman PC (2009) Myosin motor proteins in the cell biology of axons and other neuronal compartments. Results Probl Cell Differ . doi:10.1007/400_2009_10

Brooks AS, Bertoli-Avella AM, Burzynski GM, Breedveld GJ, Osinga J, Boven LG, Hurst JA, Mancini GM, Lequin MH, de Coo RF et al (2005) Homozygous nonsense mutations in KIAA1279 are associated with malformations of the central and enteric nervous systems. Am J Hum Genet 77:120–126

Brough D, Schell MJ, Irvine RF (2005) Agonist-induced regulation of mitochondrial and endoplasmic reticulum motility. Biochem J 392:291–297

Cai Q, Gerwin C, Sheng ZH (2005) Syntabulin-mediated anterograde transport of mitochondria along neuronal processes. J Cell Biol 170:959–969

Cameron HA, Kaliszewski CK, Greer CA (1991) Organization of mitochondria in olfactory bulb granule cell dendritic spines. Synapse 8:107–118

Chada SR, Hollenbeck PJ (2003) Mitochondrial movement and positioning in axons: the role of growth factor signaling. J Exp Biol 206:1985–1992

Chada SR, Hollenbeck PJ (2004) Nerve growth factor signaling regulates motility and docking of axonal mitochondria. Curr Biol 14:1272–1276

Chan DC (2006) Mitochondrial fusion and fission in mammals. Annu Rev Cell Dev Biol 22:79–99

Chang DT, Reynolds IJ (2006) Mitochondrial trafficking and morphology in healthy and injured neurons. Prog Neurobiol 80:241–268

Chang DT, Honick AS, Reynolds IJ (2006) Mitochondrial trafficking to synapses in cultured primary cortical neurons. J Neurosci 26:7035–7045

Chicurel ME, Harris KM (1992) Three-dimensional analysis of the structure and composition of CA3 branched dendritic spines and their synaptic relationships with mossy fiber boutons in the rat hippocampus. J Comp Neurol 325:169–182

Colin E, Zala D, Liot G, Rangone H, Borrell-Pages M, Li XJ, Saudou F, Humbert S (2008) Huntingtin phosphorylation acts as a molecular switch for anterograde/retrograde transport in neurons. EMBO J 27:2124–2134

Cox RT, Spradling AC (2006) Milton controls the early acquisition of mitochondria by Drosophila oocytes. Development 133:3371–3377

Danial NN, Gramm CF, Scorrano L, Zhang CY, Krauss S, Ranger AM, Datta SR, Greenberg ME, Licklider LJ, Lowell BB et al (2003) BAD and glucokinase reside in a mitochondrial complex that integrates glycolysis and apoptosis. Nature 424:952–956

Das S, Boczan J, Gerwin C, Zald PB, Sheng ZH (2003) Regional and developmental regulation of syntaphilin expression in the brain: a candidate molecular element of synaptic functional differentiation. Brain Res Mol Brain Res 116:38–49

De Vos K, Goossens V, Boone E, Vercammen D, Vancompernolle K, Vandenabeele P, Haegeman G, Fiers W, Grooten J (1998) The 55-kDa tumor necrosis factor receptor induces clustering of mitochondria through its membrane-proximal region. J Biol Chem 273:9673–9680

De Vos K, Severin F, Van Herreweghe F, Vancompernolle K, Goossens V, Hyman A, Grooten J (2000) Tumor necrosis factor induces hyperphosphorylation of kinesin light chain and inhibits kinesin-mediated transport of mitochondria. J Cell Biol 149:1207–1214

De Vos KJ, Sable J, Miller KE, Sheetz MP (2003) Expression of phosphatidylinositol (4, 5) bisphosphate-specific pleckstrin homology domains alters direction but not the level of axonal transport of mitochondria. Mol Biol Cell 14:3636–3649

Deacon SW, Serpinskaya AS, Vaughan PS, Lopez Fanarraga M, Vernos I, Vaughan KT, Gelfand VI (2003) Dynactin is required for bidirectional organelle transport. J Cell Biol 160:297–301

Denton RM (2009) Regulation of mitochondrial dehydrogenases by calcium ions. Biochim Biophys Acta doi: S0005-2728(09)00012-7 [pii] 10.1016/j.bbabio.2009.01.00, (in press)

Detmer SA, Chan DC (2007) Functions and dysfunctions of mitochondrial dynamics. Nat Rev Mol Cell Biol 8:870–879

Echeverri CJ, Paschal BM, Vaughan KT, Vallee RB (1996) Molecular characterization of the 50-kD subunit of dynactin reveals function for the complex in chromosome alignment and spindle organization during mitosis. J Cell Biol 132:617–633

Fagarasanu A, Rachubinski RA (2007) Orchestrating organelle inheritance in *Saccharomyces cerevisiae*. Curr Opin Microbiol 10:528–538

Fahim MA, Robbins N (1982) Ultrastructural studies of young and old mouse neuromuscular junctions. J Neurocytol 11:641–656

Falnikar A, Baas PW (2009) Critical roles for microtubules in axonal development and disease. Results Probl Cell Differ . doi:10.1007/400_2009_2

Fichera M, Lo Giudice M, Falco M, Sturnio M, Amata S, Calabrese O, Bigoni S, Calzolari E, Neri M (2004) Evidence of kinesin heavy chain (KIF5A) involvement in pure hereditary spastic paraplegia. Neurology 63:1108–1110

Finsterer J (2006) Central nervous system manifestations of mitochondrial disorders. Acta Neurol Scand 114:217–238

Fransson A, Ruusala A, Aspenstrom P (2003) Atypical Rho GTPases have roles in mitochondrial homeostasis and apoptosis. J Biol Chem 278:6495–6502

Fransson S, Ruusala A, Aspenstrom P (2006) The atypical Rho GTPases Miro-1 and Miro-2 have essential roles in mitochondrial trafficking. Bioch Biophys Res Com 344:500–510

Frederick RL, McCaffery JM, Cunningham KW, Okamoto K, Shaw JM (2004) Yeast Miro GTPase, Gem1p, regulates mitochondrial morphology via a novel pathway. J Cell Biol 167:87–98

Frederick RL, Okamoto K, Shaw JM (2008) Multiple pathways influence mitochondrial inheritance in budding yeast. Genetics 178:825–837

Gilbert SL, Zhang L, Forster ML, Anderson JR, Iwase T, Soliven B, Donahue LR, Sweet HO, Bronson RT, Davisson MT et al (2006) Trak1 mutation disrupts GABA(A) receptor homeostasis in hypertonic mice. Nat Genet 38:245–250

Gindhart JG Jr, Desai CJ, Beushausen S, Zinn K, Goldstein LS (1998) Kinesin light chains are essential for axonal transport in *Drosophila*. J Cell Biol 141:443–454

Giot L, Bader JS, Brouwer C, Chaudhuri A, Kuang B, Li Y, Hao YL, Ooi CE, Godwin B, Vitols E et al (2003) A protein interaction map of Drosophila melanogaster. Science 302:1727–1736

Glater EE, Megeath LJ, Stowers RS, Schwarz TL (2006) Axonal transport of mitochondria requires milton to recruit kinesin heavy chain and is light chain independent. J Cell Biol 173:545–557

Goldstein LS (2001) Kinesin molecular motors: transport pathways, receptors, and human disease. Proc Natl Acad Sci USA 98:6999–7003

Goldstein LS (2003) Do disorders of movement cause movement disorders and dementia? Neuron 40:415–425

Gorska-Andrzejak J, Stowers RS, Borycz J, Kostyleva R, Schwarz TL, Meinertzhagen IA (2003) Mitochondria are redistributed in Drosophila photoreceptors lacking milton, a kinesin-associated protein. J Comp Neurol 463:372–388

Gross SP (2003) Dynactin: coordinating motors with opposite inclinations. Curr Biol 13:15

Gross SP, Welte MA, Block SM, Wieschaus EF (2000) Dynein-mediated cargo transport in vivo. A switch controls travel distance. J Biol 148:945–956

Gross SP, Tuma MC, Deacon SW, Serpinskaya AS, Reilein AR, Gelfand VI (2002a) Interactions and regulation of molecular motors in *Xenopus* melanophores. J Cell Biol 156:855–865

Gross SP, Welte MA, Block SM, Wieschaus EF (2002b) Coordination of opposite-polarity microtubule motors. J Cell Biol 156:715–724

Gross SP, Guo Y, Martinez JE, Welte MA (2003) A determinant for directionality of organelle transport in *Drosophila* embryos. Curr Biol 13:1660–1668

Gross SP, Vershinin M, Shubeita GT (2007) Cargo transport: two motors are sometimes better than one. Curr Biol 17:R478–486

Guo X, Macleod GT, Wellington A, Hu F, Panchumarthi S, Schoenfield M, Marin L, Charlton MP, Atwood HL, Zinsmaier KE (2005) The GTPase dMiro is required for axonal transport of mitochondria to Drosophila synapses. Neuron 47:379–393

Habermann A, Schroer TA, Griffiths G, Burkhardt JK (2001) Immunolocalization of cytoplasmic dynein and dynactin subunits in cultured macrophages: enrichment on early endocytic organelles. J Cell Sci 114:229–240

Haghnia M, Cavalli V, Shah SB, Schimmelpfeng K, Brusch R, Yang G, Herrera C, Pilling A, Goldstein LS (2007) Dynactin is required for coordinated bidirectional motility, but not for dynein membrane attachment. Mol Biol Cell 18:2081–2089

Hajnoczky G, Robb-Gaspers LD, Seitz MB, Thomas AP (1995) Decoding of cytosolic calcium oscillations in the mitochondria. Cell 82:415–424

He Y, Francis F, Myers KA, Yu W, Black MM, Baas PW (2005) Role of cytoplasmic dynein in the axonal transport of microtubules and neurofilaments. J Cell Biol 168:697–703

Hirokawa N (1982) Cross-linker system between neurofilaments, microtubules, and membranous organelles in frog axons revealed by the quick-freeze, deep-etching method. J Cell Biol 94:129–142

Hirokawa N, Yorifuji H (1986) Cytoskeletal architecture of reactivated crayfish axons, with special reference to crossbridges among microtubules and between microtubules and membrane organelles. Cell Motil Cytoskeleton 6:458–468

Hirokawa N, Takemura R (2005) Molecular motors and mechanisms of directional transport in neurons. Nat Rev Neurosci 6:201–214

Hirokawa N, Noda Y (2008) Intracellular transport and kinesin superfamily proteins, KIFs: structure, function, and dynamics. Physiol Rev 88:1089–1118

Hollenbeck PJ (1996) The pattern and mechanisms of mitochondrial transport in axons. Front Biosci 1:d91–d102

Hook P, Vallee RB (2006) The dynein family at a glance. J Cell Sci 119:4369–4371

Hubley MJ, Locke BR, Moerland TS (1996) The effects of temperature, pH, and magnesium on the diffusion coefficient of ATP in solutions of physiological ionic strength. Biochim Biophys Acta 1291:115–121

Hurd DD, Saxton WM (1996) Kinesin mutations cause motor neuron disease phenotypes by disrupting fast axonal transport in *Drosophila*. Genetics 144:1075–1085

Iyer SP, Akimoto Y, Hart GW (2003) Identification and cloning of a novel family of coiled-coil domain proteins that interact with O-GlcNAc transferase. J Biol Chem 278:5399–5409

Jung D, Filliol D, Miehe M, Rendon A (1993) Interaction of brain mitochondria with microtubules reconstituted from brain tubulin and MAP2 or TAU. Cell Motil Cytoskeleton 24:245–255

Kageyama GH, Wong-Riley MT (1984) The histochemical localization of cytochrome oxidase in the retina and lateral geniculate nucleus of the ferret, cat, and monkey, with particular reference to retinal mosaics and ON/OFF-center visual channels. J Neurosci 4:2445–2459

Kanai Y, Okada Y, Tanaka Y, Harada A, Terada S, Hirokawa N (2000) KIF5C, a novel neuronal kinesin enriched in motor neurons. J Neurosci 20:6374–6384

Kang JS, Tian JH, Pan PY, Zald P, Li C, Deng C, Sheng ZH (2008) Docking of axonal mitochondria by syntaphilin controls their mobility and affects short-term facilitation. Cell 132:137–148

Kann O, Kovacs R (2007) Mitochondria and neuronal activity. Am J Physiol Cell Physiol 292:C641–657

Kim Y, Sung JY, Ceglia I, Lee KW, Ahn JH, Halford JM, Kim AM, Kwak SP, Park JB, Ho Ryu S et al (2006) Phosphorylation of WAVE1 regulates actin polymerization and dendritic spine morphology. Nature 442:814–817

King SJ, Schroer TA (2000) Dynactin increases the processivity of the cytoplasmic dynein motor. Nat Cell Biol 2:20–24

King MJ, Atwood HL, Govind CK (1996) Structural features of crayfish phasic and tonic neuromuscular terminals. J Comp Neurol 372:618–626

Koushika SP, Schaefer AM, Vincent R, Willis JH, Bowerman B, Nonet ML (2004) Mutations in *Caenorhabditis elegans* cytoplasmic dynein components reveal specificity of neuronal retrograde cargo. J Neurosci 24:3907–3916

Krendel M, Sgourdas G, Bonder EM (1998) Disassembly of actin filaments leads to increased rate and frequency of mitochondrial movement along microtubules. Cell Motil Cytoskeleton 40:368–378

Kumar J, Yu H, Sheetz MP (1995) Kinectin, an essential anchor for kinesin-driven vesicle motility. Science 267:1834–1837

Kural C, Kim H, Syed S, Goshima G, Gelfand VI, Selvin PR (2005) Kinesin and dynein move a peroxisome in vivo: a tug-of-war or coordinated movement? Science 308:1469–1472

Lai C, Lin X, Chandran J, Shim H, Yang WJ, Cai H (2007) The G59S mutation in p150(glued) causes dysfunction of dynactin in mice. J Neurosci 27:13982–13990

LaMonte BH, Wallace KE, Holloway BA, Shelly SS, Ascano J, Tokito M, Van Winkle T, Howland DS, Holzbaur EL (2002) Disruption of dynein/dynactin inhibits axonal transport in motor neurons causing late-onset progressive degeneration. Neuron 34:715–727

Langford GM (2002) Myosin-V, a versatile motor for short-range vesicle transport. Traffic 3:859–865

Lao G, Scheuss V, Gerwin CM, Su Q, Mochida S, Rettig J, Sheng ZH (2000) Syntaphilin: a syntaxin-1 clamp that controls SNARE assembly. Neuron 25:191–201

Lee CW, Peng HB (2006) Mitochondrial clustering at the vertebrate neuromuscular junction during presynaptic differentiation. J Neurobiol 66:522–536

Leterrier JF, Rusakov DA, Linden M (1994a) Statistical analysis of the surface distribution of microtubule-associated proteins (MAPs) bound in vitro to rat brain mitochondria and labelled by 10 nm gold-coupled antibodies. Bull Assoc Anat (Nancy) 78:47–51

Leterrier JF, Rusakov DA, Nelson BD, Linden M (1994b) Interactions between brain mitochondria and cytoskeleton: evidence for specialized outer membrane domains involved in the association of cytoskeleton-associated proteins to mitochondria in situ and in vitro. Microsc Res Tech 27:233–261

Levy JR, Holzbaur EL (2006) Cytoplasmic dynein/dynactin function and dysfunction in motor neurons. Int J Dev Neurosci 24:103–111

Li SH, Gutekunst CA, Hersch SM, Li XJ (1998) Interaction of huntingtin-associated protein with dynactin P150Glued. J Neurosci 18:1261–1269

Li Z, Okamoto K, Hayashi Y, Sheng M (2004) The importance of dentritic mitochondria in the morphogenesis and plasticity of spines and synapses. Cell 119:873–887

Ligon LA, Steward O (2000a) Movement of mitochondria in the axons and dendrites of cultured hippocampal neurons. J Comp Neurol 427:340–350

Ligon LA, Steward O (2000b) Role of microtubules and actin filaments in the movement of mitochondria in the axons and dendrites of cultured hippocampal neurons. J Comp Neurol 427:351–361

Ligon LA, Tokito M, Finklestein JM, Grossman FE, Holzbaur EL (2004) A direct interaction between cytoplasmic dynein and kinesin I may coordinate motor activity. J Biol Chem 279:19201–19208

Lin MT, Beal MF (2006) Mitochondrial dysfunction and oxidative stress in neurodegenerative diseases. Nature 443:787–795

Linden M, Nelson BD, Leterrier JF (1989a) The specific binding of the microtubule-associated protein 2 (MAP2) to the outer membrane of rat brain mitochondria. Biochem J 261:167–173

Linden M, Nelson BD, Loncar D, Leterrier JF (1989b) Studies on the interaction between mitochondria and the cytoskeleton. J Bioenerg Biomembr 21:507–518

Louie K, Russo GJ, Salkoff DB, Wellington A, Zinsmaier KE (2008) Effects of imaging conditions on mitochondrial transport and length in larval motor axons of *Drosophila*. Comp Biochem Physiol A Mol Integr Physiol 151:159–172

Lyons DA, Naylor SG, Mercurio S, Dominguez C, Talbot WS (2008) KBP is essential for axonal structure, outgrowth and maintenance in zebrafish, providing insight into the cellular basis of Goldberg-Shprintzen syndrome. Development 135:599–608

MacAskill AF, Brickley K, Stephenson FA, Kittler JT (2009a) GTPase dependent recruitment of Grif-1 by Miro1 regulates mitochondrial trafficking in hippocampal neurons. Mol Cell Neurosci 40:301–312

MacAskill AF, Rinholm JE, Twelvetrees AE, Arancibia-Carcamo IL, Muir J, Fransson A, Aspenstrom P, Attwell D, Kittler JT (2009b) Miro1 is a calcium sensor for glutamate receptor-dependent localization of mitochondria at synapses. Neuron 61:541–555

Mallik R, Petrov D, Lex SA, King SJ, Gross SP (2005) Building complexity: an in vitro study of cytoplasmic dynein with in vivo implications. Curr Biol 15:2075–2085

Martin M, Iyadurai SJ, Gassman A, Gindhart JG Jr, Hays TS, Saxton WM (1999) Cytoplasmic dynein, the dynactin complex, and kinesin are interdependent and essential for fast axonal transport. Mol Biol Cell 10:3717–3728

Mather WH, Fox RF (2006) Kinesin's biased stepping mechanism: amplification of neck linker zippering. Biophys J 91:2416–2426

Mattson MP (2007) Mitochondrial regulation of neuronal plasticity. Neurochem Res 32:707–715

Mattson MP, Gleichmann M, Cheng A (2008) Mitochondria in neuroplasticity and neurological disorders. Neuron 60:748–766

Miller KG, Sheetz MP (2004) Axonal mitochondrial transport and potential are correlated. J Cell Sci 117:2791–2804

Mironov SL (2006) Spontaneous and evoked neuronal activities regulate movements of single neuronal mitochondria. Synapse 59:403–411

Misgeld T, Kerschensteiner M, Bareyre FM, Burgess RW, Lichtman JW (2007) Imaging axonal transport of mitochondria in vivo. Nat Methods 4:559–561

Mok H, Shin H, Kim S, Lee JR, Yoon J, Kim E (2002) Association of the kinesin superfamily motor protein KIF1Balpha with postsynaptic density-95 (PSD-95), synapse-associated protein-97, and synaptic scaffolding molecule PSD-95/discs large/zona occludens-1 proteins. J Neurosci 22:5253–5258

Morris RL, Hollenbeck PJ (1993) The regulation of bidirectional mitochondrial transport is coordinated with axonal outgrowth. J Cell Sci 104:917–927

Morris RL, Hollenbeck PJ (1995) Axonal transport of mitochondria along microtubules and F- actin in living vertebrate neurons. J Cell Biol 131:1315–1326

Muller MJ, Klumpp S, Lipowsky R (2008) Tug-of-war as a cooperative mechanism for bidirectional cargo transport by molecular motors. Proc Natl Acad Sci USA 105:4609–4614

Nan X, Sims PA, Chen P, Xie XS (2005) Observation of individual microtubule motor steps in living cells with endocytosed quantum dots. J Phys Chem B 109:24220–24224

Nangaku M, Sato-Yoshitake R, Okada Y, Noda Y, Takemura R, Yamazaki H, Hirokawa N (1994) KIF1B, a novel microtubule plus end-directed monomeric motor protein for transport of mitochondria. Cell 79:1209–1220

Ong LL, Lim AP, Er CP, Kuznetsov A, Yu H (2000) Kinectin-kinesin binding domains and their effects on organelle motility. J Biol Chem 275:32854–32860

Overly CC, Rieff HI, Hollenbeck PJ (1996) Organelle motility and metabolism in axons vs dendrites of cultured hippocampal neurons. J Cell Sci 109(Pt 5):971–980

Pannese E, Ledda M (1991) Ribosomes in myelinated axons of the rabbit spinal ganglion neurons. J Submicrosc Cytol Pathol 23:33–38

Pannese E, Procacci P, Ledda M, Arcidiacono G, Frattola D, Rigamonti L (1986) Association between microtubules and mitochondria in myelinated axons of Lacerta muralis. A quantitative analysis. Cell Tissue Res 245:1–8

Peters A, Palay S, Webster H (1991) The fine structure of the nervous system: the neurons and supporting cells. Oxford University Press, New York

Pilling AD, Horiuchi D, Lively CM, Saxton WM (2006) Kinesin-1 and Dynein are the primary motors for fast transport of mitochondria in *Drosophila* motor axons. Mol Biol Cell 17:2057–2068

Plitz T, Pfeffer K (2001) Intact lysosome transport and phagosome function despite kinectin deficiency. Mol Cell Biol 21:6044–6055

Price RL, Lasek RJ, Katz MJ (1991) Microtubules have special physical associations with smooth endoplasmic reticula and mitochondria in axons. Brain Res 540:209–216

Reeve AK, Krishnan KJ, Turnbull D (2008) Mitochondrial DNA mutations in disease, aging, and neurodegeneration. Ann N Y Acad Sci 1147:21–29

Rintoul GL, Filiano AJ, Brocard JB, Kress GJ, Reynolds IJ (2003) Glutamate decreases mitochondrial size and movement in primary forebrain neurons. J Neurosci 23:7881–7888

Rodionov V, Yi J, Kashina A, Oladipo A, Gross SP (2003) Switching between microtubule- and actin-based transport systems in melanophores is controlled by cAMP levels. Curr Biol 13:1837–1847

Ross JL, Wallace K, Shuman H, Goldman YE, Holzbaur EL (2006) Processive bidirectional motion of dynein-dynactin complexes in vitro. Nat Cell Biol 8:562–570

Rowland KC, Irby NK, Spirou GA (2000) Specialized synapse-associated structures within the calyx of Held. J Neurosci 20:9135–9144

Russo GJ, Louie K, Wellington A, Macleod GT, Hu F, Panchumarthi S, Zinsmaier KE (2009) *Drosophila* Miro is required for both anterograde and retrograde axonal mitochondrial transport. J Neurosci 29:5443–5455

Ruthel G, Hollenbeck PJ (2003) Response of mitochondrial traffic to axon determination and differential branch growth. J Neurosci 23:8618–8624

Safieddine S, Ly CD, Wang YX, Wang CY, Kachar B, Petralia RS, Wenthold RJ (2002) Ocsyn, a novel syntaxin-interacting protein enriched in the subapical region of inner hair cells. Mol Cell Neurosci 20:343–353

Salinas S, Bilsland LG, Schiavo G (2008) Molecular landmarks along the axonal route: axonal transport in health and disease. Curr Opin Cell Biol 20:445–453

Santama N, Er CP, Ong LL, Yu H (2004) Distribution and functions of kinectin isoforms. J Cell Sci 117:4537–4549

Saotome M, Safiulina D, Szabadkai G, Das S, Fransson A, Aspenstrom P, Rizzuto R, Hajnoczky G (2008) Bidirectional Ca^{2+}-dependent control of mitochondrial dynamics by the Miro GTPase. Proc Natl Acad Sci USA 105:20728–20733

Saxton WM, Hicks J, Goldstein LSB, Raff EC (1991) Kinesins heavy chain is essential for viability and neuromuscular functions in *Drosophila*, but mutants show no defects in mitosis. Cell 64:1093–1102

Scheffler IE (2008) Mitochondria, 2nd edn. J. Wiley and Sons, Inc., Hoboken, New Jersey

Schroer TA (2004) Dynactin. Annu Rev Cell Dev Biol 20:759–779

Sheetz MP, Dai J (1996) Modulation of membrane dynamics and cell motility by membrane tension. Trends Cell Biol 6:85–89

Shepherd GM, Greer CA (1988) In: Lasek RS, Black MM (eds) Intrinsic determinants of neuronal form and function. Liss, New York, pp 245–262

Shepherd GM, Harris KM (1998) Three-dimensional structure and composition of CA3—>CA1 axons in rat hippocampal slices: implications for presynaptic connectivity and compartmentalization. J Neurosci 18:8300–8310

Smith DS, Jarlfors U, Cayer ML (1977) Structural cross-bridges between microtubules and mitochondria in central axons of an insect (Periplaneta americana). J Cell Sci 27:255–272

Smith GA, Gross SP, Enquist LW (2001) Herpesviruses use bidirectional fast-axonal transport to spread in sensory neurons. Proc Natl Acad Sci USA 98:3466–3470

Stowers RS, Megeath LJ, Gorska-Andrzejak J, Meinertzhagen IA, Schwarz TL (2002) Axonal transport of mitochondria to synapses depends on Milton, a novel *Drosophila* protein. Neuron 36:1063–1077

Su Q, Cai Q, Gerwin C, Smith CL, Sheng ZH (2004) Syntabulin is a microtubule-associated protein implicated in syntaxin transport in neurons. Nat Cell Biol 6:941–953

Sung JY, Engmann O, Teylan MA, Nairn AC, Greengard P, Kim Y (2008) WAVE1 controls neuronal activity-induced mitochondrial distribution in dendritic spines. Proc Natl Acad Sci USA 105:3112–3116

Suomalainen M, Nakano MY, Keller S, Boucke K, Stidwill RP, Greber UF (1999) Microtubule-dependent plus- and minus end-directed motilities are competing processes for nuclear targeting of adenovirus. J Cell Biol 144:657–672

Tanaka Y, Kanai Y, Okada Y, Nonaka S, Takeda S, Harada A, Hirokawa N (1998) Targeted disruption of mouse conventional kinesin heavy chain, kif5B, results in abnormal perinuclear clustering of mitochondria. Cell 93:1147–1158

Toyoshima I, Yu H, Steuer ER, Sheetz MP (1992) Kinectin, a major kinesin-binding protein on ER. J Cell Biol 118:1121–1131

Twig G, Hyde B, Shirihai OS (2008) Mitochondrial fusion, fission and autophagy as a quality control axis: the bioenergetic view. Biochim Biophys Acta 1777:1092–1097

Vale RD (2003) The molecular toolbox for intracellular transport. Cell 112:467–480

Valetti C, Wetzel DM, Schrader M, Hasbani MJ, Gill SR, Kreis TE, Schroer TA (1999) Role of dynactin in endocytic traffic: effects of dynamitin overexpression and colocalization with CLIP-170. Mol Biol Cell 10:4107–4120

Verburg J, Hollenbeck PJ (2008) Mitochondrial membrane potential in axons increases with local nerve growth factor or semaphorin signaling. J Neurosci 28:8306–8315

Wagner OI, Lifshitz J, Janmey PA, Linden M, McIntosh TK, Leterrier JF (2003) Mechanisms of mitochondria-neurofilament interactions. J Neurosci 23:9046–9058

Wang Z, Sheetz MP (1999) One-dimensional diffusion on microtubules of particles coated with cytoplasmic dynein and immunoglobulins. Cell Struct Funct 24:373–383

Wang X, Schwarz TL (2009) The mechanism of Ca^{2+}-dependent regulation of kinesin- mediated mitochondrial motility. Cell 136:163–174

Waterman-Storer CM, Karki SB, Kuznetsov SA, Tabb JS, Weiss DG, Langford GM, Holzbaur EL (1997) The interaction between cytoplasmic dynein and dynactin is required for fast axonal transport. Proc Natl Acad Sci USA 94:12180–12185

Wattenberg B, Lithgow T (2001) Targeting of C-terminal (tail)-anchored proteins: understanding how cytoplasmic activities are anchored to intracellular membranes. Traffic 2:66–71

Welte MA (2004) Bidirectional transport along microtubules. Curr Biol 14:13

Welte MA, Gross SP, Postner M, Block SM, Wieschaus EF (1998) Developmental regulation of vesicle transport in *Drosophila* embryos: forces and kinetics. Cell 92:547–557

Welte MA, Cermelli S, Griner J, Viera A, Guo Y, Kim DH, Gindhart JG, Gross SP (2005) Regulation of lipid-droplet transport by the perilipin homolog LSD2. Curr Biol 15:1266–1275

Wozniak MJ, Melzer M, Dorner C, Haring HU, Lammers R (2005) The novel protein KBP regulates mitochondria localization by interaction with a kinesin-like protein. BMC Cell Biol 6:35

Xia CH, Roberts EA, Her LS, Liu X, Williams DS, Cleveland DW, Goldstein LS (2003) Abnormal neurofilament transport caused by targeted disruption of neuronal kinesin heavy chain KIF5A. J Cell Biol 161:55–66

Yi M, Weaver D, Hajnoczky G (2004) Control of mitochondrial motility and distribution by the calcium signal: a homeostatic circuit. J Cell Biol 167:661–672

Yildiz A, Tomishige M, Vale RD, Selvin PR (2004) Kinesin walks hand-over-hand. Science 303:676–678

Zhao C, Takita J, Tanaka Y, Setou M, Nakagawa T, Takeda S, Yang HW, Terada S, Nakata T, Takei Y et al (2001) Charcot-Marie-Tooth disease type 2A caused by mutation in a microtubule motor KIF1Bbeta. Cell 105:587–597

# NGF Uptake and Retrograde Signaling Mechanisms in Sympathetic Neurons in Compartmented Cultures

Robert B. Campenot

Abstract Many neurons depend for their survival on retrograde signals to their cell bodies generated by nerve growth factor (NGF) or other neurotrophins at their axon terminals. Apoptosis resulting from the loss of retrograde NGF signaling contributes to the elimination of excess and misconnected neurons during development and to the death of neurons during the course of neurodegenerative diseases. Possible mechanisms of retrograde signaling include (1) retrograde transport of signaling endosomes, carrying NGF bound to activated TrkA, (2) retrograde transport of signaling molecules downstream of TrkA, and (3) retrograde propagation of a phosphorylation signal without transport of signaling molecules. Evidence is also described, which indicates that two or more retrograde signaling mechanisms exist to regulate neuronal survival, including recent evidence that withdrawal of NGF from distal axons produces a retrograde apoptotic signal, which is transported to the cell bodies, where it initiates the apoptotic program, leading to the death of the neuron.

1 Introduction

The function of the nervous system depends upon neuronal circuits composed of precisely connected nerve fibers. Many of these nerve fibers are long axons, extending over many centimeters and, in some cases, a meter or more from their cell bodies. Axons originate as motile outgrowths from the cell bodies, which actively seek and find the target cells that they innervate. This task is less formidable than it may seem at first; as the embryonic and postnatal organisms are much smaller than the adults, the distances are shorter, and trial-and-error is a more tractable process. The developmental "strategy" seems to overproduce neurons initially, and cull those that make inappropriate connections along with many of those that make appropriate

R.B. Campenot
Department of Cell Biology, University of Alberta, Edmonton, AB T6G 2H7, Canada
e-mail: bob.campenot@ualberta.ca

connections along with many of those that make appropriate connections, but are in excess of what is required to adequately innervate the targets. In this way, about 50% of the neurons initially produced are eliminated by apoptosis (Oppenheim 1991). In addition to eliminating entire neurons, the axonal connections of the surviving neurons can be modified by pruning axon branches that have made inappropriate connections along with some axon branches that have made appropriate connections, but are in excess of what is required to innervate the targets. Thus, axonal guidance, retrograde support of neuronal survival, and retrograde support of axonal survival are the three major players in constructing and preserving neuronal circuitry during development. Of these, support of neuronal survival and support of axonal survival both involve retrograde signaling from the axon terminals, and, in many cases, the retrograde signal is produced by nerve growth factor (NGF) or other neurotrophins produced in the target tissues interacting with Trk receptors (i.c., a family of transmembrane tyrosine kinases) on the axon terminals.

Neuronal and axonal survival support by neurotrophins remains functional in the adult organism. Axonal and neuronal degeneration occurring in neurodegenerative diseases, and after neurotrauma, is believed to arise partly from failure of the neurons to obtain neurotrophin-induced survival signals from the cells they innervate (Salehi et al. 2006; Bronfman et al. 2007). This could conceivably occur from the dysfunction in the retrograde signaling mechanism itself, or from loss of the ability to obtain neurotrophic signals from the target cells because the axon terminals have degenerated or are compromised by another mechanism. Thus, the mechanisms involved in retrograde neurotrophic signaling are of central importance for understanding the neurobiology of development, of neurodegenerative disease, and of neurotrauma.

2 Conceptual Framework Attendant to Retrograde Signaling

Since axons are structures that conduct signals along their length from one end to the other, their ability to transmit retrograde signals from the axon terminals to the cell body should not be surprising. The primary signaling mechanism in axons is, of course, the action potential, which is a bioelectrical signal that is based on changes in ionic permeability of the axonal membrane that propagates along the axon, and does not entail a longitudinal transfer of signaling molecules. There is no a priori reason why a chemical reaction, such as a phosphorylation signaling reaction, could not propagate along an axon without an actual translocation of the relevant phosphorylated molecules. However, the discovery over 30 years ago that [^{125}I]NGF presented at axon terminals in vivo is retrogradely transported intact along the axons to the cell bodies focused thinking on the transport of NGF as an essential component of retrograde survival signaling. The statement that "… NGF is able to exert its specific effect … once it has been brought to the cell body by retrograde axonal transport." (Hendry et al. 1974) was immediately accepted, and although the TrkA receptor was not identified as the receptor for the survival effects of NGF for another 15 years, it took only 3 years for the concept of the signaling endosome to be articulated in the literature: "If this membrane contains NGF receptors on its surface... then these surface receptors may be on the internal surface of the vesicles. These receptors would be stimulated by the NGF trapped within the vesicle...." (Hendry 1977).

Thus, over 30 years ago, the NGF retrograde signal was hypothesized to be carried by NGF internalized into signaling endosomes, where it remained bound to receptors that continued to generate signals, as the signaling endosome was carried to the cell body along the microtubule-based retrograde transport system. This hypothesis seems to be based partly on the conviction that there must be a reason to transport intact NGF all the way back to the cell bodies, other than just to degrade it. However, there is no compelling reason that a retrograde signal generated by NGF at the axon terminals could not be in the form of vesicles carrying signaling molecules that are activated downstream of the TrkA receptor, rather than carrying NGF bound to activated TrkA.

We now know a lot about the mechanism of transport along axons. It involves the movement of vesicles and membraneous organelles, powered by molecular motor proteins along microtubules that translocate in different directions, in which most kinesins move in an anterograde direction and cytoplasmic dyneins move in a retrograde direction (Vallee and Bloom 1991). A trafficking mechanism exists, therefore, to specify that a signal travels in a preferred direction. A view has long prevailed that transport of molecules along axons is a special problem, requiring an extraordinary mechanism. It is important to recognize that microtubule-based transport is the mechanism by which vesicles move in a vectorial manner through the Golgi stacks, and it occurs in all vertebrate cells. It did not evolve solely as an adaptation to long-range transport. Cells possess a very well developed microtubule-based transport system, making translocation extensive, targeted, and efficient. From this vantage point, transporting molecules from axon terminals to the cell bodies expressly for degradation is certainly plausible. In fact, as neurons and axon terminals are viewed as competing for the limited supply of NGF from the target tissues, which determines which neurons and axon terminals survive, it is possible that a major function of uptake, retrograde transport, and degradation of NGF is to restrict support for survival among competing neurons.

Thus, there are three possible categories of retrograde NGF signaling: (1) signaling endosomes carrying NGF bound to activated TrkA; (2) vesicles carrying activated signaling molecules downstream of TrkA; and (3) propagated signals not involving the transport of signaling molecules. Recently, we have obtained evidence that NGF at axon terminals supports neuronal survival by suppressing a retrograde death signal that is activated in the axons that are deprived of NGF. This finding not only controverts the prevailing classical view, but it also leads to a logical inversion of it; that is, the retrograde survival signal is actually the *absence of a retrograde death signal*. Before considering the evidence for these mechanisms, we need to describe the properties of the compartmented culture experimental system, which has been widely used in the investigations of retrograde neurotrophin signaling.

3 Properties of Compartmented Cultures

In compartmented cultures (Fig. 1), the cell bodies, the dendrites, and the proximal segments of the axons of cultured neurons reside in a separate compartment from the compartment containing the distal axon segments (Campenot 1977). This is established by plating the neurons into a proximal compartment formed by a Teflon divider seated on the surface of the culture dish with silicone grease. Axonal elongation is guided to the left and right by a series of parallel scratches made in the collagen-coated dish floor, which confines the growth cones to the strips of collagen between the scratches. The growth cones are thereby guided between the scratches to grow beneath silicone grease barriers to enter the left and right distal compartments. The distal axons of sympathetic neurons cultured from the superior cervical ganglia of newborn rats are well established in the distal compartments within 5–7 days in culture.

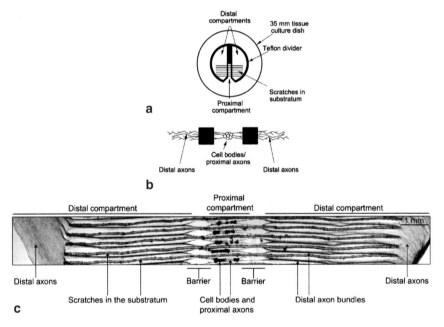

Fig. 1 Compartmented cultures. (**a**) A three-compartmented culture dish. (**b**) Neurons on a single track. (**c**) Image of 7-day-old MTT-stained rat sympathetic neurons (MacInnis and Campenot 2005) in a three-compartmented culture. Silicone grease holds a Teflon divider (**a**) in place on a collagen-coated culture dish, forming fluid barriers separating cell bodies/proximal axons in the center proximal compartment from distal axons in the left and right distal compartments (**b, c**). Axons in the proximal compartment extend on collagen tracks between parallel scratches scored on the floor of the dish and cross under the grease barrier, and into the distal compartments where they form bundles of axons extending to the end of the scratches and then fan out to the edge of the divider. Note that axons hug the scratches as they cross under the barriers (**c**), creating clear regions free of axons on each collagen track under the barriers. Each culture has 18–20 tracks of neurons, 5 of which are shown in (**c**). Scale bar: 1 mm

3.1 Proximal Compartments Always Contain Proximal Axons

Readers of our papers will note that we always refer to center compartments containing cell bodies as containing "cell bodies and proximal axons." Importantly, when treatments such as NGF are applied in the proximal compartment, proximal axons are always exposed to the NGF along with the cell bodies, and when cell extracts are prepared from proximal compartments, they always include the constituents of the proximal axons along with the cell bodies. This important qualification tends to get ignored, especially when the gel lanes prepared from cell extracts of proximal compartments are labeled with the designation "cell bodies." The importance of this distinction is clear from a consideration of cell geometry. A rat sympathetic neuron cell body is approximately a 35-μm-diameter sphere with a surface area of about 3,847 μm^2 and a volume of about 22,438 μm^3. The axon can be approximated by a 1-μm-diameter cylinder, with a surface area of 3,140 μm^2 mm^{-1} of length and a volume of 785 μm^3 mm^{-1} of length. If there is 1 mm of proximal axon per neuron in the proximal compartment, the proximal axons would contain 45% of total surface membrane and 3% of the total cytoplasm in the proximal compartment. Since axons elongate at about 1 mm per day and form branches during their initial growth in the proximal compartment, it is a reasonable assumption that there are several millimeter of proximal axon per neuron in the proximal compartment. Thus, a majority of the surface membrane in the proximal compartments likely belongs to the proximal axons, and any molecules in the axonal cytoplasm that scale with the surface membrane will also be dominantly represented. Any cytoplasmic proteins that are concentrated in axons, such as microtubules and their associated proteins, will also be strongly represented in the proximal compartments, and even cytoplasmic proteins that scale with volume would represent a significant percentage if there are several millimeters of proximal axon per neuron in the proximal compartment (dendrites are ignored in this analysis, but as dendrites are at most 50 μm long (Landis 1976), including them would not invalidate this analysis). Thus, the fraction of cellular material in proximal compartments belonging to axons and the exposure of axons to any treatments applied to the cell bodies in the proximal compartments cannot be ignored.

3.2 Proximal Compartments May Contain Distal Axons

The goal of constructing compartmented cultures for studies of retrograde signaling is to have cell bodies and only proximal axons in the proximal compartments, and all distal axons in the distal compartments. This situation is only approximately accurate as axonal growth is primarily directed away from the cell bodies along the linear tracks, but proximal axons can reverse direction, remaining within the proximal compartment, and distal axons can reverse direction in the distal compartments, reentering the proximal compartment (Campenot 1982a). To minimize axonal branching within the proximal compartments, to maximize it in the distal compartments, and to counteract reentry into the proximal compartments, a lower concentration of NGF (10 ng ml^{-1}) is provided in the proximal compartment and a higher concentration of NGF (50 ng ml^{-1}) in the distal compartments. Once distal axons are well established (5–7 days in culture), NGF is withdrawn from the proximal compartment, which terminates the growth of any distal axons remaining in the proximal compartment and causes them to degenerate (Campenot 1882a, 1982b).

In some of the works utilizing compartmented cultures reported in the literature, it is evident that the axons in the proximal compartment have escaped from the guiding influence of the scratches, and many have turned away from the barriers rather than crossing into distal compartments (e.g., Ye et al. 2003). In this case, the proximal compartment contains cell bodies, proximal axons, and the subset of distal axons that have not crossed the barriers. The distal compartments contain the subset of distal axons that have crossed the barriers. When NGF is withdrawn from the proximal compartments in cultures, in which a significant fraction of distal axons have not crossed into distal compartments, it results in the withdrawal of NGF from cell bodies, proximal axons, and a significant fraction of the distal axon branches of these neurons. This must be considered in interpreting results.

4 Mechanisms of Retrograde NGF Signaling

The evidence for the signaling endosome hypothesis of retrograde NGF signaling has been covered in several reviews (Howe and Mobley 2004; Ibanez 2007; Cosker et al. 2008). A comprehensive review of that evidence is not presented here. Rather, the issue addressed here is whether the evidence can be reasonably viewed as ruling out all other possible mechanisms of retrograde signaling.

4.1 Retrograde TrkA Phosphorylation Appears within 1 min of Applying NGF to Distal Axons

Our investigations of the retrograde transport of NGF revealed little or no detectable [^{125}I]NGF transported to the cell bodies/proximal axons of rat sympathetic neurons in compartmented cultures within 1 min after supplying [^{125}I] NGF to the distal axons, and little detectable transport occurring within 60 min. However, phosphorylated TrkA appears in the cell bodies/proximal axons within 1 min of supplying NGF to distal axons (Senger and Campenot 1997). We estimate the velocity of the retrograde transport of NGF in compartmented cultures to be in the range of 10–20 mm h^{-1} (Ure and Campenot 1997). At a retrograde transport velocity of 20 mm h^{-1}, it would take 3 minutes for a signaling endosome carrying NGF bound to phosphorylated TrkA to cross the 1-mm barrier into the proximal compartment. While the possibility cannot be ruled out that some signaling endosomes carrying [^{125}I]NGF bound to phosphorylated TrkA may have traveled faster and escaped detection, it seems more likely that the production of a signal strong enough to be detected on an immunoblot would depend on the substantial contributions of TrkA from the axons distributed in the distal compartments. Moreover, in studies in which quantum dots bearing single NGF dimers were supplied to distal axons of dorsal root ganglion neurons in compartmented cultures, it took about 40 min for the first quantum dots to appear in axons in the proximal compartments (Cui et al. 2007), suggesting that even a single NGF dimer bound to activated TrkA could not arrive within 1 min. Thus, the phosphorylated TrkA that we observe by immunoblot analysis in the cell bodies/proximal axons within 1 min of supplying NGF to the distal axons is very unlikely to represent the retrograde transport of phosphorylated TrkA from the distal axons.

Our results suggest, instead, that the retrograde phosphorylation of TrkA can occur by another mechanism, in which TrkA, already present in the cell bodies and proximal axons and not bound to NGF, is phosphorylated in response to NGF activation in the distal axons. As all retrogradely transported molecules are likely to travel at similar velocities, it is also unlikely that TrkA in the cell bodies/proximal axons was phosphorylated by a signaling molecule downstream of TrkA that was activated in the distal axons and then transported retrogradely to the cell bodies. Another possibility is that a phosphorylation signal propagates from phosphorylated TrkA bound to NGF in the distal axons to mediate phosphorylation of unoccupied TrkA receptors in the cell bodies. Also, theoretical analysis of long-range signaling via waves of protein phosphorylation has been presented with the suggestion that this type of mechanism could be involved in retrograde NGF signaling (Markevich et al. 2006). Additional experiments, moreover, indicated that phosphorylation of TrkA in the cell bodies is not required for NGF at the distal axons to produce a retrograde signal that supports neuronal survival (see Sect. 4.2). As a result, the mechanism or biological significance of the fast retrograde TrkA phosphorylation by NGF has not been further investigated at this time.

4.2 NGF Deprivation Initiates Apoptotic Signaling in the Cell Bodies Before Internalized NGF is Depleted from the Neurons

Our analysis of the retrograde transport of [^{125}I]NGF in compartmented cultures of rat sympathetic neurons indicates that NGF is taken up and retrogradely transported intact to the cell bodies, as postulated by the signaling endosome hypothesis. In the cell bodies, the radioiodinated NGF is broken down and [^{125}I] is released into the medium in low molecular weight form, presumably iodotyrosine (Ure and Campenot 1994). During retrograde transport, [^{125}I] was not substantially released by axons spanning an intermediate compartment, suggesting that NGF is not broken down in axons (Ure and Campenot 1997). These results support the conclusion that most of the [^{125}I] that is retrogradely transported to the cell bodies arrives in the form of intact NGF, which could be involved in signaling. These results also suggest that the mechanisms for degrading NGF are localized in the cell bodies.

When [^{125}I]NGF is removed from distal axons and chased with unlabeled NGF, half of the internalized [^{125}I] is released in 6 h into the medium bathing the cell bodies and proximal axons, and 75% is released in 10 h (Ure and Campenot 1997). If the distal axons were removed when the [^{125}I]NGF was withdrawn, half of the [^{125}I] remaining in the neurons was released in only 3 h, indicating that the 6 h half-life in intact neurons included time required for retrograde transport and processing the NGF that was in the distal axons at the time of NGF withdrawal. Phosphorylation and nuclear accumulation of c-jun is an early indication of apoptosis in cell bodies of NGF-deprived sympathetic neurons. In neurons supported by NGF only at the distal axons, nuclear accumulation of phosphorylated c-jun occurs within 6 h of NGF withdrawal (Mok and Campenot 2007), when half of the internalized NGF is still within the neurons.

Neuronal survival is well supported by 5 ng ml^{-1} NGF and 50 ng ml^{-1} NGF supplied to distal axons. Neurons given 50 ng ml^{-1} [^{125}I]NGF at their distal axons transported about 4.5 times more [^{125}I]NGF than neurons given 5 ng ml^{-1} [^{125}I]NGF(MacInnis and Campenot 2002); therefore, when they initiated the apoptotic mechanism after 6 h of NGF deprivation, they likely still contained more than twice the internalized NGF than neurons continuously supplied with 5 ng ml^{-1} NGF. If the internalized NGF resides in signaling endosomes enroute to the cell bodies, these neurons should not begin to activate their apoptotic programs until much later.

It is, of course, possible to hypothesize additional features of NGF processing that could account for the rapid effects of NGF deprivation; for example, possibly only a fraction of the internalized, intact NGF resides in signaling endosomes, and, possibly, this fraction is somehow degraded more rapidly after NGF withdrawal. Nonetheless, this result is not what would be predicted by the signaling endosome-only hypothesis; therefore, an alternative explanation must be sought.

4.3 The Retrograde Transport of NGF is not Required for Retrograde Survival Signaling

As a direct test of the requirement that retrograde transport of NGF serves as a retrograde survival signal, NGF covalently linked to polystyrene beads was applied to distal axons to determine if the application could produce a retrograde survival signal by activating axonal surface TrkA receptors in the absence of internalization and retrograde transport (MacInnis and Campenot 2002). We utilized the fact that after 30 h of retrograde transport of [^{125}I]NGF, the vast majority of the [^{125}I] transported has been released into the proximal compartment medium and only a small steady-state level remains in the cell bodies/proximal axons. Thus, the [^{125}I] present in the medium after 30 h is an accurate estimate of the amount of NGF retrogradely transported. We found that [^{125}I]NGF, covalently linked to 1 μm-diameter polystyrene beads to prevent internalization, activated TrkA on the neurons and supported neuronal survival when supplied to distal axons nearly as effectively as 50 ng ml^{-1}-free NGF.

To evaluate the possibility that neuronal survival could have been supported by [^{125}I]NGF that had been released from the beads, taken up by the distal axons, and retrogradely transported to the cell bodies, distal axons were supplied with [^{125}I] NGF beads for 30 h, and the amount of [^{125}I] released into the medium bathing the cell bodies/proximal axons during the incubation was determined, along with the fraction of neurons surviving at the end of the incubation. Results were compared with the survival and [^{125}I] released by cell bodies/proximal axons of neurons that had been given free [^{125}I] NGF at concentrations ranging over six orders of magnitude.

[^{125}I]NGF beads applied to distal axons supported the survival of 84% of the neurons, while the retrograde transport of [^{125}I] was barely detectable. A similar level of survival support was obtained in cultures given 5 ng ml^{-1} [^{125}I] NGF, which, however, transported at least 250 times the [^{125}I]NGF that could have been transported by neurons supplied with [^{125}I]NGF beads. The lack of correlation between the level of [^{125}I]NGF transport and neuronal survival was further demonstrated in neurons that were given 0.5 ng ml^{-1} of [^{125}I]NGF, which transported over 20 times more [^{125}I] NGF than could have been transported from [^{125}I]NGF beads, but in which survival (i.e., 29%) was not well supported. This is likely to be a substantial underestimation of initial transport, which was inadequate to support neuronal survival, inasmuch as 71% of the neurons died during the incubation, and therefore, did not transport NGF during the full 30 h. These results show that when the retrograde transport of NGF was nearly or completely eliminated, activation of TrkA receptors on the surfaces of the distal axons produced a retrograde signal that almost fully supported neuronal survival. This result is strong evidence for the existence of a retrograde signal involved in supporting neuronal survival that does not involve the retrograde transport of NGF-containing signaling endosomes.

Explanations have been offered, which interpret the aforementioned results within the context of the NGF-containing signaling endosome-only hypothesis (Ye et al. 2003); however, they need to be examined critically from an analytical standpoint. Thus, it has been argued that barely detectable amounts of [^{125}I]NGF must have been released from the beads, taken up in signaling endosomes, and retrogradely transported to the neuronal cell bodies, producing neuronal survival. To account for the inability of 0.5 ng ml^{-1} free [^{125}I]NGF to support neuronal survival, it was further proposed that even though the [^{125}I]NGF retrogradely transported in this group was more than 20 times the transport that could have occurred from beads, it must not have been carried by signaling endosomes that were functional. The basis provided for the latter assertion is that the low concentration of 0.5 ng ml^{-1} NGF is insufficient to activate TrkA receptors (Ye et al. 2003). However, if 0.1 ng ml^{-1} NGF produced half-maximal phosphorylation of TrkA in PC12 cells (Kaplan et al. 1991), then five times more certainly it would have produced phosphorylation of TrkA. Moreover, the transport of [^{125}I]NGF was specific and receptor-mediated, as it was competed in cultures provided with and excess of unlabeled NGF. Had the endocytosis and retrograde transport of [^{125}I]NGF been by means of bulk uptake of the medium, at 0.5 ng ml^{-1} NGF, 24 µl of medium would have to be transported per culture to account for the transport of 12 pg of NGF. This is 536 times the volume of the approximately 2,000 neuronal cell bodies present in a culture. At this rate, bulk endocytosis and transport of medium containing 12 pg of NGF would replace the cell body volume every 3.4 min during the 30 h incubation. These considerations indicate that the [^{125}I]NGF transported by cultures that were given 0.5 ng ml^{-1} NGF did not arise from nonspecific, bulk endocytosis, but by binding to receptors that it presumably activated. To explain how a barely detectable transport of [^{125}I]NGF in cultures exposed to [^{125}I]NGF beads could produce a retrograde survival signal, it was proposed that the massive activation of surface TrkA by bead-bound NGF somehow promoted the uptake and loading into "competent" NGF-containing signaling endosomes of a small amount of [^{125}I]NGF, hypothesized to have been spontaneously released from the surface of the beads (Ye et al. 2003).

The possibility that one NGF dimer bound to a single pair of TrkA receptors can be taken up into a functional signaling endosome has been suggested in another study. It was shown, using compartmented cultures of dorsal root ganglion neurons, that supplying NGF bound to quantum dots results in the retrograde transport of vesicles containing individual quantum dots; each quantum dot presumably bearing an individual NGF dimer (Cui et al. 2007). The size of the quantum dot-containing endosomes ranged from 50 to 100 nm, and the average velocity of retrograde transport of quantum dot-containing vesicles varied among axons, ranging from 0.93 to 2.29 μm s^{-1}. The internalization of quantum dot-NGF (QD-NGF) was blocked by K252a applied to distal axons, which blocks TrkA phosphorylation (see below), suggesting that activation of one TrkA dimer at the cell surface could promote internalization, further implying that endosomes containing a single NGF dimer are functional signaling endosomes.

Supporting this conclusion was the association of TrkA and phosphorylated Erk1/2 with some QD-NGF-containing endosomes. The analysis of quantum dots showed a frequency of 14 quantum dots per millimeter in proximal axons of dorsal root ganglion neuron cultures, which had 0.2 nM QD-NGF (equivalent to 2.6 ng ml^{-1} NGF) applied at their distal axons. There were 83 quantum dots per millimeter in cultures exposed to 2 nM QD-NGF (equivalent to 26 ng ml^{-1} NGF) at their distal axons, and 252 quantum dots per millimeter in cultures given 20 nM QD-NGF (equivalent to 260 ng ml^{-1} NGF). This is a very low level of NGF transport when considering the effect that equivalent concentrations of free NGF have (see below), which suggests that NGF bound to a quantum dot likely changes the kinetics of NGF receptor binding and internalization. As the quantum dots were 5 nm in diameter, steric hindrance might interfere with the simultaneous activation of nearby TrkA receptors, and prevent the clustering of NGF-bound TrkA, which could lead to decreased uptake. Moreover, not enough is known about these mechanisms to predict what effects this might have on the biological activity.

QD-NGF was shown to be biologically active in the PC12 cell line, in which it induced neurite outgrowth; however, neurite outgrowth could have been induced as well by activation of TrkA on the cell surfaces. Moreover, it was not reported whether QD-NGF applied to distal axons of dorsal root ganglion neurons was able to support neuronal survival. Even if survival could be supported by QD-NGF applied to distal axons, the distal axons were intensely labeled with quantum dots over their entire lengths, while only a few quantum dots were transported. It is clear from results of our [^{125}I]NGF bead experiments that a retrograde signal downstream of TrkA activated on the surfaces of the distal axons could have supported neuronal survival. Nevertheless, not reporting retrograde survival effects of QD-NGF in neurons is a serious omission. As it was concluded that a few NGF dimers bound to TrkA may have significant biological function, it would be important to know if QD-NGF applied to distal axons supported survival or produced any other biological effects. Biological effects were shown exclusively by direct application of QD-NGF to PC12 cells, which, like the distal axons exposed to QD-NGF in compartmented cultures, would be expected to have QD-NGF bound over their entire surfaces. This would likely produce a much larger TrkA signal than could have been produced by a few signaling endosomes traveling along an axon.

It would be instructive to draw a comparison between the effects of free and bead-bound NGF for the purposes of a realistic overview of the potential significance of the two modes of action. While [^{125}I] labeling studies cannot track individual [^{125}I] NGF dimers, the total transport of [^{125}I]NGF during an incubation can be calculated, providing a basis for comparison with the quantum dot data. Neurons exposed to 50 ng ml^{-1} [^{125}I]NGF at their distal axons transported 657 pg of NGF per culture during the 30 h incubation, equivalent to 15.2×10^9 NGF dimers per culture (MacInnis and Campenot 2002). Assuming 2,000 neurons per culture, this amounts to 7.6×10^6 NGF dimers transported per neuron during the 30 h incubation period. At this rate, each neuron would have transported 70.3 dimers per second. Assuming an average transport velocity of 2 μm s^{-1}, a 1 mm length of axon will deliver its NGF-containing vesicles to the cell body in 500 s. Therefore, in the case of neurons, in which 50 ng ml^{-1} NGF is applied at their distal axons, the initial segment of axon extending from the cell body to the first branch point would contain 35,200 NGF dimers per millimeter of length. This would exceed the density of quantum dots in axons supplied with 260 ng ml^{-1} QD-NGF by 140×. If one NGF dimer was contained in a single 100 nm-diameter signaling endosome, and the endosomes were linearly distributed in close proximity to each other within the axon, the endosome column would be 3.52 mm long. Clearly, this amount of NGF is not transported in endosomes, in which each would contain a single or just a few NGF dimers.

In the case of neurons given 5 ng ml^{-1} [^{125}I]NGF, the lowest concentration tested that supported neuronal survival, the analysis shows that NGF would be delivered at a rate of 10.7 dimers per neuron per second, and the initial segment of the axon would contain 5,350 dimers per millimeter of axon. In the case of neurons given 0.5 ng ml^{-1} NGF, a concentration that did not support neuronal survival, the NGF dimers would be delivered at a rate of 1.29 dimers per second, and the initial axon segment would contain 645 dimers per millimeter. This is a substantial underestimation of the actual level of transport in neurons, given an insufficient survival concentration of 0.5 ng ml^{-1} [^{125}I]NGF, because the majority of neurons die during the incubation and would not have transported for the entire 30 h. Thus, assuming that even if a single NGF dimer can be internalized into a functional signaling endosome, it seems very unlikely that the retrograde transport of just a few NGF dimers can support the survival of sympathetic neurons. If the low level of retrograde transport observed in cultures given [^{125}I]NGF beads represents transport of signaling endosomes, this would amount to the delivery of 0.06 NGF dimers per second, requiring a density of 30 dimers per millimeter in the initial segment of the axon. While this is in the range observed in the quantum dot study (Cui et al. 2007), the ability of these low levels of NGF transport to support neuronal survival was not examined.

Attempts to explain how NGF beads produce NGF-containing signaling endosomes seem to be based on the assumption that very low amounts of NGF, even possibly single dimers, can produce functional signaling endosomes, while under other conditions, relatively massive amounts of retrogradely transported NGF are not carried by functional signaling endosomes at all. Rather than supporting the signaling endosome hypothesis, these ad hoc explanations seem to stem from the conviction that NGF-containing signaling endosomes are the only credible mechanism for retrograde NGF signaling. We have preferred to make the more straight-forward interpretation; namely, that activation of TrkA on distal axons without significant internalization, and retrograde transport of NGF still produces a retrograde signal that supports neuronal survival, which must, therefore, be mediated by another mechanism. As pointed out by Thomas Kuhn, scientific hypotheses that are tenaciously held are highly resistant to changes and only do so in the face of multiple lines of evidence (Kuhn 1962). Let us now turn to a second requirement of the signaling endosome-only hypothesis – the appearance of phosphorylated TrkA in the cell bodies.

4.4 Retrograde Appearance of Phosphorylated TrkA

As required by the signaling endosome hypothesis, application of NGF to the distal axons of rat sympathetic neurons in compartmented cultures results in the appearance of phosphorylated TrkA in the cell bodies/proximal axons (Senger and Campenot 1997). To determine if the retrograde appearance of phosphorylated TrkA is required for retrograde survival signal, inhibitors of TrkA phosphorylation were applied in the proximal compartments to block any retrograde phosphorylated TrkA signal that may be occurring. We used the traditional Trk phosphorylation blocker, K252a (MacInnis et al. 2003), as well as a more recently identified blocker, Gö6976 (Mok and Campenot 2007). Both drugs completely blocked the phosphorylation of TrkA in the compartment in which they were applied. Because NGF-induced neuronal survival requires the NGF-induced phosphorylation of TrkA, each drug induced apoptosis as expected when applied to the distal axons of rat sympathetic neurons in the compartmented cultures supported by NGF exposure only to distal axons. However, the appearance of phosphorylated TrkA in the cell bodies/proximal axons was completely prevented without diminishing the survival of the neurons by application of K252a or Gö6976 to the cell bodies and proximal axons of neurons treated with NGF only at their distal axons. Thus, it appears that blocking the accumulation of phosphorylated TrkA in the cell bodies/proximal axons did not prevent or even diminish the potency of the retrograde survival signal induced by NGF provided to the distal axons. In addition, we found that in neurons that are given NGF at their distal axons, blocking of TrkA phosphorylation in the cell bodies/proximal axons with K252a or Gö6976 did not block the retrograde phosphorylation of downstream signaling molecules, Akt and CREB. These results controvert the signaling endosome-only hypothesis because inhibition of TrkA signaling at the cell bodies should abolish any signal carried by NGF bound to phosphorylated TrkA in signaling endosomes. Although there is a report that K252a applied to the cell bodies of rat sympathetic neurons in compartmented cultures does inhibit NGF retrograde survival signaling (Ye et al. 2003), it may be accounted for by differences in the cultures (see Sect. 5).

Alternative explanations do not seem adequate to explain the effects of the TrkA inhibitors. Like all pharmacological agents, these inhibitors can have multiple effects. However, in drug-only control experiments, Gö6976 did not promote the survival of NGF-deprived neurons when applied to the cell bodies and proximal axons. This indicates that Gö6976 did not support neuronal survival by another mechanism at the cell bodies, while blocking the function of signaling endosomes arriving from the distal axons. K252a did produce a partial survival effect on its own when applied to cell bodies and proximal axons of NGF-deprived neurons, but not the full survival effect that was present when the neurons were given NGF at their distal axons. The well-known survival-promoting effect of K252a was, in fact, the reason we undertook experiments with Gö6976. Gö6976 completely blocked the retrograde survival effect of NGF when applied to the distal axons, while not blocking retrograde survival at all when applied to the cell bodies and proximal axons. It is interesting that K252a also produced its survival effect when applied to the distal axons of NGF-deprived neurons. It is unlikely that membrane-permeant K252a entering the distal axons is retrogradely transported to the cell bodies to exert survival-promoting effects, because the drug would diffuse out of the neuron and equilibrate with the vastly greater volume of the bathing medium to achieve a virtually zero intracellular concentration. Thus, it seems likely that K252a perturbs a signaling system in the distal axons that can produce a retrograde signal, which partially supports neuronal survival. This suggests that a retrograde survival signal can be produced by altering signaling mechanisms in the distal axons without producing retrograde transport of NGF and phosphorylated TrkA in signaling endosomes.

In summary, three lines of evidence are contrary to the predictions of the signaling endosome-only hypothesis: (1) neurons display the nuclear accumulation of activated c-jun after NGF withdrawal, while they still contain substantial amounts of internalized NGF; (2) bead-bound NGF supplied to distal axons supports neuronal survival; and (3) block of TrkA phosphorylation in the cell bodies and proximal axons does not block the ability of NGF at distal axons to support neuronal survival.

5 Evidence for a Retrograde Apoptotic Signal

Thomas Kuhn observed that replacement of a scientific theory requires not only evidence against the theory, but also evidence for an alternative theory (Kuhn 1962). New evidence suggests that in addition to producing retrograde NGF survival signals, distal axons possess a mechanism that can generate an apoptotic signal that travels retrogradely to the cell bodies, where it triggers the initiation of the apoptotic program that kills the neuron. A death signaling mechanism intrinsic to the distal axon would have countervailing implications for the concept of an NGF retrograde survival signal. Although perhaps counterintuitive, the retrograde survival signal would comprise the absence of an active retrograde transport process that mediates survival. Rather, it would entail a local mode of action by NGF on the distal axon to suppress the generation of a retrograde death signal.

Our first indication of the existence of a retrograde apoptotic signal generated in NGF-deprived distal axons was the observation that withdrawal of NGF from distal axons produced the nuclear accumulation of activated c-jun in the cell bodies within 6 h in virtually all the neurons, even under conditions in which cell bodies and proximal axons were provided with 50 ng ml^{-1} NGF and the neurons survived (Mok et al. 2009). It seemed unlikely that the nuclear accumulation of activated c-jun in neurons treated under this protocol resulted from the loss of NGF-induced retrograde survival signals due to NGF deprivation of the distal axons. This inference is based on the efficacy of NGF acting on the cell bodies and proximal axons to provide more potent NGF-induced changes in gene expression than can be achieved by NGF applied only at the distal axons (Toma et al. 1997). The experimental findings would be consistent with a pro-apoptotic signal generated in NGF-deprived distal axons that traveled retrograde to the cell bodies to initiate the apoptotic response. The absence of apoptosis, however, would be explained by an interruption occurring downstream of c-jun activation, due to survival signals produced by TrkA activation in cell bodies supplied with NGF.

To further determine if retrograde c-jun activation arising from NGF-deprived distal axons resulted from the loss of a retrograde survival signal or the production of a retrograde apoptotic signal, axonal transport was blocked with colchicine. Treatment of distal axons with colchicine was effective because it virtually abolished the retrograde transport of fluorescent, cy3-labeled NGF. Colchicine applied to NGF-deprived distal axons blocked the retrograde accumulation of phosphorylated c-jun in the cell bodies, indicating that apoptosis, produced by NGF-deprivation in distal axons, was initiated by a retrogradely transported "apoptotic" signal traveling to the cell bodies. This result clearly indicates that loss of NGF signaling in the distal axons impacts the cell bodies by mechanisms distinct from the loss of NGF retrograde survival signals reaching the cell bodies. Since c-jun is pro-apoptotic in rat sympathetic neurons (Palmada et al. 2002), the putative signal blocked by colchicine appears to be a pro-apoptotic signal generated in NGF-deprived distal axons.

Disassembly of microtubules with colchicine, not surprisingly, causes axons to disintegrate when applied for 18 h; therefore, it was not possible to determine if a 24-h treatment of NGF-deprived distal axons could block apoptosis. However, a screen of kinase inhibitors revealed that rottlerin or chelerythrine blocked the activation of c-jun, pro-caspase-3 cleavage, and apoptosis when applied to distal axons of NGF-deprived neurons, but not when applied to the cell bodies/proximal axons (Mok et al. 2009). This suggests that a target of rottlerin and chelerythrine in the distal axons is involved in generating and/or transmitting a retrograde signal to the cell bodies that activates c-jun and pro-caspase-3, which results in apoptosis. siRNA knock-down experiments suggested that block of apoptosis by rottlerin applied to distal axons required the expression of GSK3, one of the targets of rottlerin and chelerytherine. Thus, GSK3 in distal axons may play a role in the generation and/or transport of the retrograde apoptotic signal.

A retrograde apoptotic signal may explain the results of a study that obtained different results from ours in investigations of K252a. In that study, K252a, applied to cell bodies and proximal axons of rat sympathetic neurons in compartmented cultures, blocked the ability of NGF acting on distal axons to activate survival signaling in the cell bodies and to support neuronal survival, as predicted by the signaling endosome-only hypothesis (Ye et al. 2003). However, inspection of a photomicrograph of the neurons used in that study reveals that most of the distal axons were misdirected during outgrowth, so as to remain within the proximal compartment. This was apparently a general phenomenon, because immunoblots of total TrkA and molecular markers used as loading controls showed far greater abundance in the proximal compartment cell extracts than in the distal compartment extracts. Thus, it was clear that most of the distal axons did not cross the barriers into the distal compartments, remaining instead within the proximal compartments.

In retrograde signaling experiments, NGF is withdrawn from the proximal compartments to establish the condition where survival support is obtained entirely from NGF at the distal axons. Therefore, cultures in which most of the distal axons remain in the proximal compartments, withdrawal of NGF from the cell bodies and proximal axons also deprives most of the distal axons of NGF. This would be expected to send retrograde apoptotic signals to the cell bodies. In our cultures, withdrawal of NGF from the cell bodies and proximal axons by replacing it with anti-NGF does not induce the activation and nuclear accumulation of c-jun; there is no detectable axonal retrograde death signal produced when NGF is withdrawn from the cell bodies and proximal axons. Possibly, neurons, in which death signaling has been activated, do require that signaling endosomes, containing NGF and activated TrkA, arrive from the subpopulation of distal axons in the distal compartment to overcome the death signal.

There is evidence that activation of the p75 neurotrophin receptor in sympathetic neurons by the neurotrophins, BDNF and NT-4, which do not activate TrkA, produces an apoptotic signal that kills the neuron. Activation of TrkA by NGF blocks the p75NTR death signal, and this mechanism is implicated in regulating sympathetic neuron survival during development (Bamji et al. 1998; Majdan et al. 2001; Deppmann et al. 2008). In all the published work of which we are aware, BDNF or NT-4 must be added to the culture medium to produce apoptosis by this mechanism, and so it appears that the retrograde apoptotic signal generated by NGF-deprived distal axons operates through a different mechanism.

In conclusion, our results suggest that in addition to generating retrograde survival signals that travel to the cell bodies to activate survival signaling molecules, such as Akt and CREB, NGF at distal axons suppresses the generation of a retrograde apoptotic signal that is activated in NGF-deprived distal axons and travels along axonal microtubules to the cell bodies, where it triggers the activation and nuclear accumulation of c-jun, setting in motion the apoptotic mechanism.

6 Conclusions

While the focus of the field has been on retrograde survival signaling by NGF, NGF produces many changes in gene expression to produce and maintain the differentiation of the neuron. Also, this does not just apply to NGF and the other members of the neurotrophin family that produce retrograde signals. For example, the cytokine, LIF, applied at distal axons of rat sympathetic neurons produces a retrograde signal that induces cholinergic neurotransmitter function (Ure et al. 1992; Ure and Campenot 1994).

What has evolved in neurons is an intracellular mass transit system, based on a local network of roads and highways made of microtubules that are traveled in both directions by motor-driven organelles, which are capable of going through short distances between Golgi stack membranes or long distances between the cell body and axon terminals. In principle, signaling molecules with the proper "ticket" can be targeted anywhere in the neuron, just as easily as a vesicle taking a short hop between Golgi cysterns. There is no reason to believe that all the different factors that produce retrograde signals and all the different neuronal functions they support can only operate by one specified mechanism; namely, the retrograde transport of ligand and activated receptor to the cell bodies in signaling endosomes. In fact, given the complexity of biological systems, this seems in principle to be highly unlikely.

Given the current status of the field, no plausible mechanism of retrograde signaling should be ruled out. A retrograde signal could hypothetically be carried by the retrograde transport of neurotrophin bound to its receptor in signaling endosomes, by a propagated mechanism without the retrograde transport of signaling molecules, or by the retrograde transport of downstream signaling molecules. In the case of NGF, there is evidence that retrograde signals are likely carried by NGF bound to phosphorylated TrkA in signaling endosomes, which has been discussed in several reviews (Howe and Mobley 2004; Ibanez 2007; Cosker et al. 2008). However, the evidence supporting the signaling endosome-only hypothesis does not seem compelling. The possibility of retrograde signals being carried by propagated phosphorylation of TrkA explains the rapidity of retrograde TrkA phosphorylation, but needs to be further explored. There is strong evidence that retrograde signals can be carried by the retrograde transport of molecules downstream of TrkA; specifically, an apoptotic retrograde signal generated in axons deprived of NGF.

A most intriguing question is why would NGF withdrawal from distal axons produce a retrograde apoptotic signal to the cell bodies when the loss of the NGF retrograde survival signal would kill the neuron anyway. The existence of survival and death signals generated in distal axons opens up the interesting possibility that the survival of the neuron could be determined by the balance between survival signals from the subset of its axon terminals that have connected to reliable sources of neurotrophin and apoptotic signals coming from the subset of its axon terminals that have not been connected to reliable sources of neurotrophin. This could allow for greater precision in life–death decisions, and since the function of the nervous system is based upon the development of exquisitely complex circuitry, precision in the editing process that removes 50% of the original population of neurons during development seems highly appropriate.

Axons seem poised to send a death signal to the cell bodies, which could be unleashed during neurodegenerative disease or after neurotrauma. Investigations of the possible roles of axonal death signals may lead to new insights into ways to minimize or eliminate damage to the nervous system in disease and after physical insult.

Acknowledgments Our investigations of retrograde NGF signaling in compartmented cultures represent the work of four excellent Ph.D. students: Daren Ure, Donna Senger, Bronwyn MacInnis, and Sue-Ann Mok. Excellent technical support was provided by Karen Lund, Grace Martin, Norma Jean Valli, and Megan Blacker. Financial support has been provided by the Canadian Institutes of Medical Research, The Alberta Heritage Foundation, The Alberta Paraplegic Foundation, and the Rick Hansen Man in Motion Foundation.

References

Bamji SX, Majdan M, Pozniak CD, Belliveau DJ, Aloyz R, Kohn J, Causing CG, Miller FD (1998) The p75 neurotrophin receptor mediates neuronal apoptosis and is essential for naturally occurring sympathetic neuron death. J Cell Biol 140:911–923

Bronfman FC, Escudero CA, Weis J, Kruttgen A (2007) Endosomal transport of neurotrophins: roles in signaling and neurodegenerative diseases. Dev Neurobiol 67:1183–1203

Campenot RB (1977) Local control of neurite development by nerve growth factor. Proc Natl Acad Sci USA 74:4516–4519

Campenot RB (1982a) Development of sympathetic neurons in compartmentalized cultures. I Local control of neurite growth by nerve growth factor. Dev Biol 93: 1–12

Campenot RB (1982b) Development of sympathetic neurons in compartmentalized cultures. II. Local control of neurite survival by nerve growth factor. Dev Biol 93: 13–21

Cosker KE, Courchesne SL, Segal RA (2008) Action in the axon: generation and transport of signaling endosomes. Curr Opin Neurobiol 18:270–275

Cui B, Wu C, Chen L, Ramirez A, Bearer EL, Li WP, Mobley WC, Chu S (2007) One at a time, live tracking of NGF axonal transport using quantum dots. Proc Natl Acad Sci USA 104:13666–13671

Deppmann CD, Mihalas S, Sharma N, Lonze BE, Niebur E, Ginty DD (2008) A model for neuronal competition during development. Science 320:369–373

Hendry IA (1977) The effect of the retrograde axonal transport of nerve growth factor on the morphology of adrenergic neurones. Brain Res 134:213–223

Hendry IA, Stockel K, Thoenen H, Iversen LL (1974) The retrograde axonal transport of nerve growth factor. Brain Res 68:103–121

Howe CL, Mobley WC (2004) Signaling endosome hypothesis: A cellular mechanism for long distance communication. J Neurobiol 58:207–216

Ibanez CF (2007) Message in a bottle: long-range retrograde signaling in the nervous system. Trends Cell Biol 17:519–528

Kaplan DR, Hempstead BL, Martin-Zanca D, Chao MV, Parada LF (1991) The trk proto-oncogene product: A signal transducing receptor for nerve growth factor. Science 252:554–558

Kuhn TS (1962) The structure of scientific revolutions. University of Chicago Press, Chicago

Landis SC (1976) Rat sympathetic neurons and cardiac myocytes developing in microcultures: Correlation of the fine structure of endings with neurotransmitter function in single neurons. Proc Natl Acad Sci USA 73:4220–4224

MacInnis BL, Campenot RB (2002) Retrograde support of neuronal survival without retrograde transport of nerve growth factor. Science 295:1536–1539

MacInnis BL, Campenot RB (2005) Regulation of Wallerian degeneration and nerve growth factor withdrawal-induced pruning of axons of sympathetic neurons by the proteasome and the MEK/Erk pathway. Mol Cell Neurosci 28:430–439

MacInnis BL, Senger DL, Campenot RB (2003) Spatial requirements for TrkA kinase activity in the support of neuronal survival and axon growth in rat sympathetic neurons. Neuropharmacology 45:995–1010

Majdan M, Walsh GS, Aloyz R, Miller FD (2001) TrkA mediates developmental sympathetic neuron survival in vivo by silencing an ongoing p75NTR-mediated death signal. J Cell Biol 155:1275–1285

Markevich NI, Tsyganov MA, Hoek JB, Kholodenko BN (2006) Long-range signaling by phosphoprotein waves arising from bistability in protein kinase cascades. Mol Syst Biol 2:61

Mok SA, Campenot RB (2007) A nerve growth factor-induced retrograde survival signal mediated by mechanisms downstream of TrkA. Neuropharmacology 52:270–278

Mok SA, Lund K, Campenot RB (2009) A retrograde apoptotic signal originating in NGF-deprived distal axons of rat sympathetic neurons in compartmented cultures. Cell Res doi:10.1038/cr.2009.11.

Oppenheim RW (1991) Cell death during development of the nervous system. Annu Rev Neurosci 14:453–501

Palmada M, Kanwal S, Rutkoski NJ, Gustafson-Brown C, Johnson RS, Wisdom R, Carter BD (2002) c-jun is essential for sympathetic neuronal death induced by NGF withdrawal but not by p75 activation. J Cell Biol 158:453–461

Salehi A, Delcroix JD, Belichenko PV, Zhan K, Wu C, Valletta JS, Takimoto-Kimura R, Kleschevnikov AM, Sambamurti K, Chung PP, Xia W, Villar A, Campbell WA, Kulnane LS, Nixon RA, Lamb BT, Epstein CJ, Stokin GB, Goldstein LS, Mobley WC (2006) Increased App expression in a mouse model of Down's syndrome disrupts NGF transport and causes cholinergic neuron degeneration. Neuron 51:29–42

Senger DL, Campenot RB (1997) Rapid retrograde tyrosine phosphorylation of trkA and other proteins in rat sympathetic neurons in compartmented cultures. J Cell Biol 138:411–421

Toma JG, Rogers D, Senger DL, Campenot RB, Miller FD (1997) Spatial regulation of neuronal gene expression in response to nerve growth factor. Dev Biol 184:1–9

Ure DR, Campenot RB (1994) Leukemia inhibitory factor and nerve growth factor are retrogradely transported and processed by cultured rat sympathetic neurons. Dev Biol 162:339–347

Ure DR, Campenot RB (1997) Retrograde transport and steady-state distribution of 125I-nerve growth factor in rat sympathetic neurons in compartmented cultures. J Neurosci 17:1282–1290

Ure DR, Campenot RB, Acheson A (1992) Cholinergic differentiation of rat sympathetic neurons in culture: Effects of factors applied to distal neurites. Dev Biol 154:388–395

Vallee RB, Bloom GS (1991) Mechanisms of fast and slow axonal transport. Annu Rev Neurosci 14:59–92

Ye H, Kuruvilla R, Zweifel LS, Ginty DD (2003) Evidence in support of signaling endosome-based retrograde survival of sympathetic neurons. Neuron 39:57–68

The Paradoxical Cell Biology of α-Synuclein

Subhojit Roy

Abstract Synucleinopathies are a group of neurodegenerative diseases characterized by accumulation and aggregation of the protein α-synuclein in neuronal perikarya and processes. In contrast to the proximal localization of α-synuclein in diseased states, under physiologic conditions, the bulk of α-synuclein is present in distant presynaptic terminals. Thus, pathologic conditions lead to mislocalization and aggregation of α-synuclein in neuronal cell bodies, and an outstanding question relates to the cell-biological mechanisms that can lead to such mislocalization. Like most other synaptic proteins, α-synuclein is synthesized in the neuronal perikarya and then transported into axons and synaptic domains. Accordingly, it has been hypothesized that disturbances in biogenesis/axonal transport or presynaptic targeting of α-synuclein can lead to its mislocalization in diseased states. In this chapter, key observations that lead to this hypothesis are presented in addition to a review of some recent literature that has directly addressed this issue. Finally, conflicting results that have resulted from such studies are also highlighted, and a view is offered to reconcile these controversies.

1 Introduction

Synucleinopathies or Lewy body (LB) diseases are a heterogenous group of neurodegenerative conditions characterized by movement disorders and/or dementias, and are second only to Alzheimer's disease (AD) in its prevalence. Common examples of these conditions include Parkinson's disease (PD) and dementia with Lewy bodies (DLB). A key neuropathologic hallmark of these diseases is the perikaryal and neuritic accumulation and aggregation of the small, 14-kDa protein α-synuclein into insoluble, fibrillar structures called Lewy bodies (LBs) and Lewy neurites (LN), where the wild-type (wt) protein undergoes posttranslational modifications.

S. Roy
Department of Neurosciences, University of California, San Diego, 92037, CA, USA
e-mail: s1roy@ucsd.edu

However, while the accumulation/aggregation of α-synuclein is seen in proximal neuronal compartments, under physiologic conditions, the bulk of α-synuclein is localized to distant presynaptic terminals. Thus, pathologic events lead to both accumulation/aggregation and mislocalization of presynaptic α-synuclein in diseased states. The paradoxical localization of α-synuclein in diseased states gives rise to an interesting problem in neuronal cell biology; namely, how does the presynaptic protein α-synuclein mislocalize and accumulate/aggregate proximally in these diseases? Considering that α-synuclein is synthesized in the neuronal perikaryon and is transported along the axons, much like other presynaptic proteins, subsequently localizing to presynaptic terminals, a leading hypothesis is that impairments in biogenesis/axonal transport of α-synuclein cargoes, and/or disruptions in presynaptic targeting may be early events leading to mislocalization, and subsequent accumulation of α-synuclein in synucleinopathies. While recent studies have attempted to resolve this issue, the results have been controversial, and often contradictory. In this chapter, the pathologic features of LB diseases and the evidence linking α-synuclein to these diseases are reviewed. It is then followed by a review of our current knowledge of the mechanisms underlying the biogenesis/axonal transport/targeting of wt and pathologically altered α-synuclein, and a discussion of the controversies in the field.

2 LB Diseases and the Role of α-Synuclein

LB diseases encompass a variety of conditions and a reasonable view of the diverse disease entities is briefly offered here. Patients with LB diseases often present with variable symptoms, ranging from a pure "Parkinsonian" movement abnormality ("pure" PD) to full-blown dementias ("pure" DLB) that resemble AD. The entities defining LB diseases are complex, largely a result of the heterogeneity in the clinical and pathologic presentation of these patients, and the diversity of nomenclature attributed to these diseases by clinicians over the years. The focus here is on the common neuropathologic feature in all of these diseases; namely, the proximal accumulation of α-synuclein, and the main issue is that perikaryal α-synuclein accumulations in diseased states is almost universal, and the mislocalization of α-synuclein does not depend on how these entities are segregated.

Substantial evidence links α-synuclein to disease pathogenesis in LB diseases. First, as mentioned above, perikaryal aggregates of α-synuclein are diagnostic hallmarks of LB diseases (McKeith et al. 2004). Though many other proteins are also found in LBs, including neurofilaments and other synaptic proteins, α-synuclein is one of its most abundant constituents. In addition to the aggregates in the cell bodies, most cases have marked "neuritic" pathology, and several studies have now shown that many of these "neurites" are axonal accumulations of α-synuclein (Braak et al. 1999; Saito et al. 2003; Orimo et al. 2008). Examples of α-synuclein pathology in a human brain with LB disease are shown in Fig. 1. In addition, autosomal-dominant mutations in the α-synuclein gene itself are found in familial

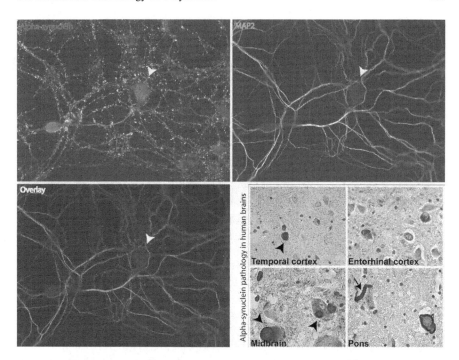

Fig. 1 Localization of α-synuclein in mature (14 DIV) cultured hippocampal neurons and in a human brain with LB disease. Cultures were stained with an antibody to α-synuclein and microtubule associated protein – MAP2 (see *red/green* overlay in ebook). Note that the vast majority of α-synuclein is localized to presynaptic boutons in these cultures, and very little protein can be discerned in the perikarya (*white arrowhead*). The *lower right panel* shows α-synuclein immunohistochemistry in various brain regions of a patient with LB disease. Note the presence of large inclusions of α-synuclein within the neuronal cell bodies (*black arrowheads*), as well as the pathology within neuritic processes (*black arrow*)

forms of LB diseases. Three of these (A30P, A53T, and E46K) representing mutations in the respective amino acid positions have been described in several families (McKeith and Mosimann 2004). Though rare, combined with the other evidence linking α-synuclein to these diseases, the mutants offer a window into the specific mechanisms by which changes in the α-synuclein protein can lead to pathology. Besides these mutations, it was recently found that some families with these diseases have a duplication or triplication of the α-synuclein gene itself (Singleton et al. 2003). Such triplication leads to a doubling of the protein levels of α-synuclein, suggesting that even a modest increase in protein levels can lead to pathology.

In addition to this compelling evidence from human genetics implicating α-synuclein in the pathogenesis of such dementias, several mouse models expressing wt, posttranslationally modified α-synucleins, as well as familial LB disease mutants replicate key features of these diseases (Masliah et al. 2000; Kahle et al. 2001; Giasson et al. 2002; Lee et al. 2002; Tofaris et al. 2006), and neurodegeneration occurs in fly and other α-synuclein animal models as well (Feany and Bender 2000; Link 2001; Lakso et al. 2003; Kuwahara et al. 2006). Thus, collectively, the immunohistochemical, biochemical, genetic, and animal model data have established α-synuclein as a major player in the pathogenesis of LB diseases.

3 The Neuronal Cell Biology of α-Synuclein

In mature neurons growing under physiologic conditions, the bulk of α-synuclein localizes to presynaptic regions. Accordingly, in dissociated cultured hippocampal neurons, the vast majority of α-synuclein in mature neurons is localized to the presynaptic terminals (Fig. 1), while in immature neurons, punctate α-synuclein staining is seen in both axons and dendrites. It was reported by the Banker laboratory that in neurons obtained from embryonic (E17–18) mice, the expression of α-synuclein is selectively suppressed until about 7 days in vitro (DIV), a time when such neurons begin to robustly generate new synapses (Withers et al. 1997). Based on this, it was speculated that α-synuclein may have a role in synapse maturation and synaptic plasticity, and not necessarily in synapse biogenesis. Indeed, the notion that α-synuclein may be involved in synaptic plasticity is also supported by the profound increase of α-synuclein expression in brains of songbirds, precisely at a time when they are actively learning new songs. However, we have found that the expression of α-synuclein in hippocampal cultures is largely dependent on the maturity of neurons at the time of plating. While E18 neurons only express α-synuclein after 7 days in culture as described by Withers et al., hippocampal neurons cultured from postnatal (P1–2) mice express α-synuclein immediately after plating, and α-synuclein expression can be seen even in neurons that grow in complete isolation (see Fig. 2). Thus, it is likely that the absence of α-synuclein expression in 1–7 DIV neurons seen by Withers et al. reflects the immature state of these cells themselves, and not the effects of synaptogenesis on the expression of α-synuclein. Curiously, in addition to synapses, α-synuclein expression is usually seen in the nucleus as well (arrowhead in Fig. 2). Though the nuclear localization has been less well studied, α-synuclein has been proposed to play a role in inhibiting histone acetylation (Kontopoulos et al. 2006).

Previous studies have revealed many key biochemical features of wt and disease-associated α-synuclein. Briefly, monomeric wt-α-synuclein is a small (140 aa), natively unfolded protein that binds to acidic phospholipids (Uversky 2007). Though α-synuclein is known to bind to certain vesicles, it lacks a transmembrane domain or a GPI anchor, and though the precise mechanisms by which α-synuclein binds to vesicles is still debated, it most likely involves the α-helical N-terminus. In diseased states, however, α-synuclein is thought to self-associate, and this self-association appears to stabilize the protein, leading to formation of oligomers and subsequent fibrillar forms, as seen in LBs and LNs in neurons. Within such LBs and LNs α-synuclein is found to undergo extensive changes, including posttranslational

The Paradoxical Cell Biology of α-Synuclein 163

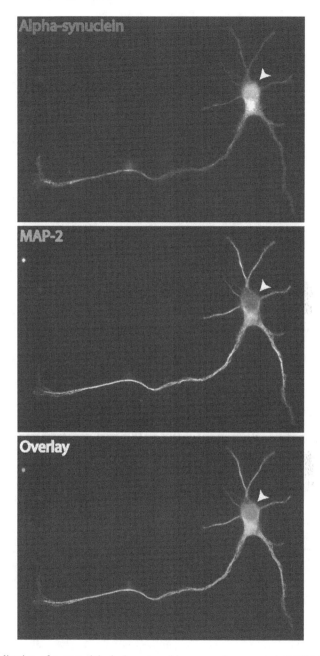

Fig. 2 Localization of α-synuclein in immature hippocampal neurons. A 3-DIV hippocampal neuron obtained from postnatal (P1) mouse brain is stained with an antibody to α-synuclein and MAP2 (see *red/green* overlay in ebook). Note that at this stage of development, punctate α-synuclein staining is seen in the axon as well as the dendrites. Also note that the α-synuclein expression in these neurons is not dependent on the connectivity of the cells or the presence of synapses. The *arrowhead* depicts the nuclear localization of α-synuclein in these neurons

modifications like phosphorylation and nitration, as well as C-terminal truncations. Many of these modifications lead to an acceleration in the rate of aggregation of the protein in in vitro biochemical experiments (reviewed in Mukaetova-Ladinska and McKeith 2006). These disease-associated modifications may play a pathogenic role, as animal models expressing these disease-associated modifications replicate some features of the disease including neuronal loss, LB-like α-synuclein accumulations, and behavioral abnormalities (Chen and Feany 2005; Tofaris et al. 2006; Gorbatyuk et al. 2008; Zhou et al. 2008). Thus, collectively, existing data suggest that specific alterations of wt-α-synuclein may play an important role in the transformation of α-synuclein into fibrillar LBs and LNs. Similarly, familial disease-associated mutations in α-synuclein also lead to acceleration in the fibril formation of the protein, and animal models expressing these mutations replicate some features of the disease.

4 Mislocalization of Presynaptic α-Synuclein in LB Dementias

Despite all this available information on the steps involved in the transformation of α-synuclein into fibrillar LBs and LNs, an outstanding question in the field is how α-synuclein, a protein that is normally localized to distal presynaptic terminals/boutons (Withers et al. 1997; Kahle et al. 2000), accumulates and aggregates into LBs and LNs that are seen in proximal neuronal compartments. One possibility is that in diseased states, pathologically altered α-synucleins fail to localize to presynaptic domains, leading to its gradual proximal accumulation, subsequent self-association, and eventual aggregation. This model of time-dependent accumulation is supported by studies in sympathetic axons and their cell bodies that are affected in LB diseases, suggesting that α-synuclein accumulation occurs in a centripetal fashion, with the axonal accumulation preceding the perikaryal accumulation/aggregation (Orimo et al. 2008). The notion that α-synuclein accumulation in the neuronal perikarya occurs gradually is also supported by immunohistochemical data from human brains with LB diseases showing that α-synuclein staining in these cases ranges from a mild diffuse perikaryal staining to a more robust pattern of staining (called "pale bodies"), in addition to the typical donut-shaped LBs that are strongly positive for α-synuclein (Wakabayashi et al. 2007). These data suggest that the variable patterns of α-synuclein pathology may represent evolving stages of LB formation.

What are the specific mechanisms that lead to the mislocalization of α-synuclein in proximal neuronal compartments? Considering that α-synuclein is synthesized in the neuronal perikarya and then transported into the axons, eventually localizing to presynaptic terminals or boutons (Jensen et al. 1999; Li et al. 2004; Saha et al. 2004), it seems reasonable to hypothesize that defects in either the presynaptic targeting and/or axonal transport/biogenesis of the protein can lead to its accumulation in diseased states. Another possibility is that the proximal accumulation of α-synuclein is a result of the failure of degradation of α-synuclein in diseased states, as suggested by other groups (Cuervo et al. 2004). This idea is not necessarily mutually exclusive of the notion of transport/targeting abnormalities in LB diseases. For example, it is conceivable that the abnormally targeted/transported α-synuclein may present the cell with problems of degrading the stalled α-synuclein, or abnormally targeted/transported α-synuclein may facilitate the formation of pathologic α-synuclein species that are difficult to degrade.

5 Biogenesis and Axonal Transport of wt and Pathologically Altered α-Synuclein

In this section, our current knowledge about mechanisms involved in the biogenesis and axonal transport of α-synuclein in physiologic and pathologic states is reviewed. As a first step, it is important to understand if α-synuclein localizes to a specific subcellular structure/organelle within the neuron. While α-synuclein's presynaptic localization suggests that it may be localized to synaptic vesicles (at least within synapses), α-synuclein's colocalization with vesicles has been difficult to demonstrate biochemically. While α-synuclein can associate with isolated purified vesicles in vitro, it behaves as a largely soluble protein in extracts from brains (Kahle et al. 2000). Ultrastructurally, while α-synuclein is clearly present in synaptic domains, it does not appear to be tightly associated with synaptic vesicles (Clayton and George 1998). Thus, the consensus view is that α-synuclein is transiently associated with vesicles at synapses, and that this interaction is dynamically regulated. Though α-synuclein is present as puncta along axons (see Fig. 2), the subcellular structure/organelle to which it binds during transport, if any, is unknown.

The biogenesis of α-synuclein is not well studied. Being largely a cytosolic protein, α-synuclein is not likely to traffic via the endoplasmic reticulum (ER) →Golgi secretory pathway. Surprisingly, however, when expressed in yeast, wt-α-synuclein localizes robustly to the plasma membrane, and this localization is likely to involve the secretory pathway (Dixon et al. 2005). Notably, in the yeast model, although the A53T mutation also localizes to the plasma membrane like the wt protein, the A30P mutation behaves like a soluble protein (Cooper et al. 2006; Soper et al. 2008). Though this result has always been interpreted as indicating the failed localization of the A30P mutant to the membrane, it could also reflect defects in ER→Golgi→plasma membrane trafficking of the mutant protein. Some evidence from nonneuronal cells also suggests a Golgi localization of α-synuclein (Tompkins et al. 2003), but we have seen that the localization of α-synuclein in neuronal cell bodies is unchanged by experimental manipulations that tend to cluster, or disperse other Golgi-derived proteins (our unpublished observations). Thus we feel that data from yeast model systems that seems to process α-synuclein very differently from neurons, must be interpreted with caution. Biogenesis of cytosolic proteins is difficult to study, and it is still unclear how α-synuclein is synthesized or how pathologic forms of α-synuclein may contribute to defects in this process.

In contrast, the overall mechanics of axonal transport of α-synuclein are better understood. In general, axonal transport conveys proteins synthesized in the neuronal cell bodies to distant sites, including presynaptic terminals. This is a constitutively active mechanism that occurs throughout the life of the neuron. Previous studies have shown that proteins are transported along axons in two overall groups. While one group of proteins was rapidly transported along axons at speeds of 50–400 mm/day, the other was transported much more slowly, at rates of only 0.2–8 mm/day. These two groups were called the fast component and the slow component (SC), respectively (reviewed in Roy et al. 2005). Closer examination of the SC revealed that it consisted of two, fairly nonoverlapping groups of proteins – one consisting of the cytoskeleton, namely, the microtubules and the neurofilaments, and the other, a slightly faster-moving group consisting of a heterogeneous collection of over 200 diverse proteins. They were called SCa and SCb, respectively.

Previous studies using the classic pulse-chase paradigm have also shown that the bulk of α-synuclein moves in slow axonal transport, in the sub-group called SCb using the classic pulse-chase paradigm (Jensen et al. 1999; Li et al. 2004). In these studies, radiolabeled amino acids were injected in the vicinity of neuronal cell bodies of living animals. The injected label was incorporated into the newly synthesized proteins in the perikarya, and these pulse-labeled proteins, including labeled α-synuclein, were transported along the axons. As the wave of radiolabeled α-synuclein moved along the axon, the composition of the labeled α-synuclein at any given time along the axon varied according to time and distance from the cell body. Thus, by studying the α-synuclein composition at various points along the axon, serial "snapshots" of the transported α-synuclein were obtained, showing that α-synuclein specifically moved in SCb. It was also reported that a small fraction (≈15%) of α-synuclein moves in the fast component as well (Jensen et al. 1999), though this was later refuted by a subsequent study (Li et al. 2004). However, it is noteworthy that small fractions (10–15%) of other SCb proteins like synapsin and heat shock proteins are also conveyed in the fast component, and the reasons for this peculiar kinetics of SCb proteins is unclear.

A limitation of the radiolabeling technique is that it can only detect bulk transport of the entire population of proteins moving in slow or fast transport and individual cargoes cannot be visualized. Thus, while the overall movement of wt-α-synuclein in SCb was well established by the above-mentioned studies, it was not clear how individual α-synuclein cargoes were transported. Recently, we designed a live-cell model-system where it is possible to visualize individual α-synuclein and SCb cargoes in living neurons (Roy et al. 2007). Using this model, we found that individual α-synuclein, and other SCb cargoes move rapidly in axons, at instantaneous velocities similar to those of the fast component. However, when compared directly to various fast component proteins, like synaptophysin and amyloid precursor protein, slow moving α-synuclein cargoes often underwent pauses during transit, and the movements were infrequent, unlike the fast component proteins that were transported much more frequently and persistently. These pauses in transit of individual slow cargoes made movement of the entire population slow, when averaged over time. Using this model, we also showed that α-synuclein cargoes were cotransported with other SCb proteins in axons (Roy et al. 2007), and that this movement was microtubule-dependent (Roy et al. 2008). More recent unpublished work from my laboratory shows that the movement of SCb proteins is an interplay of diffusion and motor-driven transport, leading to a biased transit of bulk cytosolic cargoes. These findings will be reported elsewhere shortly.

While the aforementioned studies have focused on the axonal transport of wt-α-synuclein, two previous studies have specifically looked at axonal transport of disease-associated familial α-synuclein mutants with seemingly contradictory results. While one study reported that the two familial disease-associated α-synuclein mutants A30P and A53T were transported more slowly than the wt protein (Saha et al. 2004), a subsequent study reported no differences in the transport of the wt and the mutant protein (Li et al. 2004). Notably, different methods were used to study axonal transport in these two reports. In Saha et al., the authors determined transport of wt and mutant α-synuclein in cultured cortical neurons by looking at bulk transport in fixed cultured neurons. Specifically, cultured neurons were transfected with the wt- or mutant green fluorescent protein (GFP)-tagged α-synuclein constructs, and then fixed at varying times posttransfection, allowing for transport of the transfected GFP-α-synucleins into the axon. The authors reported that they could locate the advancing "front" of the GFP-α-synuclein fluorescence in these fixed axons and measure the distance of this "front" from the cell body at varying times after transfection, as a measure of overall transport of the transfected wt- or mutant α-synuclein. Using these methods, they reported that the axonal transport of the α-synuclein "front" was significantly reduced when the protein had the A30P or A53T mutations. We note that these methods to study axonal transport are unconventional and are not yet validated by other laboratories (or in a variety of fast/slow cargoes). In our opinion, these techniques could lead to uncontrolled variability due to vagaries of transfection and/or protein expression, nevertheless, the results reported in these studies are intriguing.

In the study by Li et al. (2004), the authors studied axonal transport of the A30P and A53T mutant α-synuclein in vivo by the more conventional pulse-chase type of radiolabeling methods in peripheral sensory and motor nerves, and failed to see any difference in the axonal transport rates of the two mutants, compared to the wt protein. The reasons for such discrepancies in the transport of mutant α-synuclein in these studies are difficult to explain. However, differences in the methodology or the presence of endogenous mouse α-synuclein in these studies may account for some of the differences. Alternatively, differences in the synuclein proteins in the central and peripheral nervous systems may also account for these discrepancies. Indeed, the distribution of synucleins is quite different in the two systems, and some synuclein family members (e.g., persyn) are only present in the peripheral nervous system (Buchman et al. 1998). Saha et al. (2004) also reported that phosphorylation of α-synuclein at Ser-129 also led to a reduction in the transport of α-synuclein in their system, but this modification was not studied by Li et al. in their system. Though the underlying reasons for these discrepancies are unclear, the lack of consensus in the literature highlights the need for further studies with model-systems that can (a) directly visualize the transport of pathologic α-synuclein cargoes, (b) perform such studies in a α-synuclein null background, and (c) focus on α-synuclein targeting/transport in central nervous system neurons.

6 Mechanisms of Targeting wt and Pathologic α-Synuclein to Synapses

Proteins synthesized in the perikarya are transported into axons and synapses. Recent data clearly indicates that neurons have evolved specific mechanisms to "capture" the transiting vesicular cargoes into synapses when needed, so physiologic functioning of synapses that are distally situated would not be dependent on perikaryal protein synthesis, (reviewed in Levitan 2008). Though such mechanisms have not been shown for SC synaptic proteins yet, it is likely that these do exist for SC proteins, including α-synuclein as well. In that regard, one possible mechanism of mislocalization of α-synuclein in diseased states is that the disease-associated mutant proteins fail to localize to presynaptic domains after reaching their ultimate destination. To test this idea, many studies have compared the membrane interactions of the familial disease-associated α-synuclein mutants A30P and A53T to that of the wt protein, in *in-vitro* preparations, where such binding can be documented readily. However, the results have been controversial. While some studies showed decreased membrane-binding of the human A30P and the A53T mutants (Iwai et al. 1995; Jensen et al. 1998; Jo et al. 2000, 2002), others reported an increase in membrane-binding with the A53T mutant (Sharon et al. 2001; Lotharius et al. 2002), or no change in membrane-binding of these mutants at all (McLean et al. 2000; Perrin et al. 2000).

Most of these studies measured the ability of purified human α-synuclein to insert spontaneously into lipids, and did not take into account cytosolic factors that can regulate this interaction, including chaperones/cofactors normally present in the cell. However, a recent study adopted a novel biochemical approach by incubating membrane preparations from mice expressing wt and the A30P and A53T mutant human α-synucleins (donor) with brain cytosol from α-synuclein null mice (acceptor) (Wislet-Gendebien et al. 2006). The rationale was to evaluate the interactions of the mutant α-synucleins with membranes in the presence of cytosolic factors, thereby mimicking the in vivo situation. The rationale for using the α-synuclein null brains as the "acceptor" was to eliminate possible confounding factors introduced by endogenous mouse α-synuclein. These studies showed that both the human mutations decreased the association of α-synuclein with membranes compared to the wt protein, but remarkably, *only in the presence of brain cytosol*. These studies suggest that under conditions that mimic the *in-vivo* situation in the brain, both human α-synuclein mutants tend to show reduced association with synaptic structures, conferring a pathologic signature common to both human mutants.

Two studies have also evaluated A30P and A53T mutant human α-synucleins in cultured hippocampal neurons from wt mice, expressing endogenous mouse α-synuclein (Fortin et al. 2004, 2005). Using GFP-tagged human α-synucleins transfected in cultured hippocampal neurons from wt mice, these studies showed that while the A30P mutant was mislocalized to synapses, there was no difference in the presynaptic targeting of the A53T mutant, compared to the wt protein. These results in hippocampal cultures seem to contradict the biochemical studies by Wislet-Gendebien et al. (2006) mentioned above, though both seemingly replicated an "in vivo like" situation. Intriguingly, however, the one difference between these two studies is that while the biochemical study used cytosol from α-synuclein null mice, the experiments with cultured hippocampal neurons were done in wt-neurons that express the endogenous mouse α-synuclein as well. Thus, it is possible that the presence of endogenous mouse α-synuclein may have confounded some of the observations in the cultured neurons (Fortin et al. 2004). Further studies in α-synuclein null neurons are likely to resolve this outstanding issue. Besides these mutants, little is known about the presynaptic targeting of other pathologic α-synucleins *in-vivo*.

7 Conclusions and Perspectives

In this chapter, current understanding of the mechanisms underlying defective biogenesis/axonal transport/targeting of α-synuclein in diseased states are discussed. It is clear that despite significant effort by researchers towards understanding the mechanisms of α-synuclein in health and disease, many basic questions regarding its biology and pathology remain unanswered. One of the major problems has been the chameleon-like nature of the α-synuclein protein itself, which not only behaves diversely in in vivo and in vitro settings, but also tends to localize, and behave anomalously in non-neuronal cells. Unfortunately, the majority of the cell biological studies to date have been performed in a variety of non-neuronal cells, making it difficult to interpret and synthesize the available data. Finally, as noted above, some of these controversies may be due, at least in part, to the confounding effects of the endogenous mouse α-synuclein. In summary, though there is tantalizing data supporting the hypothesis that axonal transport and/or targeting of α-synuclein is disrupted in diseased states, the lack of systematic studies of transport/targeting of wt and disease-associated α-synucleins in a suitable model-system is lacking. Future studies focusing on these outstanding issues may eventually resolve these controversies and provide a mechanistic basis for mislocalization of this protein in diseased states.

Acknowledgment Work in our laboratory is supported by grants from the Larry Hillblom foundation, The Alzheimer's Association, the American Parkinson's Disease Association, the National Institute for Aging, and a generous donation to the UCSD Alzheimer's Center by Darlene and Donald Shiley.

References

Braak H, Sandmann-Keil D, Gai W, Braak E (1999) Extensive axonal Lewy neurites in Parkinson's disease: a novel pathological feature revealed by alpha-synuclein immunocytochemistry. Neurosci Lett 265:67–69

Buchman VL, Hunter HJ, Pinon LG, Thompson J, Privalova EM, Ninkina NN, Davies AM (1998) Persyn, a member of the synuclein family, has a distinct pattern of expression in the developing nervous system. J Neurosci 18:9335–9341

Chen L, Feany MB (2005) Alpha-synuclein phosphorylation controls neurotoxicity and inclusion formation in a Drosophila model of Parkinson disease. Nat Neurosci 8:657–663

Clayton DF, George JM (1998) The synucleins: a family of proteins involved in synaptic function, plasticity, neurodegeneration and disease. Trends Neurosci 21:249–254

Cooper AA, Gitler AD, Cashikar A, Haynes CM, Hill KJ, Bhullar B, Liu K, Xu K, Strathearn KE, Liu F, Cao S, Caldwell KA, Caldwell GA, Marsischky G, Kolodner RD, Labaer J, Rochet JC, Bonini NM, Lindquist S (2006) Alpha-synuclein blocks ER-Golgi traffic and Rab1 rescues neuron loss in Parkinson's models. Science 313:324–328

Cuervo AM, Stefanis L, Fredenburg R, Lansbury PT, Sulzer D (2004) Impaired degradation of mutant alpha-synuclein by chaperone-mediated autophagy. Science 305:1292–1295

Dixon C, Mathias N, Zweig RM, Davis DA, Gross DS (2005) Alpha-Synuclein targets the plasma membrane via the secretory pathway and induces toxicity in yeast. Genetics 170:47–59

Feany MB, Bender WW (2000) A Drosophila model of Parkinson's disease. Nature 404:394–398

Fortin DL, Troyer MD, Nakamura K, Kubo S, Anthony MD, Edwards RH (2004) Lipid rafts mediate the synaptic localization of alpha-synuclein. J Neurosci 24:6715–6723

Fortin DL, Nemani VM, Voglmaier SM, Anthony MD, Ryan TA, Edwards RH (2005) Neural activity controls the synaptic accumulation of alpha-synuclein. J Neurosci 25:10913–10921

Giasson BI, Duda JE, Quinn SM, Zhang B, Trojanowski JQ, Lee VM (2002) Neuronal alpha-synucleinopathy with severe movement disorder in mice expressing A53T human alpha-synuclein. Neuron 34:521–533

Gorbatyuk OS, Li S, Sullivan LF, Chen W, Kondrikova G, Manfredsson FP, Mandel RJ, Muzyczka N (2008) The phosphorylation state of Ser-129 in human alpha-synuclein determines neurodegeneration in a rat model of Parkinson disease. Proc Natl Acad Sci USA 105:763–768

Iwai A, Masliah E, Yoshimoto M, Ge N, Flanagan L, de Silva HA, Kittel A, Saitoh T (1995) The precursor protein of non-A beta component of Alzheimer's disease amyloid is a presynaptic protein of the central nervous system. Neuron 14:467–475

Jensen PH, Li JY, Dahlstrom A, Dotti CG (1999) Axonal transport of synucleins is mediated by all rate components. Eur J Neurosci 11:3369–3376

Jensen PH, Nielsen MS, Jakes R, Dotti CG, Goedert M (1998) Binding of alpha-synuclein to brain vesicles is abolished by familial Parkinson's disease mutation. J Biol Chem 273:26292–26294

Jo E, McLaurin J, Yip CM, George-Hyslop P, Fraser PE (2000) Alpha-Synuclein membrane interactions and lipid specificity. J Biol Chem 275:34328–34334

Jo E, Fuller N, Rand RP, George-Hyslop P, Fraser PE (2002) Defective membrane interactions of familial Parkinson's disease mutant A30P alpha-synuclein. J Mol Biol 315:799–807

Kahle PJ, Neumann M, Ozmen L, Muller V, Jacobsen H, Schindzielorz A, Okochi M, Leimer U, der PH Van, Probst A, Kremmer E, Kretzschmar HA, Haass C (2000) Subcellular localization of wild-type and Parkinson's disease-associated mutant alpha-synuclein in human and transgenic mouse brain. J Neurosci 20:6365–6373

Kahle PJ, Neumann M, Ozmen L, Muller V, Odoy S, Okamoto N, Jacobsen H, Iwatsubo T, Trojanowski JQ, Takahashi H, Wakabayashi K, Bogdanovic N, Riederer P, Kretzschmar HA, Haass C (2001) Selective insolubility of alpha-synuclein in human Lewy body diseases is recapitulated in a transgenic mouse model. Am J Pathol 159:2215–2225

Kontopoulos E, Parvin JD, Feany MB (2006) Alpha-synuclein acts in the nucleus to inhibit histone acetylation and promote neurotoxicity. Hum Mol Genet 15:3012–3023

Kuwahara T, Koyama A, Gengyo-Ando K, Masuda M, Kowa H, Tsunoda M, Mitani S, Iwatsubo T (2006) Familial Parkinson mutant alpha-synuclein causes dopamine neuron dysfunction in transgenic Caenorhabditis elegans. J Biol Chem 281:334–340

Lakso M, Vartiainen S, Moilanen AM, Sirvio J, Thomas JH, Nass R, Blakely RD, Wong G (2003) Dopaminergic neuronal loss and motor deficits in Caenorhabditis elegans overexpressing human alpha-synuclein. J Neurochem 86:165–172

Lee MK, Stirling W, Xu Y, Xu X, Qui D, Mandir AS, Dawson TM, Copeland NG, Jenkins NA, Price DL (2002) Human alpha-synuclein-harboring familial Parkinson's disease-linked Ala-53Thr mutation causes neurodegenerative disease with alpha-synuclein aggregation in transgenic mice. Proc Natl Acad Sci USA 99:8968–8973

Levitan ES (2008) Signaling for vesicle mobilization and synaptic plasticity. Mol Neurobiol 37:39–43

Li W, Hoffman PN, Stirling W, Price DL, Lee MK (2004) Axonal transport of human alpha-synuclein slows with aging but is not affected by familial Parkinson's disease-linked mutations. J Neurochem 88:401–410

Link CD (2001) Transgenic invertebrate models of age-associated neurodegenerative diseases. Mech Ageing Dev 122:1639–1649

Lotharius J, Barg S, Wiekop P, Lundberg C, Raymon HK, Brundin P (2002) Effect of mutant alpha-synuclein on dopamine homeostasis in a new human mesencephalic cell line. J Biol Chem 277:38884–38894

Masliah E, Rockenstein E, Veinbergs I, Mallory M, Hashimoto M, Takeda A, Sagara Y, Sisk A, Mucke L (2000) Dopaminergic loss and inclusion body formation in alpha-synuclein mice: implications for neurodegenerative disorders. Science 287:1265–1269

McKeith I, Mintzer J, Aarsland D, Burn D, Chiu H, Cohen-Mansfield J, Dickson D, Dubois B, Duda JE, Feldman H, Gauthier S, Halliday G, Lawlor B, Lippa C, Lopez OL, Carlos MJ, O'Brien J, Playfer J, Reid W (2004) Dementia with Lewy bodies. Lancet Neurol 3:19–28

McKeith IG, Mosimann UP (2004) Dementia with Lewy bodies and Parkinson's disease. Parkinsonism Relat Disord 10(Suppl 1):S15–S18

McLean PJ, Kawamata H, Ribich S, Hyman BT (2000) Membrane association and protein conformation of alpha-synuclein in intact neurons. Effect of Parkinson's disease-linked mutations. J Biol Chem 275:8812–8816

Mukaetova-Ladinska EB, McKeith IG (2006) Pathophysiology of synuclein aggregation in Lewy body disease. Mech Ageing Dev 127:188–202

Orimo S, Uchihara T, Nakamura A, Mori F, Kakita A, Wakabayashi K, Takahashi H (2008) Axonal alpha-synuclein aggregates herald centripetal degeneration of cardiac sympathetic nerve in Parkinson's disease. Brain 131:642–650

Perrin RJ, Woods WS, Clayton DF, George JM (2000) Interaction of human alpha-Synuclein and Parkinson's disease variants with phospholipids. Structural analysis using site-directed mutagenesis. J Biol Chem 275:34393–34398

Roy S, Zhang B, Lee VM, Trojanowski JQ (2005) Axonal transport defects: a common theme in neurodegenerative diseases. Acta Neuropathol 109:5–13

Roy S, Winton MJ, Black MM, Trojanowski JQ, Lee VM (2007) Rapid and intermittent cotransport of slow component-b proteins. J Neurosci 27:3131–3138

Roy S, Winton MJ, Black MM, Trojanowski JQ, Lee VM (2008) Cytoskeletal requirements in axonal transport of slow component-b. J Neurosci 28:5248–5256

Saha AR, Hill J, Utton MA, Asuni AA, Ackerley S, Grierson AJ, Miller CC, Davies AM, Buchman VL, Anderton BH, Hanger DP (2004) Parkinson's disease alpha-synuclein mutations exhibit defective axonal transport in cultured neurons. J Cell Sci 117:1017–1024

Saito Y, Kawashima A, Ruberu NN, Fujiwara H, Koyama S, Sawabe M, Arai T, Nagura H, Yamanouchi H, Hasegawa M, Iwatsubo T, Murayama S (2003) Accumulation of phosphorylated alpha-synuclein in aging human brain. J Neuropathol Exp Neurol 62:644–654

Sharon R, Goldberg MS, Bar-Josef I, Betensky RA, Shen J, Selkoe DJ (2001) Alpha-Synuclein occurs in lipid-rich high molecular weight complexes, binds fatty acids, and shows homology to the fatty acid-binding proteins. Proc Natl Acad Sci USA 98:9110–9115

Singleton AB et al (2003) alpha-Synuclein locus triplication causes Parkinson's disease. Science 302:841

Soper JH, Roy S, Stieber A, Lee E, Wilson RB, Trojanowski JQ, Burd CG, Lee VM (2008) {alpha}-Synuclein-induced aggregation of cytoplasmic vesicles in Saccharomyces cerevisiae. Mol Biol Cell 19:1093–1103

Tofaris GK, Garcia Reitbock P, Humby T, Lambourne SL, O'Connell M, Ghetti B, Gossage H, Emson PC, Wilkinson LS, Goedert M, Spillantini MG (2006) Pathological changes in dopaminergic nerve cells of the substantia nigra and olfactory bulb in mice transgenic for truncated human alpha-synuclein(1–120): implications for Lewy body disorders. J Neurosci 26:3942–3950

Tompkins MM, Gai WP, Douglas S, Bunn SJ (2003) Alpha-synuclein expression localizes to the Golgi apparatus in bovine adrenal medullary chromaffin cells. Brain Res 984:233–236

Uversky VN (2007) Neuropathology, biochemistry, and biophysics of alpha-synuclein aggregation. J Neurochem 103:17–37

Wakabayashi K, Tanji K, Mori F, Takahashi H (2007) The Lewy body in Parkinson's disease: molecules implicated in the formation and degradation of alpha-synuclein aggregates. Neuropathology 27:494–506

Wislet-Gendebien S, D'Souza C, Kawarai T, St George-Hyslop P, Westaway D, Fraser P, Tandon A (2006) Cytosolic proteins regulate alpha-synuclein dissociation from presynaptic membranes. J Biol Chem 281:32148–32155

Withers GS, George JM, Banker GA, Clayton DF (1997) Delayed localization of synelfin (synuclein, NACP) to presynaptic terminals in cultured rat hippocampal neurons. Brain Res Dev Brain Res 99:87–94

Zhou W, Milder JB, Freed CR (2008) Transgenic mice overexpressing tyrosine-to-cysteine mutant human {alpha}-Synuclein: a progressive neurodegenerative model of diffuse lewy body disease. J Biol Chem 283:9863–9870

Organized Ribosome-Containing Structural Domains in Axons

Edward Koenig

Abstract *Periaxoplasmic ribosomal plaques* (PARPs) are systematically recurring ribosome-containing structural domains located in the F-actin-rich periphery of axoplasm in myelinated fibers. In contrast, *endoaxoplasmic ribosomal plaques* (EARPs) are small, oval-shaped ribosomal aggregate structures randomly dispersed within the axoplasm of unmyelinated squid giant axons. Ribosomes are attached to a superficial plaque-like structural matrix, which "caps" the domain at the outer cortical margin and appears fragmented in subcortical axoplasm. As such, the matrix represents a novel hallmark of PARP domains. Molecular markers concentrated in PARP domains include β-actin mRNA, ZBP-1, SRP54, myosin Va and kinesin II molecular motor proteins. Rapid axoplasmic transport of microinjected heterologous radiolabeled BC1 RNA to putative PARP domains, mediated *pari passu* by microtubule- and F-actin-dependent systems, suggests that translation machinery, anchored by the matrix could provide targeted destinations for RNA trafficking. As distributed local centers of protein synthesis along axons, PARPs are likely to share modes of expression in common with other translational subdomains in neurons.

1 Introduction

An apparent lack of microscopically visible ribosomes in mature axons, which are the structural *sine qua non* of translational machinery was long touted as a hallmark of axons. It gave rise to a fallacious, long-entrenched dogma in neurobiology that axons do not have an endogenous capacity to synthesize proteins. This review focuses on systematic specialized ribosome-containing structural subdomains in vertebrate and invertebrate axonal compartments that conventional microscopic studies had earlier failed to discern.

E. Koenig
Department of Physiology and Biophysics, SUNY at Buffalo School of Medicine,
Buffalo, NY14214-3013, USA
e-mail: ekoenig@buffalo.edu

1.1 Initial Efforts to Document RNA in Vertebrate Model Axons

Methods to determine RNA content and nucleotide base composition in the picogram range developed by Edström were essential to documenting RNA in axoplasm and myelin sheath segments microdissected from formaldehyde-fixed goldfish spinal cord Mauthner fibers (Edström et al. 1962). Later analysis by Edström (1964a) indicated that RNA concentration in Mauthner axons in the spinal cord varied along the axon (i.e., highest at rostral and caudal ends, and lowest in the middle region), while nucleotide base composition exhibited high proportions of guanine and cytosine, consistent with those of ribosomal RNA (rRNA). A similar RNA base composition was observed in axoplasm translated out of denatured myelinated fibers of spinal accessory nerve root from cat (Koenig 1965).

Spinal cord transection in the goldfish (Edström 1964b) induced parallel increases in RNA content in axoplasm and surrounding myelin sheath. It also led to an unexpected parallel elevation in the adenine/guanine (A/G) ratio of RNA in both myelin sheath and axon over a post transection period of 20–30 days. The higher ratio reflected a preferential shift in synthesis and enrichment of nonribosomal RNA transcripts during this period. The results raised the possibility of a common source of RNA (e.g., periaxonal glial cell bodies) in sheath and subjacent axon (for discussion of glial-axonal RNA transfer, see Giuditta et al. 2008).

2 Ribosomal RNA (rRNA) in Model Axons

The important issue of ribosomal RNA (rRNA) in axons could not be addressed by a method that used ribonuclease (RNase) digestion to extract RNA from isolated cellular samples. Another methodology developed later to extract and to analyze undegraded RNA from microscopic samples showed that *native* axoplasm isolated from Mauthner fibers contained cytoribosomal RNA components typical of goldfish (Koenig 1979). Earlier, Giuditta et al. (1977) reported that multiple translational cofactors were present in the squid giant axon, and Giuditta et al. (1980) later documented cytoribosomal RNA components in isolated squid axoplasm.

Representative electrophoretic profiles of total RNA extracts from Mauthner (Koenig 1979), and from squid giant fibers (Giuditta et al. 1980) are shown in Fig. 1. A distinctive feature of axoplasmic RNA profiles of both axons is a disproportionately large $4S_E$ RNA relative to rRNA components, in contrast to RNA extracted from cell-rich tissues. The relatively large proportion of $4S_E$ RNA can be explained by assuming that the tRNA concentration must be large enough to satisfy local requirements for protein synthesis in a large axonal compartmental volume, in which there may be a low or modest content of ribosomes. Although the squid giant axon diameter exceeds that of the Mauthner axon by a factor of 10, size difference alone is not sufficient to account for the strikingly large differences between the two axonal profiles. Indeed, differences in organizational characteristics of ribosomal domains and the frequency of ribosome occurrence are factors that contribute to profile differences (e.g., see Sects. 5.1, 5.2, and 8).

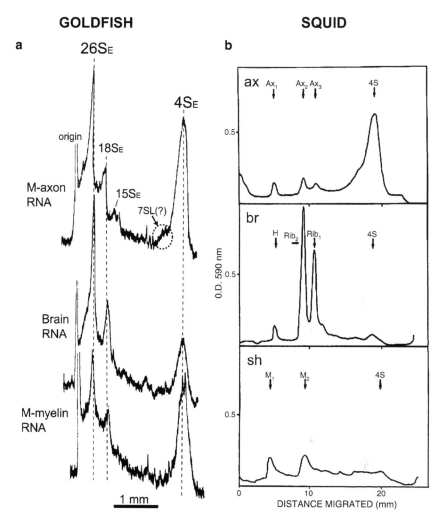

Fig. 1 Representative electrophoretic profiles of RNA extracted from axoplasm of (**a**) goldfish Mauthner fiber, and (**b**) squid giant fiber. A shoulder (*dashed oval*) on the trailing 4S peak in the "M-axon RNA" profile may be an unresolved SRP 7SL RNA (see text, and Sect. 5.3.1). Note the large disproportional 4S peak in relation to ribosomal peaks in both axon types, compared to corresponding RNA profiles from cell-rich tissues (**a, b**, *middle profiles*). Note also the disproportional 4S peak in the squid axon relative to rRNA peaks (a, ax) is much larger in comparison to that of Mauthner axon (a, M-axon RNA; see text for discussion). Mauthner myelin sheath RNA (a, M-myelin RNA) also contains rRNA peaks, and a disproportional large 4S peak similar to the M-axon. In contrast, the cell-rich squid fiber sheath (b, sh) has a typical small 4S peak. Abbreviations in (**b**) ax: axoplasm; br: brain; sh: sheath. Records modified from Koenig (1979), and from Giuditta et al. (1980) are shown with permission from Wiley-Blackwell

An apparent $15S_E$ peak in Mauthner (M-) axon RNA profile was probably derived from mitochondrial rRNA. Another consistent characteristic was a trailing "shoulder" of the $4S_E$ peak (Fig. 1a), which may be an unresolved 7SL RNA derived from the signal recognition particle (SRP). Reasons for this inference are discussed in Sect. 5.3.1.

It is also noteworthy that the myelin sheath of the Mauthner fiber, which is free of glial cell bodies, contains rRNA components, and a large $4S_E$ RNA (Fig. 1a). In contrast, the unmyelinated fiber sheath surrounding the squid giant axon contains glial cells, and has a typical small 4S peak (Fig. 1b).

3 Early Random Sightings of Ribosomes in Myelinated Axons

Although a lack of ribosomes was considered the hallmark of axons, there were early reports of ribosomes in initial axon segments of vertebrate myelinated axons based on electron microscopy (EM) (Conradi 1966; Palay et al. 1968; Peters et al. 1968). Later, EM inspection of dorsal roots and ganglia of rat by Zelenà (1972) revealed ribosomes in the initial axon segment, and ribosomal clusters in axons of dorsal root fibers; however, the probability of finding ribosomes was greatest within the intraganglionic portion of dorsal root fibers. Systematic examination of serial sections extending over tens-of-micrometers of dorsal root fibers documented ribosome distributions over lengths of 6.4, 26.2, and 58.8 mm in 3 of the 198 fibers (Pannese and Ledda (1991).

When ribosomes were observed, they were usually located in the peripheral zone near the axonal membrane in close proximity to the mitochondria (Zelenà 1972; Pannese and Ledda 1991). The latter observations are noteworthy in the light of recent data that show the importance of local extramitochondrial protein synthesis for mitochondrial function and viability of growing axons (Hillefors et al. 2007). A final point of interest is that Pannese and Ledda (1991) reported a low incidence of ribosomes (i.e., 0.4%) bound to a tubular endoplasmic reticulum (see Sect. 5.3.1).

4 Alternative Approaches to Investigate Axonal Ribosomes

Although rRNA was extracted from isolated Mauthner axoplasm, EM examination of random sections yielded no evidence of ribosomes (Koenig 1979), nor did an earlier EM survey by Edström and Sjöstrand (1969). *If Mauthner axoplasm contains rRNA, then why have conventional histological approaches at light and ultrastructural levels failed to identify significant ribosomal content?* This apparent paradox prompted an unconventional experimental approach to investigate the question, utilizing the isolated axoplasmic whole-mount as the principal experimental preparation, and electron spectroscopic imaging (ESI) to remove ambiguity in the identification of ribosomes at an EM level.

4.1 Axoplasmic Whole-Mounts

Native axoplasm behaves as a visco-elastic, easily disrupted gel (Sato et al. 1984). Tensile strength of the axoplasmic core and the ease of translating axoplasm from its ensheathment improve with an increase in diameter and neurofilament content. The visco-elastic gel is transformed into a plastic solid by zinc denaturation, which enhances tensile strength, and isolation of axoplasmic whole-mounts from small diameter fibers. The isolated preparation (Koenig 1965, 1979, 1991) offered advantages, and the large myelinated Mauthner axon, in which rRNA content was documented, clearly was the experimental model of choice. Moreover, the acutely isolated preparation provided a global overview of the axon compartment, in which the distribution and the localization of RNA could be rapidly evaluated by simple epifluorescence microscopy after staining with a high affinity fluorescent nucleic acid binding dye, such as YOYO-1.

Shearing of intercellular adhesive interactions/cross-linking between surfaces of the axon and the surrounding myelin sheath during translation, however, imposed limitations for purposes of evaluating localization and distribution of ribosomes because the latter are located in the cortical zone; i.e., inadvertent loss of peripheral axoplasmic integrity could result in loss of ribosomes. Preservation of cortical integrity during translation was variable, or indeterminable due to inconstant physical properties among experimental fiber preparations. Despite this shortcoming, the approach yielded unexpected and novel findings.

4.2 Electron Spectroscopic Imaging (ESI) of Ribosomes

Conventional electron microscopy depends on heavy metal staining to obtain visible contrast. Staining, however, is nonspecific; it can be capricious by failing to stain, and structures can only be identified by morphological criteria alone. ESI provides an objective physical basis for identifying ribosomes.

In principle, ESI requires an electron microscope equipped with an energy spectrometer, and appropriate lens systems to image inelastically scattered electrons emitted from a specimen in the image plane (Ottensmeyer 1986). A high-energy electron beam that irradiates an ultrathin (10–20 nm) plastic-embedded, unstained tissue section produces elastically and inelastically scattered electrons. The latter arise from energy loss (i.e., ΔE) that occurs at the $L_{2,3}$ inner shell of electrons of an element in a specific region of the energy loss spectrum. Ribosomes, which are particles that have almost 7,000 nucleotides with a corresponding number of phosphorus (P) atoms, produce a bright ribosome P signal of 25 nm in size in a low contrast background image, which is readily detectable by ESI (Korn et al. 1983). The ionization threshold for phosphorus (P) is $\Delta E = 132$ eV (i.e., phosphorus edge). Above the P edge, the relative brightness of the P signal is greatly augmented in intensity.

In practice, an energy window of $\Delta E \geq 150$ eV is selected for inspection. For a bright P signal to be specific, it must disappear into the low contrast background when viewed in an energy window below the P edge (e.g., $\Delta E = 110$ eV). Thus, a bright signal above the P edge that fades completely below it is a simple, reliable test that confirms the specificity of the P signal. Should a bright signal diminish, but not disappear, it may be a result of mass rather than P (see Door et al. 1997). An imperative for ESI analysis is exclusion of heavy metal staining, which precludes fixation of myelin with OsO_4 that is ordinarily used in conventional EM. Professor Rainer Martin (Universität Ulm) examined ultrathin sections of all Epon-embedded, glutaraldehyde-fixed isolated axoplasmic whole-mounts, and one case of lipid-extracted isolated myelinated fibers in which ribosomes were mapped by ESI.

5 Periaxoplasmic Ribosomal Plaque (PARP) Domains in Vertebrate Myelinated Axons

5.1 PARP Domains in Myelinated Mauthner Axons at a Light Microscope (LM) Level

Salient characteristics of ribosome-containing domains in axoplasmic whole-mounts isolated from Mauthner cell fibers (Koenig and Martin 1996) are illustrated in Fig. 2. YOYO-1 staining reveals discrete, intermittent fluorescent domains distributed along the axoplasmic whole-mount (Fig. 2a). Their appearance is highly variable, and pre-digestion with ribonuclease (RNase) eliminates binding of YOYO-1 (Fig. 2a1), which indicates that dye binding depends on undegraded RNA.

Native, nondenatured axoplasmic whole-mounts yield the most optimal structural detail when cortical integrity is preserved. High magnification reveals "plaque-like" surface structures with low fluorescence intensity, surrounded by highly fluorescent perimeters (Fig. 2b, b1). In addition, fluorescent punctae, which form an underlying "cloud-like" distribution, delimit the axoplasmic boundary of the domain. The plaque-like surface structure is clearly visible in DIC, or phase images of corresponding fluorescent domains (Fig. 2c–d1).

The plaque structure is a novel hallmark feature of ribosome-containing domains in axoplasmic whole-mounts from myelinated fibers, which prompted the name, *periaxoplasmic ribosomal plaques* (PARPs) (Koenig and Martin 1996). That these periodic domains along the axon's cortical zone contain ribosomes is based on experiments described in detail elsewhere (Koenig and Martin 1996; Koenig et al. 2000), and briefly reviewed here. Evidence supporting this conclusion includes: RNase-sensitive binding of high affinity RNA dyes, binding of monoclonal (mAb) Y-10B antibodies that recognizes the large rRNA subunit (Lerner et al. 1981), and binding of polyclonal antibodies against human ribosomal P proteins. Finally, ESI analysis identifies specific 25 nm rRNA phosphorus signals at an EM level (e.g., see Sect. 5.2).

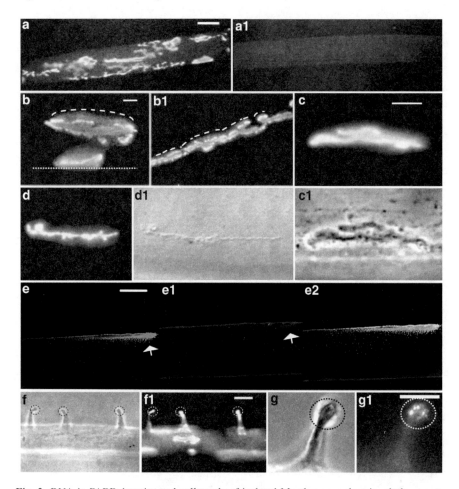

Fig. 2 RNA in PARP domains and collaterals of isolated Mauthner axoplasmic whole-mounts. (a) Fluorescent PARP domains stained by YOYO-1 are located along the surface boundary of the whole-mount (*note*: PARPs on opposite surface are *blurred*). (a1) RNase digestion eliminates YOYO-1 fluorescence staining. (b, b1) YOYO-1 fluorescent staining of PARP domains of a *native* axoplasmic whole-mount, in which fluorescent "punctae" form "cloud-like" distributions subjacent to the surface PARP structure (contoured dashed lines), which serve to "delimit" the axoplasmic boundary of the domain (*note: finely dashed line* in (b) marks lateral margin of whole-mount). (c, d) Examples of PARPs stained with YOYO-1, and (c1, d1) corresponding *en face* views of surface-localized structural correlates visible by (c1) phase-contrast, and (d1) DIC images. (e–e2) Confocal images of a PARP domain in longitudinal profile. (e) YOYO-1 fluorescence staining of RNA exhibits dense punctate distribution in the cortical and subcortical axoplasm (*arrow*). (e1) Cortical F-actin layer, and subcortical dense F-actin network (*arrow*) revealed by rhodamine phalloidin fluorescence. (e2) Merging of (e) RNA and (e1) F-actin images illustrate the overlapping co-distributions. Some scattered fluorescent punctae extend well into the axoplasmic core (e, e2). (f) Phase image of short axon collaterals with bouton-like terminals (*dashed circles*), projecting from a whole-mount segment, and (f1) corresponding epifluorescence image after YOYO-1 staining that reveals RNA localization in nearby PARPs, and in terminals. (g) High power phase image of a terminal bouton (*dashed circle*), and (g1) corresponding mAb Y-10B immuno-fluorescence image, showing punctate ribosomal aggregates. *Scale bars*: 10 μm. a–e2 from Koenig and Martin (1996), with permission from *Journal of Neuroscience*

A cortical filamentous actin layer characteristically underlies the plasma membrane of most eukaryotic cells. In the PARP domain, fluorescently labeled punctae, which are putative polyribosomes (see Sect. 5.2), appear to be an inclusion of the cortical F-actin layer, and are also codistributed with F-actin in subcortical axoplasm. This is illustrated in a longitudinal confocal image profile of a PARP domain, in which RNA is labeled by YOYO-1 (Fig. 2e), and F-actin is labeled by rhodamine–phalloidin (Fig. 2e1). RNA fluorescence (Fig. 2e) codistributes with F-actin fluorescence (Fig. 2e1) in the cortical layer, and with a dense meshwork of actin filaments below the cortical layer (Fig. 2e2). The importance of the actin cytoskeletal network in supporting translational activity in the PARP domain is underscored by strong inhibition of [^{32}S]met/cys incorporation into axoplasmic proteins in vitro caused by disruption of the actin cytoskeleton (Sotelo-Silveira et al. 2008).

The axon terminal is a specialized subdomain of the axon, in which protein synthesis may play a synapse-related functional role (reviewed by Giuditta et al. 2008). The Mauthner axon has periodic short collaterals that project through the thickness of the myelin sheath to form *en passant* synapses with a neural network mediating a reflex "C"-bend of trunk musculature. Collaterals that project from an isolated whole-mount illustrated in Fig. 2f show variable RNA fluorescence staining of bouton-like terminals with YOYO-1 (Fig. 2f1). Immuno-labeling of rRNA by mAb Y-10B, however, at high resolution (Fig. 2g) reveals punctate immunofluorescence of ribosomal aggregates (Fig. 2g1).

5.2 *ESI of PARP Domains in Mauthner Axons at an EM Level*

Selected features of the PARP domain in Mauthner axoplasmic whole-mounts as revealed by ESI are illustrated in Fig. 3. The domain region is "capped" with a fenestrated/layered matrix located at the surface margin of axoplasm. The high contrast matrix appears dark because of very low inelastic scattering of electrons in the energy window selected for phosphorus detection (Fig. 3a). The matrix is probably the structural correlate identified in phase and DIC images (see Fig. 2c1, d1). Bright ribosomal P signals above the P edge (Fig. 3a), which disappear below the P edge (Fig. 3a1), indicate that they are P-rich ribosomes. Clusters of ribosomal P signals are probably polyribosomes that correspond to fluorescent punctae at a LM level (Fig. 2b, b1, e).

(c) A PARP domain mapped above the P edge ($\Delta E = 150$ eV) in a tangentially cut section of a rabbit whole-mount from a ventral root fiber that contains ribosome aggregates (*inset, arrows*) and smaller non-ribosomal RNP P signals (*inset, arrowheads*) and endoplasmic reticulum (ER). (**c1**) Domain mapped below the P edge ($\Delta E = 110$ eV) indicate that bright signals in (**c**) are P. (**d**) A PARP domain in a lipid-extracted ventral root myelinated fiber, showing cluster complexes of ribosomal P signals (*double headed arrows*), mapped above the P edge ($\Delta E = 155$ eV), and (**d1**) mapped above the carbon edge ($\Delta E = 300$ eV) to reveal corresponding binding matrix sites (double headed arrows; see text). Scale bars: (**a, a1**) = 0.44 µm; (**b**) = 0.30 µm; (**d, d1**) = 0.15 µm. (**a-b**) Modified from Koenig and Martin (1996); (**c-d1**) modified from Koenig et al. 2000, with permission from *Journal of Neuroscience*

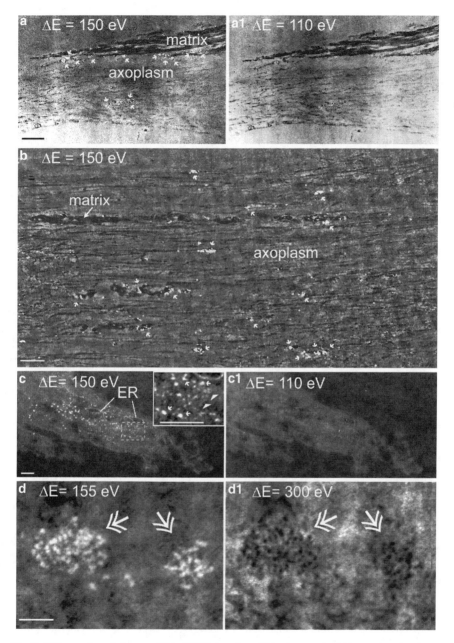

Fig. 3 ESI mapping of ribosomal P in PARP domains of Mauthner and rabbit axoplasmic wholemounts. (**a**) A grazing, tangentially cut ultrathin section mapped in an energy window above the P edge (ΔE = 150 eV), in which clusters of ribosomes (*arrows*) that probably represent polyribosomes are associated with a structural matrix, and also associated with matrix fragments that are distributed within subcortical axoplasm. (**a1**) The same PARP domain as in (**a**) mapped in an energy window below the P edge (ΔE = 110 eV) shows loss of bright ribosomal P signals. (**b**) A horizontal section through subcortical axoplasm of a PARP domain mapped above the P edge (ΔE = 150 eV), in which matrix fragments with ribosomal clusters (*arrows*) are shown in greater detail.

A fragmentary matrix is also associated with ribosomal P signals in axoplasm (Fig. 3a, c). It is unknown whether matrix fragments are fragmented, or are actually structurally continuous with periaxoplasmic PARP matrix. In any case, the frequent association of ribosomes with the matrix suggests that the latter may serve as scaffolding for translational machinery to spatially organize the PARP domain. As such it could serve as a potential target for RNA trafficking (see Sect. 6).

5.3 PARP Domains in Mammalian Myelinated Axons at the LM Level

PARP domains in mammalian myelinated axons were examined in axoplasmic whole-mounts from dorsal and ventral spinal nerve fibers of rabbit and rat (Koenig et al. 2000). Intermittent distribution of PARPs along the surface boundary of whole-mounts was similar to Mauthner axons, but different from standpoints of size and configuration complexity.

Mammalian PARPs exhibit a simpler configuration, in being narrow, elongated domains (Fig. 4). Measurements of PARP sizes in a subset of rabbit axoplasmic whole-mounts that ranged 4–12 mm in diameter were performed on 378 PARPs stained with either YOYO-1, or mAbY-10B. PARP lengths, ranging 1–45 mm, had a mean length of 9.7 ± 5.9 mm, and a mean width of 2.1 ± 0.6 mm. Long PARPs are observed especially in whole-mounts isolated from old animals (Fig. 4a1, a2).

When PARPs are viewed with either DIC, or phase optics, the plaque structure appears as an excrescence, protruding from the surface margin of the whole-mount (Fig. 4c–g) similar to that observed in the goldfish (Fig. 2c1, d1). High resolution polyribosomal punctate fluorescence (Fig. 4b, b1, d1–h1) is similar in configuration with that of the "encompassing" phase plaque structures (Fig. 4d–h1). Again, this suggests that the structural matrix provides a requisite scaffold for docking translational machinery (see Sect. 5.2).

RNA labeling in nodes of Ranvier by YOYO-1 was only infrequently observed (Fig. 4i, j1), and the pattern of fluorescence was highly variable, unlike the stereotyped pattern characteristic of PARP domains. There was also no distinctive surface structural correlate.

Kun et al. (2007) performed in situ hybridization (ISH) of rRNA, and immunocytochemistry at LM and EM levels to survey localization/distribution of ribosomes in the axons of intact rat sciatic nerve fibers. ISH of rRNA localized at the axon boundary, consistent with the location of PARP domains. Immuno-reaction deposits of polyclonal antibodies, raised against a purified rat brain ribosomal fraction, localized to similar peripheral axonal sites, and scattered discrete sites within the axoplasmic core. Occasional "multivesicular" structures spanning the myelin–axon interface boundaries, containing discrete immuno-reactive deposits at the pole protruding into the axon compartment were also described. The authors suggested these ribosome-containing multivesicular structures might be engaged in transcellular transfer of ribosomes (see Sect. 6).

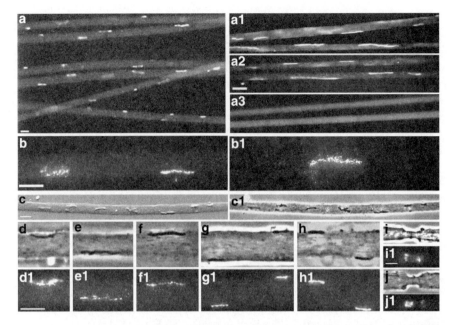

Fig. 4 Ribosome or RNA labeling of PARP domains, and in nodes of Ranvier of axoplasmic whole-mounts from rabbit ventral root fibers. (**a**) Low power image of isolated whole-mounts immuno-labeled by mAb Y-10B to illustrate random longitudinal distribution of PARP-containing ribosomes. Large PARP domains in whole-mounts of an older animal, in which (**a1**) ribosomes are immuno-labeled by mAb Y-10B, and (**a2**) RNA is labeled by YOYO-1. (**a3**) RNase digestion of the same whole-mounts shown in (**a2**) removes bound YOYO-1 fluorescence. (**b, b1**) Immuno-labeling of ribosomes by mAb Y-10B of PARP domains in high-resolution images reveal circumscribed fluorescent punctate distributions, which are probably polyribosomes (*see text*). (**c, c1**) Low power DIC, and corresponding phase images of the same whole-mount, reveal the distinctive structural excrescence that marks PARP domains at the surface boundary of the whole-mount. (**d–h**) Phase structural correlates of PARP domains, and (**d1–h1**) corresponding mAb Y-10B fluorescence labeling of ribosomal distributions show that configurations of the ribosome distributions parallel those of the structural correlates. (**i, j**) Phase images of two nodes of Ranvier, and (**i1, j1**) corresponding RNA fluorescence labeling by YOYO-1. RNA fluorescence distributions in occasionally labeled nodes are variable, and lack apparent phase structural correlates. *Scale bars*: 10 µm. (**a-h1**) from Koenig et al. (2000) with permission from *Journal of Neuroscience*

5.3.1 Evidence for Signal Recognition Particle in PARP Domains

Eukaryotic signal recognition particle (SRP) is a ribonucleoprotein complex, composed of a 7SL RNA backbone, and six associated proteins. SRP RNA, mapped by ESI, was shown to be ellipsoidal in shape that measured 15 nm (Bazzett-Jones 1988). SRP54, an associated protein, binds nascent chains of secretory/membrane polypeptides to interrupt translation transiently on ribosomes, and targets the ribosome-nascent chain complex to the ER, where it also mediates GTP dependent binding to the signal recognition site to resume translation (see Keenan et al. 2001).

A number of observations are consistent with the probability of SRP particles in the axon compartment. As noted previously (see Sect. 2), an unresolved peak in the trailing $4S_E$ peak is probably 7SL RNA (Fig. 1a), confirmed later by PCR of Mauthner whole-mount RNA (Sotelo-Silveira and Koenig, unpublished). ER bound ribosomes have been reported at an EM level (Pannese and Ledda 1991), and ESI of PARP domains shows smaller RNP P signals (Fig. 3c). Microinjection of heterologous conopressin receptor mRNA into isolated axons from *Lymmnaea* snail neurons resulted in the insertion of a functionally active translation product into the membrane (Spencer et al. 2000). In addition, local synthesis of the EphA2 receptor was up regulated, and expressed in the surface of growing axons (Brittis et al. 2002). A recent report (Merianda et al. 2008) describes ER and Golgi equivalents, and locally synthesized membrane proteins in growing axons.

The likelihood of SRPs in axons is further supported by immuno-labeling experiments in axoplasmic whole-mounts isolated from rabbit ventral root fibers (Koenig, unpublished observations). Antibodies against gel-purified SRP54 (Walter and Blobel 1983) selectively label PARP domains (Fig. 5a–a3, b1–e1). PARPs, in which structural correlates were lost during isolation, were immuno-labeled with mAb Y-10B to localize ribosomes (Fig. 5b–f), and with anti SRP54 to localize SRPs (Fig. 5b1–f1) revealed that residual SRP signals codistributed with ribosome distributions in merged images of the two pseudocolored distributions (Fig. 5b2–f2).

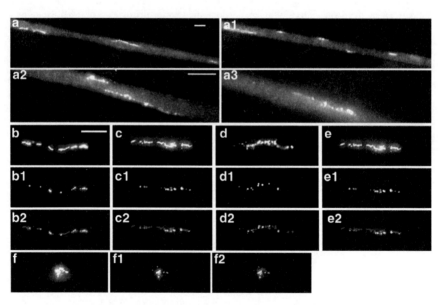

Fig. 5 SRP54 localizes in PARP domains. (**a, a1**) Low, and (**a2, a3**) high magnification images of SRP54 immuno-labeling of PARP domains in axoplasmic whole-mounts isolated from rabbit ventral root fibers. (**b–f2**) Double labeling of PARPs in isolated rabbit axoplasmic whole-mounts to illustrate (**b–f**) ribosomal distributions, labeled by mAb Y10-B, and (**b1-f1**) SRP54, labeled by a specific polyclonal antibody (Walter and Blobel, 1983; gift of Prof. Peter Walter). (**b2–f2**) Merged images of the pseudo-colored distributions, respectively, show overlapping distributions. *Scale bars*: (**a–g**) = 10 µm

5.3.2 Specific Molecular Motor Proteins Localize in PARP Domains

Localization and distribution of several molecular motor proteins were investigated in axoplasmic whole-mounts from Mauthner, rabbit and rat (Sotelo-Silveira et al. 2004). Specific immuno-labeling of myosin Va, and KIF3A heavy chain of kinesin II occurred primarily in PARP domains. In contrast, kinesin I (i.e., KIF5A/C isoforms), and dynein molecular motor proteins were diffusely distributed within axoplasm.

The functional significance of specific actin- and microtubule-based molecular motor proteins in a restricted domain such as that of a PARP is presently unknown; however, several explanations can be considered. PARP domains contain a dense actin filament network (see Fig. 2e–e2), and are likely targets for RNA trafficking in myelinated axons (see Sect. 5.4). Molecular motor proteins involved in long-range microtubule dependent, and short-range actin dependent RNA trafficking to PARP domains, exemplified by BC1 RNA transport experiments (Muslimov et al. 2002; see Sect. 6), may accumulate in PARP domains after off-loading cargoes. Molecular motor proteins within the PARP domain may mediate dynamic redistribution of machinery, and components associated with translational activity. This function could also include dynamic organizational rearrangement of PARP domains. Finally, mRNA has been identified in squid axoplasm (Gioio et al. 1994), which makes it plausible that one or more molecular motor protein(s) may be synthesized locally in the PARP domain.

5.4 ESI of PARP Domains in Rabbit Axoplasmic Whole-mounts

ESI of selected PARP domain examples in axoplasmic whole-mounts isolated from rabbit ventral root fibers (Koenig et al. 2000) are illustrated in Fig. 3. Unfortunately, the matrix at the outer PARP margins was not preserved during isolation of whole-mounts examined by ESI. Nonetheless, the ESI image pair in Fig. 3d, d1 is instructive because the image in Fig. 3d, mapped above the P edge (i.e., DE = 155 eV), shows two dense clusters of ribosomal P signals that disappear, and are replaced in Fig. 3d1 by an unsuspected ribosomal binding matrix when mapped above the carbon edge (i.e., DE = 300 eV).

6 Transcytosis of Ribosomes from Schwann Cell to the Axon

While a reasonable presumption is that all axonal ribosomes are transported from the cell body, Court et al. (2008) have now documented polyribosomes in axoplasm that originated from Schwann cells. Severed distal sciatic nerve fibers in the Wallerian degeneration slow (*Wld^s*) mouse strain, a strain in which degeneration is markedly delayed, were transfected with a construct comprising L4 ribosome protein fused with green fluorescent protein to label newly formed ribosomes in Schwann cells. After an interval of 7 days, labeled ribosomal signals were present in axoplasm along internode and paranodal regions. Axons containing polyribosomes increased by more than a hundredfold 7 days after crush. Interestingly, ribosome-containing axons in explant fibers from both wild type and *Wld^s* mice in culture for 24 h increased significantly. Thus, an equivalent short-term increase in ribosome occurrence in fibers indicated that the "reactive increase" was not unique to *Wld^s fibers*. At an EM level, several types of polyribosome-containing multimembrane bounded inclusions in axoplasm near the periaxonal margin, and within the axonal core were illustrated, suggesting transit across the myelin sheath and axonal transcytosis. Although PARP domains were not discerned, the findings indicated a trans myelin sheath-axoplasmic pathway, was likely to be mediated by multivesicular structures with ribosomal inclusions that were also reported by Kun et al. (2007; see Sect. 5.3).

The increase in A/G RNA ratio in proximal stump of Mauthner axon after transection reported by Edström (1964b; see Sect. 1.1), however, is not consistent with a reactive increase of ribosome content reported in distal stumps. Clearly, issues related to what extent the extraneuronal pathway contributes to supplying ribosomes to intact axons, and whether it contributes significantly in the post-transected proximal stump will need to be sorted out. Finally, compelling issues center on what signaling mechanisms trigger an upregulation mediated by the extraneuronal pathway, and why it is triggered in axons destined to degenerate.

7 PARPs as Targeted Subdomains for RNA Trafficking in the Axon Compartment

Experiments of BC1 RNA transport in the Mauthner neuron (Muslimov et al. 2002) suggested the likelihood that PARPs are destinations for RNA trafficking. BC1 RNA is an untranslated cytoplasmic transcript of 152 nt, specific to neurons in rodent brain that acts as a repressor of translation at the level of initiation (Wang et al. 2002). Axoplasmic transport of radiolabeled BC1 RNA was tracked systematically by autoradiography in microdissected Mauthner perikarya and isolated axoplasmic whole-mounts at progressive intervals after microinjection in vivo. The full-length BC1 RNA transcript, or a 65 nt sequence in the 5' region was transported rapidly in an equivalent manner; whereas, irrelevant RNA transcripts of comparable sizes were not transported.

After an initial diffuse wave of radioactivity was rapidly transported in axoplasm (Fig. 6a), a clearing of radioactivity occurred in its wake, leaving periodic focal concentrations of radioactivity at peripheral cortical sites (Fig. 6b, b1). Experiments further demonstrated that the rapid axial BC1 RNA transport depended on microtubules, while radial transport to the cortical zone depended on actin filaments. The focal concentrations of residual radioactivity in the cortical zone were consistent with being putative PARP domain sites.

BC1 RNA is specifically targeted to rat dendrites in culture (Muslimov et al. 1997), where it regulates translation (Wang et al. 2002), and expression of proteins

Fig. 6 Transport of radiolabeled, full-length BC1 RNA into the Mauthner axon after microinjection into the Mauthner cell body of the goldfish in vivo. (**a**) An initial diffuse wave of radioactivity is transported rapidly into the axon after microinjection into the Mauthner cell perikaryon. (**b, b1**) Focal accumulations of silver grains in peripheral loci of axoplasm remain after diffuse distribution of radioactivity from initial wave is "cleared" from axoplasm. The focal accumulations appear consistent with targeting and localization of BC1 RNA in PARP domain locations. *Scale bar* = 50 μm. Modified from Muslimov et al. (2002) with permission from *Journal of Neuroscience*

in postsynaptic cellular subdomains (see Kindler et al. 2005; Bramham and Wells 2007). In as much as heterologous BC1 targets putative PARP domains, there is a good likelihood that mRNAs are targeted there as well.

Support for this hypothesis is further provided by in situ hybridization of β-actin mRNA (Fig. 7a–b2), and immuno-labeling of zipcode-binding protein-1 (ZBP-1) (Sotelo-Silveira et al. 2008). ISH of β-actin mRNA in axoplasmic whole-mounts isolated from goldfish Mauthner, rabbit and rat nerve fibers revealed localization in PARP domains identified by phase structural correlates (Fig. 7a, b). In addition, confocal microscopy of RNA fluorescence (Fig. 7c), and of ZBP-1 immunofluorescence (Fig. 7c1) in PARP domains of Mauthner whole-mounts showed overlapping codistributions (Fig. 7c2). Codistribution and partial colocalization (see Sotelo-Silveira et al. 2008) of β-actin mRNA with its transacting binding factor, ZBP-1, in PARP domains strengthen the likelihood of RNA trafficking to PARP domains.

8 Endoaxoplasmic Ribosomal Plaques in Squid Giant Axon

There have been several reports of ribosomes based on ESI analysis in the postsynaptic region of the squid axo–axonic giant synapse (Martin et al. 1989, 1998), in polyribosomes isolated from squid axoplasm (Giuditta et al. 1991), in presynaptic terminals of synaptosome fractions from the optic lobe (Crispino et al. 1993), and from the squid brain (Crispino et al. 1997). Immunocytochemistry has also been used to survey ribosome distribution in the squid axon (Sotelo et al. 1999).

Bleher and Martin (2001) performed an in-depth analysis of ribosomal domain organization, and distribution in the proximal squid giant axon by, using mAb Y-10B immunofluorescence analysis at an LM level, and ESI at an EM level. ESI analysis documented small, oval-shaped ribosomal aggregates, 0.7–1.3 μm in diameter (Fig. 8). The latter were called *endoaxoplasmic ribosomal plaques* (EARPs) because ribosomes were associated with a matrix analogous to vertebrate PARPs. Unlike PARPs,

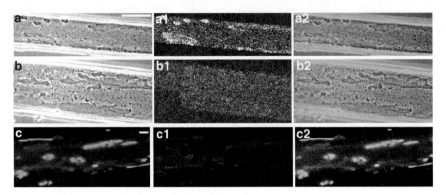

Fig. 7 Confocal images showing In situ hybridization (ISH) of β-actin mRNA, and double labeling of RNA and zip binding protein-1 (ZBP-1) in PARP domains of Mauthner whole-mounts. (**a, b**) Phase correlates that mark PARP domains. (**a1**) ISH with the β-actin mRNA antisense probe has discrete localization pattern. (**b1**) ISH with the β-actin mRNA sense probe has diffuse pattern. (**a2**) Pseudo coloring of PARPs identified by phase correlates shown in (**a**), and ISH reaction product in (**a1**) merged to show that β-actin mRNA is localized in PARP domains. (**b1, b2**) The sense probe exhibits diffuse random background labeling that does not correlate with PARP domains. (**c**) Z-projection of confocal image stack that shows YOYO-1 fluorescence labeling of PARPs in a whole-mount segment. (**c1**) The corresponding z-projection of PARPs immuno-labeled with rabbit anti chicken ZBP-1 (gift of Dr. Gary Bassell). (**c2**) Merged image of (**c**) and (**c1**) shows similar overlapping codistributions of RNA and ZBP-1 in PARPs. *Scale bars*: (**a–b2**) = 50 μm; (**c–c2**) = 12.2 μm. (**a–c2**) from Sotelo-Silveira et al. (2008) with permission from Wiley-Blackwell

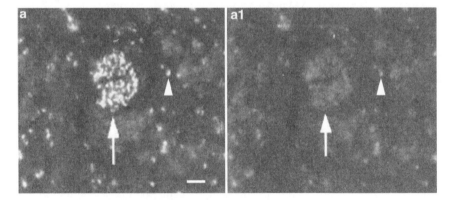

Fig. 8 Example of an *endoaxoplasmic ribosomal plaque* (EARP) in the squid giant axon. (**a**) ESI mapping of an oval-shaped EARP (*large arrow*) located in squid axoplasm 4 cm distal to the axon's emergence from the stellate ganglion in an energy window above the P edge (ΔE = 150 eV). (**a1**) Corresponding image mapped below the P edge (ΔE = 110 eV). Note the intrinsic plaque-like structure. Scattered bright signals of neurofilaments (*arrowhead*) in (**a**) that are incompletely eliminated in (**a1**), indicate contribution of mass to signal (*see text*). Scale bar = 0.1 μm. Modified from Bleher and Martin (2001) with permission from Elsevier

however, EARPs were randomly distributed throughout the axoplasm, and not preferentially localized in the cortical region as suggested by immuno-electron microscopy (Sotelo et al. 1999). The larger, more complex EARPs were located in the postsynaptic region of the giant synapse within the stellate ganglion, where the giant axon takes origin. The size and complexity of EARPs diminished with distance from the ganglion. In a parallel manner, the frequency of occurrence also diminished significantly with distance from the ganglion. The higher concentration, size and complexity of plaques in the postsynaptic region of the giant axon suggested that this subpopulation of EARPs might have special significance associated with synaptic activity.

Whether sparse, random distribution of EARPs in the squid giant axon is characteristic of unmyelinated axons in general is unknown. What is clear is that unmyelinated vertebrate axons are extremely thin (e.g., < 1.5 µm), not unlike immature, growing axons – absent glial/Schwann cell ensheathment. Growing axons have ribosomes/mRNAs that may have a preferential distribution in growth cones (Bassell et al. 1998), but are also distributed throughout axons, as indicated by ISH localization of mRNA/RNPs (Lee and Hollenbeck 2003), and by the evidence of protein synthesis throughout the isolated axonal fields (Koenig and Adams 1982; Lee and Hollenbeck 2003).

The differences in 4S RNA proportions between the squid and Mauthner axons noted previously (e.g., see Sect. 2) can now be explained by considering the differences in organization, and the frequency of occurrence of ribosome domains. Ribosome-containing small EARPs in the squid giant axon are very sparse, and randomly distributed in a very large compartmental volume. The Mauthner axon in comparison has a frequent occurrence of ribosome-containing PARP domains that are systematically distributed in a much smaller compartmental volume.

9 Concluding Remarks

Ribosome-containing structural domains in myelinated axons were effectively invisible to conventional histological and EM methods. The discovery of intermittent periaxoplasmic ribosomal plaques provided a long sought structural basis for local protein synthesizing machinery that biochemical and metabolic studies had foreshadowed (for review, see Giuditta et al. 2008), adding new-found complexity to the biology of the axon. While it seems likely that PARPs share with other actin-rich translational subdomains common mechanisms related to RNA trafficking, translation, and local regulation of expression, a feature that appears unique to the PARP domain is the structural matrix that "caps" it, which prompts questions about matrix composition and organization. A hypothesis, suggested by available data, is that *matrix serves as scaffolding for protein-synthesizing machinery recruited by the actin cytoskeletal system, which thereby governs the overall spatial organization of the translational domain.* Finally, important additional emergent questions relate to PARP origin. How are PARP domains formed during differentiation/maturation of the myelinated axon? What determines the location/distribution of PARP domain sites?

References

Bassell GJ, Zhang H, Byrd AL, Femino AM, Singer RH, Taneja KL, Lifshitz LM, Herman IM, Kosik KS (1998) Sorting of β-actin mRNA and protein to neurites and growth cones in culture. J Neurosci 18:251–265

Bazzet-Jones DP (1988) Phosphorus imaging of the 7-S ribonucleoprotein particle. J Ultrastr Mol Struct Res 99:59–69

Bleher R, Martin R (2001) Ribosomes in the squid giant axon. Neuroscience 107:527–534

Bramham CR, Wells DG (2007) Dendritic mRNA: transport, translation and function. Nature Rev Neurosci 8:776–787

Brittis PA, Lu Q, Flanagan JG (2002) Axonal protein synthesis provides a mechanism for localized regulation at an intermediate target. Cell 110:223–235

Conradi S (1966) Ultrastructural specialization in the initial segment of cat lumber motoneurons. Acta Soc Med Upsalien 71:281–284

Court FA, Hendriks WTJ, MacGillavry HD, Alvarez J, van Minnen J (2008) Novel understanding of the role of glia in the nervous system. J Neurosci 28:11024–1102

Crispino M, Castigli E, Perrone Capano C, Martin R, Menichini E, Kaplan BB, Giuditta A (1993) Protein synthesis in a synaptosomal fraction from squid brain. Mol Cell Neurosci 4:366–374

Crispino M, Kaplan BB, Martin R, Alvarez J, Chun JT, Benech JC, Giuditta A (1997) Active polysomes are present in the large presynaptic endings of the synaptosomal fraction from squid brain. J Neurosci 17:7694–7702

Door R, Richter K, Martin R (1997) Detection of low phosphorus contents in neurofilaments of squid axons by image-EELS contrast spectroscopy. J Microsc 188:173–181

Edström A (1964a) The ribonucleic acid in the Mauthner neuron of the goldfish. J Neurochem 11:309–314

Edström A (1964b) Effect of spinal cord transection on the base composition and content of RNA in the Mauthner nerve fibre of the goldfish. J Neurochem 11:557–559

Edström A, Sjöstrand J (1969) Protein synthesis in isolated Mauthner nerve fibre components. J Neurochem 16:67–81

Edström J-E, Eichner D, Edström A (1962) The ribonucleic acid of axons and myelin sheaths from Mauthner neurons. Biochim Biophys Acta 61:178–184

Gioio AE, Chun JT, Crispino M, Perrone Capano C, Giuditta A, Kaplan BB (1994) Kinesin mRNA is present in the squid giant axon. J Neurochem 63:3–18

Giuditta A, Metafora S, Felsani A, Del Rio A (1977) Factors for protein synthesis in the axoplasm of squid giant axons. J Neurochem 28:1393–1395

Giuditta A, Cupello A, Lazzarini G (1980) Ribosomal RNA in the axoplasm of the squid giant axon. J Neurochem 34:1757–1760

Giuditta A, Menichini E, Perrone Capano C, Langella M, Martin R, Castigli E, Kaplan BB (1991) Active polysomes in the axoplasm of the squid giant axon. J Neurosci Res 26:18–28

Giuditta A, Chun JT, Eyman M, Cefaliello C, Bruno PA, Crispino M (2008) Local gene expression in axons and nerve endings: the glia-neuron unit. Physiol Rev 88:515–555

Hillefors M, Gioio AE, Mameza MG, Kaplan BB (2007) Axon viability and mitochondrial function are dependent on local protein synthesis in sympathetic neurons. Cell Mol Neurobiol 27:701–716

Keenan RJ, Freymann DM, Stroud RM, Walter P (2001) The Signal recognition particle. 70:755–777

Kindler S, Wang H, Richter D, Tiedge H (2005) RNA transport and local control of translation. Annu Rev Cell Dev Biol 21:223–245

Koenig E (1965) Synthetic mechanisms in the axon. Part II: RNA in myelin-free axons of the cat. J Neurochem 12:357–361

Koenig E (1979) Ribosomal RNA in Mauthner axon: implications for a protein synthesizing machinery in the myelinated axon. Brain Res 174:95–107

Koenig E (1991) Evaluation of local synthesis of axonal proteins in goldfish Mauthner cell axon and axons of dorsal and ventral roots of the rat in vitro. Mol Cell Neurosci 2:384–394

Koenig E, Adams P (1982) Local protein synthesizing activity in axonal fields regenerating in vitro. J Neurochem 39:386–400

Koenig E, Martin R (1996) Cortical plaque-like structures identify ribosome- containing domains in the Mauthner axon. J Neurosci 16:1400–1411

Koenig E, Martin R, Titmus M, Sotelo-Silveira JR (2000) Cryptic peripheral ribosomal domains distributed intermittently along mammalian myelinated axons. J Neurosci 20:8390–8400

Korn AP, Spitnik-Elson P, Elson D (1983) Specific visualization of ribosomal RNA in intact ribosomes by electron spectroscopic imaging. Eur J Cell Biol 31:334–340

Kun A, Otero L, Sotelo-Silveira JR, Sotelo JR (2007) Ribosomal distributions in axons of mammalian myelinated fibers. J Neurosci Res 85:2087–2098

Lee S-K, Hollenbeck PJ (2003) Organization and translation of mRNA in sympathetic axons. J Cell Sci 116:4467–4478

Lerner EA, Lerner MR, Janeway CA Jr, Steitz JA (1981) Monoclonal antibodies to nucleic acid-containing cellular constituents: probes for molecular biology and autoimmune disease. Proc Natl Acad Sci U S A 78:2737–2741

Martin R, Fritz W, Giuditta A (1989) Visualization of polyribosomes in the postsynaptic area of the squid giant synapse by electron spectroscopic imaging. J Neurocytol 18:11–18

Martin R, Vaida B, Bleher R, Crispino M, Giuditta A (1998) Protein synthesizing units in presynaptic and postsynaptic domains of squid neurons. J Cell Sci 111:3157–3166

Merianda TT, Lin A, Lam J, Vuppalanchi D, Willis DE, Karin N, Holt CE, Twiss JL (2008) A functional equivalent of endoplasmic reticulum and Golgi in axons for secretion of locally synthesized proteins. Mol Cell Neurosc doi:10.1016/j.mcn.2008.09.008

Muslimov IA, Santi E, Homel P, Perini S, Higgins D and Tedge H (1997) RNA transport in dendrites: cis-acting targeting element is contained within neuronal BC1 RNA. J Neurosci 17:4722-4733

Muslimov IA, Titmus M, Koenig E, Tiedge H (2002) Transport of neuronal BC1 RNA in Mauthner axons. J Neurosci 22:4293–4301

Ottensmeyer FP (1986) Elemental mapping by energy filtration: advantages, limitations and compromises. Ann NY Acad Sci 483:339–353.

Palay SL, Sotelo C, Peters A, Orkland PM (1968) The axon hillock and the initial segment. J Cell Biol 38:193–201

Pannese E, Ledda M (1991) Ribosomes in myelinated axons of the rabbit spinal ganglion neurons. J Submicrosc Cytol Pathol 23:33–38

Peters A, Proskauer CG, Kaiserman-Abramof IR (1968) The small pyramidal neuron of the cat cerebral cortex. The axon hillock and the initial segment. J Cell Biol 39:604–619

Sato M, Wong TZ, Brown DT, Allen RD (1984) Rheological properties of living cytoplasm: a preliminary investigation of squid axoplasm (Loligo pealei). Cell Motil 4:7–23

Sotelo JR, Kun A, Benech JC, Giuditta A, Morillas J, Benech CR (1999) Ribosomes and polyribosomes are present in the squid giant axon: an immunocytochemical study. Neuroscience 90:705–715

Sotelo-Silveira JR, Calliari A, Cárdenas M, Koenig E, Sotelo JR (2004) Myosin Va and kinesin II motor proteins are concentrated in ribosomal domains (periaxoplasmic ribosomal plaques) of myelinated axons. J Neurobiol 60:187–196

Sotelo-Silveira J, Crispino M, Puppo A, Sotelo JR, Koenig E (2008) Myelinated axons contain b-actin mRNA and ZBP-1 in periaxoplasmic ribosomal plaques and depend on cyclic AMP and F-actin integrity for in vitro translation. J Neurochem 104:545–557

Spencer GE, Syed NI, van Kesteren E, Lukowiak K, Geraerts WP, van Minnen J (2000) Synthesis and functional integration of a neurotransmitter receptor in isolated invertebrate axons. J Neurobiol 44:72–81

Walter P, Blobel G (1983) Subcellular distribution signal recognition particle and 7SL-RNA determined with polypeptide-specific antibodies and complementary DNA probe. J Cell Biol 97:1693–1699

Wang H, Iacoangeli A, Popp A, Muslimov IA, Imataka H, Sonenberg N, Lomakin IB, Tiedge H (2002) Dendritic BC1 RNA: functional role in regulation of translation initiation. J Neurosci 22:10232–10241

Zelenà J (1972) Ribosomes in myelinated axons of dorsal root ganglia. Z Zellforsch 124:217–229

Regulation of mRNA Transport and Translation in Axons

Deepika Vuppalanchi, Dianna E. Willis, and Jeffery L. Twiss

Abstract Movement of mRNAs into axons occurs by active transport by microtubules through the activity of molecular motor proteins. mRNAs are sequestered into granular-like particles, referred to as transport ribonucleoprotein particles (RNPs) that mediate transport into the axonal compartment. The interaction of mRNA binding proteins with targeted mRNA is a key event in regulating axonal mRNA localization and subsequent localized translation of mRNAs. Several growth-modulating stimuli have been shown to regulate axonal mRNA localization. These do so by activating specific intracellular signaling pathways that converge upon RNA binding proteins and other components of the transport RNP to regulate their activity specifically. Transport can be both positively and negatively regulated by individual stimuli with regard to individual mRNAs. Consequently, there is exquisite specificity for regulating the axon's composition of mRNAs and proteins that control expression in the axon. Finally, recent studies indicate that axotomy can also trigger changes in axonal mRNA composition by specifically shifting the populations of mRNAs that are transported into distal axons.

1 Both Developing and Adult Neurons Transport mRNAs and Protein Synthesis Machinery into their Axons

mRNAs are actively transported to peripheral regions of the cytoplasm of polarized cells, allowing subcellular domains to autonomously regulate protein levels over space and time (Besse and Ephrussi 2008). A classic setting for mRNA localization is during embryonic development, in which movement of mRNAs to different

---

D. Vuppalanchi and J.L. Twiss (✉)
Department of Biological Sciences, University of Delaware, Newark, DE, USA
e-mail: twiss@medsci.udel.edu

D.E. Willis and J.L. Twiss
Center for Translational Neurobiology, Nemours Biomedical Research,
Alfred I. duPont Hospital for Children, Wilmington, DE, USA

regions of the early embryo is needed to establish polarity (St Johnston 2005). A recent high throughput analysis of RNA localization in the early stages of Drosophila embryogenesis showed that more than two-thirds of approximately 3,500 transcripts preferentially localized to subcellular sites (Lecuyer et al. 2007). A clear implication of this study is that mRNA trafficking and subcellular localization are much more likely to be a global mechanism than studies limited to one or a few transcripts have led us to believe.

Neurons are undoubtedly the most polarized of cells, with their axonal processes sometimes extending distances of more than 1,000-fold the diameter of their cell body. Axons are very much longer than dendrites, but mRNA transport and localized protein synthesis in dendrites have received considerably more attention from the neuroscience community. Studies from Levy and Steward (1982) initially showed that ribosomes are concentrated at the base of dendritic spines in hippocampal neurons (Steward and Levy 1982). However, ribosomes were not seen in the axons of the mature CNS neurons of the mammalian hippocampus used in these ultrastructural analyses. This and subsequent studies of CNS neurons gave rise to the notion that ribosomes, translation factors, and mRNAs are excluded from the axonal compartment (Twiss and van Minnen 2006). However, as noted below and in other chapters in this volume, a steadily growing number of studies show that axons of many different neuronal types contain mRNAs and the requisite machinery to synthesize proteins de novo, at least under some circumstances. In this chapter, we focus on two principal aspects of protein expression in axons: (1) how mRNAs are transported into axons, and (2) how localized translation of mRNAs is regulated within axons. Since much more experimental effort has been devoted to dendritic mRNAs, particularly with regard to the nature of the transported RNA structure, we will also include references to the dendritic RNA transport/translation literature, which largely centers on localized protein synthesis related to synaptic plasticity. Such studies become particularly relevant in the context of the axon, and the axonal biologists can draw experimental design and reagents from this more developed field of dendritic protein synthesis. Before addressing the main subject areas, we briefly highlight some early literature that influenced our and others' approaches to the mechanisms governing axonal RNA transport and localized protein synthesis.

Early studies from Koenig dating from 1965 to 1967 were the first to provide biochemical data on protein synthesis in axons (Koenig 1965a, b, 1967a, b). These studies stemmed from earlier observations on recovery of specific acetylcholinesterase activity in peripheral nerves from cholinergic neurons after systemic irreversible inactivation (Koenig and Koelle 1960, 1961). Koenig (1967b) suggested that this recovery of neuronal acetylcholinesterase activity was due to the generation of new protein within the distal nerve, probably within the axons themselves. However, the methodologies available at the time could not completely rule out nonneuronal components of the nerve as sources of newly synthesized proteins for the axons (Edström 1966; Fischer and Litvak 1967).

Studies of invertebrate axons, in particular the squid giant axon where axoplasmic contents could be isolated with much less potential for glial contamination, pointed to axonal protein synthesis activity (Giuditta et al. 1968). However, reports of glial transfer of invertebrate proteins and even some nucleic acids to axons suggested that metabolic labeling studies of the squid system were actually detecting glial, rather than axonal, synthetic capacity (Tytell and Lasek 1984; Sheller et al. 1995). A few recent publications indicate that there can be transfer of mammalian proteins, RNAs, and even ribosomes between cells, including transfer from glia to axons (Twiss and Fainzilber 2009). Nonetheless, the early experiments that were performed long before the sensitivity of molecular biology rose to meet the challenges of this quest were indeed correct, and now there is also irrefutable evidence that both invertebrate and vertebrate axons actively synthesize proteins. In the sections below, we try to highlight how this field has now blossomed largely through advances in molecular and cellular experimental technologies as well as renewed interest from the neuroscience community.

2 A Complex Population of mRNAs Is Transported into Axons

mRNA localization has now been recognized as a common mechanism for spatially and temporally regulating protein levels in subcellular domains (Besse and Ephrussi 2008). Advances in molecular biology for detecting and amplifying RNAs have made neurons a very appealing model system for the identification of localized mRNAs. The neuron's dendrites and axons essentially represent widely separated subcellular compartments, where localization can be conclusively demonstrated. With the ability to isolate large quantities of axoplasm, the squid giant axon provided the first insight into the complexity of the axonal mRNA population. By analyzing RNA–cDNA hybridization kinetics, Capano et al. (1987) estimated that the squid giant axon RNA could code for 200 different polypeptides. Subsequent identification of axonal mRNAs from this preparation yielded those encoding β-actin, β-tubulin, neurofilaments, and kinesin heavy chain (Capano et al. 1987; Gioio et al. 1994; Chun et al. 1996). Moreover, at least some of these axonal mRNAs were associated with ribosomes (i.e., "polysomes"), indicating that they were translationally active in the squid axoplasm (Giuditta et al. 1991; Crispino et al. 1997). Many of the identified mRNAs encode cytoskeletal proteins. As outlined below, modification of the cytoskeleton is a major role for locally synthesized proteins in vertebrate axons. However, sensory axons from *Lymnaea* contained large amounts of neuropeptide-encoding mRNAs (van Minnen 1994), and bag cells of *Aplysia* locally generate and secrete egg laying hormones (ELH) (Lee and Wayne 2004).

These observations argue that localized protein synthesis in invertebrate axons can fulfill more functions than just modifying the cytoskeleton. Consistent with this notion of complexity, a brute force sequencing of a cDNA library generated from *Aplysia* sensory axons identified approximately 200 different mRNAs (Moccia et al. 2003). Both this and a cDNA library from *Lymnaea* sensory axons (van Kesteren et al. 2006) indicate that cytoskeletal encoding mRNAs represent only a fraction of the transcripts that localize in invertebrate axons. A potential concern in interpreting these studies is that invertebrate axons can also show dendritic features (Martin et al. 2000). This "dendritic ambiguity" has been used to argue that neuronal polarity is not complete in these less advanced organisms when compared with mammalian neurons. Thus, the mRNA localization in invertebrates was suggested to be a feature of incomplete polarity.

Early studies in vertebrate neurons detected vasopressin, oxytocin, and tyrosine hydroxylase (TH) mRNAs in axons of hypothalamic neurons that project to the posterior pituitary (Jirikowski et al. 1990; Mohr et al. 1991; Trembleau et al. 1994; Wensley et al. 1995). Olfactory axons were also shown to contain mRNAs encoding odorant receptors and olfactory marker protein (OMP) (Ressler et al. 1994). However, no translational machinery was detected in these axons, so there was doubt that these mRNAs could be utilized for protein production (Mohr and Richter 1992). Furthermore, there was also evidence that vasopressin and TH mRNAs could be retrogradely transported from posterior pituitary to the cell bodies in the hypothalamus (Jirikowski et al. 1992; Maciejewski-Lenoir et al. 1993). It is counterintuitive to consider that a neuron would transport an mRNA into distal processes only to send the transcript back to the cell body at a later time. However, there is evidence of specificity for the localization of mRNAs into the neurohypophysis, since conditions of hypoosmolality increase vasopressin but not oxytocin mRNA in the hypothalamic axons (Svane et al. 1995). As discussed below, studies in cultured neurons have shown that the movement of mRNAs most typically occurs by an active microtubule-based transport. Thus, even though the mRNAs mentioned above are abundant transcripts in these neurons, their axonal localization is likely the result of specific targeting rather than passive diffusion.

Preparations of cultured neurons, where neuronal processes can be visibly isolated from glial cells, have been used to demonstrate mRNAs and translational machinery in axons. Olink-Coux and Hollenbeck (1996) used fluorescent in situ hybridization (FISH) to visualize polyadenylated RNA, undoubtedly mRNAs, in the axons of cultured chick sympathetic neurons. Using axonal processes that were manually dissected from the cell bodies in explant cultures, these authors' also initially detected mRNAs encoding cytoskeletal proteins by reverse transcriptase-coupled polymerase chain reaction (RT-PCR). Bassell et al. (1998) later used FISH techniques to show that β-actin mRNA but not γ-actin mRNA localizes to the axonal growth cones of cultured CNS neurons. In retrospect, this should have quelled any suggestions that abundant mRNAs could nonselectively diffuse into axons, since β-actin and γ-actin mRNAs are both quite abundant, and, similar to fibroblasts and myoblasts (Hill and Gunning 1993; Kislauskis et al. 1993), these cortical neurons were selectively localizing only the β-actin mRNA. Importantly, Bassell and colleagues showed that ribosomes were also present in these cortical growth cones, putting the mRNAs and ribosomes in the same subcellular space (Bassell et al. 1998).

Techniques to physically isolate axonal processes from cell bodies and nonneuronal cells have provided much insight into the complexity of axonal mRNA content and protein synthesis. These approaches merit inclusion here since they have shown the diversity of axonally synthesized proteins, and emphasize that neurons somehow select, which mRNAs are to be sent into the axons. Furthermore, the knowledge gleaned from recognizing the complexities of axonal mRNA populations' offers insight into how mRNA transport is regulated. Using a compartmentalized culture system, or a "Campenot culture," Eng et al. (1999) not only showed that axonal mRNAs are translated into new proteins locally, but also emphasized that only a small fraction of β-actin protein is generated in the axonal compartment of sympathetic neurons. Lee et al. (2003) later quantitated general protein synthesis in sympathetic axons and concluded that 5–10% of total cellular protein synthesis can occur in the axonal compartment of cultured neurons. Thus, the assertions that axonal proteins are synthesized in the cell body and anterogradely transported into the axonal compartment are correct, at least for the 90% of axonal proteins in cultured sympathetic neurons.

Experiments in physically severed axons prove the importance of the fraction of axonally synthesized proteins. Campbell and Holt (2001) used axons severed from the cell body to show that turning in response to some guidance cues requires newly synthesized proteins in the growth cone. Zheng et al. (2001) similarly showed that regenerating sensory axons require localized protein synthesis to maintain the growth cone after severing from their cell body. Although recent work from the Letourneau lab suggests that the axon may not always utilize locally synthesized proteins for these responses (Roche et al. 2009), it is clear that, at least under some conditions, this small amount of protein synthesized in the axonal compartment serves essential functions in the distal axon.

Severed axons have been used to determine the breadth of proteins that can be synthesized in axons. Willis et al. (2005) used metabolic labeling of isolated adult rat sensory axons to identify axonally synthesized proteins by MALDI–TOF/TOF mass spectroscopy. Although this approach was limited to the most abundant axonally synthesized polypeptides, and any with significant hydrophobic regions (e.g., integral membrane proteins) could not be well resolved by the two-dimensional gels utilized for this study, Willis and colleagues increased the number of proteins known to be axonally synthesized by more than fivefold. Importantly, they showed that cytoskeletal proteins constitute only a fraction of the axonally generated proteins, similar to the observations in invertebrate axonal processes (Moccia et al. 2003; van Kesteren et al. 2006).

Transcript profiling of these same adult sensory axons was used in an attempt to identify some of the less abundant axonally synthesized proteins and to determine if mRNAs encoding membrane and secreted proteins might localize to axons (Willis et al. 2007). Indeed, more than 200 different mRNAs were identified in the dorsal root ganglion (DRG) axons, and the functions of the encoded proteins are quite diverse (Fig. 1). These sensory neurons cultured from DRG have limited polarity since they are pseudounipolar in vivo, and they extend multiple axons but no dendrites in culture (Zheng et al. 2001).

Analyses of axonal mRNAs in more polarized neurons (e.g., cortical neurons) have been hindered by the lack of any means to separate axonal and dendritic processes. Jeon and colleagues developed a microfluidic device to overcome this limitation (Taylor et al. 2005). This device essentially guides axons to a separate compartment

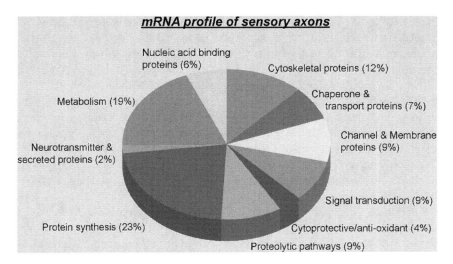

Fig. 1 Axonally synthesized proteins have many different functions. A gene ontology analysis of proteins encoded by mRNA identified in axons of sensory neurons cultured from adult rats is shown. This analysis is based on ~200 axonal mRNAs identified by Willis et al. (2007) using isolated axons that were tested for purity based on exclusion of γ-actin and MAP2 mRNAs

from the cell bodies and dendrites, providing a means to isolate only axons. Taylor et al. (2009) used this approach to profile the axonal mRNAs from cultured hippocampal neurons after these neurons achieved full polarity with the potential to establish synaptic contacts. They detected a few hundred more mRNAs than were seen in the sensory neurons, but there were many mRNAs in common between the two preparations.

It important to note, that the above array analyses may underrepresent axonal mRNA content, because there is always ambiguity over the hybridization intensity level to call a "hit." In the studies by Willis et al. (2007), the cutoff for presence of an mRNA in the axons was based on their previous proteomics studies (Willis et al. 2005). Arrays from Taylor et al. (2009) provide a more comprehensive picture; however, even their absolute number of axonal hits varied when the differences between mature and regenerating cortical axons were compared (see Sect. 6). All in all, array technology is best used for a differential comparison of the two or more populations. For example, a recent transcript profiling study of cellular protrusion from fibroblasts showed that approximately 50 mRNAs differentially accumulated in the pseudopodial protrusions, when compared with the cell body in response to lysophosphatidic acid and fibronectin (Mili et al. 2008). This study illustrates the benefits of array analyses for studying the asymmetric distribution of RNAs but not for the conclusive demonstration of whether a specific mRNA is present within a particular subcellular region. Thus, one must remain aware that studies on the enrichment of mRNAs in neuronal regions under different conditions do not answer the absolute question of whether an mRNA localizes or not. This is particularly true when comparing mRNA samples from processes vs. total or cell body fractions, as was done in profiling the

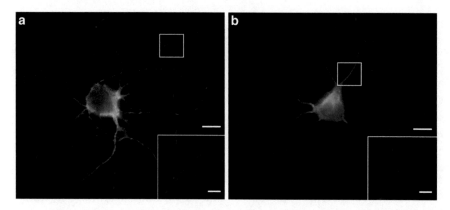

Fig. 2 RNA FISH shows specificity of RNA localization. Micrographs of cultured DRG neurons from adult rats that were immunostained for neurofilament proteins (*green*) and hybridized for β-actin (**a**) and γ-actin (**b**) mRNAs are shown. The *large panel* shows low magnification images, where the signals for neurofilament proteins (*green*) are merged with those for the actin mRNAs (*red*). The *inset panels* at lower right of each image are higher power images of the *boxed regions* with only the *red* mRNA channel shown. The RNA probes used to generate these images consisted of three synthetic antisense 50–60′mer oligonucleotides for each transcript that were labeled with digoxigenin at four sites each; a Cy3-conjugated antidigoxigenin secondary was used to visualize the mRNAs. Note that β-actin mRNA extends into the distal axonal processes of these cultured DRG neurons, while γ-actin is restricted to the proximal segment of the axon. Additionally, the images of β-actin mRNA show that the majority of the FISH signal is restricted to the cell body, and only a fraction of the total cellular β-actin mRNA pool is transported into the axonal compartment. See Willis et al. (2005, 2007) for technical details on methods and probe sequence. [*Scale bars* = 25 μm; *Inset scale bars* = 5 μm]

RNA content of hippocampal neurons (Poon et al. 2006). Indeed, based on the growing number of axonal mRNAs that we have examined in the DRG system, the vast majority of individual transcripts remain in the cell body (Fig. 2).

2.1 Neurons Choose Which mRNAs to Transport into Axons Based on RNA Structure

RNA transport is of particular importance to neurons due to the complexity of their multiple functional compartments and the use of localized translation for regulation of growth and synaptic strength. As discussed in the chapter by Rossoll and Bassell (2009), alterations in RNA transport have been seen in animal models of spinal muscular atrophy (SMA). Altered RNA transport into dendrites is also seen in experimental models of fragile X mental retardation (Dictenberg et al. 2008; Estes et al. 2008). Modifying the transport of mRNAs into axons is one way to regulate localized protein synthesis. One can postulate that deficits in mRNA localization, either through disruption of targeting or transport mechanisms, could occur in many other neurodegenerative diseases. The loss of capacity to locally express specific proteins in neuronal processes might also contribute to the pathogenesis of disease in the nervous system. Although there have been no mechanistic studies in humans, mutations in tyrosyl and glycyl tRNA synthetases (YARS and GARS, respectively) in axons can lead to Charcot Marie Tooth type peripheral neuropathies (Antonellis et al. 2006; Jordanova et al. 2006; Seburn et al. 2006; Nangle et al. 2007). Both GARS and YARS proteins localize to axons, and drosophila and murine neurons carrying human disease-causing GARS mutations show decreased protein synthesis in axons (Chihara et al. 2007). Thus, the axonally synthesized proteins likely support long-term health of distal axons.

Long-range movement of mRNAs in neuronal processes occurs through microtubule-based transport mechanisms. In axons, kinesin is the anterograde transport motor, and kinesin isoforms also provide mRNA movement in dendrites. However, there is also evidence for short-range movement on microfilaments, undoubtedly through myosin-based motors. This latter mechanism was shown to account for the ligand-dependent redistribution of β-actin mRNAs within growth cones (Leung et al. 2006; Yao et al. 2006a). Interestingly, the periaxoplasmic ribosomal domains that Koenig and colleagues have detected in myelinated axons contain kinesin and myosin motor proteins (Sotelo-Silveira et al. 2004) (see chapter by Koenig (2009)). Thus, it is possible that mRNAs, or, more likely, the ribonucleoprotein particles (RNPs) may be handed off from kinesin to myosin for fine-tuning of their localization within subdomains within the axon.

Targeting of mRNAs to subcellular regions is highly selective, since it is clear that not all neuronal mRNAs extend into the axonal or dendritic compartments. In neurons, β-actin and γ-actin mRNAs encode remarkably similar proteins but only β-actin mRNA is transported into neuronal processes, while γ-actin mRNA is confined to the cell body (Fig. 2) (Bassell et al. 1998). Tau mRNA selectively localizes into axons, but not dendrites (Litman et al. 1993, 1994; Behar et al. 1995). Similarly, mRNAs encoding microtubule-associated protein 2 (MAP2) and the α-subunit of Ca-Calmodulin kinase II (CAMKIIα) selectively localize into dendrites and not into axons (Garner et al. 1988; Burgin et al. 1990). However, some mRNAs can be transported into both axonal and dendritic compartments. For example, The Bassell group has shown that β-actin mRNA transport into dendrites, and this dendritic localization of β-actin mRNA can be regulated by depolarizing stimuli (Tiruchinapalli et al. 2003). Early work from the Eberwine lab showed that growing dendrites contain several mRNAs that are associated with growth-related activities (e.g., mRNAs encoding GAP-43 and neurofilament subunits) (Crino and Eberwine 1996). Some of these mRNAs also localize into sensory axons (Willis et al. 2007). The neuron may differentially target mRNAs to growing vs. mature functional dendrites. The growing dendrite likely shares some features with axons in its need for localized protein synthesis during elongation.

The selective localization of some mRNAs into axons vs. dendrites outlined above suggests that the neuron employs selective mechanisms for transport of individual mRNAs to specific neuronal subdomains. Subcellular localization of mRNAs is controlled by *cis*-acting elements, which are usually found in the 3′ untranslated region (3′ UTR). Such localizing mRNA regions have also been termed "zip-codes," targeting elements, and localization sequences. These RNA elements are recognized by RNA-binding proteins (RBPs). The mRNA-protein complex is then transported as a ribonucleoprotein complex or RNP particle. We will reserve the term "zip-code" to refer to β-actin mRNA's *cis*-element that is recognized by the "zip-code binding proteins" (see below). The 3′ UTRs of CaMKIIα, brain-derived neurotrophic factor (BDNF), and Arc mRNAs are sufficient to target transcripts into dendrites (Mayford et al. 1996; Kobayashi et al. 2005; An et al. 2008). β-actin mRNA's 3′ UTR is sufficient for targeting into axons and dendrites of neurons and into the leading lamellipodial pole of fibroblasts (Kislauskis et al. 1993; Zhang et al. 2001; Tiruchinapalli et al. 2003). Conservation of RNA localization mechanisms between insects, yeasts, and vertebrates (Jansen 2001; Besse and Ephrussi 2008) may provide evidence for evolutionary pressure to conserve these UTR structures.

Among the many mRNAs that are now known to localize into axons, β-actin mRNA is the best characterized. Subcellular localization of β-actin mRNA, both in fibroblasts and neurons, is imparted by a ~50 nucleotide long "zip-code" in its 3′ UTR (Kislauskis et al. 1994; Zhang et al. 2001). The zip-code binding protein 1 (ZBP1) binds to β-actin's 3′ UTR zip-code element, and this interaction is needed for subcellular localization of β-actin mRNA. Treatment of neuronal cultures with antisense oligonucleotides directed at the zip-code region blocks localization of β-actin mRNA in neurons and fibroblasts, resulting in growth cone retraction and altered polarity of cell migration, respectively (Shestakova et al. 2001; Zhang et al. 2001). The 3′ UTR of β-actin mRNA also confers localization in the peripheral nervous system (PNS) neurons of mouse and rat (Willis et al. 2007). Despite this clear role for the zip-code in directing subcellular transport of β-actin mRNA, knowledge of its primary and secondary structure has not been helpful in determining the localizing elements of other axonal mRNAs. In analyzing 3′ UTRs of other axonal mRNAs for homology to the 3′UTR of β-actin using either the 54 nucleotide zip-code from rat β-actin mRNA (Willis et al. 2007) or the minimal zip-code consensus sequence (Farina et al. 2003), we have not found any clear homologies with more than 200 transcripts (S McCahan and DE Willis, unpublished observations). Studies in other cellular systems have shown that muscle-like binding protein 1 (MLP1), which binds to and peripherally localizes integrin α3 mRNA, also binds to the zip-code of β-actin mRNA (Adereth et al. 2005). Despite having similar functions, MLP1 and ZBP1 do not show any clear homology to each other by BLAST alignment tools (http://blast.ncbi.nlm.nih.gov), which show only two short peptides (≤7 amino acids each) that align between human MLP1 and ZBP1. Thus, multiple RBPs can recognize the zip-code element. ZBP1 may indeed interact with many other axonal mRNAs through elements with related secondary structures or protein-protein interactions. In support of the latter notion, recent ribonomics analysis of proteins showed that a few hundred mRNAs copurify with IMP1, the human ortholog of ZBP1, and most did not have an identifiable zip-code element (Jonson et al. 2007). Some of these IMP1-associated mRNAs have also been detected in sensory axon RNA profiling studies (Willis et al. 2007).

The 3′ UTRs of many other mRNAs have been shown to be necessary and sufficient for axonal RNA localization. A 240-nucleotide span of Tau mRNA's 3′ UTR was shown to confer axonal localization in differentiated P19 cells, a pluripotent teratocarcinoma cell line that can be induced to differentiate into polarized neurons (Aronov et al. 2001). Just as with β-actin's zip-code element, the mechanisms conferring localization are conserved between species. The 3′ UTR of Tau mRNA is sufficient to localize a reporter mRNA to the vegetal hemisphere in *Xenopus* oocytes, but reporter constructs containing only the coding region of Tau mRNA (i.e., without the 3′ UTR element) were uniformly distributed in the oocyte (Litman et al. 1996). Within an organism, lineage-specific expression of the target mRNA and corresponding RBP determine which mRNAs can be localized.

Work from the Jaffrey group showed that RhoA and CREB mRNAs are transported into developing axons, and the 3′ UTRs of these mRNAs are sufficient to confer axonal localization to fluorescent reporter mRNAs (Wu et al. 2005; Cox et al. 2008). Additionally, the 3′ UTRs of RhoA and CREB provide ligand-dependent translation in the developing axons. Different mRNAs generated from the same gene can also give rise to differential RNA sorting in neurons. Yudin et al. (2008) recently showed that sensory neurons express two mRNAs for Ran binding protein 1 (RanBP1); these mRNAs have identical coding regions but 3′ UTRs that differ in length due to differential transcriptional termination sites. Only the RanBP1 with the longer 3′ UTR localizes into sensory axons, indicating that the localization element is in the terminal 208 nucleotides (Yudin et al. 2008). A similar differential localization is seen with BDNF mRNA variants, where only the transcripts that use a distal transcription termination site localize to dendrites (An et al. 2008). Neither of these UTRs shows homology to β-actin mRNA's zip-code element. While the majority of transcripts studied to date contain localization elements within their 3′ UTRs, the 3′ UTR is not always sufficient for localization. Recent analyses of κ-opioid receptor mRNA shows, that its localization in primary sensory axons requires both 3′ and 5′ UTR *cis*-elements (Bi et al. 2006).

The technical approaches to test for localization capabilities of potential RNA *cis*-acting elements bear mention here. To show that an element is sufficient for localization, it must be able to confer localization to a normally nonlocalizing mRNA. Investigators have cleverly adapted fluorescent reporter systems, of which many were originally pioneered in dendritic analyses, to prove that an engineered mRNA can localize. While in situ hybridization methods allow visualization of the mRNA in fixed systems, the ability to visualize movement and translation of the mRNA in living cells has provided means to dissect both transport and translational control mechanisms. RNA tagging with fluorescent RNA binding proteins has been used for visualizing mRNA transport. The best characterized system is the MS2 bacteriophage RNA binding protein; MS2-GFP fusion proteins show sequence-specific binding to a prokaryotic RNA element that can be cloned into eukaryotic expression constructs (Querido and Chartrand 2008). An appeal of this approach is that it allows one to observe the actual mRNA in transit and single transcript sensitivity can be approached by increasing the number of MS2 recognition loops in the target mRNA (Fusco et al. 2003; Shav-Tal et al. 2004). A second RNA tagging system, the λB box binding peptide (Daigle and Ellenberg 2007), may offer the potential to visualize movement of multiple mRNAs simultaneously since this peptide recognizes an entirely different mRNA sequence element (Lange et al. 2008). Tagging the cDNAs of fluorescent reporters with potential RNA localization elements has been used to test for both RNA localizing capability and localized translation. This requires the ability to distinguish locally synthesized reporter (i.e., in distal axons) from cell body-synthesized reporter that may diffuse or be transported anterogradely down the axon. Tagging the reporter with membrane localizing elements has dramatically decreased passive diffusion of fluorescent reporters like eGFP (Aakalu et al. 2001).

Both photobleaching and photoconvertible forms of fluorescent reporter proteins have been used to differentiate cell body-derived vs. locally synthesized fluorescent reporters. For the photobleaching approach (i.e., FRAP technique), the distal axonal segment is photobleached with high laser power and then monitored for recovery of fluorescence over time at a much lower laser power. Significant attenuation of recovery by translational inhibition proves that the recovery requires protein synthesis, which is highly suggestive of axonal localizing capability of the element. Yudin et al. (2008) used this approach to show that the distal segment of the RanBP1 mRNA's 3′ UTR is sufficient and necessary for axonal localization of the transcript. The Jaffrey group has elegantly used a photoconvertible fluorescence protein, which changes from green to red emission upon exposure to violet light, to demonstrate localizing activity of CREB mRNA's 3′ UTR (Cox et al. 2008). An advantage of this reporter is that the fate of the axonally synthesized protein can be monitored by including critical coding sequences from the axonal mRNA. With this approach, Cox et al. (2008) showed that the protein product of axonal CREB mRNA is retrogradely transported after stimulating translation with nerve growth factor (NGF). These photobleaching/conversion approaches are currently limited to cultured cells and expose the cells to high laser power, which can be detrimental to axonal health. Better methodologies are needed to monitor protein synthesis in vivo. Brittis et al. (2002) used a fluorescent reporter that converts from green to red emission as the protein ages to show that the 3′ UTR of EphA2 mRNA confers axonal localization in the developing spinal cord (Brittis et al. 2002). Such naturally converting and short-lived fluorescent reporters hold promise for in vivo studies of axonal protein synthesis.

The sequence elements within mRNAs that RBPs recognize are often composed of secondary structures (and possibly tertiary structures), which makes predicting *cis*-elements quite difficult (Hamilton and Davis 2007; Jambhekar and Derisi 2007). Presently, there is no reliable means to predict RNA localization elements. The structures that RBPs recognize are likely secondary structures derived from Watson–Crick base pairing, creating double stranded stems and single-stranded loops. Recent bioinformatic tools that allow large-scale comparison of RNA secondary structures do hold promise for determining common structural motifs in localizing mRNAs (Yao et al. 2006b; Hamilton et al. 2009). However, biological testing of individual mRNAs for localizing ability is currently more feasible than any in silico analyses for common structures using computing tools. Our attempts to use motif-searching algorithms to find potential common sequence motifs in the UTRs of axonal mRNAs have largely failed, despite quite protracted computing sessions (DE Willis, S McCahan, J Coleman, and JL Twiss, unpublished observations). In our experience, searching for evolutionarily conserved 3′ UTR segments between species has proven more reliable when using primary RNA structure as a query.

Several small noncoding RNAs, including tRNAs and the 7S RNA component of the signal recognition particle (SRP), also localize to axons (Black and Lasek 1977; Merianda et al. 2009). The mechanisms used to transport these into the axonal compartment have not been explored. MicroRNAs (miRNA) are particularly interesting because these provide means for the cell to regulate levels and/or translation of specific mRNAs (Cannell et al. 2008). Components of the RNA interference (RNAi) machinery, including argonaute-3, argonaute-4, Dicer, and the fragile X mental retardation protein (FMRP), have been detected in the mammalian PNS axons (Hengst et al. 2006; Murashov et al. 2007).

Synthetic RNAi substrates (i.e., "siRNA") have been used to deplete axons of RhoA and tubulin mRNAs (Hengst et al. 2006; Murashov et al. 2007). Thus, the RNAi machinery that was detected in axons appears to be functional even in vivo. The possibility that naturally occurring miRNAs might localize to axons is an appealing mechanism for temporal and spatial control of axonal mRNA and protein levels. miRNAs have been detected in dendrites, where they can modulate synaptic plasticity (Schratt et al. 2006; Kye et al. 2007). Aschrafi et al. (2008) showed that miR-338 modulates expression of CoxIV in sympathetic neurons by targeting its mRNA locally at the axons. Importantly, localized translation of CoxIV supports mitochondrial respiration in axons and survival of these sympathetic neurons (Hillefors et al. 2007; Aschrafi et al. 2008) (see chapter by Kaplan et al. 2009). There has been some evidence for transfer of proteins, small RNAs, and ribosomes from glia to axons (Eyman et al. 2007; Court et al. 2008). However, Aschrafi et al. (2008) used sympathetic axons that were physically separated from the nonneuronal cells, so miR-338 must have derived from anterograde transport rather than any localized glial transfer. Further studies are clearly needed to specifically characterize the targeting mechanisms used to get endogenous miRNAs into axons.

The overwhelming majority of studies on axonal mRNA localization discussed here have been limited to cultured neurons. The field needs to develop better techniques to visualize mRNAs and locally synthesized proteins in axons that still reside in their natural environments in vivo. Work from the Fainzilber lab indicates that Importin β1, vimentin, and RanBP1 mRNAs localize to axons of the PNS in vivo (Hanz et al. 2003; Perlson et al. 2005; Yudin et al. 2008). Sotelo-Silveira et al. (2008) have also recently visualized β-actin mRNA in myelinated PNS axons using in situ hybridization. Other mRNAs will undoubtedly localize to axons in vivo, likely including CNS axons under some conditions. A difficulty with many of the axonal mRNAs known to date is that they are also expressed in adjacent non-neuronal cells. However, confocal microscopy techniques can be used to distinguish axonal and glial signals when fluorescent RNA probes are combined with immuno-

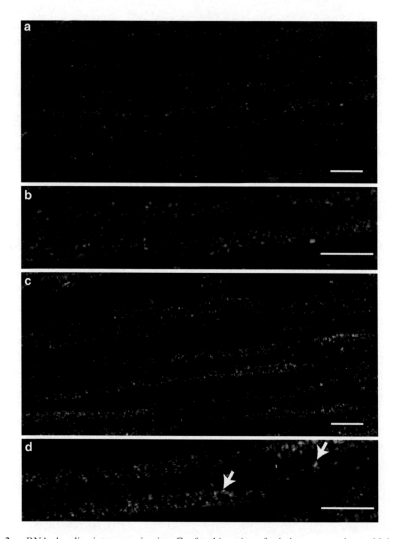

Fig. 3 mRNAs localize into axons in vivo. Confocal imaging of sciatic nerve sections with immunofluorescent signals for neurofilament (**a**, **b**) and S100 (**c**, **d**) combined with fluorescence in situ hybridization for calreticulin mRNA is shown. The third channel is calreticulin protein (**a**, **b**) or ribophorin II (**c**, **d**). The methods for FISH and probe preparation are based on those detailed in Fig. 2 with modifications for tissue sections (Muddashetty et al. 2007). Panels (**a**) and (**b**) each show a single optical XY plane with calreticulin mRNA (*red*), calreticulin protein (*green*), and neurofilament protein signals (*blue*) merged. Granular-appearing signals for calreticulin mRNA overlap the neurofilament immunoreactivity. Panel (**b**) is high magnification of a single axon in longitudinal section through the center of the axoplasm. The punctate appearance of calreticulin protein signals suggests localization into an ER-like vesicular compartment, as we have recently demonstrated for cultured neurons (Merianda et al. 2009). In panels (**c**) and (**d**), S100 staining (*red*) was used to distinguish Schwann cell cytoplasm, and ribophorin II (*green*) was used to highlight RER. Single optical planes of longitudinal sections of sciatic nerve are shown in panels (**a**) and (**b**). As expected, the nonneuronal calreticulin mRNA (*red*) is largely explained by signals in the Schwann cells. In the high-magnification image shown in panel, calreticulin mRNA is clearly seen in what would be axonal profiles (i.e., between S100 positive regions), and there is focal overlap with ribophorin II signals (*arrows*). The latter overlap may indeed represent axonal calreticulin mRNA that is being actively translated on ER-bound ribosomes [*Scale bars* = 5 μm]

fluorescence for axonal and glial markers (Fig. 3). Still, this cannot fully address the mechanisms of RNA transport. We will need to develop means to tag mRNAs or use fluorescent reporters similar to those outlined above to determine functionality of RNA cis-elements in vivo.

3 Axonal mRNAs are Transported as Ribonucleoprotein Particles

As suggested above, the process of mRNA localization in neurons, as well as in other cells, is complex and involves multiple RBPs. hnRNP R, hnRNP Q, and SMN proteins have been shown to associate with β-actin mRNA in motor axons (Rossoll et al. 2002). Tau mRNA-containing RNPs from P19 cells contain the Elav protein HuD, IMP1 (i.e., ZBP1), a ras regulating protein (G3BP), and the motor protein KIF3A (Atlas et al. 2004). κ-opioid receptor mRNA-containing RNPs include HuR, the coatamer protein Copb1, and Grb7. HuR appears to stabilize the mRNA, Cop1b is needed for localization of the mRNA, and Grb7 plays a role in translational repression of the mRNA (Tsai et al. 2006; Bi et al. 2007). Imaging studies of neurons and other cell types indicate that mRNAs are transported in granular-appearing structures that contain both mRNAs and RBPs. These proteins that bind to Tau, κ-opioid receptor, and β-actin mRNAs likely constitute part of their structural transport complex. These granular RNP profiles have been referred to as "RNA granules" – in this chapter, the term "transport RNP" will be used exclusively to avoid any confusion with other RNA granules, such as stress granules (Kiebler and Bassell 2006). However, it should be noted that, as outlined below, the transport RNPs show some similarities to stress granules in that both structures seem to function as storage compartments for mRNAs (Anderson and Kedersha 2006). Additionally, there is evidence that neuronal transport RNPs share some features with "processing bodies," or "P-bodies," that have been associated with RNA decay (Zeitelhofer et al. 2008). It is quite possible that some of what has been designated as "transport RNPs" in axons will become better differentiated in terms of their exact nature in the coming years.

mRNA localization is initiated by association of the transcript with RNA binding proteins that ultimately generate a transport RNP in the cytoplasm. In Xenopus oocytes, an mRNA initially interacts with RBPs in the nucleus, and the mRNA-RBP complex is subsequently remodeled in the cytoplasm prior to interacting with motor proteins for transport (Kress et al. 2004). This has not been extensively tested in other cell types, nor have many mRNAs been subjected to such analysis. For β-actin, the KSRP splicing factor homolog ZBP2 initially binds to newly transcribed β-actin mRNA in the nucleus, and this interaction facilitates the transcript's subsequent interaction with ZBP1(Pan et al. 2007). The Achsel lab recently showed that a dendritic RNP containing Lsm1 and CBP80, which is formed on nascent transcripts in the nucleus, is also transported into spinal cord axons (di Penta et al. 2009). One implication of this nuclear determination of localizing capability for mRNAs is that the fate of an individual mRNA may be determined shortly after it is transcribed. RBPs that bind within the nucleus can also function in the export of the mRNAs into the cytoplasm and are needed for subsequent localization (Nakielny and Dreyfuss 1999; Schnapp 1999). Although we tend to think of transcription, mRNA processing, nuclear export, and subcellular localization as independent steps in the life of an mRNA, for some cell types, this may represent a sequential series of mechanistically linked events that share common mediators. Thus, it is possible that the fate of an mRNA could be determined cotranscriptionally. However, this may not always be the case in neurons, since Willis et al. (2007) showed that sensory neurons can mobilize transport of mRNAs in response to trophic and tropic agents, even in the absence of ongoing transcription. This raises the possibility that neurons can store "localizable mRNAs" within the cell body, using them for either cell body or localized protein synthesis. Additionally, the rate limiting steps for RNA localization have yet to be established. Since only a fraction of many mRNAs are localized into the axonal compartment (see Fig. 2), some components of the transport system are likely limited.

Axonal RNPs can exhibit dynamic movement, with long-range anterograde motility that occurs at rates comparable to the well-established rates of fast axonal transport (2.2–4.6 μm s^{-1}) (Kiebler and Bassell 2006). In live cell imaging experiments, movement of RBPs fused to fluorescent proteins or mRNAs "tagged" by fluorescent methods show that RNPs exhibit oscillatory motion and, occasionally, short-range retrograde movement. It is not entirely clear whether such retrograde movements reflect microtubule-based transport through dynein or transition to microfilament-based transport through myosin. Microfilament-based transport of RNPs is employed for localization in budding yeasts and fibroblasts (Sundell and Singer 1991; Bertrand et al. 1998), and movement within axonal growth cones can be microfilament based (Leung et al. 2006; Yao et al. 2006a). At least for the La RBP (also known as SSB), long-range retrograde movement is microtubule-dependent through interactions with dynein, which requires a posttranslational modification (i.e., sumoylation) of La (van Niekerk et al. 2007). Whether other RBPs show such long-range retrograde movement remains to be determined. The retrograde transport of La is likely used to return the protein to the cell body or nucleus to be "recycled." The possibility that other RBPs are similarly recycled has not been tested, but this could provide a means for the distal axon to signal to the cell body. Moreover, if an RBP level is limited; recycling could effectively increase the available pool of individual mRNAs that can be localized.

4 Transport RNPs also Provide Translational Control

A common feature for the transport RNPs is that the mRNAs within these structures appear to be translationally silent. Analyses of the protein composition of the biochemically isolated RNPs show that they contain a complex population of proteins and multiple mRNAs. Krichevsky and Kosik used classical gradient centrifugation methods to isolate RNPs from neurons (Krichevsky and Kosik 2001). Kanai et al. (2004) isolated mobile neuronal RNPs based on their affinity for conventional kinesin (Kif5). The transport RNPs from both groups contained multiple mRNAs, translation factors, RBPs, and ribosome components. Ribosome subunits were detected in the RNPs isolated by density gradient centrifugation, but these appeared to be translationally inactive (Krichevsky and Kosik 2001). Ribosomal components were also identified in other RNP purification schemes (Kanai et al. 2004; Elvira et al. 2006). It is likely that each of the purification schemes above have identified multiple transport RNPs, and this may account for the complexity of the protein and mRNA composition. Recent work from the Sossin laboratory emphasizes the complexity of dendritic and axonal RNPs where several different transport RNPs could be visualized in living neurons based on their composition of different DEAD box RBPs (Miller et al. 2009).

Translational silencing of RNPs has been effectively shown when testing individual RNPs or mRNAs. When β-actin mRNA is bound to ZBP1 protein, the transcript is translationally repressed. Tyrosine phosphorylation of ZBP1 by Src family kinase(s) causes the RBP to dissociate from the mRNA, freeing it for translational initiation (Huttelmaier et al. 2005). This provides a means to spatially restrict translation of β-actin mRNA to sites of Src kinase activity. It is not known if ZBP1 inhibits translation of the other target mRNAs that have been identified as copurifying with IMP1 (Jonson et al. 2007). The cytoplasmic element binding protein (CPEB) similarly restricts translation of target mRNAs by forming a protein complex that blocks access to the mRNA cap structure (Huang and Richter 2004). Phosphorylation of CPEB disrupts the protein complex, freeing the 5' mRNA cap structure for recognition by eIF4 complex and initiation of its translation (Mendez and Richter 2001). The role of CPEB in neuronal mRNA transport and translational control is well established in dendrites (Wells 2006). CPEB clearly plays a role in *Aplysia* axons (Si et al. 2003), and there is evidence that CPEB can play a role in vertebrate axonal development, particularly for translation-dependent axonal guidance (Brittis et al. 2002; Lin et al. 2009). Thus far, CPEB has been demonstrated only in developing spinal cord commissural axons, so its role in other vertebrate axons remains unknown. Also, drawing an analogy to dendritic protein synthesis, recent work from the Fallon lab indicates that RNAs containing the FMRP and fragile X-related proteins 1 and/or 2 (FXR1, FXR2, respectively) are transported into the presynaptic compartment of developing neurons (Christie et al. 2009). FMRP has been localized to PNS axons, where it is argued to be part of the RNAi machinery (Murashov et al. 2007). Thus, it is possible that axons and dendrites may have more overlapping mechanisms of RNA transport and translational control than currently appreciated.

As previously noted (see Sect. 3), there are also similarities between transport RNPs, stress granules, and P-bodies. P-bodies are sites for RNA degradation and transient storage of untranslated mRNA derived from disassembled polysomes (Parker and Sheth 2007). In addition to decapping enzymes and exonucleases, P-body-like structures are seen in dendrites, where they show activity-dependent changes in their composition (Cougot et al. 2008). Recent work from Zeitelhofer et al. (2008) suggests transport RNPs can dock with P-bodies in dendrites. P-bodies also contain components of the RNAi machinery, so they likely contribute to miRNA-mediated degradation (Liu et al. 2005). Considering that PNS axons also contain RNAi machinery (Hengst et al. 2006; Murashov et al. 2007), axonal transport RNPs could share similar interactions with P-bodies. Stress granules are classically used to store mRNAs in response to different types of cellular stress; this allows the cell to selectively choose to synthesize proteins needed to respond to specific stress by sequestering other mRNAs in stress granules (Anderson and Kedersha 2006). Stress granules have been demonstrated in dendrites and axons (Vessey et al. 2006; Tsai et al. 2009). Interestingly, Tsai et al. (2009) suggest that dynein motor protein activity is needed to assemble and dissemble stress granules, which raises the interesting possibility that axonal stress granules could be derived from distal components.

5 Regulation of Axonal RNA Transport by Extracellular Cues

Most of the analyses of axonal protein synthesis in developing neurons have focused on regulation by guidance cues and tropic stimuli. Localized translation of RhoA, cofilin, CREB, and β-actin mRNAs in axons can be triggered by different ligands that are physiologically relevant for developmental axonal growth (Wu et al. 2005; Leung et al. 2006; Piper et al. 2006; Yao et al. 2006a; Cox et al. 2008). Li et al. (2004a) showed that stimulation with semaphorin, ligands that regulate axonal translation of RhoA (Wu et al. 2005), also triggered an increase in both retrograde and anterograde axonal transport. Interestingly, this increase in transport required activation of downstream protein kinases and localized translation (Li et al. 2004a). Earlier work in cultured neurons using fluorescent RNA dyes (Sytox®) showed that neurotrophin-3 (NT3) increases movement of RNPs into neuronal processes (Knowles and Kosik 1997). The Bassell lab went on to show that NT3 specifically increases localization of β-actin mRNA in developing cortical axons, using a quantitative in situ hybridization approach (Zhang et al. 1999). This effect was mimicked by elevating cAMP levels and required β-actin mRNA interaction with ZBP1 (Zhang et al. 1999, 2001). This suggests that NT3 and cAMP signaling pathways converge on the RNA transport machinery (e.g., through ZBP1). These reports of mRNA transport and translation were limited by studies of single mRNAs or nonspecific staining of all RNAs, so they did not address the question of whether there is specificity of transport at the level of individual mRNAs.

To determine if ligand stimulation of axonal processes generates changes in levels of multiple mRNAs, we have taken advantage of our enlarging database of axonal RNA populations for quantitative RT-PCR analyses of axonal RNA levels (Willis et al. 2007). By locally stimulating the axonal compartment of dissociated sensory neuron cultures with growth-promoting (neurotrophins) vs. growth-inhibiting stimuli (myelin-associated glycoprotein and semaphorin 3A), we have shown that neurons can both positively and negatively regulate the transport rates of individual mRNAs. There is a surprising level of specificity in how mRNAs respond, both in the RNA populations that change and in the magnitude of responses for the same mRNA to different ligands. Importantly, these experiments were performed in the presence of transcriptional inhibitors that blocked more than 95% of the new RNA synthesis. Thus, changes in axonal mRNA levels in these analyses were the result of ligand-dependent mobilization of an existing pool of mRNAs in the perikaryon (Willis et al. 2005, 2007). For each transcript tested, cell body mRNA levels decreased or increased conversely in response to increase or decrease in axonal mRNA levels. These findings are consistent with a pool of mRNAs in the cell body that can be selectively mobilized in response to ligand stimulation in the axonal compartment. In analyzing this large cohort of mRNAs for transport regulation, families of mRNAs, coding for proteins that in general have, like functions tended to show like responses (Fig. 4). The neurotrophins also tended to regulate mRNAs distinctly from myelin-associated glycoprotein and semaphorin, which cause growth cone retraction or turning in cultured sensory neurons. Even when responses to individual neurotrophins were examined, transcripts showed differential regulation by different signaling pathways. Inhibition of downstream signaling pathways indicated that ligands can actively downregulate transport of mRNAs that would normally localize in the absence of ligand stimulation. Interestingly, there were many mRNAs whose axonal levels were not affected by these ligands, suggesting that their transport into the axons is constitutive or that other stimuli are needed to regulate their transport (Willis et al. 2007). As noted below, injury itself may be one such stimulus.

The exquisite specificity of axonal mRNA transport regulation is likely conferred by ligand dependent activation of signaling pathways that converge on

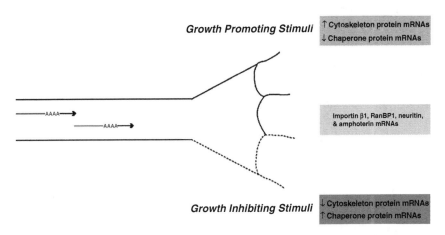

Fig. 4 Tropic stimuli modulate transport of RNA groups with like functions. Summary of shifts in axonal mRNA levels in response to positive (neurotrophins) and negative cues (myelin-associated glycoprotein and semaphorin 3A) in cultures of injury-conditioned DRG neurons summarized from Willis et al. (2007). *Upgoing arrows* indicate increased transport, and *downgoing arrows* indicate decreased transport

RBPs, or other components of the transport RNPs. This would be consistent with the findings that retrograde transport is required for modulation of RNA transport in response to neurotrophins (Willis et al. 2007). Of course, this does not exclude the possibility of localized responses within the axon, where signaling mechanisms may converge on transport machinery already within the axonal compartment. Indeed, depletion of β-actin mRNA from sites of myelin-associated glycoprotein stimulation along sensory axons was shown to be independent of retrograde transport (Willis et al. 2007). Application of netrin-1 and BDNF gradients to growing axons triggers an asymmetrical redistribution of β-actin mRNA within growth cones towards the ligand source (Leung et al. 2006; Yao et al. 2006a). Although such analyses have not been performed for other mRNAs, this elegant work from the Zheng and Holt labs emphasizes that localized signaling mechanisms instruct the neuron exactly where to send mRNAs within the growth cone by acting upon RBPs and components of the transport RNP (Leung et al. 2006; Yao et al. 2006a; see chapter by Yoon et al. 2009). It is likely that such redistribution of RNPs may also occur along the axon shaft in response to guidance cues and/or other ligands.

In addition to ligand-stimulated mRNA transport, some mRNAs are targeted to specific regions of the axon as a way to properly localize their protein products to domains or subdomains within the axon. Several mRNAs encoding membrane proteins have been identified in the axons of PNS and CNS neurons (Willis et al. 2007; Piper et al. 2008; Taylor et al. 2009). Both transfected and endogenously encoded membrane proteins that are locally translated can be inserted into the axonal plasma membrane (Spencer et al. 2000; Brittis et al. 2002; Merianda et al. 2009). Likewise, a recent publication from Toth et al. (2009) is highly suggestive of secretion of axonally synthesized calcitonin gene-related peptide (CGRP) from regenerating PNS axons, with the secreted CGRP inducing Schwann cell migration into the injured segment. Although a classic ultrastructural-appearing rough endoplasmic reticulum (RER) and Golgi apparatus have not been seen in axons, Merianda et al. (2009) have shown that components of these organelles needed for cotranslational protein secretion are present in rat, mouse, and *Xenopus* axons. mRNAs encoding membrane and secreted proteins must find their way to outposts of RER within the axonal compartment. One way to achieve this would be to target the mRNAs to the correct region of the axon. Interestingly, Cop1b protein is apparently needed for axonal localization of κ-opioid receptor mRNA (Bi et al. 2007). Cop1b is a "coatamer" protein that is part of a protein complex that functions in bidirectional transfer of vesicles between the endoplasmic reticulum (ER) and Golgi (Lee et al. 2004). It is surprising that a protein involved in membrane sorting plays a role in targeting an mRNA at subcellular compartments. Cop1b may position κ-opioid receptor mRNA into local regions of the axon that contain RER and Golgi.

There are mechanistic connections between the secretory machinery and RNA transport machinery in nonneuronal systems. Localization of drosophila Gurken mRNA to subcellular regions is needed to correctly target this protein into a secretory pathway (Herpers and Rabouille 2004). Without this targeting, Gurken protein does not generate a correct gradient in the embryo. Trailer hitch protein is required for proper targeting of Gurken mRNA to outposts of RER and Golgi in the embryo. Trailer hitch is part of an RNP that includes mRNAs encoding proteins needed for ER-to-Golgi progression (Wilhelm et al. 2005), similar to Cop1b's role in vesicle transport between the ER and Golgi. As more is learned of mRNA sorting within the axon, we are likely to see more evidence for targeting within the axonal subdomains. Several axonal mRNAs encode for components of the translational machinery (Willis et al. 2007; Taylor et al. 2009), and these may also need differential sorting compared with mRNAs that encode axoplasmic contents needed for growth.

6 Injury as a Stimulus to Shift Axonal mRNA Transport Capacity

Observations in injured neurons pointed us to a potential role for protein synthesis in neurite regeneration. Indeed, NGF-primed PC12 cells can rapidly regenerate sheared neurites through translational control of mRNAs that existed prior to the injury (Twiss and Shooter 1995). Adult rodent sensory neurons that are "injury-conditioned" show rapid regenerative growth when cultured 3–28 days after an in vivo axonal crush injury (Lankford et al. 1995, 1998; Smith and Skene 1997). This rapid axonal regrowth is independent of new mRNA synthesis but requires new protein synthesis (Smith and Skene 1997; Twiss et al. 2000). Zheng et al. (2001) went on to show that some of these newly synthesized proteins are generated directly in the axons. When axons growing from unconditioned or "naïve" neurons are compared to those of 7-day injury-conditioned neurons, injury appears to dramatically increase the protein synthetic capacity of the axonal compartment (Willis et al. 2005). These observations have led us to hypothesize that injury reactivates the developmental mechanism of axonal growth, allowing (perhaps, even encouraging) localization of mRNAs and translational machinery into the axonal compartment. However, as noted below, works from other groups have shown that axonal injury can recruit dormant mRNAs to the protein synthesis machinery in axons, and these proteins play a critical role in regeneration.

Injury of axons is well known to stimulate retrograde transport. Indeed, this is one way the neuronal cell body senses that the axon is injured, and such retrograde signals initiate a regenerative response in the cell body by altering programs of gene expression (Hanz and Fainzilber 2006). Hanz et al. (2003) showed that crush injury of PNS axons generates a retrograde signaling complex through localized protein synthesis (Hanz et al. 2003). Both Importin β1 and RanBP1 mRNAs are locally translated in axons in response to injury, a mechanism that apparently requires elevated Ca^{2+} (Hanz et al. 2003; Yudin et al. 2008). With the introduction of RanBP1 protein into the axoplasm, the newly produced Importin β1 generates a cargo-carrying complex with resident Importin α3 protein that signals injury to the nucleus, likely by triggering a growth-associated transcriptional program (Yudin and Fainzilber 2009).

The role of Ca^{2+} in the injury-induced translational response is intriguing. In *Aplysia* neurites, Ca^{2+} influx and subsequent Ca^{2+}-activated calpain-dependent proteolysis are needed to form a growth cone following injury (Ziv and Spira 1997; Gitler and Spira 2002). Increased Ca^{2+} is also needed for membrane sealing after injury of mammalian neurites (Detrait et al. 2000). Verma et al. (2005) directly linked localized protein synthesis to growth cone formation by showing that the severed axons of cultured mammalian neurons do not form a growth cone when localized mRNA translation is inhibited. Several studies have implicated changes in intra-axonal Ca^{2+} levels for outgrowth and guidance of axons (for review see Zheng and Poo 2007). Thus far, the only conclusive evidence for calcium-dependent mRNA localization is in the relocalization of β-actin mRNA-ZBP1 complex in response to BDNF discussed above (Yao et al. 2006a). However, since the Importin β1 and RanBP1 mRNAs appear to be dormant or inefficiently utilized in the mature, uninjured sciatic nerve axons, injury may trigger a relocalization of these mRNAs to axonal regions where the protein synthesis machinery is concentrated. Additional studies will be needed to determine how such dormant mRNAs are recruited in response to injury. Convergence of signaling cascades on RNPs or individual RBPs to modify their activity is a likely mechanism. Interestingly, increase in cytoplasmic Ca^{2+} typically blocks protein synthesis by inactivating translation factors (Gmitter et al. 1996; Brostrom and Brostrom 1998). Translation of some mRNAs is paradoxically increased by elevated Ca^{2+} (Brostrom et al. 1996; Brostrom and Brostrom 1998, 2003), including those encoding ER chaperone proteins that have been detected in axons (e.g., calreticulin and grp78/BiP) (Willis et al. 2005, 2007). Thus, local Ca^{2+} concentration may provide a means to regulate specificity of protein synthesis in the axonal compartment.

Our studies point to an injury-induced elevation of axonal protein synthesizing capacity (Zheng et al. 2001; Willis et al. 2005), including altered aggregation of cotranslational secretory machinery in axons (Merianda et al. 2009). The possibility that axonal injury can alter the levels or activities of RBPs and other constituents of the RNA transport apparatus is an appealing hypothesis to explain shift in this axonal translational capacity. Consistent with this possibility, Toth et al. (2009) very recently showed that peripheral nerve injury results in increased axonal localization and translation of the CGRP mRNA. In culture, axons of the injury conditioned sensory neurons have 2.5-fold more CGRP mRNA than those from naïve animals, and the mRNA concentrates in the growth cones (Fig. 5). Importantly, this shift in localization occurs despite falling CGRP mRNA and protein levels in the cell body (Li et al. 2004b). Toth et al. (2009) also proved the importance of this increased localized synthesis of CGRP for axonal regeneration in vivo, using siRNA technologies to selectively deplete axons of CGRP mRNA or Schwann cells of the CGRP receptor.

It is highly likely that transport of groups of axonal mRNAs will be regulated in parallel by axotomy. Taylor et al. (2009) have recently used microfluidic cultures of CNS neurons to test this hypothesis. As previously noted (see Sect. 2), the microfluidic device allows for isolation of axons from cell bodies, nonneuronal cells, and

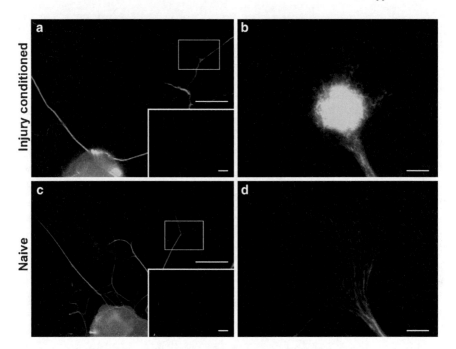

Fig. 5 Transport of CGRP mRNA into axons is increased by injury conditioning. Epifluorescent images for CGRP mRNA hybridization (*red*) and neurofilaments (*green*) from cultures of 7-day injury-conditioned (**a, b**) vs. naïve (**c, d**) DRG neurons are shown. The boxed regions in panels (**a**) and (**c**) are shown at high power in panels (**b**) and (**c**). CGRP mRNA clearly extends into the axonal compartment and concentrates in the growth cone. Note that signal for CGRP mRNA is clearly granular as seen with other localized mRNAs. Image pairs (**a** and **c**, **b** and **d**) are matched for exposure, offset, gain, and postprocessing. Note that CGRP mRNA is dramatically enriched in the axons from the injury-conditioned DRG neurons compared with the naïve neurons. Methods for in situ hybridization, probe design, as well as probe sequence are published (Toth et al. 2009). [*Scale bars*: **a, c** = 25 μm; **b, d** = 5 μm]

dendrites (Taylor et al. 2005). Using array technology, Taylor and colleagues showed that roughly half of 866 mRNAs showing greater than 20% difference between naïve and regenerating axons were increased, and the other half were decreased in regenerating axons (Taylor et al. 2009). There are clearly differences in the injury of cultured neurons used by Taylor et al. (2009) compared to the injury conditioning produced in vivo for the PNS axonal mRNA studies discussed above. Also, one must not presume that the uninjured CNS axons in these microfluidic chambers are comparable to mature axons in vivo. Nonetheless, these CNS axonal mRNA studies clearly show that injury itself can trigger a broad shift in the axonal transport of entire families of mRNAs. Interestingly, gene ontology analyses showed that the mRNAs with increased levels in regenerating axons have functions in axonal targeting, synaptogenesis, and synaptic function (Taylor et al. 2009). Thus, such shifts in mRNA transport are likely used by the neuron to match the localized mRNA population in response to the environmental and metabolic needs of the distal axon.

7 Concluding Remarks

In this chapter, we have tried to summarize the current literature on regulation of axonal mRNA transport, emphasizing the limitations of available data and identifying directions for future research. Despite evidence in the literature for translational activity and RNAs in axons from the 1960s and early 1990s, most neuroscientists did not appreciate until recently the potential for protein synthetic activity in the axon. The schematic presented in Fig. 6 attempts to outline the current state of

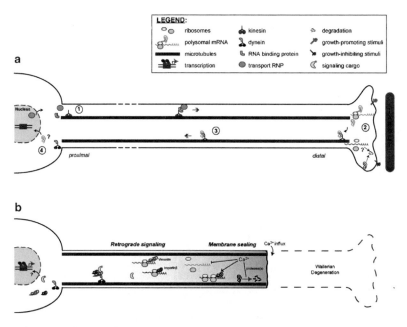

Fig. 6 Summary of mechanisms to regulate axonal mRNA translation. Schematic of the potential mechanisms that are used to regulate axonal mRNA transport and translation in intact (**a**) and injured (**b**) axons are depicted with components of transport and translational regulation indicated in the legend. (**a**) Axonal mRNA targeting requirement is initiated by mRNA-protein interactions, which ultimately generate and transport RNP containing multiple mRNAs, RNA binding proteins and other constituents to link the RNP to motor proteins for microtubule-based transport (1). Once an mRNA reaches the correct locale, its translation can be regulated by stimuli including growth promoting and growth inhibiting ligands (2). Ligand stimulation of the distal axon can also generate retrograde signals (3) that can further modulate localized protein synthesis by modifying RNA transport through transcriptional and posttranscriptional means (4). Some ligands can trigger proteolysis in axons, including proteolysis of recently synthesized proteins. (**b**) With axotomy, the distal severed axon undergoes Wallerian degeneration with loss of target contact. Generation of a new growth cone by the proximal stump requires axonal protein synthesis and degradation. Interestingly, proteolysis can generate peptides with distinct activities from the parent protein; for vimentin, this peptide fragment contributes to retrograde signaling. A transient elevation of Ca^{2+} in the injured axon is needed to trigger translation of injury-associated mRNAs. This likely occurs through posttranslational modification of initiation factors and RNA binding proteins, which ultimately recruits dormant mRNAs into translation. Some of the locally synthesized proteins generate a retrograde signaling complex that tells the cell body the axon has been injured. Other locally synthesized proteins are used to locally initiate regenerative growth

knowledge on axonal transport in developing injured/regenerating axons. Although we tend to regard RNA transport and translation as distinct steps, the mechanisms governing transport RNP formation, actual transport, and localized translation are undoubtedly a continuum, where each step is mechanistically linked. We also frequently refer to localized protein synthesis providing autonomy to the distal axon, but this protein synthesis requires a coordinated interaction with the cell body, with locally synthesized proteins signaling to the cell body. Thus, despite the tremendous distances that can separate the distal axon from the cell body, they truly are inseparable.

Acknowledgements The authors were supported by funds from NIH (R01-NS049041), Christopher and Dana Reeve Foundation (TCB2-0602), Adelson Medical Research Foundation, and Nemours. Drs. Dianna Willis, Cory Toth, and Douglas Zochodne provided unpublished data for Fig. 5. The Nemours Histotechnology laboratory provided technical assistance for tissue preparation for Fig. 3.

References

Aakalu G, Smith WB, Nguyen N, Jiang C, Schuman EM (2001) Dynamic visualization of local protein synthesis in hippocampal neurons. Neuron 30:489–502

Adereth Y, Dammai V, Kose N, Li R, Hsu T (2005) RNA-dependent integrin alpha3 protein localization regulated by the Muscleblind-like protein MLP1. Nat Cell Biol 7:1240–1247

An JJ, Gharami K, Liao GY, Woo NH, Lau AG, Vanevski F, Torre ER, Jones KR, Feng Y, Lu B, Xu B (2008) Distinct role of long 3¢ UTR BDNF mRNA in spine morphology and synaptic plasticity in hippocampal neurons. Cell 134:175–187

Anderson P, Kedersha N (2006) RNA granules. J Cell Biol 172:803–808

Antonellis A, Lee-Lin SQ, Wasterlain A, Leo P, Quezado M, Goldfarb LG, Myung K, Burgess S, Fischbeck KH, Green ED (2006) Functional analyses of glycyl-tRNA synthetase mutations suggest a key role for tRNA-charging enzymes in peripheral axons. J Neurosci 26:10397–10406

Aronov S, Aranda G, Behar L, Ginzburg I (2001) Axonal tau mRNA localization coincides with tau protein in living neuronal cells and depends on axonal targeting signal. J Neurosci 21:6577–6587

Aschrafi A, Schwechter AD, Mameza MG, Natera-Naranjo O, Gioio AE, Kaplan BB (2008) MicroRNA-338 regulates local cytochrome c oxidase IV mRNA levels and oxidative phosphorylation in the axons of sympathetic neurons. J Neurosci 28:12581–12590

Atlas R, Behar L, Elliott E, Ginzburg I (2004) The insulin-like growth factor mRNA binding-protein IMP-1 and the Ras-regulatory protein G3BP associate with tau mRNA and HuD protein in differentiated P19 neuronal cells. J Neurochem 89:613–626

Bassell GJ, Zhang H, Byrd AL, Femino AM, Singer RH, Taneja KL, Lifshitz LM, Herman IM, Kosik KS (1998) Sorting of beta-actin mRNA and protein to neurites and growth cones in culture. J Neurosci 18:251–265

Behar L, Marx R, Sadot E, Barg J, Ginzburg I (1995) cis-acting signals and trans-acting proteins are involved in tau mRNA targeting into neurites of differentiating neuronal cells. Int J Dev Neurosci 13:113–127

Bertrand E, Chartrand P, Schaefer M, Shenoy SM, Singer RH, Long RM (1998) Localization of ASH1 mRNA particles in living yeast. Mol Cell 2:437–445

Besse F, Ephrussi A (2008) Translational control of localized mRNAs: restricting protein synthesis in space and time. Nat Rev Mol Cell Biol 9:971–980

Bi J, Tsai NP, Lin YP, Loh HH, Wei LN (2006) Axonal mRNA transport and localized translational regulation of kappa-opioid receptor in primary neurons of dorsal root ganglia. Proc Natl Acad Sci U S A 103:19919–19924

Bi J, Tsai NP, Lu HY, Loh HH, Wei LN (2007) Copb1-facilitated axonal transport and translation of kappa opioid-receptor mRNA. Proc Natl Acad Sci U S A 104:13810–13815

Black MM, Lasek RJ (1977) The presence of transfer RNA in the axoplasm of the squid giant axon. J Neurobiol 8:229–237

Brittis PA, Lu Q, Flanagan JG (2002) Axonal protein synthesis provides a mechanism for localized regulation at an intermediate target. Cell 110:223–235

Brostrom CO, Brostrom MA (1998) Regulation of translational initiation during cellular responses to stress. Prog Nucleic Acid Res Mol Biol 58:79–125

Brostrom MA, Brostrom CO (2003) Calcium dynamics and endoplasmic reticular function in the regulation of protein synthesis: implications for cell growth and adaptability. Cell Calcium 34:345–363

Brostrom CO, Prostko CR, Kaufman R, Brostrom MA (1996) Inhibition of translational initiation by activators of the glucose-regulated stress response and heat shock protein stress response systems. J Biol Chem 271:24995–25002

Burgin KE, Waxham MN, Rickling S, Westgate SA, Mobley WC, Kelly PT (1990) In situ hybridization histochemistry of Ca^{2+}/Calmodulin-dependent protein kinase in developing rat brain. J Neurosci 10:1788–1798

Campbell D, Holt C (2001) Chemotropic responses of reginal growth cones mediated by rapid local protein synthesis and degradation. Neuron 32:1013–1016

Cannell IG, Kong YW, Bushell M (2008) How do microRNAs regulate gene expression? Biochem Soc Trans 36:1224–1231

Capano CP, Giuditta A, Castigli E, Kaplan BB (1987) Occurrence and sequence complexity of polyadenylated RNA in squid axoplasm. J Neurochem 49:698–704

Chihara T, Luginbuhl D, Luo L (2007) Cytoplasmic and mitochondrial protein translation in axonal and dendritic terminal arborization. Nat Neurosci 10:828–837

Christie SB, Akins MR, Schwob JE, Fallon JR (2009) The FXG: a presynaptic fragile X granule expressed in a subset of developing brain circuits. J Neurosci 29:1514–1524

Chun J-T, Gioio A, Crispino M, Capano C, Giuditta A, Kaplan B (1996) Differential compartmentalization of mRNAs in squid giant axon. J Neurochem 67:1806–1812

Cougot N, Bhattacharyya SN, Tapia-Arancibia L, Bordonne R, Filipowicz W, Bertrand E, Rage F (2008) Dendrites of mammalian neurons contain specialized P-body-like structures that respond to neuronal activation. J Neurosci 28:13793–13804

Court FA, Hendriks WTJ, Mac Gillavry HD, Alvarez J, van Minnen J (2008) Schwann cell to axon transfer of ribosomes: toward a novel understanding of the role of glia in the nervous system. J Neurosci 28:11024–11029

Cox LJ, Hengst U, Gurskaya NG, Lukyanov KA, Jaffrey SR (2008) Intra-axonal translation and retrograde trafficking of CREB promotes neuronal survival. Nat Cell Biol 10:149–159

Crino PB, Eberwine J (1996) Molecular characterization of the dendritic growth cone: regulated mRNA transport and local protein synthesis. Neuron 17:1173–1187

Crispino M, Kaplan BB, Martin R, Alvarez J, Chun JT, Benech JC, Giuditta A (1997) Active polysomes are present in the large presynaptic endings of the synaptosomal fraction from squid brain. J Neurosci 17:7694–7702

Daigle N, Ellenberg J (2007) LambdaN-GFP: an RNA reporter system for live-cell imaging. Nat Methods 4:633–636

Detrait E, Yoo S, Eddleman C, Fukada M, Bittner G, Fishman H (2000) Plasmalemmal repair of severed neurites of PC12 cells requires Ca^{2+} and synaptotagmin. J Neurosci Res 62:566–573

di Penta A, Mercaldo V, Florenzano F, Munck S, Ciotti MT, Zalfa F, Mercanti D, Molinari M, Bagni C, Achsel T (2009) Dendritic LSm1/CBP80-mRNPs mark the early steps of transport commitment and translational control. J Cell Biol 184:423–435

Dictenberg JB, Swanger SA, Antar LN, Singer RH, Bassell GJ (2008) A direct role for FMRP in activity-dependent dendritic mRNA transport links filopodial-spine morphogenesis to fragile X syndrome. Dev Cell 14:926–939

Edström A (1966) Amino Acid Incorporation in isolated Mauthner nerve fibre components. J Neurochem 13:315–321

Elvira G, Wasiak S, Blandford V, Tong XK, Serrano A, Fan X, del Rayo Sanchez-Carbente M, Servant F, Bell AW, Boismenu D, Lacaille JC, McPherson PS, DesGroseillers L, Sossin WS (2006) Characterization of an RNA granule from developing brain. Mol Cell Proteomics 5:635–651

Eng H, Lund K, Campenot RB (1999) Synthesis of beta-tubulin, actin, and other proteins in axons of sympathetic neurons in compartmented cultures. J Neurosci 19:1–9

Estes PS, O'Shea M, Clasen S, Zarnescu DC (2008) Fragile X protein controls the efficacy of mRNA transport in Drosophila neurons. Mol Cell Neurosci 39:170–179

Eyman M, Cefaliello C, Ferrara E, De Stefano R, Lavina ZS, Crispino M, Squillace A, van Minnen J, Kaplan BB, Giuditta A (2007) Local synthesis of axonal and presynaptic RNA in squid model systems. Eur J Neurosci 25:341–350

Farina KL, Huttelmaier S, Musunuru K, Darnell R, Singer RH (2003) Two ZBP1 KH domains facilitate beta-actin mRNA localization, granule formation, and cytoskeletal attachment. J Cell Biol 160:77–87

Fischer S, Litvak S (1967) The incorporation of microinjected 14C-amino acids into TCA insoluble fractions of the giant axon of the squid. J Cell Physiol 70:69–74

Fusco D, Accornero N, Lavoie B, Shenoy SM, Blanchard JM, Singer RH, Bertrand E (2003) Single mRNA molecules demonstrate probabilistic movement in living Mammalian cells. Curr Biol 13:161–167

Garner C, Tucker R, Matus A (1988) Selective localization of messenger RNA for MAP2 in dendrites. Nature 336:674–677

Gioio A, Chun J-T, Crispino M, Capano C, Giuditta A, Kaplan B (1994) Kinesin mRNA is present in the squid giant axon. J Neurochem 63:13–18

Gitler D, Spira M (2002) Short window of opportunity for Calpain induced growth cone formation after axotomy in *Aplysia* neurons. J Neurobiol 52:267–279

Giuditta A, Dettbarn W-D, Brzin M (1968) Protein synthesis in the isolated giant axon of the squid. Proc Natl Acad Sci U S A 59:1284–1287

Gmitter D, Brostrom CO, Brostrom MA (1996) Translational suppression by Ca^{2+} ionophores: reversibility and roles of Ca^{2+} mobilization, Ca^{2+} influx, and nucleotide depletion. Cell Biol Toxicol 12:101–113

Guiditta A, Menichini E, Capano C, Langella M (1991) Active polysomes in the axoplasm of the squid giant axon. J Neurosci Res 28:18–28

Hamilton RS, Davis I (2007) RNA localization signals: deciphering the message with bioinformatics. Semin Cell Dev Biol 18:178–185

Hamilton RS, Hartswood E, Vendra G, Jones C, Van De Bor V, Finnegan D, Davis I (2009) A bioinformatics search pipeline, RNA2DSearch, identifies RNA localization elements in Drosophila retrotransposons. RNA 15:200–207

Hanz S, Fainzilber M (2006) Retrograde signaling in injured nerve - the axon reaction revisited. J Neurochem 99:13–19

Hanz S, Perlson E, Willis D, Zheng JQ, Massarwa R, Huerta JJ, Koltzenburg M, Kohler M, vanMinnen J, Twiss JL, Fainzilber M (2003) Axoplasmic importins enable retrograde injury signaling in lesioned nerve. Neuron 40:1095–1104

Hengst U, Cox LJ, Macosko EZ, Jaffrey SR (2006) Functional and selective RNA interference in developing axons and growth cones. J Neurosci 26:5727–5732

Herpers B, Rabouille C (2004) mRNA localization and protein sorting mechanisms dictates the usage of tER-golgi units involved in Gurken transport in Drosophila oocytes. Mol Biol Cell 15:5306–5317

Hill MA, Gunning P (1993) Beta and gamma actin mRNAs are differentially located within myoblasts. J Cell Biol 122:825–832

Hillefors M, Gioio A, Mameza M, Kaplan B (2007) Axon viability and mitochondrial function are dependent on local protein synthesis in sympathetic neurons. Cell Mol Neurobiol 27:701–706

Huang YS, Richter JD (2004) Regulation of local mRNA translation. Curr Opin Cell Biol 16:308–313

Huttelmaier S, Zenklusen D, Lederer M, Dictenberg J, Lorenz M, Meng X, Bassell GJ, Condeelis J, Singer RH (2005) Spatial regulation of beta-actin translation by Src-dependent phosphorylation of ZBP1. Nature 438:512–515

Jambhekar A, Derisi JL (2007) cis-acting determinants of asymmetric, cytoplasmic RNA transport. RNA 13:625–642

Jansen RP (2001) mRNA localization: message on the move. Nat Rev Mol Cell Biol 2:247–256

Jirikowski GF, Sanna PP, Bloom FE (1990) Messenger RNA coding for oxytocin is present in axons of the hypothalamoneurohypophysial tract. Proc Natl Acad Sci U S A 87:7400–7404

Jirikowski GF, Sanna PP, Maciejewski-Lenoir D, Bloom FE (1992) Reversal of diabetes insipidus in Brattleboro rats: intrahypothalamic injection of vasopressin mRNA. Science 255:996–998

Jonson L, Vikesaa J, Krogh A, Nielsen LK, Hansen T, Borup R, Johnsen AH, Christiansen J, Nielsen FC (2007) Molecular composition of IMP1 ribonucleoprotein granules. Mol Cell Proteomics 6:798–811

Jordanova A, Irobi J, Thomas FP, Van Dijck P, Meerschaert K, Dewil M, Dierick I, Jacobs A, De Vriendt E, Guergueltcheva V, Rao CV, Tournev I, Gondim FA, D'Hooghe M, Van Gerwen V, Callaerts P, Van Den Bosch L, Timmermans JP, Robberecht W, Gettemans J, Thevelein JM, De Jonghe P, Kremensky I, Timmerman V (2006) Disrupted function and axonal distribution of mutant tyrosyl-tRNA synthetase in dominant intermediate Charcot-Marie-Tooth neuropathy. Nat Genet 38:197–202

Kanai Y, Dohmae N, Hirokawa N (2004) Kinesin transports RNA: isolation and characterization of an RNA-transporting granule. Neuron 43:513–525

Kaplan BB, Gioio AE, Hillefors M, Aschrafi A (2009) Axonal protein synthesis and the regulation of local mitochondrial function. Results Probl Cell Differ. doi: 10.1007/400_2009_1

Kiebler MA, Bassell GJ (2006) Neuronal RNA granules: movers and makers. Neuron 51:685–690

Kislauskis EH, Li Z, Singer RH, Taneja KL (1993) Isoform-specific 3′-untranslated sequences sort alpha-cardiac and beta-cytoplasmic actin messenger RNAs to different cytoplasmic compartments. J Cell Biol 123:165–172

Kislauskis EH, Zhu X, Singer RH (1994) Sequences responsible for intracellular localization of beta-actin messenger RNA also affect cell phenotype. J Cell Biol 127:441–451

Knowles RB, Kosik KS (1997) Neurotrophin-3 signals redistribute RNA in neurons. Proc Natl Acad Sci U S A 94:14804–14808

Kobayashi H, Yamamoto S, Maruo T, Murakami F (2005) Identification of a cis-acting element required for dendritic targeting of activity-regulated cytoskeleton-associated protein mRNA. Eur J Neurosci 22:2977–2984

Koenig E (1965a) Synthetic mechanisms in the axon - II: RNA in myelin-free axons of the cat. J Neurochem 12:357–361

Koenig E (1965b) Synthetic mechanisms in the axon - I: local axonal synthesis of acetylcholinesterase. J Neurochem 12:343–355

Koenig E (1967a) Synthetic mechanisms in the axon - IV: in vitro incorporation of [3H]precursors into axonal protein and RNA. J Neurochem 14:437–446

Koenig E (1967b) Synthetic mechanisms in the axon - III: stimulation of acetylcholinesterase synthesis by actinomycin-D in the hypoglossal nerve. J Neurochem 14:429–435

Koenig E, Koelle GB (1960) Acetylcholinesterase regeneration in peripheral nerve after irreversible inactivation. Science 132:1249–1250

Koenig E, Koelle GB (1961) Mode of regeneration of acetylcholinesterase in cholinergic neurons following irreversible inactivation. J Neurochem 8:169–188

Koenig E (2009) Organized ribosome-containing structural domains in axons. Results Probl Cell Differ, doi: 10.1007/400_2008_29

Kress TL, Yoon YJ, Mowry KL (2004) Nuclear RNP complex assembly initiates cytoplasmic RNA localization. J Cell Biol 165:203–211

Krichevsky A, Kosik K (2001) Neuronal RNA granules: a link between RNA localization and stimulation-dependent translation. Neuron 32:683–696

Koenig E (2009) Organized ribosome-containing structural domains in axons. Results Probl Cell Differ, doi:10.1007/400_2008_29

Kye MJ, Liu T, Levy SF, Xu NL, Groves BB, Bonneau R, Lao K, Kosik KS (2007) Somatodendritic microRNAs identified by laser capture and multiplex RT-PCR. RNA 13:1224–1234

Lange S, Katayama Y, Schmid M, Burkacky O, Brauchle C, Lamb DC, Jansen RP (2008) Simultaneous transport of different localized mRNA species revealed by live-cell imaging. Traffic 9:1256–1267

Lankford KL, Rand MN, Waxman SG, Kocsis JD (1995) Blocking Ca2 + mobilization with thapsigargin reduces neurite initiation in cultured adult rat DRG neurons. Brain Res Dev Brain Res 84:151–163

Lankford K, Waxman S, Kocsis J (1998) Mechanisms of enhancement of neurite regeneration in vitro following a conditioning sciatic nerve lesion. J Comp Neurol 391:11–29

Lecuyer E, Yoshida H, Parthasarathy N, Alm C, Babak T, Cerovina T, Hughes TR, Tomancak P, Krause HM (2007) Global analysis of mRNA localization reveals a prominent role in organizing cellular architecture and function. Cell 131:174–187

Lee, S, Hollenbeck P (2003) Organization and translation of mRNA in sympathetic axons. J Cell Sci 116:4467–4478

Lee W, Wayne NL (2004) Secretion of locally synthesized neurohormone from neurites of peptidergic neurons. J Neurochem 88:532–537

Lee MC, Miller EA, Goldberg J, Orci L, Schekman R (2004) Bi-directional protein transport between the ER and Golgi. Annu Rev Cell Dev Biol 20:87–123

Leung KM, van Horck FP, Lin AC, Allison R, Standart N, Holt CE (2006) Asymmetrical beta-actin mRNA translation in growth cones mediates attractive turning to netrin-1. Nat Neurosci 9:1247–1256

Li C, Sasaki Y, Takei K, Yamamoto H, Shouji M, Sugiyama Y, Kawakami T, Nakamura F, Yagi T, Ohshima T, Goshima Y (2004a) Correlation between semaphorin3A-induced facilitation of axonal transport and local activation of a translation initiation factor eukaryotic translation initiation factor 4E. J Neurosci 24:6161–6167

Li XQ, Verge VM, Johnston JM, Zochodne DW (2004b) CGRP peptide and regenerating sensory axons. J Neuropathol Exp Neurol 63:1092–1103

Lin AC, Tan CL, Lin CL, Strochlic L, Huang YS, Richter JD, Holt CE (2009) Cytoplasmic polyadenylation and cytoplasmic polyadenylation element-dependent mRNA regulation are involved in Xenopus retinal axon development. Neural Dev 4:8

Litman P, Barg J, Rindzoonski L, Ginzburg I (1993) Subcellular localization of tau mRNA in differentiating neuronal cell culture: implications for neuronal polarity. Neuron 10:627–638

Litman P, Barg J, Ginzburg I (1994) Microtubules are involved in the localization of tau mRNA in primary neuronal cell cultures. Neuron 13:1463–1474

Litman P, Behar L, Elisha Z, Yisraeli JK, Ginzburg I (1996) Exogenous tau RNA is localized in oocytes: possible evidence for evolutionary conservation of localization mechanisms. Dev Biol 176:86–94

Liu J, Valencia-Sanchez MA, Hannon GJ, Parker R (2005) MicroRNA-dependent localization of targeted mRNAs to mammalian P-bodies. Nat Cell Biol 7:719–723

Maciejewski-Lenoir D, Jirikowski GF, Sanna PP, Bloom FE (1993) Reduction of exogenous vasopressin RNA poly(A) tail length increases its effectiveness in transiently correcting diabetes insipidus in the Brattleboro rat. Proc Natl Acad Sci U S A 90:1435–1439

Martin KC, Barad M, Kandel ER (2000) Local protein synthesis and its role in synapse-specific plasticity. Curr Opin Neurobiol 10:587–592

Mayford M, Baranes D, Podsypanina K, Kandel ER (1996) The 3′-untranslated region of CaMKII alpha is a cis-acting signal for the localization and translation of mRNA in dendrites. Proc Natl Acad Sci U S A 93:13250–13255

Mendez R, Richter JD (2001) Translational control by CPEB: a means to the end. Nat Rev Mol Cell Biol 2:521–529

Merianda TT, Lin AC, Lam JS, Vuppalanchi D, Willis DE, Karin N, Holt CE, Twiss JL (2009) A functional equivalent of endoplasmic reticulum and Golgi in axons for secretion of locally synthesized proteins. Mol Cell Neurosci 40:128–142

Mili S, Moissoglu K, Macara IG (2008) Genome-wide screen reveals APC-associated RNAs enriched in cell protrusions. Nature 453:115–119

Miller LC, Blandford V, McAdam R, Sanchez-Carbente MR, Badeaux F, DesGroseillers L, Sossin WS (2009) Combinations of DEAD box proteins distinguish distinct types of RNA:protein complexes in neurons. Mol Cell Neurosci 40:485–495

Moccia R, Chen D, Lyles V, Kapuya E, E Y, Kalachikov S, Spahn CM, Frank J, Kandel ER, Barad M, Martin KC (2003) An unbiased cDNA library prepared from isolated *Aplysia* sensory neuron processes is enriched for cytoskeletal and translational mRNAs. J Neurosci 23:9409–9417

Mohr E, Fehr S, Richter D (1991) Axonal transport of neuropeptide encoding mRNAs within the hypothalamo-hypophysial tract of rats. EMBO 10:2419–2424

Mohr E, Richter D (1992) Diversity of Messenger RNAs in the Axonal Compartment of Peptidergic Neurons in the Rat. Eur J Neurosci 4:870–876

Muddashetty RS, Kelic S, Gross C, Xu M, Bassell GJ (2007) Dysregulated metabotropic glutamate receptor-dependent translation of AMPA receptor and postsynaptic density-95 mRNAs at synapses in a mouse model of fragile X syndrome. J Neurosci 27:5338–5348

Murashov AK, Chintalgattu V, Islamov RR, Lever TE, Pak ES, Sierpinski PL, Katwa LC, Van Scott MR (2007) RNAi pathway is functional in peripheral nerve axons. FASEB J 21:656–670

Nakielny S, Dreyfuss G (1999) Transport of proteins and RNAs in and out of the nucleus. Cell 99:677–690

Nangle LA, Zhang W, Xie W, Yang XL, Schimmel P (2007) Charcot-Marie-Tooth disease-associated mutant tRNA synthetases linked to altered dimer interface and neurite distribution defect. Proc Natl Acad Sci U S A 104:11239–11244

Olink-Coux M, Hollenbeck PJ (1996) Localization and Active Transport of mRNA in Axons of Sympathetic Neurons in Culture. J Neurosci 16:1346–1358

Pan F, Huttelmaier S, Singer RH, Gu W (2007) ZBP2 facilitates binding of ZBP1 to beta-actin mRNA during transcription. Mol Cell Biol 27:8340–8351

Parker R, Sheth U (2007) P bodies and the control of mRNA translation and degradation. Mol Cell 25:635–646

Perlson E, Hanz S, Ben-Yaakov K, Segal-Ruder Y, Segar R, Fainzilber M (2005) Vimentin-dependent spatial translocation of an activated MAP kinse in injured nerve. Neuron 45:715–726

Piper M, Anderson R, Dwivedy A, Weinl C, van Horck F, Leung KM, Cogill E, Holt C (2006) Signaling mechanisms underlying Slit2-induced collapse of Xenopus retinal growth cones. Neuron 49:215–228

Piper M, Dwivedy A, Leung L, Bradley RS, Holt CE (2008) NF-protocadherin and TAF1 regulate retinal axon initiation and elongation in vivo. J Neurosci 28:100–105

Poon MM, Choi SH, Jamieson CA, Geschwind DH, Martin KC (2006) Identification of process-localized mRNAs from cultured rodent hippocampal neurons. J Neurosci 26:13390–13399

Querido E, Chartrand P (2008) Using fluorescent proteins to study mRNA trafficking in living cells. Methods Cell Biol 85:273–292

Ressler KJ, Sullivan SL, Buck LB (1994) A molecular dissection of spatial patterning in the olfactory system. Curr Opin Neurobiol 4:588–596

Roche FK, Marsick BM, Letourneau PC (2009) Protein synthesis in distal axons is not required for growth cone responses to guidance cues. J Neurosci 29:638–652

Rossoll W, Bassell GJ (2009) Spinal muscular atrophy and a model for survival of motor neuron protein function in axonal RNP complexes. Results Probl Cell Differ. doi: 10.1007/400_2009_4

Rossoll W, Kroning AK, Ohndorf UM, Steegborn C, Jablonka S, Sendtner M (2002) Specific interaction of Smn, the spinal muscular atrophy determining gene product, with hnRNP-R and gry-rbp/hnRNP-Q: a role for Smn in RNA processing in motor axons? Hum Mol Genet 11:93–105

Schnapp BJ (1999) A glimpse of the machinery. Curr Biol 9:R725–R727

Schratt GM, Tuebing F, Nigh EA, Kane CG, Sabatini ME, Kiebler M, Greenberg ME (2006) A brain-specific microRNA regulates dendritic spine development. Nature 439:283–289

Seburn KL, Nangle LA, Cox GA, Schimmel P, Burgess RW (2006) An active dominant mutation of glycyl-tRNA synthetase causes neuropathy in a Charcot-Marie-Tooth 2D mouse model. Neuron 51:715–726

Shav-Tal Y, Darzacq X, Shenoy SM, Fusco D, Janicki SM, Spector DL, Singer RH (2004) Dynamics of single mRNPs in nuclei of living cells. Science 304:1797–1800

Sheller RA, Tytell M, Smyers M, Bittner GD (1995) Glia-to-axon communication: enrichment of glial proteins transferred to the squid giant axon. J Neurosci Res 41:324–334

Shestakova EA, Singer RH, Condeelis J (2001) The physiological significance of beta -actin mRNA localization in determining cell polarity and directional motility. Proc Natl Acad Sci U S A 98:7045–7050

Si K, Giustetto M, Etkin A, Hsu R, Janisiewicz AM, Miniaci MC, Kim JH, Zhu H, Kandel ER (2003) A neuronal isoform of CPEB regulates local protein synthesis and stabilizes synapse-specific long-term facilitation in *Aplysia*. Cell 115:893–904

Smith DS, Skene P (1997) A transcription-dependent switch controls competence of adult neurons for distinct modes of axon growth. J Neurosci 17:646–658

Sotelo-Silveira JR, Calliari A, Cardenas M, Koenig E, Sotelo JR (2004) Myosin Va and kinesin II motor proteins are concentrated in ribosomal domains (periaxoplasmic ribosomal plaques) of myelinated axons. J Neurobiol 60:187–196

Sotelo-Silveira J, Crispino M, Puppo A, Sotelo JR, Koenig E (2008) Myelinated axons contain beta-actin mRNA and ZBP-1 in periaxoplasmic ribosomal plaques and depend on cyclic AMP and F-actin integrity for in vitro translation. J Neurochem 104:545–557

Spencer GE, Syed NI, van Kesteren E, Lukowiak K, Geraerts WP, van Minnen J (2000) Synthesis and functional integration of a neurotransmitter receptor in isolated invertebrate axons. J Neurobiol 44:72–81

St Johnston D (2005) Moving messages: the intracellular localization of mRNAs. Nat Rev Mol Cell Biol 6:363–375

Steward O, Levy WB (1982) Preferential localization of polyribosomes under the base of dendritic spines in granule cells of the dentate gyrus. J Neurosci 2:284–291

Sundell CL, Singer RH (1991) Requirement of microfilaments in sorting of actin mRNAs. Science 253:1275–1277

Svane PC, Thorn NA, Richter D, Mohr E (1995) Effect of hypoosmolality on the abundance, poly(A) tail length and axonal targeting of arginine vasopressin and oxytocin mRNAs in rat hypothalamic magnocellular neurons. FEBS Lett 373:35–38

Taylor AM, Blurton-Jones M, Rhee SW, Cribbs DH, Cotman CW, Jeon NL (2005) A microfluidic culture platform for CNS axonal injury, regeneration and transport. Nat Methods 2:599–605

Taylor AM, Berchtold NC, Perreau VM, Tu CH, Jeon NL, Cotman CW (2009) Axonal mRNA in un-injured and regenerating cortical mammalian axons. J Neurosci 29:4697-4707

Tiruchinapalli DM, Oleynikov Y, Kelic S, Shenoy SM, Hartley A, Stanton PK, Singer RH, Bassell GJ (2003) Activity-dependent trafficking and dynamic localization of zipcode binding protein 1 and beta-actin mRNA in dendrites and spines of hippocampal neurons. J Neurosci 23:3251–3261

Toth CC, Willis D, Twiss JL, Walsh S, Martinez JA, Liu WQ, Midha R, Zochodne DW (2009) Locally synthesized calcitonin gene-related peptide has a critical role in peripheral nerve regeneration. J Neuropathol Exp Neurol 68:326-337

Trembleau A, Morales M, Bloom F (1994) Aggregration of vasopressin mRNA in a subset of axonal swellings of the median eminence and posterior pituitary: light and electron microscopic evidence. J Neurosci 14:39–53

Tsai NP, Bi J, Loh HH, Wei LN (2006) Netrin-1 signaling regulates de novo protein synthesis of kappa opioid receptor by facilitating polysomal partition of its mRNA. J Neurosci 26:9743–9749

Tsai NP, Tsui YC, Wei LN (2009) Dynein motor contributes to stress granule dynamics in primary neurons. Neuroscience 159:647-656

Twiss JL, Shooter EM (1995) Nerve growth factor promotes neurite regeneration in PC12 cells by translational control. J Neurochem 64:550–557

Twiss JL, van Minnen J (2006) New insights into neuronal regeneration: the role of axonal protein synthesis in pathfinding and axonal extension. J Neurotrauma 23:295–308

Twiss JL, Fainzilber M (2009) Ribosomes in axons – scrounging from the neighbors? Trends Cell Biol 19:236–243

Twiss J, Smith D, Chang B, Shooter E (2000) Translational control of ribosomal protein L4 is required for rapid neurite extension. Neurobiol Dis 7:416–428

Tytell M, Lasek R (1984) Glial polypeptides transferred into the squid giant axon. Brain Res 324:223–232

van Kesteren RE, Carter C, Dissel HM, van Minnen J, Gouwenberg Y, Syed NI, Spencer GE, Smit AB (2006) Local synthesis of actin-binding protein beta-thymosin regulates neurite outgrowth. J Neurosci 26:152–157

van Minnen J (1994) RNA in the axonal domain: a new dimension in neuronal functioning? Histochem J 26:377–391

van Niekerk EA, Willis DE, Chang JH, Reumann K, Heise T, Twiss JL (2007) Sumoylation in axons triggers retrograde transport of the RNA-binding protein La. Proc Natl Acad Sci U S A 104:12913–12918

Verma P, Chierzi S, Codd AM, Campbell DS, Meyery RL, Holt CE, Fawcett JW (2005) Axonal protein synthesis and degradation are necessary for efficient growth cone regeneration. J Neurosci 25:331–342

Vessey JP, Vaccani A, Xie Y, Dahm R, Karra D, Kiebler MA, Macchi P (2006) Dendritic localization of the translational repressor Pumilio 2 and its contribution to dendritic stress granules. J Neurosci 26:6496–6508

Wells DG (2006) RNA-binding proteins: a lesson in repression. J Neurosci 26:7135–7138

Wensley CH, Stone DM, Baker H, Kauer JS, Margolis FL, Chikaraishi DM (1995) Olfactory marker protein mRNA is found in axons of olfactory receptor neurons. J Neurosci 15:4827–4837

Wilhelm JE, Buszczak M, Sayles S (2005) Efficient protein trafficking requires trailer hitch, a component of a ribonucleoprotein complex localized to the ER in Drosophila. Dev Cell 9:675–685

Willis DE, Li KW, Zheng J-Q, Smit AB, Kelly TK, Merianda TT, Sylvester J, van Minnen J, Twiss JL (2005) Differential transport and local translation of cytoskeletal, injury-response, and neurodegeneration protein mRNAs in axons. J Neurosci 25:778–791

Willis DE, van Niekerk EA, Sasaki Y, Mesngon M, Merianda TT, Williams GG, Kendall M, Smith DS, Bassell GJ, Twiss JL (2007) Extracellular stimuli specifically regulate localized levels of individual neuronal mRNAs. J Cell Biol 178:965–980

Wu KY, Hengst U, Cox LJ, Macosko EZ, Jeromin A, Urquhart ER, Jaffrey SR (2005) Local translation of RhoA regulates growth cone collapse. Nature 436:1020–1024

Yao J, Sasaki Y, Wen Z, Bassell GJ, Zheng JQ (2006a) An essential role for beta-actin mRNA localization and translation in Ca(2+)-dependent growth cone guidance. Nat Neurosci 9:1265–1273

Yao Z, Weinberg Z, Ruzzo WL (2006b) CMfinder–a covariance model based RNA motif finding algorithm. Bioinformatics 22:445–452

Yoon BC, Zivraj KH, Holt CE (2009) Local translation and mRNA trafficking in axon pathfinding. Results Probl Cell Differ. doi: 10.1007/400_2009_5

Yudin D, Hanz S, Yoo S, Iavnilovitch E, Willis D, Segal-Ruder Y, Vuppalanchi D, Ben-Yaakov K, Hieda M, Yoneda Y, Twiss J, Fainzilber M (2008) Localized regulation of axonal RanGTPase controls retrograde injury signaling in peripheral nerve. Neuron 59:241–252

Yudin D, Fainzilber M (2009) Ran on tracks – cytoplasmic roles for a nuclear regulator. J Cell Sci 122:587–593

Zeitelhofer M, Karra D, Macchi P, Tolino M, Thomas S, Schwarz M, Kiebler M, Dahm R (2008) Dynamic interaction between P-bodies and transport ribonucleoprotein particles in dendrites of mature hippocampal neurons. J Neurosci 28:7555–7562

Zhang HL, Singer RH, Bassell GJ (1999) Neurotrophin regulation of beta-actin mRNA and protein localization within growth cones. J Cell Biol 147:59–70

Zhang HL, Eom T, Oleynikov Y, Shenoy SM, Liebelt DA, Dictenberg JB, Singer RH, Bassell GJ (2001) Neurotrophin-induced transport of a beta-actin mRNP complex increases beta-actin levels and stimulates growth cone motility. Neuron 31:261–275

Zheng JQ, Poo MM (2007) Calcium signaling in neuronal motility. Annu Rev Cell Dev Biol 23:375–404

Zheng J-Q, Kelly T, Chang B, Ryazantsev S, Rajasekaran A, Martin K, Twiss J (2001) A functional role for intra-axonal protein synthesis during axonal regeneration from adult sensory neurons. J Neurosci 21:9291–9303

Ziv N, Spira M (1997) Localized and transient elevations of intracellular Ca^{2+} induce dedifferentiation of axonal segments into growth cones. J Neurosci 17:3568–3579

Axonal Protein Synthesis and the Regulation of Local Mitochondrial Function

Barry B. Kaplan, Anthony E. Gioio, Mi Hillefors, and Armaz Aschrafi

Abstract Axons and presynaptic nerve terminals of both invertebrate and mammalian SCG neurons contain a heterogeneous population of nuclear-encoded mitochondrial mRNAs and a local cytosolic protein synthetic system. Nearly one quarter of the total protein synthesized in these structural/functional domains of the neuron is destined for mitochondria. Acute inhibition of axonal protein synthesis markedly reduces the functional activity of mitochondria. The blockade of axonal protein into mitochondria had similar effects on the organelle's functional activity. In addition to mitochondrial mRNAs, SCG axons contain approximately 200 different microRNAs (miRs), short, noncoding RNA molecules involved in the posttranscriptional regulation of gene expression. One of these miRs (miR-338) targets cytochrome c oxidase IV (COXIV) mRNA. This nuclear-encoded mRNA codes for a protein that plays a key role in the assembly of the mitochondrial enzyme complex IV and oxidative phosphorylation. Over-expression of miR-338 in the axon markedly decreases COXIV expression, mitochondrial functional activity, and the uptake of neurotransmitter into the axon. Conversely, the inhibition of endogenous miR-338 levels in the axon significantly increased mitochondrial activity and norepinephrine uptake into the axon. The silencing of COXIV expression in the axon using short, inhibitory RNAs (siRNAs) yielded similar results, a finding that indicated that the effects of miR-338 on mitochondrial activity and axon function were mediated, at least in part, through local COXIV mRNA translation. Taken together, recent findings establish that proteins requisite for mitochondrial activity are synthesized locally in the axon and nerve terminal, and call attention to the intimacy of the relationship that has evolved between the distant cellular domains of the neuron and its energy generating systems.

B.B. Kaplan (✉), A.E. Gioio, M. Hillefors, and A. Aschrafi
Laboratory of Molecular Biology, National Institute of Mental Health,
NIH, Bethesda, MD, 20892-1381, USA
e-mail: kaplanb@mail.nih.gov

1 Introduction

One of the central tenets in neuroscience has been that the protein constituents of the distal structural/functional domains of the neuron (e.g., the axon and presynaptic nerve terminal) are synthesized in the nerve cell body, and are subsequently transported to their ultimate sites of function. Although the majority of neuronal mRNAs are indeed translated in the neuronal cell soma, increasing attention is being focused on that subset of the transcriptome that is selectively transported to the distal domains of the neuron. The local translation of these mRNAs plays a key role in the development of the neuron and the function of the axon to include navigation of the axon (Campbell and Holt 2001; Campbell et al. 2001; Ming et al. 2002; Zhang and Poo 2002; Wu et al. 2005; Leung et al. 2006), synthesis of membrane receptors employed as axon guidance molecules (Brittis et al. 2002), axon regeneration (Zhang et al. 2001; Hanz et al. 2003; Verma et al. 2005), axon transport (Li et al. 2004), synapse formation (Schacher and Wu 2002), axon viability (Hu et al. 2003), neuronal survival (Cox et al. 2008), and activity-dependent synaptic plasticity (Martin et al. 1997; Casadio et al. 1999; Beaumont et al. 2001; Liu et al. 2003; Si et al. 2003).

Initial estimates of the number of mRNAs that are present in axon derived from invertebrate model systems are approximately 200–400 different mRNAs (Perrone-Capano et al. 1987; Moccia et al. 2003). These messengers code for a diverse population of proteins that include cytoskeletal proteins, translation factors, ribosomal proteins, molecular motors, chaperone proteins, and metabolic enzymes (for review, see Giuditta et al. 2008, 2009). Results of early quantitative RT-PCR analyses established that the relative abundance of the mRNAs present in the axon differed markedly from that in the cell soma, a finding that suggested that these gene transcripts were being differentially transported into the axonal compartment (Chun et al. 1996). One surprising feature of the axonal mRNA population was the presence of a significant number of nuclear-encoded mitochondrial mRNAs (Gioio et al. 2001; Hillefors et al. 2007). Based in part upon this observation, it was hypothesized that proteins requisite for the maintenance of mitochondrial function are synthesized locally in the axon and presynaptic nerve terminal. In this communication, we review the evidence that both invertebrate and mammalian axons contain a heterogeneous population of nuclear-encoded mitochondrial mRNAs, and that the local translation of these mRNAs plays a critical role in the regulation of the functional activity of the local mitochondrial population.

2 Nuclear-Encoded Mitochondrial mRNAs are Present in Invertebrate Axons and Nerve Endings

The initial evidence that mRNAs coding for nuclear-encoded mitochondrial proteins were present in the distal structural domains of the neuron derived from the squid giant axon and the large presynaptic nerve terminals of squid retinal photoreceptor neurons (Gioio et al. 2001, 2004). These structures were being employed as components of model invertebrate motor and sensory neurons, respectively.

In these early studies, differential mRNA display was used to compare the mRNAs present in the axon and neuronal cell soma. This comparison yielded approximately 150 sequences, which manifested a higher relative abundance in the axon. These mRNA fragments were subsequently screened for evolutionary sequence conservation by dot-blot hybridization using cDNA prepared from mouse brain mRNA. Fifty of the squid sequences cross-hybridized to mouse brain cDNA and were subsequently subjected to DNA sequence analyses. Surprisingly, we were able to establish the identity of four nuclear-encoded mRNAs coding for mitochondrial proteins (COX 17, propionyl-OcA carboxylase, dihydrolipoamide dehydrogenase and CoQ7). In addition, mRNAs encoding the molecular chaperones Hsp70 and Hsp90, proteins facilitating the import of preproteins into the mitochondria, were also identified. The association of these mRNAs with polysomes prepared from presynaptic nerve terminals of photoreceptor neurons was subsequently established by RT-PCR and provided evidence to suggest that the mRNAs were being actively translated (Fig. 1).

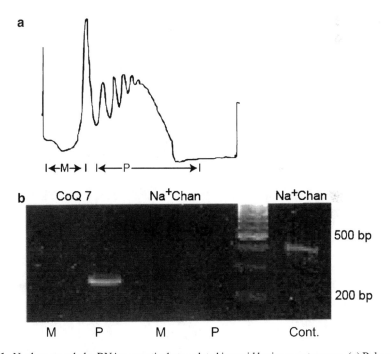

Fig. 1 Nuclear-encoded mRNAs are actively translated in squid brain synaptosomes. (**a**) Polysomes were prepared from squid optic lobe synaptosomes and were displayed on linear sucrose density gradients. Gradients were divided into monosome (M) and polysome (P) fractions. UV absorbance of RNA was monitored continuously at 254 nm. (**b**) RT-PCR analysis of monosome and polysome fractions using gene-specific primer sets for the nuclear-encoded mitochondrial protein CoQ7 and Na^+ channel used as an internal control. PCR products were fractionated on agarose gels and visualized by ethidium bromide staining. The absence of amplicons generated from the Na^+ channel primers indicates that the polysome fraction is devoid of RNA contamination from the neuronal cell soma. *Cont* Na^+ channel amplicons obtained from total RNA prepared from the optic lobe, *bp* base-pairs, *MW* molecular weight. Reproduced from Gioio et al. (2001)

To further evaluate the hypothesis that nuclear-encoded mitochondrial proteins were being locally synthesized, synaptosomes prepared from squid brain were incubated with [^{35}S]methionine, and the amount of newly synthesized protein associated with purified mitochondria was determined. Pretreatment of the synaptosomes with the antibiotic chloramphenicol was used to inhibit endogenous mitochondrial translational activity. Surprisingly, 20–25% of the total translational activity of squid brain synaptosomes was associated with the mitochondrial fraction. Hence, it appears that one of the major functions of the local protein synthetic system is to maintain the energy generating system present in the axon and nerve endings.

The results of a subsequent independent analysis of protein synthesized in squid brain synaptosomes confirmed the local synthesis of nuclear-encoded mitochondrial proteins (Jimenez et al. 2002). In this study, [^{35}S]methionine-labeled synaptosomal proteins were fractionated by two-dimensional gel electrophoresis, and peptides generated by in-gel tryptic digestion were identified by mass spectrometry. Essentially, this proteomics study established the de novo synthesis of about 80 different proteins, several of which were nuclear-encoded mitochondrial proteins, as well as the molecular chaperone, Hsp70. Interestingly, in a high-throughput proteomic analysis of proteins synthesized in synaptoneurosomes derived from cultured rat cortical neurons, Liao et al. (2007) reported that brain-derived neurotrophic factor (BDNF) upregulated the synthesis of approximately 200 different proteins. Eleven percent of these proteins were associated with mitochondria.

3 Axons of Sympathetic Neurons Contain a Heterogeneous Population of Nuclear-Encoded Mitochondrial mRNAs

To assess the general applicability of the early findings derived from invertebrate model systems, RT-PCR analyses and in situ hybridization histochemistry was employed to identify nuclear-encoded mitochondrial mRNAs in mammalian axons. In this study, primary sympathetic neurons were prepared from rat superior cervical ganglia (SCG) and were cultured in Campenot multicompartment chambers (Campenot and Martin 2001). The use of this cell culture system allows one to plate dissociated SCG neurons in the center compartment and grow axons into the two side compartments, providing pure axonal populations in both side compartments. One advantage of this culture system is that it permits the manipulation of axons in one of the side compartments, while axons in the opposite (contralateral) side compartment of the same culture dish serves as experimental controls (Hillefors et al. 2007).

Consistent with the findings derived from the squid giant axon and presynaptic nerve endings, SCG axons contain numerous nuclear-encoded mitochondrial mRNAs to include ATP synthase, cytochrome c oxidase IV and Va, DNA polymerase γ, as well as the molecular chaperones Hsp70 and Hsp90 (e.g., see Fig. 2a, b). The axonal localization of these mRNAs was confirmed by in situ hybridization (Fig. 2c).

In contrast to these findings, we were unable to detect the presence of several nuclear-encoded mitochondrial mRNAs in the SCG axons, which were readily detected

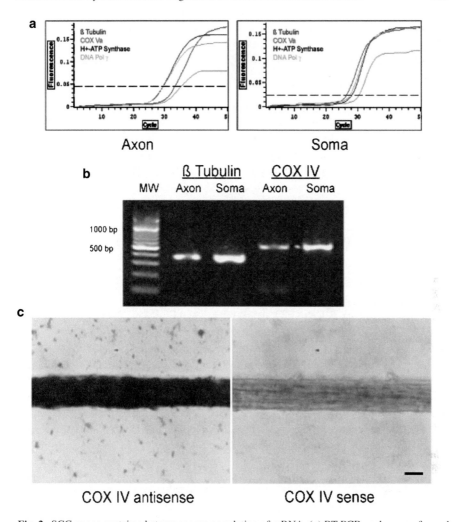

Fig. 2 SCG axons contain a heterogeneous population of mRNA. (a) RT-PCR analyses performed on total RNA isolated from SCG axons and somas. Note the shift in abundance of β-tubulin mRNA relative to H$^+$-ATP synthase and DNA polymerase γ mRNAs between axon and soma. (b) PCR products for β-tubulin and COXIV, a nuclear-encoded mitochondrial protein, were size-fractionated by agarose gel electrophoresis and visualized by ethidium bromide staining. MW, molecular weight. (c) COX IV subunit mRNA is present in SCG axons, as demonstrated with in situ hybridization histochemistry. Representative phase contrast photomicrographs of SCG axons after fixation and hybridization with antisense (COX IV antisense) and sense (Control; COX IV sense) riboprobes. *Bar* = 10 μm. From Hillefors et al. (2007)

in the cell somas located in the central compartment of the Campenot chambers (e.g., Mitochondrial topoisomerase 1, Cytochrome P450, Citrate lyase B subunit, and Timm 10, a mitochondrial inner membrane import translocase subunit). These findings indicate that only a subset of the nuclear-encoded mitochondrial mRNAs that are expressed in the neuronal cell soma are translocated to the axon (Aschrafi et al. 2008).

The application of quantitative RT-PCR methodology allowed us to compare the relative abundance of different mRNAs present in the distal SCG axons, present in the side compartments, to that of the mRNAs present in the proximal axons and neuronal cell somas, located in the central compartment. In somas and proximal axons, β-tubulin mRNA was more abundant than the mRNAs encoding ATP synthase and DNA polymerase γ, while the opposite was the case in the distal axon (Fig. 2a). The alterations in the relative abundance of these mRNAs in the different compartments of the neurons suggest that there is a differential transport of nuclear-encoded mitochondrial mRNAs into SCG axons, an observation that is consistent with early findings reported in the squid giant axon (Chun et al. 1996). In this regard, mammalian axons may prove quite similar to those of cephalopod mollusks.

4 Axonal Protein Synthesis and Mitochondrial Activity

4.1 SCG Axons Contain a Local Protein Synthetic System

To test the hypothesis that SCG axons contain a local protein synthetic system, distal axons were exposed to low concentrations of the antibiotics cycloheximide, chloramphenicol, or emetine prior to the addition of [^{35}S]methionine. After a 4-h labeling period, the amount of newly synthesized protein was assessed in the distal axons. The use of cycloheximide, emetine, and chloramphenicol provides one the ability to differentiate between the axoplasmic and endogeneous mitochondrial translational activity, respectively. Cycloheximide and emetine markedly decreased the incorporation of [^{35}S]methionine into protein (approximately 70–80%). In contrast, chloramphenicol had only modest effects on axonal protein synthesis, a finding that suggests that mitochondria, per se, contribute approximately 20–25% of the total protein synthetic activity in the axon. Once again, the data obtained in a model mammalian neuron system were consistent with the results obtained from the early metabolic labeling studies conducted in squid brain synaptosomes (Gioio et al. 2001; Jimenez et al. 2002).

4.2 Mitochondrial Membrane Potential and ATP Synthetic Capacity is Dependent on Local Protein Synthesis and Import of Axonal Protein

Mitochondrial viability and function are dependent on proteins synthesized from the cellular genome, and these nuclear-encoded proteins are imported into the organelle from the cytosol (for review, see Bauer and Hoffman 2006). The import of mitochondrial preproteins utilizes the molecular chaperones Hsp70 and Hsp90, and a translocase complex situated on the outer membrane (TOM). The TOM70 receptor is located on the cytosolic side of the TOM complex, and disruption of the interaction of Hsp90 with the receptor inhibits the import of proteins into the

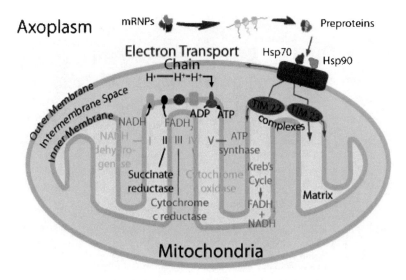

Fig. 3 Model of a mitochondrion showing the outer and inner membrane, intramembrane space, and matrix. The Kreb's cycle takes place in the matrix and the electron transport chain, generating ATP, is located in the inner membrane. Molecular chaperones such as Hsp70 and Hsp90 help facilitate the import of preproteins synthesized in the axoplasm into the organelle via the translocase of outermembrane (TOM) complex, which is located on the outer membrane of mitochondrion

mitochondria (Young et al. 2003). These events and intermolecular interactions are schematized in Fig. 3. To evaluate the potential significance of axonal protein synthesis on mitochondrial function, we employed both protein synthesis inhibitors and a competitive inhibitor of the TOM70 receptor. The results of these experiments are summarized briefly below:

4.2.1 Protein Synthesis Inhibition

The findings that SCG axons contain both nuclear-encoded mitochondrial mRNAs and a protein synthetic system raised the possibility that proteins requisite for mitochondrial function are locally synthesized. To explore this postulate, axonal protein synthesis was inhibited by either cycloheximide or emetine, and mitochondrial membrane potential was monitored using the mitochondrial-specific fluorescent dyes, JC-1 and TMRE. The fluorescence intensity of these molecular probes is proportional to the membrane potential of the organelle. Acute inhibition of local protein synthesis by either cycloheximide or emetine (3 h) decreased mitochondrial membrane potential (Fig. 4a, b). Brief exposure to the protein synthesis inhibitors also inhibited the SCG axon's ability to restore ATP levels after a depolarization stress induced by KCl treatment (Fig. 4c). These results suggest that the maintenance of axonal mitochondrial membrane potential and the organelle's ability to generate ATP is dependent, at least in part, on locally synthesized proteins.

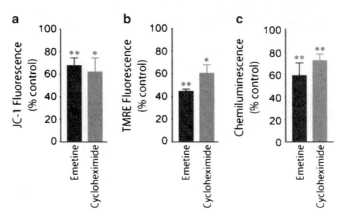

Fig. 4 Inhibition of local protein synthesis decreases axonal mitochondrial membrane potential and generation of ATP. Axons in one side compartment of the Campenot cell culture chamber were exposed to the protein synthesis inhibitors emetine and cycloheximide for 3 h and mitochondrial membrane potential assessed by subsequent treatment with the mitochondrial-specific fluorescent dyes JC-1 (**a**) or TMRE (**b**). Distal axons present in the contralateral side compartment served as vehicle-treated controls. Values for mitochondrial membrane potentials, as determined with each dye, are given as Mean ± S.E.M. (*$P < 0.05$, **$P < 0.01$) (**c**) Protein synthesis inhibitors impede the recovery of depolarization-induced decreases in axonal ATP levels. Depolarization was induced by exposing untreated SCG axons and axons exposed to emetine or cycloheximide for 3 h followed by a 5 min exposure to 50 mM KCl. Axonal ATP levels were measured using ATPlite 1step kit and a microplate reader. Data shown are provided as the Mean ± S.E.M. (**$P < 0.01$). Reproduced with modification from Hillefors et al. (2007)

4.2.2 Blockade of Mitochondrial Protein Import

One means used to disrupt the import of axoplasmic proteins into mitochondria is to expose the TOM70 receptor to the carboxy terminal portion of Hsp90 (C90; Young et al. 1998). Although the C90 fragment can bind to the receptor, it is nonfunctional in that it cannot bind cargo proteins. When introduced into the axon, C90 molecules act as competitive inhibitors preventing full-length functional Hsp90 molecules from binding to the TOM70 receptor. To facilitate the detection of mitochondria with C90 molecules bound to their TOM70 receptors, C90 was labeled with fluorescent Alexa Fluor 488, and the inhibitor lipofected directly into the distal axons in one side compartment of the Campenot chambers. Axons in the contralateral compartment were treated with the reagent vehicle and served as controls. Alterations in mitochondrial membrane potential were subsequently employed to evaluate the effects of decreased protein import into the organelle. After 4 h of exposure to C90, the membrane potential of Alexa-labeled mitochondria was reduced to 35% control values, indicating that mitochondria present in the distal axon require the import of local protein to maintain their functional integrity.

5 MicroRNAs in the Axon

MicroRNAs (miRs) are small (21–23 nucleotides), single-stranded, noncoding RNAs involved in the posttranscriptional regulation of gene expression. They are highly conserved and play a key role in the regulation of numerous biological processes such as cell proliferation and differentiation (for a neuronal perspective, see Kosik 2006). Results of a recent microarray analysis indicate that in the axons of primary sympathetic neurons there are approximately 200 different miRs. The axonal localization of twenty of these miRs was subsequently verified by quantitative RT-PCR methodology. Here, we focus on a brain-specific microRNA, miR-338, which modulates the expression of cytochrome c oxidase IV (COXIV), a nuclear-encoded protein that plays an important role in the assembly of the mitochondrial enzyme complex IV and oxidative phosphorylation (Fig. 3).

5.1 MicroRNAs Target Nuclear-Encoded Mitochondrial mRNAs

As mentioned earlier, COXIV mRNA is one of the several nuclear-encoded mRNAs that is present in the axons of sympathetic neurons (Fig. 2). Sequence analysis of the relatively short 3' untranslated region (3' UTR) of this mRNA revealed the presence of a putative target site for miR-338 (MTS). A RNA secondary structural analysis of the region indicated that the miR-338 MTS was positioned on a hairpin-loop structure, in an exposed position, that might facilitate miR accessibility. The transfection of axons of cultured SCG neurons with chimeric reporter gene constructs containing the COXIV 3' UTR with or without the putative MTS established that miR-338 could specifically target COXIV mRNA (Aschrafi et al. 2008).

To explore the possibility that mature miRs might function in the axon, the presence of Dicer and the RNA-induced silencing complex (RISC) component eIF2c was visualized in the distal axons by immunocytochemistry. Immunofluorescence of Dicer and eIF2c antibodies revealed the presence of granule-like structures along the entire length of the axon. These findings support the hypothesis that microRNAs play a role in the regulation of mRNA levels in the axon and are consistent with the report of Hengst et al. (2006), in which the latter authors demonstrated that axons of dorsal root ganglion neurons were capable of autonomously silencing a gene without the contribution of the cell body.

In addition, SCG axons were transfected with precursor miR-338 (pre miR-338) and levels of mature miR-338 determined by RT-PCR using specific primers for the mature form of the molecule. Transfection of axons with pre miR-338 resulted in a 42-fold increase in mature miR-338 levels within 4 h. After 24 h, a 42,000-fold increase in mature miR-338 levels was observed in transfected axons compared to endogenous levels in sham-transfected control axons. Clearly, distal axons of SCG neurons contain Dicer and RISC and have the capability of processing miR precursors to form the mature form of the molecule.

5.2 Regulation of Axonal COXIV Expression by miR-338

To evaluate whether miR-338 can modulate COXIV expression in the distal axon, COXIV mRNA and protein levels were monitored after transfecting SCG axons with pre miR-338. COXIV mRNA and COXIV protein decreased by 80% and 60% of control values 24 h after transfection, respectively (Fig. 5a, b). Conversely, the inhibition of endogenous miR-338 by transfection of the axons with a competitor molecule (anti miR-338) resulted in a 3.5-fold increase in axonal COXIV mRNA levels 24 h after transfection (Fig. 5c). Twofold increases in axonal COXIV protein levels were observed 24 h after transfection with anti miR-338 (Fig. 5d).

The specificity of the COXIV response to anti miR-338 was evaluated by assessing COXII mRNA levels by quantitative RT-PCR after transfection of the axons with anti miR-338. This mRNA codes for one of the three subunits of the enzyme complex IV encoded by the mitochondrial genome. In contrast to the alterations in COXIV expression induced by anti miR-338, no differences in COXII mRNA levels were observed 24 h after transfection with anti miR-338 (Fig. 5e).

5.3 miR-338 Mediates Axonal Mitochondrial Activity

Alteration in the levels of miR-338 in the axon has a profound effect on the activity of the axonal mitochondrial population and the basal metabolic rate of the axon. For example, the over-expression of miR-338 in the axon significantly reduces mitochondrial oxygen consumption, as demonstrated by the reduction of Alamar Blue (AB), a redox dye used to assess the metabolic activity of mitochondria (Fig. 6a). Inhibition of the miR-338 activity in the axon by the introduction of a competitor RNA (anti miR-338) directly into the axon by lipofection increased oxygen consumption by 50–60% within 24 h (Fig. 6b). Consistent with this finding, axonal ATP levels were elevated by 50% after the inhibition of endogenous axonal miR-338 (Fig. 6c).

To delineate the impact of the modulation of axonal miR-338 levels on ATP synthesis and respiration on the function of the axon and presynaptic nerve terminal, the distal axons present in the side compartments of the Campenot chambers were exposed to the neurotransmitter, [^{3}H]norepinephrine (NE), and the neurotransmitter uptake was assessed after introducing precursor miR-338 or anti miR-338 directly into the distal axons. Under the experimental cell culture conditions employed in these experiments, desiprimine, a powerful NE uptake inhibitor, reduced the amount of [^{3}H]NE in the distal axons by 85–95%. The introduction of anti miR-338 or pre miR-338 into the distal axons resulted in a 50% increase or decrease in the uptake of NE, respectively (Fig. 6d, e). These findings clearly established that modulation of axonal miR-338 levels has marked effects on axonal ATP synthesis, metabolic rate, and axonal function, as judged by catecholamine neurotransmitter uptake.

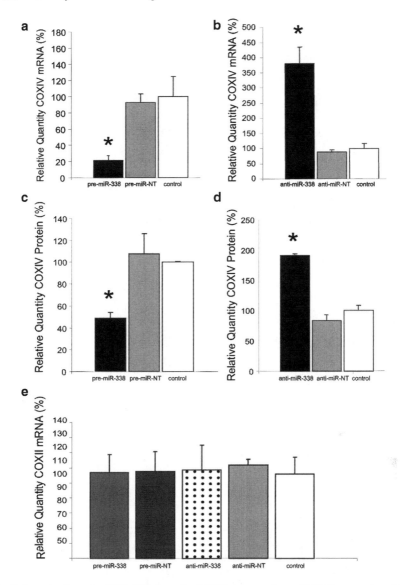

Fig. 5 Introduction of miRNA-338 into distal SCG axons reduces COXIV expression. Quantification of COXIV mRNA levels in the distal axons of SCG neurons transfected with pre miR-338 (**a**) or anti miR-338 (**b**), as determined by quantitative RT-PCR 24 h after oligonucleotide transfection. COXIV mRNA levels are expressed relative to β-actin mRNA. Values given are the Mean ± S.E.M. (*$P < 0.05$). The introduction into the distal axons of pre miR-338 (**c**) or anti miR-338 (**d**) also altered axon COXIV protein levels 24 h after transfection as determined by immunoblot analyses. Values shown are Mean ± S.E.M. (*$P < 0.05$). Transfection of SCG axons with either precursor or anti miR-338 oligonucleotides did not affect the relative abundance of COXII mRNA in the axon as shown by quantitative RT-PCR (**e**). Reproduced from Aschrafi et al. (2008) with modification

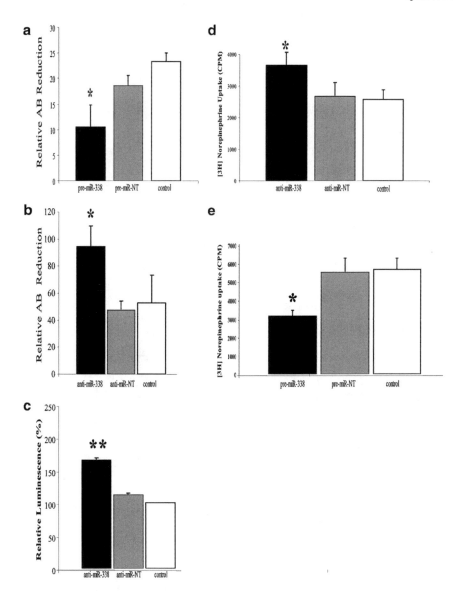

Fig. 6 MiR-338 mediated alteration in COXIV levels modulates metabolic activity of mitochondria and neurotransmitter uptake into the distal axons of sympathetic neurons. SCG axons were transfected with either precursor miR-338 (**a**), anti miR-338 (**b**), or nontargeting short oligonucleotides and metabolic activity of the axons assessed using the redox dye Alomar Blue (AB). Data represent Mean ± S.E.M. (*$P < 0.05$). ATP levels were measured in anti miR-338 transfected axons using the luciferase cell viability assay (**c**). Values are given in arbitrary luminescence units and represent the Mean ± S.E.M. (**$P < 0.0001$). Distal axons were also transfected with either anti miR-338 (**d**), pre miR-338 (**e**), or nontargeting short oligonucleotides (NT) and were subsequently exposed to radiolabeled norepinephrine (NE). NE uptake into treated distal axons was measured by liquid scintillation spectrometry. Data represent Mean ± S.E.M. (*$P < 0.002$). Reproduced with modification from Aschrafi et al. (2008)

5.4 Knockdown of Local COXIV Expression Alters Mitochondrial Activity

Although the modulation of miR-338 levels in the axon has significant effects on axonal metabolic rate and function, this microRNA may have multiple target gene transcripts, and it is unclear whether the effects described above were derived from the miR-338-mediated regulation of COXIV expression per se. To further explore this phenomenon, short inhibitory RNAs (siRNA), designed to specifically silence COXIV expression, were introduced directly into the axons of primary sympathetic neurons cultured in Campenot chambers by lipofection. The introduction of these siRNAs into the axon resulted in marked reductions in the levels of COXIV mRNA and COXIV protein (Fig. 7a, b). Identical to the findings obtained with the

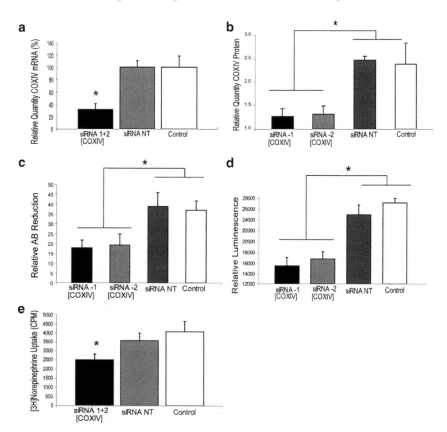

Fig. 7 Knock-down of local COXIV levels decreases axonal respiration and ATP levels, and diminishes NE uptake into distal axons. Two independent siRNA oligonucleotides targeted against COXIV mRNA were introduced into distal axons of SCG neurons by lipofection. COXIV mRNA (**a**) and protein levels (**b**) were quantitated 24 h later by quantitative RT-PCR and immunoblot analyses, respectively. Knock-down of axonal COXIV expression reduced basal oxygen consumption (**c**), ATP levels (**d**), and [^{3}H]NE uptake (**e**). Values are expressed as Mean ± S.E.M and statistical significance evaluated using one-way ANOVA (*$P < 0.03$). From Aschrafi et al. (2008)

over-expression of miR-338, reduction in axonal COXIV expression by RNA interference resulted in significant decrements in basal oxygen consumption (Fig. 7c) and axonal ATP levels (Fig. 7d). In contrast, the introduction into distal axons of a nontargeting siRNA (NT) had no effect on these experimental outcomes. Moreover, the knockdown of local COXIV expression reduced axonal function by 25–30%, as judged by the reduction of [^{3}H]NE uptake into the distal axons (Fig. 7e). These findings provide persuasive evidence that the effects of miR-338 on mitochondrial activity and axonal function are mediated, at least in part, by translational inhibition of COXIV mRNA.

6 Conclusion

The mitochondrial genome encodes only a few proteins (approximately 13); thereby, requiring the mitochondria to import several hundred nuclear-encoded proteins to maintain its structure, function, and replicative activity. For example, 10 of the 13 subunits of enzyme complex IV, the proteins involved in the electron transport chain and ATP synthesis, derive from the nuclear genome (Heales et al. 2006). Numerous studies have demonstrated that mitochondrial functional activity requires a continuous import of nuclear-encoded proteins.

In neurons, it is well established that mitochondria are translocated via fast axonal transport from the cell soma to the axon and presynaptic nerve terminals (Hollenbeck and Saxton, 2005), and that mitochondrial trafficking can be regulated by neuronal activity (e.g., see Chang et al. 2006). However, little is known about the half-life or fate of these organelles once they reach their ultimate sites of function. If axons and nerve terminals lacked a cytosolic protein synthetic system, then the resupply of protein for the local mitochondria must by default differ significantly from that in the cell soma. Alternatively, the organelles present in the distal domains of the neuron could be relatively short-lived and continuously resupplied. The weight of the recent evidence would indicate that the mitochondria present in the axon and nerve endings are being sustained by local protein synthesis. Most significantly, it has recently been reported that mitochondria can replicate in the axon and nerve ending (Amiri and Hollenbeck 2008). The finding that the nuclear-encoded mitochondrial DNA polymerase γ mRNA is present in the axons of sympathetic neurons is consistent with this report (Hillefors et al. 2007), and raises the possibility that the local synthesis of nuclear-encoded mitochondrial proteins also plays an important role in the biogenesis of this organelle.

The local synthesis of both nuclear-encoded mitochondrial proteins and the molecular chaperones necessary for their import into the organelle suggests that a substantial portion of the protein synthesized in the axon and nerve terminals are directed toward fulfilling the biological demands of its energy generating and calcium buffering system. As mentioned earlier (Sect. 2), 20–25% of the total protein synthesized in these regions of the neuron is destined for the mitochondria. One surprising feature of the relationship between the axon and its mitochondrial population is the apparent rapid turnover rate of the locally synthesized proteins. For example, only brief time periods of protein synthesis inhibition (3–4 h) or the blockade of local protein import into the organelle was sufficient to induce mitochondrial membrane depolarization and diminish the organelle's ability to generate ATP (Hillefors et al. 2007). These observations indicate that a significant portion of mitochondrial protein is being rapidly turned over in the axon, and raises the possibility that one of the major functions of the local protein synthetic system is to replace these highly labile constituents of this organelle. Interestingly, Li et al. (2006) demonstrated that COXIV plays a rate-limiting role in the assembly of enzyme complex IV, and that dysfunctional COXIV resulted in a compromised mitochondrial membrane potential, as well as decreased ATP levels and respiration. Such deficits could ultimately have deleterious effects on axon function and the ability of the neuron to respond to prolonged periods of activation or stress.

In summary, the findings reviewed here support the hypothesis that protein synthesis in the distal regions of the neuron is vital for mitochondrial activity, and suggests a new model for the maintenance of mitochondrial function in the axon and nerve terminal (Fig. 8). In this schema, nuclear-encoded mitochondrial mRNAs are transported to the distal structural/functional domains of the neuron at a rapid rate in the

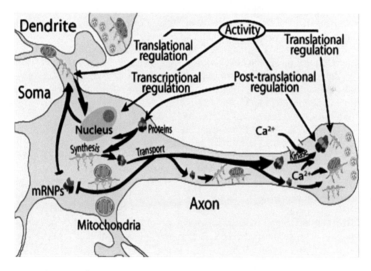

Fig. 8 Model of neuronal protein synthesis. Protein synthesis occurs in multiple compartments within neurons to include the dendrite, axon, and nerve terminal. Key features of the model include the rapid and selective transport of stable messenger ribonucleoprotein complexes (mRNPs) to the neuronal periphery and the local regulation of mRNA translation in response to neuronal activity. The model also shows that synthesis of proteins occurs in the vicinity, or on the surface of mitochondria in the distal parts of the neuron. The activity in the neuron can be modulated by transcriptional regulation in the soma, as well as translational and posttranslational regulation in dendrites, axons, and nerve terminals. The down regulation of gene expression in the distal structural/functional domains of the neuron can also be effected by microRNAs. From Hillefors et al. (2007)

form of relatively stable ribonucleoprotein particles (mRNPs). These complexes can serve as templates for the synthesis of mitochondrial preproteins, and their translation can be regulated by multiple mechanisms to include microRNAs at their site of function in response to local neuronal activity. In this regard, it has been reported that ribosome clusters (i.e., polysomes) in the squid giant axon are often visualized in association with mitochondria (Bleher and Martin 2001), and that the close approximation of mRNA to the vicinity of mitochondria is essential for respiratory function (Margeot et al. 2005). Hence, in this model, it is envisaged that preproteins are synthesized juxtaposed to, or on, the surface of axonal mitochondria, and that this relationship plays a key role in the biogenesis, maintenance, and functional activity of the organelle.

Acknowledgements Work in the laboratory of B.B. Kaplan was supported by the Division of Intramural Research Programs of the National Institute of Mental Health. The expert technical assistance of Ms. Orlangie Natera-Naranjo is greatly appreciated.

References

Amiri M, Hollenbeck PJ (2008) Mitochondrial biogenesis in the axons of vertebrate peripheral neurons. Dev Neurobiol 68:1348–1361
Aschrafi A, Schwechter AD, Mameza MG, Natera-Naranjo O, Gioio AE, Kaplan BB (2008) MicroRNA-338 regulates local cytochrome c oxidase IV mRNA levels and oxidative phosphorylation in the axons of sympathetic neurons. J Neurosci 28:12581–12590
Bauer MF, Hofman S (2006) Import of mitochondrial proteins. In: Shapira AHV (ed) Mitochondrial function and dysfunction. Academic, San Diego, CA, pp 57–90
Beaumont V, Zhong N, Fletcher R, Froemke RC, Zucker RS (2001) Phosphorylation and local presynaptic protein synthesis in calcium- and calcincurin-dependent induction of crayfish long-term facilitation. Neuron 32:489–501
Bleher R, Martin R (2001) Ribosomes in the squid giant axon. Neuroscience 107:527–534
Brittis PA, Lu Q, Flanagan JG (2002) Axonal protein synthesis provides a mechanism for localized regulation at an intermediate target. Cell 110:223–235
Campbell DS, Holt CE (2001) Chemotropic responses of retinal growth cones mediated by rapid local protein synthesis and degradation. Neuron 32:1013–1026
Campbell DS, Regan AG, Lopez JS, Tannahill D, Harris WA, Holt CE (2001) Semaphorin 3A elicits stage-dependent collapse, turning, and branching in Xenopus retinal growth cones. J Neurosci 21:8538–8547
Campenot RN, Martin G (2001) Construction and use of compartmented cultures for studies of cell biology in neurons. In: Federoff S, Richardson A (eds) Protocols for neural cell culture. Humana, Totowa, NJ, pp 49–57
Casadio A, Martin KC, Giustetto M, Zhu H, Chen M, Bartsch D, Bailey CH, Kandel ER (1999) A transient, neuron-wide form of CREB-mediated long-term facilitation can be stabilized at specific synapses by local protein synthesis. Cell 99:221–237
Chang DTW, Honick AS, Reynolds IJ (2006) Mitochondrial trafficking to synapses in cultured primary cortical neurons. J Neurosci 26:7035–7045
Chun JT, Gioio AE, Crispino M, Giuditta A, Kaplan BB (1996) Differential compartmentalization of mRNAs in squid giant axon. J Neurochem 67:1806–1812
Cox LJ, Hengst U, Gurskaya NG, Lukyanov KA, Jaffrey SR (2008) Intra-axonal translation and retrograde trafficking of CREB promotes neuronal survival. Nat Cell Biol 10:149–159

Gioio AE, Eyman M, Zhang H, Lavina ZS, Giuditta A, Kaplan BB (2001) Local synthesis of nuclear-encoded mitochondrial proteins in the presynaptic nerve terminal. J Neurosci Res 64:447–453

Gioio AE, Lavina ZS, Jurkovicova D, Zhang H, Eyman M, Giuditta A, Kaplan BB (2004) Nerve terminals of squid photoreceptor neurons contain a heterogeneous population of mRNAs and translate a transfected reporter mRNA. Eur J Neurosci 20:865–872

Giuditta A et al (2009) Protein synthesis in nerve terminals and the glia-neuron unit. Results Probl Cell Differ doi: 10.1007/400_2009_9

Giuditta A, Chun JT, Eyman M, Cefaliello C, Bruno AP, Crispino M (2008) Local gene expression in axons and nerve endings: the glia-neuron unit. Physiol Rev 88:515–555

Hanz S, Perlson E, Willis D, Zheng JQ, Massarwa R, Huerta JJ, Koltzenburg M, Kohler M, van-Minnen J, Twiss JL, Fainzilber M (2003) Axoplasmic importins enable retrograde injury signaling in lesioned nerve. Neuron 40:1095–1104

Heales SJR, Gegg ME, Clark JB (2006) Oxidative phosphorylation: structure, function, and Intermediary metabolism. In: Shapira AHV (ed) Mitochondrial function and dysfunction. Academic, San Diego, CA, pp 25–56

Hengst U, Cox LJ, Macosko EZ, Jaffrey SR (2006) Functional and selective RNA interference in developing axons and growth cones. J Neurosci 26:5727–5732

Hillefors M, Gioio AE, Mameza MG, Kaplan BB (2007) Axon viability and mitochondrial function are dependent on local protein synthesis in sympathetic neurons. Cell Mol Neurobiol 27:701–716

Hollenbeck PJ, Saxton WM (2005) The axonal transport of mitochondria. J Cell Sci 118:5411–5419

Hu JY, Meng X, Schacher S (2003) Redistribution of syntaxin mRNA in neuronal cell bodies regulates protein expression and transport during synapse formation and long-term synaptic plasticity. J Neurosci 23:1804–1815

Jimenez CR, Eyman M, Lavina ZS, Gioio A, Li KW, van der Schors RC, Geraerts WP, Giuditta A, Kaplan BB, van Minnen J (2002) Protein synthesis in synaptosomes: a proteomics analysis. J Neurochem 81:735–744

Kosik KS (2006) The neuronal microRNA system. Nat Rev Neurosci 7:911–920

Leung KM, van Horck FP, Lin AC, Allison R, Standart N, Holt CE (2006) Asymmetrical β-actin mRNA translation in growth cones mediates attractive turning to netrin-1. Nat Neurosci 9:1247–1256

Li Y, Park JS, Deng JH, Bai Y (2006) Cytochrome c oxidase subunit IV is essential for assembly and respiratory function of the enzyme complex. J Bioenerg Biomembr 38:283–291

Li C, Sasaki Y, Takei K, Yamamoto H, Shouji M, Sugiyama Y, Kawakami T, Nakamura F, Yagi T, Ohshima T, Goshima Y (2004) Correlation between semaphorin3A-induced facilitation of axonal transport and local activation of a translation initiation factor eukaryotic translation initiation factor 4E. J Neurosci 24:6161–6170

Liao L, Pilotte J, Xu T, Wong CCL, Edelman GM, Vanderklish P, Yates JR III (2007) BDNF induces widespread changes in synaptic protein content and up-regulates components of the translational machinery: an analysis using high-throughput proteomic. J Proteome Res 6:1059–1071

Liu K, Hu JY, Wang D, Schacher S (2003) Protein synthesis at synapse versus cell body: enhanced but transient expression of long-term facilitation at isolated synapses. J Neurobiol 56:275–286

Margeot A, Garcia M, Wang W, Tetaud E, di Rago JP, Jacq C (2005) Why are many mRNAs translated to the vicinity of mitochondria: a role in protein complex assembly? Gene 354:64–71

Martin KC, Casadio A, Zhu H, Yaping E, Rose JC, Chen M, Bailey CH, Kandel ER (1997) Synapse-specific, long-term facilitation of aplysia sensory to motor synapses: a function for local protein synthesis in memory storage. Cell 91:927–938

Ming GL, Wong ST, Henley J, Yuan XB, Song HJ, Spitzer NC, Poo MM (2002) Adaptation in the chemotactic guidance of nerve growth cones. Nature 417:411–418

Moccia R, Chen D, Lyles V, Kapuya E, E Y, Kalachikov S, Spahn CM, Frank J, Kandel ER, Barad M, Martin KC (2003) An unbiased cDNA library prepared from isolated Aplysia sensory neuron processes is enriched for cytoskeletal and translational mRNAs. J Neurosci 23:9409–9417

Perrone Capano C, Giuditta A, Castigli E, Kaplan BB (1987) Occurrence and sequence complexity of polyadenylated RNA in squid axoplasm. J Neurochem 49:698–704

Schacher S, Wu F (2002) Synapse formation in the absence of cell bodies requires protein synthesis. J Neurosci 22:1831–1839

Si K, Giustetto M, Etkin A, Hsu R, Janisiewicz AM, Miniaci MC, Kim JH, Zhu H, Kandel ER (2003) A neuronal isoform of CPEB regulates local protein synthesis and stabilizes synapse-specific long-term facilitation in aplysia. Cell 115:893–904

Verma P, Chierzi S, Codd AM, Campbell DS, Meyer RL, Holt CE, Fawcett JW (2005) Axonal protein synthesis and degradation are necessary for efficient growth cone regeneration. J Neurosci 25:331–342

Wu KY, Hengst U, Cox LJ, Macosko EZ, Jeromin A, Urquhart ER, Jaffrey SR (2005) Local translation of RhoA regulates growth cone collapse. Nature 436:1020–1024

Young JC, Obermann WM, Hartl FU (1998) Specific binding of tetratricopeptide repeat proteins to the C-terminal 12-kDa domain of hsp90. J Biol Chem 273:18007–18010

Young JC, Hoogenraad NJ, Hartl FU (2003) Molecular chaperones Hsp90 and Hsp70 deliver preproteins to the mitochondrial import receptor Tom70. Cell 112:41–50

Zhang X, Poo MM (2002) Localized synaptic potentiation by BDNF requires local protein synthesis in the developing axon. Neuron 36:675–688

Zheng JQ, Kellyc TK, Chang B, Ryazantsev S, Rajasekaran AK, Martin KC, Twiss JL (2001) A functional role for intra-axonal protein synthesis during axonal regeneration from adult sensory neurons. J Neurosci 21:9291–9303

Protein Synthesis in Nerve Terminals and the Glia–Neuron Unit

Marianna Crispino, Carolina Cefaliello, Barry Kaplan, and Antonio Giuditta

Abstract The progressive philogenetic lengthening of axonal processes and the increase in complexity of terminal axonal arborizations markedly augmented the demands of the neuronal cytoplasmic mass on somatic gene expression. It is proposed that in an adaptive response to this challenge, novel gene expression functions developed in the axon compartment, consisting of axonal and presynaptic translation systems that rely on the delivery of transcripts synthesized in adjacent glial cells. Such intercellular mode of gene expression would allow more rapid plastic changes to occur in spatially restricted neuronal domains, down to the size of individual synapses. The cell body contribution to local gene expression in well-differentiated neurons remains to be defined. The history of this concept and the experimental evidence supporting its validity are critically discussed in this article. The merit of this perspective lies with the recognition that plasticity events represent a major occurrence in the brain, and that they largely occur at synaptic sites, including presynaptic endings.

Abbreviations

CAP	Chloramphenicol
CXM	Cycloheximide
EB	Ethidium bromide
EM	Electron microscopy
ESI	Electron spectroscopic imaging

M. Crispino, C. Cefaliello, and A. Giuditta (✉)
Department of Biological Sciences, University of Naples "Federico II",
Naples, Italy
e-mail: giuditta@unina.it

B. Kaplan
Laboratory of Molecular Biology, National Institute of Mental Health, NIH,
Bethesda, MD 20892-1381, USA
e-mail: kaplanb@mail.nih.gov

MW   Molecular weight
NF   Neurofilament
RNA bases   *A* adenine, *C* cytosine, *G* guanine, *U* uracil
RNP   Ribonucleoprotein particle
TH   Tyrosine hydroxylase

1 Introduction

The philogenetic development of axonal processes allowed neurons to relay signals among body and brain regions. In the body, axons attained considerable length, and in the brain, distal axonal fields formed complex arborizations, in which terminal boutons connected with specialized postsynaptic sites. Both terminal and preterminal features markedly increased the cytoplasmic mass of the axon domain, with the obvious consequence of proportionally decreasing the relative mass of the cell body. In spite of this basic consideration, trophic needs of axons and nerve terminals were initially assumed to depend entirely on the neuron soma, in view of the consistent degeneration of decentralized axons, and of later evidence of the distal axonal transport of proteins and organelles delivered from the cell body (Grafstein and Forman 1980). As a necessary corollary, axons and nerve terminals were assumed to lack ribosomes and protein synthesis. This long-lasting dogma was, however, challenged by theoretical considerations, and was finally overturned by experimental data demonstrating the local synthesis of axonal and presynaptic proteins (for reviews, see Koenig and Giuditta 1999; Alvarez et al. 2000; Giuditta et al. 2002; 2008; Piper and Holt 2004).

Additional problems arose with regard to the cellular origin of the RNAs involved in peripheral translation. In view of their transcription on nuclear DNA, they were assumed to derive exclusively from the cell body, notwithstanding that analogous theoretical difficulties concerning the somatic origin of axonal proteins are encountered. Here, we emphasize the likelihood that supporting trophic capacities of the neuron soma may satisfy the needs of cytoplasm upto a threshold mass. Beyond that limit, which may be exceeded in differentiated neurons, trophic needs of axon and nerve endings require the additional contribution provided by periaxonal and perisynaptic glia. The trophic role of the latter cells may also be relevant to the selective delivery of macromolecules to single nerve endings engaged in plastic changes (Martin et al. 1997; Grosche et al. 1999). Indeed, a local mechanism would bypass the requirement of specific targeting instructions needed for the supply of soma-derived macromolecules to correct distal destinations. In addition, a local glial supply would represent a viable alternative to tagging activated synapses (Frey and Morris 1997; Casadio et al. 1999), as it would bypass any metabolic load imposed on the neuron soma, and it would account for selective forms of plasticity implemented in different nerve terminals of the same neuron.

Our review article will discuss the evidence, the properties, and the role of protein synthesis in nerve endings (see Sect. 2), and the composition and cellular origin of the population of RNAs present in axons and nerve endings (see Sects. 3 and 4).

2 Protein Synthesis in Nerve Terminals

The study of protein synthesis in nerve terminals has been largely carried out in brain synaptosomal fractions (Whittaker 1993). Synaptosomes are subcellular vesicles enclosed by a plasma membrane, mostly consisting of pinched-off presynaptic, and postsynaptic elements subjected to mild tissue homogenization. During disruption, membranes reseal and intrinsic components are segregated from the bathing medium. As a result, the synaptosomal capacity to incorporate radiolabeled amino acids into proteins reflects the activities of included mitochondrial and cytoplasmic systems of protein synthesis, which are readily distinguished by use of selective inhibitors. Synaptosomal fractions are generally enriched in synaptosomes derived from nerve endings, which are identified by inclusion of synaptic vesicles and mitochondria. The variable occurrence of synaptosomes derived from post-synaptic elements and glial processes has led to opposing interpretations as to the prevalent localization of the synaptosomal system of cytoplasmic protein synthesis (Giuditta et al. 2002, 2008; see Sect. 2.1).

2.1 Vertebrate Nerve Endings

The presence of protein synthesis in purified mammalian brain synaptosomes was initially demonstrated in the late 1960s by the incorporation of radiolabeled amino acids into protein (Austin and Morgan 1967; Morgan and Austin 1968, 1969). The preparation did not require addition of exogenous soluble factors, or energy sources; it was insensitive to ribonuclease, but was strongly inhibited by specific inhibitors of eukaryotic protein synthesis, such as cycloeximide (CXM), and partially inhibited by inhibitors of mitochondrial protein synthesis, such as chloramphenicol (CAP). Synaptosomal protein synthesis was strongly inhibited by ouabain, a specific inhibitor of the Na^+/K^+-ATPase (Verity et al. 1979), which induced the lowering of intrasynaptosomal K^+ with the resulting inhibition of both CXM- and CAP-sensitive protein synthesis. At least 50% of synaptosomes containing newly synthesized proteins were derived from nerve endings, as shown by ultrastructural autoradiographic methods (Cotman and Taylor 1971; Gambetti et al. 1972). Remaining structures were derived from glial cells and postsynaptic elements.

A different experimental approach was used to demonstrate the local presynaptic synthesis of tyrosine hydroxylase (TH), a marker of catecholaminergic neurons. In the heart, which lacks catecholaminergic cell bodies (Thoenen et al. 1970), TH activity significantly increased a few days after administration of reserpine that depletes nerve terminals of biogenic amine neurotransmitters. This effect was blocked by inhibitors of cytoplasmic protein synthesis and could not be attributed to TH delivery by axonal transport. These observations indicated that TH mRNA was present in heart nerve terminals and was actively engaged in translation.

More recently, local protein synthesis was demonstrated in structures (neuromas) that sprout from the proximal stump of centrally disconnected peripheral nerves when the target tissue is not accessible. The rate of protein synthesis is much higher in neuromas than in nerves, and many of its features are comparable to those of nerve terminals. Some of the neuroma proteins expressed at high level are neuron-specific (Huang et al. 2008).

2.2 Invertebrate Nerve Endings

The heterogeneous composition of mammalian synaptosomal fractions led to contrasting interpretations of data regarding the subsynaptic derivation of actively translating synaptosomes (Alvarez et al. 2000). The problem was circumvented in studies of more homogeneous synaptosomal preparations from squid optic lobes, in which the most abundant components are large synaptosomes derived from the presynaptic terminals of retinal photoreceptor neurons (Cohen 1973). Detailed investigations of this preparation provided conclusive evidence that nerve endings contain an active system of cytoplasmic protein synthesis (Hernandez et al. 1976; Crispino et al. 1997; Martin et al. 1998).

The purity of the squid preparation was initially suggested by the marked differences exhibited by the electrophoretic pattern of newly synthesized proteins in comparison with those of neuronal cell bodies and glial cells. This was confirmed by electron microscopic (EM) analyses that demonstrated the prevalence of unusually large synaptosomes in the synaptosomal fraction (Crispino et al. 1993b). Autoradiographic ultrastructural studies demonstrated the nearly exclusive localization of newly synthesized synaptosomal proteins in the large synaptosomes (Crispino et al. 1997). In addition, polysomes purified from optic lobe synaptosomes incubated with [^{35}S]methionine contained nascent peptide chains (Crispino et al. 1993b), and were shown not to be contaminated by microsomal polysomes (Gioio et al. 2001). Polysomes were also visualized with specific methods within the large optic lobe synaptosomes, and the carrot-shaped photoreceptor nerve endings in squid optic lobes (Crispino et al. 1997; Martin et al. 1998).

Proteins synthesized by optic lobe synaptosomes, or cell-free synaptosomal polysomes, include members of the neurofilament (NF) family (Crispino et al. 1993a), and calexcitin, a learning-related signaling protein (Eyman et al. 2003). In addition, RT-PCR analyses of synaptosomal polysomes demonstrated their association with mRNAs coding for cytoskeletal and ribosomal proteins, translation factors, and, most notably, mitochondrial proteins encoded by nuclear DNA (Gioio et al. 2001, 2004). The data indicated translation of presynaptic mRNAs. This was confirmed by 2-D electrophoretic analysis of newly synthesized synaptosomal proteins, which included about 80 different proteins, several of which were shown by mass spectroscopy to be mitochondrial proteins encoded by nuclear DNA (Jimenez et al. 2002). Hence, maintenance and new genesis of presynaptic mitochondria were largely independent from the supply of the corresponding nuclear encoded mitochondrial, and chaperon proteins from the cell body (see Kaplan et al. 2009), which demonstrated the marked degree of autonomy of nerve terminals.

The presence and role of protein synthesis in nerve endings were also shown in cultures of *Aplysia* neurons, in which the bifurcated axon of a sensory neuron established separate synapses with two different motor neurons. The repeated administration of serotonin to one nerve terminal resulted in the long-term facilitation (LTF) of that synapse, which allowed the study of LTF by comparison with the other synapse that was used as a control. Under such experimental conditions, protein synthesis markedly increased in the nerve terminal involved in LTF, and its selective inhibition completely abolished LTF (Martin et al. 1997; Casadio et al. 1999; Martin 2004). These studies provided support to the tagging hypothesis (Frey and Morris 1997) that had been advanced to explain plastic modifications of individual synapses in the face of cell-wide release of soma-derived mRNAs and proteins. One tagging component synthesized in the nerve ending of the facilitated synapse was identified as the neuron-specific isoform of the cytoplasmic polyadenylation element binding protein (CPEB), a translational regulator which activates latent mRNAs by elongation of their poly(A) tails (Si et al. 2003a). Interestingly, the N-terminal peptide of *Aplysia* CPEB shares prion features that allow its persistent activity.

The formation of specific synaptic connections also requires the selective presynaptic accumulation of the mRNA coding for sensorin, a neuron-specific peptide (Schacher et al. 1999). This event parallels the distribution of sensorin and indicates the requirement of mRNA translation at presynaptic sites (Lyles et al. 2006). The aforementioned data and more recent results (Hu et al. 2007; Chihara et al. 2007; Puthanveettil et al. 2008) demonstrate that long-term plasticity changes at synapses are implemented by presynaptic translation processes. Later bouts of presynaptic protein synthesis are likewise required for the stabilization of the new synaptic asset (Bekinschtein et al. 2008; Lee et al. 2008; Lu et al. 2008; Miniaci et al. 2008) and for the reconsolidation process (Lee 2008; Smith and Kelly 1988).

3 mRNAs in Axons and Nerve Terminals

3.1 Early Studies

Early investigations of axonal and presynaptic RNAs were strictly intertwined with initial evidence of local protein synthesis. In the late 1960s, protein synthesis was demonstrated in the isolated squid giant axon by incorporation of radiolabeled amino acids into axonal proteins (Fisher and Litvak 1967; Giuditta et al. 1968). The process was strongly inhibited by CXM, which indicated the presence of a eukaryotic translation system. The cellular localization of this system in the axon, or in periaxonal glia was soon addressed and was initially resolved in favor of periaxonal glia (Gainer 1978). This conclusion was based on the presumed absence of translational machinery in the axon, a conclusion that was invalidated in later experiments by evidence that all components of the translation machinery were present in the axoplasm (Giuditta et al. 2002), including the occurrence of a population of about 200 mRNAs (Giuditta et al. 1986; Perrone Capano et al. 1987) and active polysomes (Giuditta et al. 1991). The local synthesis of axonal protein in the M-neuron had been previously demonstrated using comparable methods (Edström 1966). Interestingly, in the same axon, inhibition of protein synthesis by actinomycin D indicated that local translation depended on local glial transcription, and suggested that the coding mRNAs were undergoing a marked turnover (Edström 1967).

The Mauthner cell axon has been a useful model system also with regard to the identification of the peripheral RNAs. While RNA was believed to disappear from the axon during development (Hughes and Flexner 1956), a drastic change occurred when RNA was identified in adult axons (Edström et al. 1962; Koenig 1965), and its total content appeared to be several times higher than in cell bodies. Several RNA species were identified in M-axons by electrophoretic fractionation (Koenig 1979), including a prevalent 4S RNA, cytoplasmic 26S, and 18S rRNA, and minor amounts of unidentified 15S and 8S components. Similar data were obtained in squid giant axons (Giuditta et al. 1980). In both model systems, the 4S RNA peak included other small RNAs, such as 7S and 5S RNA.

3.2 Components of the Axonal mRNA Population

The first direct estimates of the diversity of the mRNA population present in axons derived from RNA–cDNA kinetic hybridization analysis conducted on polyadenylated mRNA obtained from the squid giant axon (Perrone Capano et al. 1987). The sequence complexity of the mRNA population was sufficient to encode approximately 200 different mRNAs averaging 1,500–2,000 nucleotides in length. In contrast, the mRNA transcriptome of the parental cell bodies of the giant axon was composed of approximately 15,000 different polyadenylated RNA transcripts. Follow-up in situ hybridization analysis using radiolabeled poly(U) as a generic mRNA hybridization probe facilitated the visualization of mRNA in the axon. These findings clearly established that the axonal mRNA population represented a relatively small and highly selective subset of the transcriptome.

The construction of a cDNA library to mRNA present in the squid giant axon and the sequencing of several randomly selected clones subsequently led to the identification of several axonal mRNAs to include β-actin and β-tubulin (Kaplan et al. 1992), enolase (Chun et al. 1995), the heavy chain of kinesin (Gioio et al. 1994),

MAP H1 (Chun et al. 1996), and a LDL receptor adaptor-like protein (Chun et al. 1997). The axonal localization of all these mRNAs was established by in situ hybridization histochemistry. In addition, the presence of neurofilament mRNA in the squid giant axon was established by in situ hybridization histochemistry, using a heterologous neurofilament cDNA as a hybridization probe (Giuditta et al. 1991). Interestingly, an early comparison of the relative abundance of these gene transcripts in the giant axon, and parental cell bodies by quantitative RT-PCR analyses revealed marked differences in relative distribution of these mRNAs. For example, in the cell soma, the amount of these gene transcripts varied over a fourfold range, with β-tubulin being the most abundant mRNA species. In contrast, kinesin mRNA was most abundant in the axon with individual mRNA levels varying 15-fold.

Results of this quantitative analysis suggested that the levels of some mRNAs in the axonal domain (e.g., kinesin and MAP H1) may be comparable to, or even greater than the levels in the cell soma, and indicated that specific mRNAs were being differentially transported into the axon (Chun et al. 1996). This postulate is consistent with the findings that local treatment of dorsal root ganglion axons with neurotrophins can induce a twofold to fivefold increase in axonal levels of mRNAs encoding select cytoskeletal proteins (Willis et al. 2005).

Studies using the freshwater snail, *Lymnea stagnalis*, established that axons of neuropeptidergic neurons contain significant amounts of neuropeptide encoding mRNAs (Dirks et al. 1993; van Minnen et al. 1997). The axonal localization of sensorin mRNA has also been reported (Schacher et al. 1999; Moccia et al. 2003).

More recently, the sequencing of clones from an unbiased cDNA library prepared from *Aplysia* sensory neurites led to the identification of over 200 distinct mRNAs (Moccia et al. 2003). Similar to the situation in the squid giant axon, the *Aplysia* axonal mRNA population is enriched for mRNA encoding cytoskeletal proteins, ribosomal proteins, and components of the translational machinery. Interestingly, as noted by Martin and colleagues, many of the ribosomal mRNAs present in the axon encode proteins located on the surface of the ribosome, raising the possibility that the local synthesis of these proteins mediates the assembly and/or translational competence of the ribosome.

Consistent with the above findings, the nerve terminal of squid photoreceptor neurons also contains a heterogeneous population of mRNAs, and biologically active polysomes (Crispino et al. 1993b, 1997). The results of RT-PCR analyses clearly established the presence of both β-actin and synapsin mRNA sequences in polysomes purified from these presynaptic nerve terminals (Gioio et al. 2004). An independent proteomic analysis conducted on this synaptosomal preparation revealed the local synthesis of approximately 80 different proteins to include several cytoskeletal proteins, molecular chaperones, and nuclear-encoded mitochondrial proteins (Jimenez et al. 2002).

A partial listing of mRNAs present in invertebrate axons is provided in Table 1. These gene transcripts can be organized into several broad functional categories, raising the possibility that the axon can synthesize key components of its own cytoskeleton and local translational machinery, molecular motors, chaperone molecules, neuropeptides, and components of its energy metabolic system (for discussion of

Table 1 mRNAs identified in invertebrate axons

Cytoskeletal proteins	Nuclear-encoded mitochondrial proteins
β-actin[a]	Cytochrome C oxidase assembly protein COX17[b]
α and β tubulin[a,c]	Cytochromosome C oxidase subunit Vb[d]
β-spectrin[d]	Ubiquinone biosynthesis protein CoQ7[b]
β-thymosin[c]	Dihydrolipoamide dehydrogenase[b]
Neurofilament proteins[e]	
Molecular motors	**Neuropeptides**
Kinesin[g]	Sensorin[c,f]
MAP H1[i]	Caudodorsal cell hormone[h]
	APG–Wamide[j]
	Egg laying hormone[k]
Ribosome-associated proteins	**Other proteins**
S5, S6, S15, S16, S19, S29[c,d]	Enolase[l]
L7A, L8, L9, L11, L18	Fructose PTS enzyme II[d]
L22, L31, L36, L37	Heat shock protein 70 (Hsp70)[b]
Translational factors	LDL receptor adaptor protein[m]
EF 1α[c]	Nucleotide diphosphate kinase[d]
EF 2[d]	Selenoprotein W[d]
CPED[c]	Synapsin[d]
	Syntaxin[n]
	Ubiquitin[c,d]

[a]Kaplan et al. (1992); [b]Gioio et al. (2001); [c]Moccia et al. (2003); [d]Gioio et al. (2004); [e]Giuditta et al. (1991); [f]Schacher et al. (1999); [g]Gioio et al. (1994); [h]Dirks et al. (1993); [i]Chun et al. (1996); [j]van Minnen (1994); [k]van Minnen et al. (1997); [l]Chun et al. (1995); [m]Chun et al. (1997); [n]Hu et al. (2003).

mRNAs present in mammalian axons, see Willis et al. 2007, and Vuppalanchi et al. 2009). One surprising feature of the categorization is that axons contain a significant number of nuclear-encoded mitochondrial mRNAs. In fact, approximately 25% of the total proteins synthesized locally in the presynaptic nerve terminal of squid photoreceptor neurons are transported into the mitochondria (Gioio et al. 2004). This finding indicates that the local protein synthesizing system plays a key role in the maintenance, and regulation of mitochondrial activity in the distal, structural and functional domains of the neuron, and calls attention to the intimacy of the relationship that has evolved between the axon/nerve terminal, and its energy generating system (for detailed discussion, see Hillefors et al. 2007; Aschrafi et al. 2008; Kaplan et al. 2009).

3.3 Local Translation of Reporter Genes

Invertebrate model systems have also played an important role in establishing that axons and nerve terminals had the capacity to synthesize proteins *de novo* (Giuditta et al. 1991; Crispino et al. 1993b, 1997). These studies were later substantiated by the use of heterologous reporter genes. For example, van Minnen and colleagues microinjected the mRNA encoding a molluscan egg laying hormone (ELH) into isolated axons of *Lymnea stagnalis*. In this study, injection of ELH mRNA into serotonergic neurons that did not normally express ELH (pedal A neurons) resulted in the synthesis of ELH peptides. The synthesis of ELH was detectable within 2 h after mRNA injection and was blocked by anisomycin, a protein synthesis inhibitor (van Minnen et al. 1997).

Similar findings were obtained when cultured *Lymnea* motor axons were microinjected with a heterologous mRNA encoding the conopressin receptor, a homolog of the vertebrate vasopressin receptor. Using immunocytochemical, electrophysiological, and pharmacological techniques, Spencer et al. (2000) were able to demonstrate that the locally synthesized receptor was inserted into the axolemma and was functionally active. Subsequent ultrastructural analyses demonstrated that the polysomal machinery requisite for the synthesis of the receptor was present in cultured isolated axons. The results of this elegant study were intriguing in that they established that isolated axons could intrinsically synthesize, and integrate integral proteins into the axonal membrane in the apparent absence of a well-demarcated rough endoplasmic reticulum and Golgi apparatus.

Reporter genes have also been introduced into synaptosomes prepared from squid photoreceptor neurons (Gioio et al. 2004). These large presynaptic nerve terminals were previously shown to contain a heterogenous population of mRNAs, and biologically active polysomes (Crispino et al. 1993b, 1997; Gioio et al. 2001). The transfection of the mRNA encoding green fluorescent protein (GFP) into these isolated nerve terminals resulted in the rapid synthesis of the GFP (Gioio et al. 2004). The immunohistochemical visualization of synapsin was used to confirm the presynaptic terminal localization of the GFP translation product. These results clearly demonstrated that the isolated nerve terminal had the capacity to translate a heterologous reporter mRNA.

4 Local Glial Synthesis of Neuronal RNA

One of the main questions raised by the demonstration of protein synthesis in axons and nerve endings was with regard to the cellular origin of their RNAs. Knowledge that cytoplasmic RNA is transcribed on nuclear DNA led to the apparently reasonable assumption that peripheral RNAs were exclusively supplied by the neuron soma. But problems were soon encountered when the hypothesis was tested, as radiolabeled RNA that invaded the axon was associated (and often preceded) with a persisting wave of radiolabeled RNA precursors (Autilio-Gambetti et al. 1973; Por et al. 1978), which could be taken up by glia cells, and incorporated into RNA that eventually was transferred to the axon. Hence, the origin of radiolabeled axonal RNA from the neuron cell body could not be distinguished from its derivation from periaxonal glia. This unsolved dilemma prevented the general acceptance of the somatic source of axonal RNAs, with the only exception of 4S RNA in regenerating goldfish axons (Ingoglia 1982). More recently, heterologous BC1 RNA, microinjected into the Mauthner neuron soma, was transported distally along the axon, and targeted to periaxonal ribosomal plaques (Muslimov et al. 2002).

A related key issue concerns the intraneuronal distribution of soma-derived RNA granules, which have been demonstrated in neurites of neurons in culture and in dendrites of mature neurons (Kindler et al. 2005; Cambray et al. 2009), but have not yet been described in differentiated axons (Sotelo-Silveira et al. 2006). They are distinguished in three main types: transport RNPs which have been detected in growing neurites and mature dendrites, stress granules which contain mRNAs, and in which translation is regulated, and P-bodies which contain mRNAs that are degraded (Kiebler and Bassell 2006). The latter two kinds have been identified only in dendrites. Neuronal mRNAs are largely associated with microtubules and actin filaments (Steward and Banker 1992). In cultured neurons, they move distally in neurites in association with ribosomes (Landry and Campagnoni 1998). As to compact RNA granules from cultured embryonic neurons, and adult mouse brain, they engage in translation when assuming a looser organization. In the adult brain, they are exclusively transported into dendrites (Krichevsky and Kosik 2001; Hirokawa 2006; Martin and Zukin 2006).

Long before the aforementioned studies, neuronal cell body RNA had been shown to derive from perisomatic glial cells under conditions of enhanced neural activity or learning (see Sect. 4.1). These observations were prompted by a line of thought that more than a century ago had suggested the name of "trophocytes" (Holmgren 1904) for cells that exhibited intimate contacts with neurons (now called glial cells). Since then, the use of analytical procedures of greater power and resolution considerably extended these data (for reviews, see Kuffler and Nicholls 1966; Bunge 1993; Newman 2003; Miller 2005). Glial cells are now known to express a number of ion channels, receptors, neurotransmitters, and neuromodulators that mediate their 'cooperative dialogue' with neuronal domains, notably with mammalian synapses (Gafurov et al. 2001; Auld and Robitaille 2003; Bezzi et al. 2004; Volterra and Meldolesi 2005).

The synthesis and delivery of glial transcripts to the neuron cell body and to axonal and presynaptic domains are described in the following subsections.

4.1 Cell Body RNA

Microchemical methods that allowed determination of RNA content and base composition at picogram levels were used to demonstrate the transfer of glial RNA to neuron somas under conditions of enhanced neural activity and learning. Chronic *in vivo* stimulation for 7 days of rabbit vestibular nucleus of Deiters increased the RNA content of Deiters cell bodies and led to a comparable RNA loss in perisomatic glial cells (Hydén and Egyházi 1962a). More marked inverse changes in glia and neuron soma concerned the enzyme cytochrome oxidase, while anaerobic glycolysis underwent inverse changes in the opposite direction. As pericapillary glia were barely affected, changes were specifically related to perisomatic glia (Hamberger

1963). The data were interpreted to reflect cooperation between neurons and glial cells, which allowed neurons to receive glial RNA, and upregulate oxidative metabolism under conditions of enhanced neuronal activity (Hamberger and Hydén 1963).

Short-lasting (1 h) pharmacologic treatments of rabbits, or rats also induced comparable inverse changes in the RNA content of Deiters neurons and glial cells (Hydén and Egyházi 1968). Furthermore, the RNA guanine (G) content increased in neurons, but decreased in glia, and opposite changes occurred in cytosine (C) content. In addition, when the amount and base composition of the RNA lost by glia were compared to the RNA gained by neurons, the two RNAs turned out to be almost equivalent (Hydén and Lange 1965, 1966). These features strongly supported the intercellular transfer of glial RNA to the neuron cell body. On the other hand, the enhanced RNA content of Deiters cytoplasm was found to be associated with a significant loss of nuclear RNA, and a marked decrease in the uridine (U) content of neuronal nuclear RNA (Hydén and Lange 1965, 1966). The data indicated that neuronal transcription was involved. At variance with acute experiments, RNA changes in neurons and glia were absent when the pharmacological treatment was prolonged for weeks (Slagel et al. 1966), most probably because neurons had adapted to the chronic functional load (Watson 1974).

Quantitative monitoring of neural activity was reported in similar investigations of the lobster slowly adapting stretch receptor. When the receptor neuron was stimulated by stretch stimulus for periods up to 6 h, the content of neuronal RNA did not change, but the A/U ratio significantly increased (Grampp and Edström 1963). With more intense and longer stimulations, the RNA content and base composition were not modified in the neuron, but glial RNA significantly decreased. In addition, in the unstimulated preparation, the block of transcription induced similar RNA losses in neuron and glia, and similar changes in their base composition. This indicated a parallel loss of a short-lived RNA rich in A, and poor in C. These effects were markedly reduced in neurons, but were reduced to a much lower extent in glia when the stretch receptor was stimulated for the same period of time (Edström and Grampp 1965).

Inverse changes in the RNA content of stimulated neurons and glia were also observed, using cytospectrophotometric methods, which avoided possible errors arising from microdissection artifacts (Pevzner 1965). Following high frequency stimulation of the afferent nerve to the cat superior cervical ganglion for 3 h, neuronal RNA markedly increased and glial RNA decreased. This observation confirmed microchemical data and excluded the possibility of attributing glial RNA loss to dendrites contaminating the glial sample (Kuffler and Nicholls 1966). The data were confirmed and extended in additional preparations (Pevzner 1971). On the whole, the experiments indicated that glial RNA was transferred to the neuron soma under conditions of sufficiently intense and prolonged stimulation, but that RNA losses occurred in both neurons and glia when the functional load exceeded a given threshold.

Changes in the RNA base composition of neurons and perisomatic glia were also reported in rats trained for an appetitive balancing task for 8 days. In comparison with control rats exposed to vestibular stimulation, in trained rats, the A/U ratio markedly increased in Deiters nuclear RNA, while the A content of glial RNA increased and the C content decreased. Both changes indicated the enhanced synthesis of DNA-like RNA (Hydén and Egyházi 1962a, 1963). Comparable effects were observed in rats forced to use their nonpreferred paw for 4 days to retrieve food. In trained animals, cortical pyramidal neurons of the contralateral hemisphere exhibited a slight increment in the content of RNA, and a decrease in the RNA GC/AU ratio, in comparison with the corresponding contralateral neurons. This again suggested a higher synthesis of DNA-like RNA. The effect was absent in control rats, and in rats that had used their preferred paw (Hydén and Egyházi 1964).

The methodology and interpretation of the microchemical data were harshly criticized on several counts (Kuffler and Nicholls 1966), including: (1) the possibility that the increased content of neuronal RNA was not due to the transfer of glial RNA, but to the transfer of dendritic RNA contaminating the glial sample, and (2) the lack of direct monitoring of the enhanced neuronal activity. In a recent reevaluation of these criticisms (Giuditta et al. 2008), we pointed out (1) that the loss of glial RNA had been independently confirmed using morphological methods (Pevzner 1965, 1971), and (2) that direct monitoring of neuronal activity was reported in experiments that yielded comparable data (Grampp and Edström 1963; Edström and Grampp 1965; Pevzner 1965, 1971). Overall, the criticism appeared based on the implicit rejection of any type of intercellular macromolecular exchange, despite its demonstrated occurrence in a large number of cells and conditions (Motta et al. 1995; Rustom et al. 2004).

4.2 Axonal RNA

Early studies of the glial origin of neuronal RNA were not limited to the neuron cell body, but they also considered the axon. Initial evidence of the local synthesis of axonal RNA was reported 45 years ago in studies of the goldfish Mauthner (M-) neuron. In the first investigation (Edström 1964), microchemical RNA analyses of the proximal stump of transected M-fibres showed that the base composition of axonal and myelin RNA underwent remarkable parallel changes that persisted for several weeks after the lesion. Notably, the A/G ratio increased 43% in the axon, and 50% in the myelin sheath, which indicated a rapid RNA turnover in both, and further suggested a common origin of the changes, most probably in periaxonal glia. In addition, in an independent experiment, 30 min stimulation of the M-neuron *in vivo* induced an initial 40% loss of axonal RNA, which was fully recovered in the following 3 h (Jakoubek and Edström 1965). As myelin RNA underwent comparable changes, the observation also suggested their dependence on a single source. Direct evidence was later provided by incorporation of radiolabeled precursors into the axonal RNA of M-fibers (Edström et al. 1969). Interestingly, sedimentation analysis indicated high molecular weight RNA (16S and 28–30S) and 4S RNAs were synthesized when intact M-fibers contained oligodendroglial cell nuclei, but 4S RNA was the only synthesized species when glial nuclei were absent, very likely in axonal mitochondria. This observation strongly supported the conclusion that axonal RNA was transcribed on glial nuclear DNA when fibers were decentralized from Mauthner cell bodies.

Incorporation of radiolabeled precursors into axoplasmic RNA was also demonstrated in isolated squid giant axons (Fischer et al. 1969; Cutillo et al. 1983). In *Loligo vulgaris*, the disproportionate increment of newly synthesized axoplasmic RNA that was monitored after an 8 h incubation period was attributed to the transit time required by glial RNA to reach the axoplasmic core. The local synthesis of squid axoplasmic RNA also occurred in vivo (Cutillo et al. 1983). Newly synthesized axoplasmic RNA was eventually identified as tRNA, rRNAs, and poly(A)$^+$RNA (Rapallino et al. 1988), and shown to be assembled in ribonucleoprotein particles (Menichini et al. 1990). Interestingly, with short incubation times, axoplasmic RNAs of small size were synthesized at a much faster rate than in periaxonal glia, or nerve cell bodies. These observations were confirmed in experiments with internally perfused giant axons. Radiolabeled RNA appeared in the perfusate soon after addition of [^{3}H]uridine to the incubation chamber, and accumulated in the perfusate at nearly linear rate. Most of it was of small size, and partly colocalized with tRNA, but some was associated with marker subribosomal particles (Eyman et al. 2007a). The rate of delivery of radiolabeled RNA to the axon perfusate was strongly enhanced by depolarizing the axon, which suggested that the effect was mediated by neurotransmitters released by the axon, which bound to glial receptors. Stimulated giant axons were in fact known to induce glial hyperpolarization, which was mediated by the release of axonal neurotransmitters that bound to glial receptors (Lieberman et al. 1994).

This possibility was demonstrated by monitoring the effects of agonists and antagonists of glial cholinergic and glutamatergic receptors on the rate of delivery of radiolabeled RNA to the axon perfusate. The delivery process was markedly enhanced by *N*-acetylaspartylglutamate, the first neurotransmitter shown to be released by stimulated crayfish axons (Gafurov et al. 2001), as well as by glutamate, NMDA, D-aspartate, and carbachol. In addition, when antagonists of NMDA (MK801), or cholinergic receptors (d-tubocurarine) were paired with the corresponding agonists, the rate of delivery of radiolabeled RNA to the axon perfusate remained at its basal level. Together with previous data (Cutillo et al. 1983; Menichini et al. 1990; Rapallino et al. 1988), these results represented the most direct demonstration of the intercellular transfer of glia-derived RNA to the axon, and of the modulation of this process by neurotransmitters released by the axon that activated glia receptors (Eyman et al. 2007a).

Comparable incorporation experiments demonstrated the local synthesis of axonal RNA in mammalian axons (Koenig 1967). Quantitative autoradiographic methods were used to demonstrate the local synthesis of axonal RNA in proximal stumps of transected rat sciatic nerves in vivo (Benech et al. 1982). Radiolabeled RNA was more concentrated in Schwann cells than in myelin sheaths and axons, which suggested the glial origin of axonal RNA. The rate of synthesis reached a peak 1 day after axotomy. In Amphibia, decentralized brachial nerves of the newt incorporated [³H]uridine into axonal RNA, as shown by autoradiographic methods (Singer and Green 1968). Radiolabeled RNA was also present over Schwann cells and myelin, which was consistent with axonal RNA being mostly supplied by periaxonal glia. In agreement with results of the rat experiments (Benech et al. 1982), newly synthesized RNA appeared to be transferred to the axon in the regions of glial cytoplasm abutting the axon, such as Schmidt–Lantermans incisures and paranodal regions (Bunge 1993).

The origin of axonal ribosomes from Schwann cells has been demonstrated in sciatic nerves of the mouse Wld_s mutant (Brown et al. 1994) and C57BL strain. In the mutant, 1 week after nerve decentralization, axons do not degenerate, and, actually, they contain clusters of polysomes, some of which are associated with membranes, presumably derived from periaxonal glia (Court and Alvarez 2005). This was recently confirmed in an elegant study with fluorescent antibody probes specific for ribosomes, axoplasm, and glial cytoplasm (Court et al. 2008). A similar but less pronounced process occurs in the C57BL strain. The transfer is believed to be mediated by invagination of ribosome-containing glial fingers into the axon, in which the double membrane enveloping the axoplasmic inclusion is eventually broken down. Comparable formations mediate the transfer of glial ribosomes to chicken and rat axons (Li et al. 2005; Kun et al. 2007).

4.3 Presynaptic RNA

The first indirect evidence of the local synthesis of presynaptic RNA was obtained by RT-PCR analyses of TH mRNA in brain regions lacking catecholaminergic cell bodies, such as cerebellum and striatum (Melia et al. 1994). In the cerebellum, the content of TH mRNA underwent a marked increment after reserpine treatment, while a significant decrease of the mRNA was induced in the ispilateral striatum by unilateral damage to the cathecolaminergic pathway.

Direct data were provided by analyses of newly synthesized RNA in the unusually large synaptosomes of squid optic lobes, which derive from the nerve terminals of retinal photoreceptors (Crispino et al. 1993b, 1997; Martin et al. 1998). The retinal location of their cell bodies implied that, in optic lobe slices incubated with radiolabeled precursors, the presence of newly synthesized RNA in the large synaptosomes would prove its local transcription. This experimental design yielded positive results, as newly synthesized RNA became prevalently associated with the synaptosomal fraction after incubation periods longer than half an hour (Eyman et al. 2007a). In addition, when radiolabeled slices were homogenized in hypo-osmotic buffer, newly synthesized synaptosomal RNA was almost completely released into the medium, as expected from the osmotically driven swelling and disruption of synaptosomes. Moreover, autoradiographic analyses of the synaptosomal fraction demonstrated that essentially all radiolabeled RNAs were localized in the large presynaptic synaptosomes identified by selective staining with a presynaptic marker antibody.

As with internally perfused squid giant axons, sedimentation analyses of radiolabeled presynaptic RNA indicated that it was mostly of small size, cosedimenting with marker tRNA, but that some RNA cosedimented with marker subribosomal particles (Eyman et al. 2007a). The large prevalence of small size RNA in presynaptic and axoplasmic RNA synthesized during short incubation times indicated their remarkably high turnover (Menichini et al. 1990; Eyman et al. 2007a). This may partly explain the relatively quick changes occurring in the base composition of axonal RNA following axotomy (Edström 1964), and the enhanced turnover of axonal RNA elicited by vestibular stimulation (Jakoubek and Edström 1965). As some small size RNAs may be miRNAs, they could be involved in the modulation of axonal and presynaptic protein synthesis. This possibility is supported by the selective decrement in the content of axonal tubulin and tubulin mRNA, which is induced in mouse by exposure to specific siRNA (Murashov et al. 2007). In neurons, the involvement of miRNAs in translational control has largely been related to postsynaptic regions (Lugli et al. 2008; Fiore et al. 2008).

The abundant presence of mitochondria in nerve endings prompted investigations of synaptosomal fractions from rat brain to characterize the synthesis of mitochondrial RNA in extrasomatic neuronal domains. The rate of this process was very modest in young adult rats, but was much higher in 10-day-old rats, in which most newly synthesized mitochondrial RNA included 16S, 12S, and 4S species (England and Attardi 1976). Ethidium bromide (EB), a specific inhibitor of mitochondrial transcription, strongly inhibited the synthesis of RNA larger than 4S, but inhibited to a much smaller extent the synthesis of 4S RNA. In addition, mitochondrial DNA satisfactorily annealed with RNA larger than 4S, but annealing was much less efficient with 4S RNA. Interestingly, 50% of newly synthesized EB-sensitive RNA was present in presynaptic synaptosomes, as shown by EM autoradiography. On the whole, the data established the synthesis of mitochondrial RNA in nerve terminals and suggested that synaptosomal 4S RNA might contain non-mitochondrial RNA species.

Fewer studies have been aimed at clarifying the cellular derivation of presynaptic cytoplasmic RNA in mammalian brain. Following the intracranial injection of radiolabeled precursors in young rats (DeLarco et al. 1975), the specific activity of synaptosomal RNA was initially lower than in mitochondria and polysomes, but became markedly higher with time. Notably, the specific activity of poly(A)$^+$RNA was much higher than in polysomes for several hours. Since its poly(A) tail was markedly shorter compared to axonal mRNAs (Mohr and Richter 2000), most newly synthesized synaptosomal poly(A)$^+$RNA is likely to be derived from nerve endings. In adult rats, newly synthesized synaptosomal RNA was less abundant than in mitochondria and microsomes 1 h after the intraventricular injection of [^{3}H] uridine, but the percent of radiolabeled poly(A)$^+$RNA was markedly higher than in microsomes (Cupello and Hydén 1982), which confirmed the high turnover of synaptosomal poly(A)$^+$RNA.

The transcription of presynaptic RNA in perisynaptic cells (Eyman et al. 2007a) raises the intriguing question of whether the modifications of glia RNA detected under conditions of enhanced neural activity, and learning are really reflecting the delivery of glial RNA to the neuron soma (Giuditta et al. 2008). This doubt is raised in view of the increment in rate of neuronal transcription, which might entirely explain the increment in neuronal RNA (Hydén and Egyházi 1962b). If so, the associated changes in glial RNA might imply its delivery to the numerous nerve endings that remain attached to the soma of dissected neurons (Hydén and Pigon 1960), and are included in the neuronal sample. The transfer of glial RNA to nerve endings implies targeting a plastic neuronal compartment which is capable of protein synthesis, but is devoid of transcriptional activity. Supplying glial transcripts to nerve terminals (and possibly dendrites) would allow synapses to adapt quickly to environmental conditions.

The intercellular transfer of glial RNAs and RNPs to extrasomatic neuronal domains raises the problem of the mechanism of the translocation process. Among possible alternatives (Vincent and Magee 2002; Rustom et al. 2004), we have considered most likely the axonal invagination of glial cytoplasm followed by pinching off and eventual dissolution of the enclosing plasma membranes (Giuditta et al. 2008). This mechanism has received support in recent studies (Li et al. 2005; Kun et al. 2007; Court et al. 2008).

5 Gene Expression in Axons and Nerve Terminals

As previously noted, the concept of a 'glia–neuron unit' was proposed more than a century ago to account for the intimate contacts the neuron cell body entertains with glial cells (Holmgren 1904). The concept received experimental support half a century later when RNA and protein were shown to be transferred from perisomatic glia to the neuron soma (Giuditta et al. 2008). In the same time period, most axonal RNA was found to be synthesized in periaxonal glia (Edström et al. 1969, Koenig 1967). After an intervening period of controversies (Gainer 1978), the concept was revived in experiments dealing with the glial origin of axonal, and presynaptic proteins and RNAs (Bittner 1991; Koenig and Giuditta 1999; Alvarez et al. 2000; Giuditta et al. 2002, 2008).

Overall, the data emphasized the view that proteins of well-differentiated axons and nerve endings are not exclusively supplied by the cell body, as originally believed, but are synthesized by peripheral translation machineries requiring delivery of glial transcripts. Together with the evidence that axonal translation is modulated by glial transcription (Tobias and Koenig 1975), these observations suggested that axons and nerve terminals utilize local gene expression systems for maintenance and to respond to local stimuli by appropriate plastic modifications (Eyman et al. 2007a; Giuditta et al. 2008). Peripheral systems appear to be largely independent from trophic modulation by the neuron cell body. Indeed, they supplement their role when the mass of the axonal domain exceeds the capacities of the neuron soma. The minimal conditions required to accept this proposal include: (1) the demonstration that RNAs involved in peripheral translation are transcribed in periaxonal and perisynaptic glia, and are eventually delivered to axons and nerve endings, and

(2) the demonstration that local glial transcripts control axonal or presynaptic translation. These conditions are largely satisfied by axons, in view of overwhelming evidence that axons are capable of protein synthesis (see Sect. 3.1; Vuppalanchi et al. 2009), that axonal RNA originates in periaxonal glia (see Sect. 4.2), and that glial transcription modulates local protein synthesis (Edström 1967; Tobias and Koenig 1975).

On the other hand, in nerve terminals, notably from mammalian brain, some of these conditions remain to be demonstrated, as suitable experimentation is still lacking. They include the local synthesis of presynaptic RNA, and the local control of presynaptic protein synthesis by glial transcripts. Protein synthesis has nonetheless been demonstrated in nerve terminals of vertebrate and invertebrate species (see Sect. 2), and presynaptic RNAs are synthesized in squid perisynaptic cells (see Sect. 4.3). In addition, (1) the local synthesis of presynaptic RNA is indirectly supported by evidence that the content of TH mRNA is locally modulated in mammalian brain regions lacking catecholaminergic cell bodies (Melia et al. 1994); furthermore, local transcriptional modulation of presynaptic protein synthesis is suggested by the increase in TH activity that occurs in the heart following reserpine treatment (Thoenen et al. 1970), (2) the presence and high turnover of poly(A)$^+$RNA in mammalian nerve endings are indicated by the prevalence of newly synthesized poly(A)$^+$RNA in brain synaptosomes, and by its shorter poly(A) tail (De Larco et al. 1975; Cupello and Hydén 1982), that reproduces the same unique feature of axonal mRNA (Mohr and Richter 2000), and (3) the control of presynaptic translation by locally transcribed miRNA is indirectly suggested by the prevalence of newly synthesized low MW RNA in squid nerve terminals, and by their presumed content of miRNA (Eyman et al. 2007a). This possibility is also in agreement with recent observations (Murashov et al. 2007).

Circumstantial evidence in loose agreement with the proposal of a presynaptic system of gene expression is provided by the involvement of perisynaptic glia in synaptic plasticity and sensory function (Murai and Van Meyel 2007; Bains and Olliet 2007; Bacaj et al. 2008; Flavell and Greenberg 2008; Reichenbach and Pannicke 2008; Fellin 2009), by the capacity of neuronal compartments to take up exogenous protein (Holtzman and Peterson 1969; Giuditta et al. 1971; Kolodny 1971), and by the intercellular transfer of mRNAs and miRNAs by exosomes (Valadi et al. 2007).

In the last years a number of well-designed experiments have brought to light the complex dialogue between nerve endings and neuron soma that is required to implement plastic synaptic modifications in cultured *Aplysia* neurons (Miniaci et al. 2008). The data provide strong support to the view that presynaptic events are conditioned by neuronal transcription rather than by perisynaptic glia. At present, the possibility that a comparable mechanism may also apply to mature, well differentiated neurons lacks experimental support. In this regard, we note that the axonal/presynaptic mass of cultured *Aplysia* neurons is substantially smaller than that of most well differentiated neurons. Since the trophic glial involvement is assumed to be depended on the ratio between the axonal/presynaptic mass and the trophic capacities of the cell body, cultured *Aplysia* neurons are likely to qualify well below that threshold. As a result, their plastic behavior does not appear to controvert the proposal of a presynaptic gene expression system, which only concerns differentiated neurons with ratio values above threshold. Furthermore, it is obvious that experiments involving glia-free neurons cannot yield information on the role of glial cells.

6 Conclusions

The original concept of 'glia–neuron unit', initially proposed for neuronal soma, takes on a broader, functionally more relevant meaning if applied to mature axons and nerve endings that are capable of protein synthesis but lack transcription capacity in contrast to the cell body. The delivery of glial transcripts to peripheral gene expression systems relieves the cell body of a considerable trophic load and allows peripheral translation to be locally modulated by transcription. This bypasses the slower, spatially indiscriminate mode of communication with the cell body. Such mechanism is especially valuable for synaptic regions that are continuously undergoing highly selective plastic modifications (Eyman et al. 2007b; Guan and Clark 2006; Kang and Schuman 1996; Mariucci et al. 2007; Martin et al. 1997). A comparable gene expression system might possibly apply to dendrites also; notably, those endowed with extensive arborization.

The possibility of a somatic contribution to peripheral gene expression cannot be excluded, and is actually to be expected; presumably, with modalities that will include transcellular modulation of perisomatic and peripheral glia. If such prediction will be validated, the roles of central and peripheral trophic sources should be determined under basal conditions, and during enhanced neural activity, in order to attain an overall view of neuronal gene expression.

The still lingering belief that axons and nerve terminals depend on somatic gene expression to a large, or exclusive extent (Hengst and Jaffrey 2007; Reichenbach and Pannicke 2008) should be revised on the basis of data highlighting the essential, and prompt involvement of local gene expression systems. In addition, it would seem preferable not to equate local protein synthesis with dendritic protein synthesis, and not refer to invertebrate axons as a poorly defined hybrid of dendritic and axonal features. Their receptive structures are limited to most proximal axonal locations.

References

Alvarez J, Giuditta A, Koenig E (2000) Protein synthesis in axons and terminals: significance for maintenance, plasticity and regulation of phenotype. With a critique of slow transport theory. Prog Neurobiol 62:1–62

Aschrafi A, Schwechter AD, Mameza MG, Natera-Naranjo O, Gioio AE, Kaplan BB (2008) Micro RNA-338 regulates local cytochrome oxidase IV mRNA levels and oxidative phosphorylation in the axons of sympathetic neurons. J Neurosci 19:12581–12590

Auld DS, Robitaille R (2003) Glial cells and neurotransmission: an inclusive view of synaptic function. Neuron 40:389–400

Austin L, Morgan IG (1967) Incorporation of ^{14}C-labelled leucine into synaptosomes from rat cerebral cortex in vitro. J Neurochem 14:377–387

Autilio-Gambetti L, Gambetti P, Shafer B (1973) RNA and axonal flow. Biochemical and autoradiographic study in the rabbit optic system. Brain Res 53:387–398

Bacaj T, Tevlin M, Lu Y, Shaham S (2008) Glia are essential for sensory organ function in C. elegans. Science 322:744–747

Bains JS, Oliet SH (2007) Glia: they make your memories stick! Trends Neurosci 30:417–424

Bekinschtein P, Cammarota M, Katche C, Slipczuk L, Rossato JI, Goldin A, Izquierdo I, Medina JH (2008) BDNF is essential to promote persistence of long-term memory storage. Proc Natl Acad Sci USA 105:2711–2716

Benech C, Sotelo JR Jr, Menendez J, Correa-Luna R (1982) Autoradiographic study of RNA and protein synthesis in sectioned peripheral nerves. Exp Neurol 76:72–82

Bezzi P, Gundersen V, Galbete JL, Seifert G, Steinhauser C, Pilati E, Volterra A (2004) Astrocytes contain a vesicular compartment that is competent for regulated exocytosis of glutamate. Nat Neurosci 7:613–620

Bittner GD (1991) Long-term survival of anucleate axons and its implications for nerve regeneration. Trends Neurosci 14:188–193

Brown MC, Perry VH, Hunt SP, Lapper SR (1994) Further studies on motor and sensory nerve regeneration in mice with delayed Wallerian degeneration. Eur J Neurosci 6:420–428

Bunge RP (1993) Expanding roles for the Schwann cell: ensheathment, myelination, trophism and regeneration. Curr Opin Neurobiol 3:805–809

Cambray S, Pedraza N, Rafel M, Garí E, Aldea M, Gallego C (2009) Protein kinase KIS localizes to RNA granules and enhances local translation. Mol Cell Biol 29:726–735

Casadio A, Martin KC, Giustetto M, Zhu H, Chen M, Bartsch D, Bailey CH, Kandel ER (1999) A transient, neuron-wide form of CREB-mediated long-term facilitation can be stabilized at specific synapses by local protein synthesis. Cell 99:221–237

Chihara T, Luginbuhl D, Luo L (2007) Cytoplasmic and mitochondrial protein translation in axonal and dendritic terminal arborization. Nat Neurosci 10:828–837

Chun JT, Gioio AE, Crispino M, Giuditta A, Kaplan BB (1995) Characterization of squid enolase mRNA: sequence analysis, tissue distribution, and axonal localization. Neurochem Res 20:923–930

Chun JT, Gioio AE, Crispino M, Giuditta A, Kaplan BB (1996) Differential compartmentalization of mRNAs in squid giant axon. J Neurochem 97:1806–1812

Chun JT, Gioio AE, Crispino M, Eyman M, Giuditta A, Kaplan BB (1997) Molecular cloning and characterization of a novel mRNA present in the squid giant axon. J Neurosci Res 49:144–153

Cohen AI (1973) An ultrastructural analysis of the photoreceptors of the squid and their synaptic connections. III. Photoreceptor terminations in the optic lobe. J Comp Neurol 147:399–426

Cotman CW, Taylor DA (1971) Autoradiographic analysis of protein synthesis in synaptosomal fractions. Brain Res 29:366–372

Court F, Alvarez J (2005) Local regulation of the axonal phenotype, a case of merotropism. Biol Res 38:365–374

Court FA, Hendriks WT, Macgillavry HD, Alvarez J, van Minnen J (2008) Schwann cell to axon transfer of ribosomes: toward a novel understanding of the role of glia in the nervous system. J Neurosci 28:11024–11029

Crispino M, Capano CP, Kaplan BB, Giuditta A (1993a) Neurofilament proteins are synthesized in nerve endings from squid brain. J Neurochem 61:1144–1146

Crispino M, Castigli E, Perrone Capano C, Martin R, Menichini E, Kaplan BB, Giuditta A (1993b) Protein synthesis in a synaptosomal fraction from squid brain. Mol Cell Neurosci 4:366–374

Crispino M, Kaplan BB, Martin R, Alvarez J, Chun JT, Benech JC, Giuditta A (1997) Active polysomes are present in the large presynaptic endings of the synaptosomal fraction from squid brain. J Neurosci 17:7694–7702

Cupello A, Hydén H (1982) Labeling of poly(A)-associated RNA in synaptosomes and the other subcellular fractions of rat cerebral cortex in basal conditions and during training. J Neurosci Res 8:575–579

Cutillo V, Montagnese P, Gremo F, Casola L, Giuditta A (1983) Origin of axoplasmic RNA in the squid giant fibre. Neurochem Res 8:1621–1634

DeLarco J, Nakagawa S, Abramowitz A, Bromwell K, Guroff G (1975) Polyadenylic acid-containing RNA from rat brain synaptosomes. J Neurochem 25:131–137

Dirks RW, van Dorp AGM, van Minnen J, Fransen JAM, van der Ploeg M, Rapp AK (1993) Ultrastructural evidence for the axonal localization of caudodorsal cell hormone mRNA in the central nervous system of the mollusk Lymnea stagnalis. Microsc Res Techn 25:12–18

Edström A (1964) Effect of spinal cord transection on the base composition and content of RNA in the Mauthner nerve fibre of the goldfish. J Neurochem 11:557–559

Edström A (1966) Amino acid incorporation in isolated Mauthner nerve fibre components. J Neurochem 13:315–321

Edström A (1967) Inhibition of protein synthesis in Mauthner nerve fibre components by actinomycin-D. J Neurochem 14:239–243

Edström A, Edström JE, Hökfelt T (1969) Sedimentation analysis of ribonucleic acid extracted from isolated Mauthner nerve fibre components. J Neurochem 16:53–66

Edström JE, Eichner D, Edström A (1962) The ribonucleic acid of axons and myelin sheaths from Mauthner neurons. Biochim Biophys Acta 61:178–184

Edström JE, Grampp W (1965) Nervous activity and metabolism of ribonucleic acids in the crustacean stretch receptor neuron. J Neurochem 12:735–741

England JM, Attardi G (1976) Analysis of RNA synthesized by an isolated rat brain synaptosomal fraction. J Neurochem 27:895–904

Eyman M, Crispino M, Kaplan BB, Giuditta A (2003) Squid photoreceptor terminals synthesize calexcitin, a learning related protein. Neurosci Lett 347:21–24

Eyman M, Cefaliello C, Ferrara E, De Stefano R, Scotto Lavina Z, Crispino M, Squillace A, van Minnen J, Kaplan BB, Giuditta A (2007a) Local synthesis of RNA in axons and nerve terminals. Eur J Neurosci 25:341–350

Eyman M, Ferrara E, Cefaliello C, Mandile P, Crispino M, Giuditta A (2007b) Synaptosomal protein synthesis from rat brain is selectively modulated by learning. Brain Res 1132:148–157

Fellin T (2009) Communication between neurons and astrocytes: relevance to the modulation of synaptic and network activity. J Neurochem 108:533–544

Fiore R, Siegel G, Schratt G (2008) MicroRNA function in neuronal development, plasticity and disease. Biochim Biophys Acta 1779:471–478

Fischer S, Litvak S (1967) The incorporation of micro injected ^{14}C-aminoacids into TCA insoluble fractions of the giant axon of the squid. J Cell Physiol 70:69–74

Fischer S, Gariglio P, Tarifeno E (1969) Incorporation of H^3-uridine and the isolation and the characterization of RNA from squid axon. J Cell Physiol 74:155–162 Pfeiffer BE, Huber

Flavell SW, Greenberg ME (2008) Signaling mechanisms linking neuronal activity to gene expression and plasticity of the nervous system. Annu Rev Neurosci 31:563–590

Frey U, Morris RG (1997) Synaptic tagging and long-term potentiation. Nature 385:533–536

Gafurov B, Urazaev AK, Grossfeld RM, Lieberman EM (2001) N-acetylaspartylglutamate (NAAG) is the probable mediator of axon-to-glia signaling in the crayfish medial giant nerve fiber. Neuroscience 106:227–235

Gainer H (1978) Intercellular transfer of proteins from glial cells to axons. Trends Neurosci 1:93–96

Gambetti P, Autilio-Gambetti LA, Gonatas NK, Shafer B (1972) Protein synthesis in synaptosomal fractions. Ultrastructural radioautographic study. J Cell Biol 52:526–535

Gioio AE, Chun JT, Crispino M, Perrone Capano C, Giuditta A, Kaplan BB (1994) Kinesin mRNA is present in the squid giant axon. J Neurochem 63:13–18

Gioio AE, Eyman M, Zhang H, Lavina ZS, Giuditta A, Kaplan BB (2001) Local synthesis of nuclear-encoded mitochondrial proteins in the presynaptic nerve terminal. J Neurosci Res 64:447–453

Gioio AE, Scotto Lavina Z, Jurkovicova D, Eyman M, Giuditta A, Kaplan BB (2004) Nerve terminals of squid photoreceptor neurons contain a heterogeneous population of mRNAs and translate a transfected reporter mRNA. Eur J Neurosci 20:865–872

Giuditta A, Dettbarn WD, Brzin M (1968) Protein synthesis in the isolated giant axon of the squid. Proc Natl Acad Sci USA 59:1284–1287

Giuditta A, D'Udine B, Pepe M (1971) Uptake of protein by the giant axon of the squid. Nature 229:29–30

Giuditta A, Cupello A, Lazzarini G (1980) Ribosomal RNA in the axoplasm of the squid giant axon. J Neurochem 34:1757–1760

Giuditta A, Hunt T, Santella L (1986) Messenger RNA in squid axoplasm. Neurochem Intern 8:435–442

Giuditta A, Menichini E, Perrone Capano C, Langella M, Martin R, Castigli E, Kaplan BB (1991) Active polysomes in the axoplasm of the squid giant axon. J Neurosci Res 26:18–28

Giuditta A, Kaplan BB, van Minnen J, Alvarez J, Koenig E (2002) Axonal and presynaptic protein synthesis: new insights into the biology of the neuron. Trends Neurosci 25:400–404

Giuditta A, Chun JT, Eyman M, Cefaliello C, Crispino M (2008) Local gene expression in axons and nerve endings: the glia-neuron unit. Physiol Rev 88:515–555

Grafstein B, Forman DS (1980) Intracellular transport in neurons. Physiol Rev 60:1167–1283

Grampp W, Edström JE (1963) The effect of nervous activity on ribonucleic acid of the crustacean receptor neuron. J Neurochem 10:725–731

Grosche J, Matyash V, Möller T, Verkhratsky A, Reichenbach A, Kettenmann H (1999) Microdomains for neuron-glia interaction: parallel fiber signaling to Bergmann glial cells. Nat Neurosci 2:139–143

Guan X, Clark GA (2006) Essential role of somatic and synaptic protein synthesis and axonal transport in long-term synapse-specific facilitation at distal sensomotor connections in *Aplysia*. Biol Bull 210:238–254

Hamberger A (1963) Difference between isolated neuronal and vascular glia with respect to respiratory activity. Acta Physiol Scand Suppl 58:1–58

Hamberger A, Hydén H (1963) Inverse enzymatic changes in neurons and glia during increased function and hypoxia. J Cell Biol 16:521–525

Hengst U, Jaffrey SR (2007) Function and translational regulation of mRNA in developing axons. Semin Cell Dev Biol 18:209–215

Hernandez AG, Langford GM, Martinez JL, Dowdall MJ (1976) Protein synthesis by synaptosomes from the head ganglion of the squid, *Loligo pealli*. Acta Cient Venez 27:120–123

Hillefors M, Gioio AE, Mameza MG, Kaplan BB (2007) Axon viability and mitochondrial function are dependent on local protein synthesis in sympathetic neurons. Cell Mol Neurobiol 27:701–716

Hirokawa N (2006) mRNA transport in dendrites: RNA granules, motors, and tracks. J Neurosci 26:7139–7142

Holmgren E (1904) Ueber die trophospongien der nervenzellen. Anat Anz 24:225–244

Holtzman E, Peterson ER (1969) Uptake of proteins by mammalian neurons. J Cell Biol 40:863–869

Hu JY, Meng X, Schacher S (2003) Redistribution of syntaxin mRNA in neuronal cell bodies regulates protein expression and transport during synapse formation and long-term synaptic plasticity. J Neurosci 23:1804–1815

Hu JY, Chen Y, Schacher S (2007) Protein kinase C regulates local synthesis and secretion of a neuropeptide required for activity-dependent long-term synaptic plasticity. J Neurosci 27:8927–8939

Huang HL, Cendan CM, Roza C, Okuse K, Cramer R, Timms JF, Wood JN (2008) Proteomic profiling of neuromas reveals alterations in protein composition and local protein synthesis in hyper-excitable nerves. Mol Pain 4:33

Hughes A, Flexner LB (1956) A study of the development of the cerebral cortex of the foetal guinea-pig by means of the ultraviolet microscope. J Anat 90:386–394

Hydén H, Egyházi E (1962a) Changes in the base composition of nuclear ribonucleic acid of neurons during a short period of enhanced protein production. J Cell Biol 15:37–44

Hydén H, Egyházi E (1962b) Nuclear RNA changes of nerve cells during a learning experiment in rats. Proc Natl Acad Sci USA 48:1366–1373

Hydén H, Egyházi E (1963) Glial RNA changes during a learning experiment in rats. Proc Natl Acad Sci USA 49:618–624

Hydén H, Egyházi E (1964) Changes in RNA content and base composition in cortical neurons of rats in a learning experiment involving transfer of handedness. Proc Natl Acad Sci USA 52:1030–1035

Hydén H, Egyházi E (1968) The effect of tranylcypromine on synthesis of macromolecules and enzyme activities in neurons and glia. Neurology 18:732–736

Hydén H, Lange PW (1965) A differentiation in RNA response in neurons early and late during learning. Proc Natl Acad Sci USA 53:946–952

Hydén H, Lange PW (1966) A genetic stimulation with production of adenic-uracil rich RNA in neurons and glia in learning. Naturwissen 53:64–70

Hydén H, Pigon A (1960) A cytophysiological study of the functional relationship between oliodendroglial cells and nerve cells of Deiters' nucleus. J Neurochem 6:57–72

Ingoglia NA (1982) 4S RNA in regenerating optic axons of goldfish. J Neurosci 2:331–338

Jakoubek B, Edström JE (1965) RNA changes in the Mauthner axon and myelin sheath after increased functional activity. J Neurochem 12:845–849

Jimenez CJ, Eyman M, Scotto Lavina Z, Gioio AE, Li KW, van den Schors R, Geraerts WPM, Giuditta A, Kaplan BB, J van Minnen (2002) Protein synthesis in synaptosomes: a proteomic analysis. J Neurochem 81:735–744

Kang H, Schuman EM (1996) A requirement for local protein synthesis in neurotrophin-induced hippocampal synaptic plasticity. Science 273:1402–1406

Kaplan BB, Gioio AE, Perrone Capano C, Crispino M, Giuditta A (1992) β-actin and β-tubulin are components of a heterogeneous mRNA population present in the squid giant axon. Mol Cell Neurosci 3:133–144

Kaplan BB, Gioio AE, Hillefors M, Aschrafi A, (2009) Axonal Protein Synthesis and the Regulation of Local Mitochondrial Function. Results Probl Cell Differ. doi: 10.1007/400_2009_1

Kiebler MA, Bassell GJ (2006) Neuronal RNA granules: movers and makers. Neuron 51:685–690

Kindler S, Wang H, Richter D, Tiedge H (2005) RNA transport and local control of translation. Annu Rev Cell Dev Biol 21:223–245

Koenig E (1965) Synthetic mechanisms in the axon. II. RNA in myelin-free axons of the cat. J Neurochem 12:357–361

Koenig E (1967) Synthetic mechanisms in the axon. IV. In vitro incorporation of [3H]precursors into axonal protein and RNA. J Neurochem 14:437–446

Koenig E (1979) Ribosomal RNA in Mauthner axon: implications for a protein synthesizing machinery in the myelinated axon. Brain Res 174:95–107

Koenig E, Giuditta A (1999) Protein synthesizing machinery in the axon compartment. Neuroscience 89:5–15

Kolodny GM (1971) Evidence for transfer of macromolecular RNA between mammalian cells in culture. Exp Cell Res 65:313–324

Krichevsky AM, Kosik KS (2001) Neuronal RNA granules: a link between RNA localization and stimulation-dependent translation. Neuron 32:683–696

Kuffler SW, Nicholls JG (1966) The physiology of neuroglial cells. Ergeb Physiol 57:1–90

Kun A, Otero L, Sotelo-Silveira JR, Sotelo JR (2007) Ribosomal distribution in axons of mammalian myelinated fibers. J Neurosci Res 85:2087–2098

Landry CF, Campagnoni AT (1998) Targeting of mRNAs into neuronal and glial processes: intracellular and extracellular influences. Neuroscientist 4:77–87

Lee JL (2008) Memory reconsolidation mediates the strengthening of memories by additional learning. Nat Neurosci 11:1264–1266

Lee YS, Bailey CH, Kandel ER, Kaang BK (2008) Transcriptional regulation of long-term memory in the marine snail *Aplysia*. Mol Brain 1:3

Li YC, Li YN, Cheng CX, Sakamoto H, Kawate T, Shimada O, Atsumi S (2005) Subsurface cisterna-lined axonal invaginations and double-walled vesicles at the axonal-myelin sheath surface. Neurosci Res 53:298–303

Lieberman EM, Hargittai PT, Grossfeld RM (1994) Electrophysiological and metabolic interactions between axons and glia in crayfish and squid. Prog Neurobiol 44:333–376

Lu Y, Christian K, Lu B (2008) BDNF: a key regulator for protein synthesis-dependent LTP and long-term memory? Neurobiol Learn Mem 89:312–323

Lugli G, Torvik VI, Larson J, Smalheiser NR (2008) Expression of microRNAs and their precursors in synaptic fractions of adult mouse forebrain. J Neurochem 106:650–661

Lyles V, Zhao Y, Martin KC (2006) Synapse formation and mRNA localization in cultured Aplysia neuron. Neuron 49:349–356

Mariucci G, Giuditta A, Ambrosini MV (2007) Permanent brain ischemia induces marked increments in hsp72 expression and local protein synthesis in synapses of the ischemic hemisphere. Neurosci Lett 415:77–80

Martin KC (2004) Local protein synthesis during axon guidance and synaptic plasticity. Curr Opin Neurobiol 14:305–310

Martin KC, Casadio A, Zhu H, Yaping E, Rose JC, Chen M, Bailey CH, Kandel ER (1997) Synapse-specific, long-term facilitation of *Aplysia* sensory to motor synapses: a function for local protein synthesis in memory storage. Cell 91:927–938

Martin KC, Zukin RS (2006) RNA trafficking and local protein synthesis in dendrites: an overview. J Neurosci 26:7131–7134

Martin R, Vaida B, Bleher R, Crispino M, Giuditta A (1998) Protein synthesizing units in presynaptic and postsynaptic domains of squid neurons. J Cell Sci 111:3157–3166

Melia KR, Trembleau A, Oddi R, Sanna PP, Bloom FE (1994) Detection and regulation of tyrosine hydroxylase mRNA in catecholaminergic terminal fields: possible axonal compartmentalization. Exp Neurol 130:394–406

Menichini E, Castigli E, Kaplan BB, Giuditta A (1990) Synthesis of axoplasmic RNA particles in the isolated squid giant axon. Neurosc Res Commun 7:89–96

Miller G (2005) The dark side of glia. Science 308:778–781

Miniaci MC, Kim JH, Puthanveettil SV, Si K, Zhu H, Kandel ER, Bailey CH (2008) Sustained CPEB-dependent local protein synthesis is required to stabilize synaptic growth for persistence of long-term facilitation in *Aplysia*. Neuron 59:1024–1036

Moccia R, Chen D, Lyles V, Kapuya E, Yaping E, Kalachikov S, Spahn CM, Frank J, Kandel ER, Barad M, Martin KC (2003) An unbiased cDNA library prepared from isolated *Aplysia* sensory neuron processes is enriched for cytoskeletal and translational mRNAs. J Neurosci 23:9409–9417

Mohr E, Richter D (2000) Axonal mRNAs: functional significance in vertebrates and invertebrates. J Neurocytol 29:783–791

Morgan IG, Austin L (1968) Synaptosomal protein synthesis in a cell-free system. J Neurochem 15:41–51

Morgan IG, Austin L (1969) Ion effects and protein synthesis in synaptosomal fraction. J Neurobiol 2:155–167

Motta CM, Castriota Scanderbeg M, Filosa S, Andreuccetti P (1995) Role of pyriform cells during the growth of oocytes in the lizard Podarcis sicula. J Exp Zool 273:247–256

Murai KK, Van Meyel DJ (2007) Neuron glial communication at synapses: insights from vertebrates and invertebrates. Neuroscientist 13:657–666

Murashov AK, Chintalgattu V, Islamov RR, Lever TE, Pak ES, Sierpinski PL, Katwa LC, Van Scott MR (2007) RNAi pathway is functional in peripheral nerve axons. FASEB J 21:656–670

Muslimov IA, Titmus M, Koenig E, Tiedge H (2002) Transport of neuronal BC1 RNA in Mauthner axons. J Neurosci 22:4293–4301

Newman EA (2003) New roles for astrocytes: regulation of synaptic transmission. Trends Neurosci 26:536–542

Perrone Capano C, Giuditta A, Castigli E, Kaplan BB (1987) Occurrence and sequence complexity of polyadenylated RNA in squid axoplasm. J Neurochem 49:698–704

Pevzner LZ (1965) Topochemical aspects of nucleic acid and protein metabolism within the neuron-neuroglia unit of the superior cervical ganglion. J Neurochem 12:993–1002

Pevzner LZ (1971) Topochemical aspects of nucleic acid and protein metabolism within the neuron-neuroglia unit of the spinal cord anterior horn. J Neurochem 18:895–907

Piper M, Holt C (2004) RNA translation in axons. Annu Rev Cell Dev Biol 20:505–523

Por S, Gunning PW, Jeffrey PL, Austin L (1978) Axonal transport of 4S RNA in the chick optic system. Neurochem Res 3:411–422

Puthanveettil SV, Monje FJ, Miniaci MC, Choi YB, Karl KA, Khandros E, Gawinowicz MA, Sheetz MP, Kandel ER (2008) A new component in synaptic plasticity: upregulation of Kinesin in the neurons of the gill withdrawal reflex. Cell 135:960–973

Rapallino MV, Cupello A, Giuditta A (1988) Axoplasmic RNA species synthesized in the isolated squid giant axon. Neurochem Res 13:625–631

Reichenbach A, Pannicke T (2008) Neuroscience. A new glance at glia. Science 2008 322:693–694

Rustom A, Saffrich R, Markovic I, Walther P, Gerdes HH (2004) Nanotubular highways for intercellular organelle transport. Science 303:1007–1010

Schacher S, Wu F, Panyko JD, Sun Z-Y, Wang D (1999) Expression and branch-specific export of mRNA are regulated by synapse formation and interaction with specific postsynaptic targets. J Neurosci 19:6338–6347

Si K, Lindquist S, Kandel ER (2003a) A neuronal isoform of the *Aplysia* CPEB has prion-like properties. Cell 115:879–891

Si K, Giustetto M, Etkin A, Hsu R, Janisiewicz AM, Miniaci MC, Kim JH, Zhu H, Kandel ER (2003b) A neuronal isoform of CPEB regulates local protein synthesis and stabilizes synapse-specific long-term facilitation in *Aplysia*. Cell 115:893–904

Singer M, Green M (1968) Autoradiographic studies of uridine incorporation in peripheral nerve of the newt, *Triturus*. J Morphol 124:321–344

Slagel DE, Hartman HA, Edström JE (1966) The effect of iminodiproprionitrile on the ribonucleic acid content and composition of mesencephalic V cells, anterior horn cells, glial cells, and axonal balloons. J Neuropathol Exp Neurol 25:244–253

Smith C, Kelly G (1988) Paradoxical sleep deprivation applied two days after end of training retards learning. Physiol Behav 43:213–216

Sotelo-Silveira JR, Calliari A, Kun A, Koenig E, Sotelo JR (2006) RNA trafficking in axons. Traffic 7:508–515

Spencer GE, Syed NI, van Kesteren E, Lucowiak K, Geraerts WPM, van Minnen J (2000) Synthesis and functional integration of a neurotransmitter receptor in isolated invertebrate axons. J Neurobiol 44:72–81

Steward O, Banker GA (1992) Getting the message from the gene to the synapse: sorting and intracellular transport of RNA in neurons. Trends Neurosci 15:180–186

Thoenen H, Mueller RA, Axelrod J (1970) Phase difference in the induction of tyrosine hydroxylase in cell body and nerve terminals of sympathetic neurones. Proc Natl Acad Sci USA 65:58–62

Tobias GS, Koenig E (1975) Influence of nerve cell body and neurolemma cell on local axonal protein synthesis following neurotomy. Exp Neurol 49:235–245

Valadi H, Ekström K, Bossios A, Sjöstrand M, Lee JJ, Lötvall JO (2007) Exosome-mediated transfer of mRNAs and microRNAs is a novel mechanism of genetic exchange between cells. Nat Cell Biol 9:654–659

van Minnen J (1994) RNA in the axonal domain: a new dimension in neuronal functioning? Histochem J 26:377–391 van Minnen J, Bergman JJ, van Kesteren ER, Smit AB, Geraerts WPM, Lukowiak K, Hasan SU, Syed NI (1997) De novo protein synthesis in isolated axons of identified neurons. Neuroscience 80:1–7

Verity MA, Brown WJ, Cheung MK (1979) On the mechanism of ouabain inhibition of synaptosome protein synthesis. J Neurochem 32:1295–1301

Vincent JP, Magee T (2002) Argosomes: membrane fragments on the run. Trends Cell Biol 12:57–60

Volterra A, Meldolesi J (2005) Astrocytes, from brain glue to communication elements: the revolution continues. Nature Rev Neurosci 6:626–640

Vuppalanchi et al. (2009) Results Probl Cell Differ

Watson WE (1974) Physiology of neuroglia. Physiol Rev 54:245–271

Whittaker VP (1993) Thirty years of synaptosome research. J Neurocytol 22:735–742

Willis DE, Li KW, Zheng J-Q, Chang JH, Smit A, Kelly T, Merianda TT, Sylvester J, van Minnen J, Twiss JL (2005) Differential transport and local translation of cytoskeletal, injury-response, and neurogeneration protein mRNAs in axons. J Neurosci 26:778–791

Willis DE, Niekerk EA, Sasaki Y, Mesngon M, Merianda TT, Williams GG, Kendall M, Smith DS, Bassell GJ, Twiss JL (2007) Extracellular stimuli specifically regulate localized levels of individual neuronal mRNAs. J Cell Biol 178:965–980

Local Translation and mRNA Trafficking in Axon Pathfinding

Byung C. Yoon, Krishna H. Zivraj, and Christine E. Holt

Abstract Axons and their growth cones are specialized neuronal sub-compartments that possess translation machinery and have distinct messenger RNAs (mRNAs). Several classes of mRNAs have been identified using candidate-based, as well as unbiased genome-wide-based approaches. Axonal mRNA localization serves to regulate spatially the protein synthesis; thereby, providing axons with a high degree of functional autonomy from the soma during axon pathfinding. Importantly, *de novo* protein synthesis in navigating axonal growth cones is necessary for chemotropic responses to various axon guidance cues. This chapter discusses the molecular components involved in regulating axonal mRNA trafficking, targeting, and translation, and focuses on RNA binding proteins (RNBPs) and microRNAs. The functional significance of local mRNA translation in the directional response of growth cones to a gradient is highlighted along with the downstream signaling events that mediate local protein synthesis. The view that emerges is that local translation is tightly coupled to extracellular cues, enabling growth cones to respond to new signals with exquisite adaptability and spatiotemporal control.

1 Introduction

During development, axons often navigate significant distances to their synaptic targets and are guided by multiple molecular cues, including the netrins, slits, semaphorins, and ephrins (Dickson 2002). The growth cone at the tip of the axon senses the guidance cues and steers towards or away from them (Fig. 1a, b). Given the extreme distance that can separate growth cones from their cell bodies (sometimes

---

B.C. Yoon, K.H. Zivraj, and C.E. Holt (✉)
Department of Physiology, Development and Neuroscience, University of Cambridge, Downing Street, Cambridge, CB2 3DY, UK
e-mail: ceh@mole.bio.cam.ac.uk

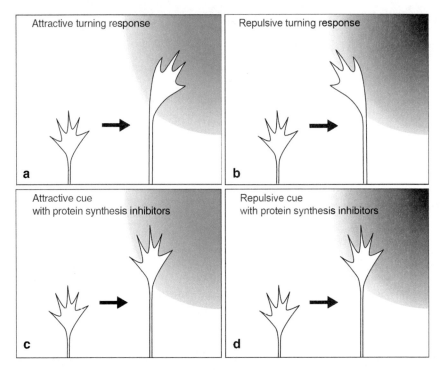

Fig. 1 Growth cone steering in response to guidance cues in the absence or presence of protein synthesis inhibitors. (**a**) Growth cone turns toward the gradient of an attractive cue. (**b**) Growth cone turns away from the gradient of a repulsive cue. (**c**) Attractive turning response is abolished when local protein synthesis is inhibited. (**d**) Repulsive turning response is also abolished when local protein synthesis is inhibited

centimeters), axon pathfinding presents a unique challenge to neurons in ensuring that growth cones respond properly and rapidly to complex extracellular stimuli encountered along the way.

Classically, it has been thought that the proteins required for directional steering originate in cell bodies, where they are synthesized and subsequently transported to growth cones to mediate responses to extracellular stimuli (for review, see Koenig and Giuditta 1999). However, a growing body of evidence challenges this view. For instance, the isolated growth cones of *Xenopus laevis* retinal ganglion cells (RGCs) not only survive for up to 3 h when detached from their cell bodies *in vitro* and *in vivo*, but also migrate and respond to guidance cues (Campbell and Holt 2001; Harris et al. 1987). Since chemotropic responses require precise regulation of protein levels and activity, the question arises as to how isolated growth cones are able to sustain their responses without contributions from their cell bodies. Recent evidence suggests that this functional autonomy seen in axonal growth cones is mediated, at least in part, by local protein synthesis. According to this view, subsets of neuronal mRNAs are trafficked into axons and growth cones, where they are locally translated independent of cell bodies. This chapter will focus on the role of local protein synthesis in axon pathfinding during development as well as the regulators involved in this process.

2 Axonal mRNA Localization and Translation: Historical Perspectives

Cytoplasmic RNA localization is an evolutionarily conserved mechanism that contributes to spatially restricted protein synthesis. The first hint of RNA localization in neurons dates back to the 1960s when ribosomes associated with endoplasmic reticulum (ER), together known as "ergastoplasm," were detected near dendritic synapses of spinal cord neurons in monkeys by electron microscopy (EM) (Bodian 1965). This anatomical finding suggested that these neurons employ protein synthesis of localized RNAs for their synaptic function. Several years later, the concept of RNA localization was established in various other cell types ranging from unicellular organisms such as yeasts to oocytes and embryos of *Drosophila* and *Xenopus* in addition to other vertebrate and invertebrate neurons (Palacios and St. Johnston 2001).

RNA localization is particularly relevant in neurons where the polarity of cell processes (dendrites and axons) is important for structure and function. Specific targeting of various molecules to dendrites and axons establishes subcellular compartmentalization within neurons. In addition to the well-known concept of targeting proteins, the identification of polyribosomes at the base of dendritic spines in the 1980s indicated that synapses have close access to protein synthesis machinery (Steward and Levy 1982). The concept of dendrites synthesizing new proteins for synaptic function was not only consistent with Bodian's early observations, but also triggered a whole new area of research for neuroscientists. Shortly following this observation, several mRNAs were identified in dendrites, and the molecular mechanisms underlying dendritic mRNA localization and translation became a subject of intense interest. Local translation of mRNAs in dendrites is now relatively well characterized and is critically important for synaptic plasticity and memory (Bramham and Wells 2007).

While postsynaptic dendritic mRNA translation enjoyed the limelight, the concept of *de novo* protein synthesis in presynaptic axons remained controversial. Although EM studies in the embryonic rabbit spinal cord axons revealed the presence of polysomes in axons (Tennyson 1970; Yamada et al. 1971; Zelenà 1970), the failure to detect biochemically the protein translation machinery in axons hampered further investigation (Alvarez et al. 2000). However, the notion of axonal localization and translation of mRNAs gathered momentum with the identification of mRNAs and ribosomes in vertebrate axons (Koenig and Giuditta 1999; Koenig and Martin 1996; Koenig et al. 2000), along with the evidence for local protein synthesis of cytoskeletal proteins such as actin, tubulin, and neurofilament in both vertebrate and invertebrate axons (Bassell et al. 1998; Eng et al. 1999; Koenig and Giuditta 1999; Olink-Coux and Hollenbeck 1996).

The first direct evidence for axonal protein synthesis came from experiments in the 1960s with the demonstration of acetylcholinesterase (AChE) resynthesis in axons of cat cholinergic neurons, in which this particular enzyme was irreversibily inactivated (Koenig and Koelle 1960). Axonal protein synthesis was subsequently shown in rabbit spinal root neurons by metabolic labeling experiments wherein neurons were incubated with [^{3}H]leucine, which was incorporated into newly synthesized proteins of isolated axons (Koenig 1967). Similar metabolic labeling experiments were done using the squid giant axon and Mauthner neurons of the carpfish (*Carassius carassius*; for review, see Giuditta et al. 2008). Moreover, isolated axons of goldfish RGCs and newborn rat sympathetic neurons are capable of synthesizing actin and β-tubulin (Koenig and Adams 1982; Eng et al. 1999). In fact, it is estimated that the axonal protein synthesis capacity of chick sympathetic axons accounts for 5% of the total protein synthesis occurring in the cells (Lee and Hollenbeck 2003). Moreover, local synthesis of a reporter protein of the receptor EphA2, and its expression at the cell surface, has been demonstrated in chick spinal cord axons, indicating that axons and their growth cones can synthesize new proteins and export them to the cell surface (Brittis et al. 2002). This finding is also supported by recent evidence showing that active ER and Golgi components in axons are required for protein synthesis and secretion, and that axons can target locally synthesized proteins into secretory pathways (Merianda et al. 2009).

Collectively, these findings raise several questions regarding mRNA localization within axons and growth cones: What are the mRNAs that are targeted to axons and growth cones? How are they selectively transported into the axons from the cell body? Does mRNA localization and translation in axons and growth cones serve a functional role in axon guidance? These questions will be discussed in the following sections.

3 Identity of Axonal mRNAs

Specific mRNAs were first identified in invertebrate axons using biochemical fractionation and metabolic labeling in the 1960s. These studies revealed the presence of ribosomal RNAs, transfer RNAs, translation initiation and elongation factor mRNAs, and mRNAs encoding three cytoskeletal elements: actin, tubulin, and neurofilament (Giuditta et al. 2008; Koenig and Giuditta 1999). In vertebrates, similar studies using rat spinal ventral roots, components of the peripheral nervous system (PNS), identified the same three cytoskeletal protein mRNAs as the major newly synthesized proteins among several other smaller molecular weight proteins that were not identified (Giuditta et al. 2008). Some of the first mRNAs identified in vertebrate central nervous system (CNS) axons were the peptide hormones oxytocin (Jirikowski et al. 1990) and vasopressin (Trembleau et al. 1996). These were localized using *in situ* hybridization (ISH) and EM studies on neurons of the rat hypothalamus. Other vertebrate CNS axonal mRNAs include cytoskeletal proteins; namely, β-actin (Bassell et al. 1998; Leung et al. 2006; Olink-Coux and Hollenbeck

1996), β-tubulin (Eng et al. 1999), microtubule-associated protein 1b (MAP1b) (Antar et al. 2006), tau (Litman et al. 1993), neurofilament (Sotelo-Silveira et al. 2000; Weiner et al. 1996), and actin depolymerizing factor ADF/cofilin (Lee and Hollenbeck 2003). mRNAs encoding other functional proteins include cell signaling molecules such as RhoA (Wu et al. 2005), transcription factors like CREB (Cox et al. 2008), and transmembrane receptors like EphB2 (Brittis et al. 2002), the κ-opioid receptor (Bi et al. 2006), capsaicin receptor VR1 (Tohda et al. 2001), and the olfactory marker protein (Wensley et al. 1995).

Generation of pure axonal preparations followed by genome-wide analysis allows for the identification of axonally localized mRNA transcripts in an unbiased manner. Proteomic analysis and reverse-transcription polymerase chain reaction (RT-PCR) of injury-conditioned or regenerating dorsal root ganglion cell (DRG) axons have revealed the localization of several different mRNAs in these axons. The mRNAs identified include many cytoskeletal encoding transcripts, and also several mRNAs encoding heat-shock proteins, endoplasmic reticulum (ER) proteins, such as calreticulin, antioxidant proteins like peroxiredoxin 1 and 6, and metabolic proteins such as phosphoglycerate kinase 1 (PGK1) (Willis et al. 2005). Moreover, microarray analysis of injury-conditioned DRG axons expanded the list of axonal mRNAs to over 200 transcripts, revealing many previously unidentified mRNAs. These include several transmembrane proteins and components of the translational machinery such as ribosomal proteins (Willis et al. 2007). This suggests that axons contain a diverse repertoire of mRNAs that can be potentially translated similar to dendrites. Increasingly sensitive methods of mRNA isolation and new sequencing technologies will doubtless reveal further complexity in the mRNA content and localization in axons.

4 Molecular Mechanism of Axonal mRNA Transport

The presence of mRNAs in axons raises the question of how they are transported. Messenger RNA localization in axons, as in nonneuronal cells and in dendrites (Kindler et al. 2005), involves *cis*-acting sequence elements within a given transcript that are recognized by specific *trans*-acting factors or RNBPs. mRNAs bound to RNBPs along with other proteins, such as motor proteins, together constitute the ribonucleoprotein particles (RNPs) or RNA granules. These RNP complexes undergo dynamic transitions ranging from stress granules, P-bodies, transport particles and microRNA RNPs (miRNPS) or RNA interference silencing complexes (RISCs) (Sossin and DesGroseillers 2006). Importantly, the mRNA in the RNA granule is usually translationally silent during transport (Krichevsky and Kosik 2001; Fig. 2). Translational repression is regulated by RNBPs themselves or by miRNAs, a class of small noncoding RNAs in axons. Although, much of our current understanding of RNPs comes from dendrites, we will consider the role of each of these molecular components in axonal mRNA targeting and translation, and draw parallels with dendritic mRNA targeting.

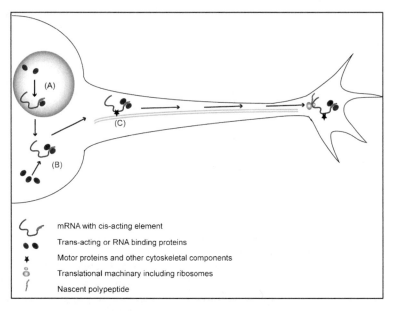

Fig. 2 Axonal mRNA localization in neurons. (a) Recognition of mRNA *cis*-acting elements (*orange*) by *trans*-acting factors (*blue ovals*) in the nucleus forming RNP complexes. (b) Nuclear export of RNP particles into the soma and recruitment of additional *trans*-acting factors (*red ovals*) into the complex. (c). Transport of remodeled RNP or RNA granule along the cytoskeleton (*light blue lines*) with the help of motor proteins (*black star*) to the axon/growth cone. Translational machinery including ribosomes (*grey ovals*) attach to the RNA granule to initiate protein synthesis. Ribosomes may even attach to the RNA granule earlier during transport but begin translation only after reaching its final destination

4.1 Cis-Acting Factors

Cis-acting elements, present in the 3′ or 5′ untranslated region (UTR) of mRNAs, exist in the form of specific nucleotide sequences, or in the form of secondary or tertiary stem-loop structures (Bassell and Kelic 2004). The most well-characterized *cis*-acting element is the 54-nucleotide (nt) sequence in the 3′UTR of β-actin mRNA, referred to as the "zipcode" (Kislauskis et al. 1994; Zhang et al. 2001). Mutational analysis identified this 54-nt zipcode as the minimum *cis*-acting element essential and sufficient to target β-actin mRNA into growth cones of chick forebrain axons (Eom et al. 2003). Antisense oligonucleotides to this zipcode, as well as mutations disrupting its secondary structure, abolish the localization of β-actin mRNA to growth cones (Zhang et al. 2001). Similarly, the *cis*-acting axonal localization element of tau mRNA, which encodes a microtubule associated protein that is exclusively localized to the axonal compartment, is a distinct 240-nt AU-rich sequence in tau mRNA 3'UTR (Aronov et al. 1999). Thus, axonal mRNA *cis*-acting elements are complex and variable and, on the basis of primary sequence alone, cannot be used to predict reliably whether a particular mRNA will be targeted to the axon.

4.2 Trans-Acting Factors, or RNBPs: Dual Function in mRNA Targeting and Translational Regulation

Regulation of mRNA localization is mediated by the recognition of *cis*-acting elements by *trans*-acting RNBPs. The most well-studied *trans*-acting proteins contain either RNA recognition motifs (RRMs) and/or hnRNP K homology (KH) domains (Bassell and Kelic 2004). One example is the zipcode-binding protein-1 (ZBP1), which binds to β-actin mRNA's zipcode and represses β-actin translation (Ross et al. 1997). ZBP1 itself is regulated by src kinase, which phosphorylates ZBP1 and disrupts its binding to β-actin mRNA (Huttelmaier et al. 2005). Fragile-X mental retardation protein (FMRP), another RNA-binding protein, interacts with MAP1b mRNA in axons and dendrites (Antar et al. 2006). FMRP acts as a translational repressor in dendrites (Zalfa et al. 2006) and is required for filopodia formation and motility in axonal growth cones (Antar et al. 2006). Similarly, the cytoplasmic polyadenylation element (CPE)-binding protein 1, CPEB1, binds to the CPE sequence (consensus UUUUUAU) in mRNAs and regulates cytoplasmic polyadenylation of transcripts in oocytes and dendrites (Huang et al. 2003; Richter 1999). Control of polyA-tail length is a conserved mechanism for mRNA-specific translational regulation, and it will be interesting to know whether it is involved in regulating translation in axons. In general, RNBPs are involved in multiple aspects of mRNA localization in axons (e.g., trafficking, targeting, and anchoring) and also help to "silence" translation during transport.

4.3 MicroRNA-Mediated Translational Repression

MicroRNAs and RNA interference (RNAi) are recently described players in the regulation of mRNA translation in neurons. Several miRNAs identified in mammalian neurons are developmentally regulated and are also associated with polyribosomes (Kim et al. 2004; Kye et al. 2007). Moreover, FMRP interacts with miRNAs and with miRNA machinery, including Dicer and the mammalian orthologue of Argonaute (Jin et al. 2004). Functional RNA interference (RNAi) machinery is also present in axons (Hengst et al. 2006), and in fact, the 3'UTR of RhoA mRNA contains putative miRNA-binding sequences (Wu et al. 2005). MicroRNAs regulate translation in dendrites; for example, miR-134 inhibits the translation of Lim kinase 1 (LIMK1), which is involved in regulating actin cytoskeletal dynamics in dendrites by suppressing cofilin activity (Schratt et al. 2006). This in turn alters dendritic spine morphology, although it remains unknown whether this particular miRNA plays a role in regulating axonal mRNA translation. Recently, miR-338, the first axonal miRNA identified, was found in axons of sympathetic neurons. miR-338 targets cytochrome c oxidase IV mRNA in sympathetic axons and controls the local synthesis of cytochrome c oxidase IV; thereby, modulating energy metabolism in these axons in a soma-independent mechanism (Aschrafi et al. 2008). These findings suggest that miRNAs play a distinct role in regulating axonal mRNA translation, and future studies may uncover a role in axon guidance.

4.4 RNA Granule Formation

As soon as mRNAs are generated in the nucleus, they associate with several RNBPs involved in mRNA processing and nuclear export. Evidence for the formation of RNP complexes in the nucleus comes from *Drosophila* and *Xenopus* oocytes (Hachet and Ephrussi 2004; Kress et al. 2004). RNBPs regulating cytoplasmic RNA localization are also localized to the nucleus, implying that these proteins may serve as nucleocytoplasmic shuttling proteins that regulate nuclear and cytoplasmic mRNA targeting (Kindler et al. 2005). For example, zipcode binding protein 2 (ZBP2), also known as KSRP, MARTA1, or FBP2, is predominantly a nuclear protein (Gu et al. 2002; Min et al. 1997; Rehbein et al. 2002). It associates with β-actin mRNA in the nucleus to form the initial R NP and shuttles out into the cytoplasm. Additional RNA binding proteins like ZBP1 are required in the cytoplasm, while ZBP2 shuttles back into the nucleus (Gu et al. 2002). Thus, an RNP formed in the nucleus is commonly remodeled with additional transport factors once outside in the cytoplasm, resulting in large RNP complexes or RNA granules (Gu et al. 2002; Kindler et al. 2005). Biochemical characterization of these granules from neuronal extracts has revealed the presence of many mRNAs, RNBPs, ribosomal proteins, proteins constituting the translational machinery, and motor proteins that move these granules along the cytoskeleton (Kanai et al. 2004). Hence, RNA granules are highly dynamic structures with multiple proteins orchestrating the targeting of a particular mRNA right from the time it is transcribed in the nucleus until its final destination in the cytoplasm.

4.5 Cytoskeletal Elements and Motor Proteins

RNA granules need to be actively transported along the cytoskeleton to reach the target site. Live-imaging studies of RNA granules in neurons show that these granules move with an average speed of 0.1–0.2 $\mu m\ s^{-1}$ (Knowles et al. 1996). While microfilaments are used to localize β-actin RNA along with ZBP1 in fibroblasts, the same RNP complex predominantly travels along microtubules in neuronal processes (Bassell et al. 1998). Microtubules are polarized polymers arranged in axons with the plus-ends oriented away from the cell body (Baas et al. 1988). Moreover, microtubules are required for stimulus-induced changes in axonal localization of certain mRNAs (Willis et al. 2007). Although the motor proteins regulating mRNA transport in axons have not been studied, it is likely that as in neuronal dendrites, mRNAs are transported by kinesin and dynein molecular motor proteins, which are components of RNP particles (Hirokawa 2006). Dendritic CPEB granules, for example, contain both kinesin and dynein motor proteins; thereby, allowing bidirectional movement along microtubules (Huang et al. 2003). While kinesins are predominantly "plus-end" motor proteins carrying their cargo in the anterograde direction, axonal retrograde transport is mediated by the "minus-end" directed dynein motor proteins (Hirokawa and Takemura 2005).

In addition to microtubules, axonal growth cones have a fine meshwork of actin microfilaments in their peripheral domain. Several RNBPs are detected in developing growth cones such as ZBP1 or Vg1RBP in *Xenopus*, and FMRP (Antar et al. 2006; Leung et al. 2006). Cytochalasin D treatment abolishes the movement of Vg1RBP granules in RGC axons and growth cones, indicating the essential role of actin microfilaments (Leung et al. 2006). Moreover, Vg1RBP co-localizes with β-actin mRNA in retinal growth cones, and netrin-1 stimulation induces the directed movement of Vg1RBP granules into filopodia (Leung et al. 2006). Hence, microtubules, as well as microfilaments, along with various molecular motors facilitate the transport of RNA granules in axons.

5 Local Protein Synthesis and Cue-Induced Chemotropic Guidance

Axonal growth cones navigate accurately over long distances through complex microenvironments. This directed migration or pathfinding requires the transduction of extracellular stimuli in the growth cone. A key question raised by the findings is: does mRNA localization and translation have any functional role in the directed migration of axons? Local protein synthesis could contribute to growth cone autonomy in responding to guidance cues, since it would allow a rapid regulation of proteins without requiring transport to and from cell bodies.

One of the earliest studies demonstrating that local protein synthesis plays a role in the cue-induced chemotropic responses of growth cones was done using developing RGCs of *Xenopus* (Campbell and Holt 2001). Global application of the guidance cues Semaphorin3A (Sema3A) or netrin-1 elicited rapid (within 10 min) protein synthesis in isolated axons, as measured by the incorporation of [^{3}H]leucine. The guidance cue stimulation also resulted in rapid phosphorylation of translation initiation factor (eIF-4E) and initiation factor binding protein (eIF-4EBP1), which mark translation initiation. This translation pathway was shown to be functionally significant, as inhibition of protein synthesis with anisomycin or cycloheximide abolished the directional turning and collapse responses to

Sema3A and the turning responses to netrin-1 in growth cones (Fig. 1c, d). The same results were obtained when the axon was cut, separating the growth cone from its soma, demonstrating that the translation was local (Campbell and Holt 2001). Subsequent studies have shown that other guidance cues, such as Engrailed-2, Slit-2, brain-derived neurotrophic factor (BDNF), and pituitary adenylate cyclase-activating polypeptide (PACAP), require local protein synthesis (Brunet et al. 2005; Guirland et al. 2003; Piper et al. 2006; Yao et al. 2006). Interestingly, 1-α-Lysophosphatidic acid (LPA)-induced growth cone collapse does not require local translation, neither ephrinB nor EphB, indicating that local protein synthesis is a differentially regulated mechanism employed by specific cues (Campbell and Holt 2001; Mann et al. 2003). These findings suggest that specific guidance cues elicit rapid, local protein synthesis in growth cones, and that this local translation is required for the chemotropic response *in vitro*.

6 Differential Translation of mRNAs

The finding that local protein synthesis is required for a growth cone's response to guidance cues raises several questions. For instance, what are the proteins that are synthesized in response to guidance cues? Are different mRNAs translated in response to different guidance cues? For chemotropic steering, growth cones need to sense a minute gradient of molecular cues across their width (Rosoff et al. 2004). Does local translation play a role in the detection of the chemotropic gradient? A recently proposed model, termed the "differential translation" model, suggests that different cues stimulate the translation of a specific subset of proteins, and that guidance cue gradients do so asymmetrically, resulting in growth cone steering (Fig. 3 and Table 1; Lin and Holt 2008).

6.1 Attractive Cues

Two recent studies support the idea of differential translation (Leung et al. 2006; Yao et al. 2006). Using immunostaining, Leung et al. (2006) showed that netrin-1, under attractive conditions, elicits a translation-dependent increase in β-actin levels locally within *Xenopus* RGC growth cones. The authors confirmed the immunostaining results using a reporter construct with the 3'-UTR of β-actin fused to the Kaede protein. Kaede is a green fluorescent protein that can be irreversibly converted to red fluorescence upon ultraviolet (UV) or violet irradiation (350–400 nm) (Ando et al. 2002). By photoconverting the pre-existing green Kaede to red and measuring the new green signal, the *de novo* Kaede signal can be quantified (Leung and Holt 2008). Upon netrin-1 application, the intensity of the green signal increased markedly, indicating that netrin-1, indeed, induces β-actin translation. In addition, preventing β-actin translation with antisense morpholino oligonucleotides

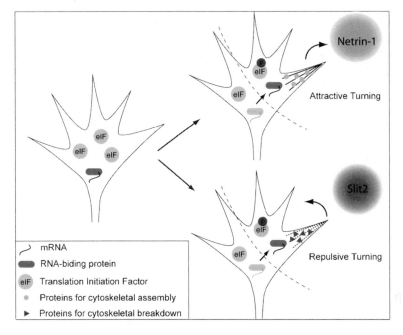

Fig. 3 "Differential translational" model describing local protein synthesis and growth cone steering. A gradient of an attractive cue (e.g., netrin-1) elicits asymmetric phosphorylation of translation initiation factor and subsequent asymmetric synthesis of proteins that build up the cytoskeleton closer to the gradient. A gradient of a repulsive cue (e.g., slit2) elicits asymmetric synthesis of proteins that break down the cytoskeleton close to the gradient, resulting in growth cone steering away from the cue. Adapted from Lin and Holt (2008)

Table 1 Differential translation of axonal mRNAs in response to distinct cues

External cue	mRNA level	Reference
Netrin-1	β-Actin ↑	Leung et al. (2006)
	Enolase ↔	
Sema3A	RhoA ↑	Wu et al. (2005)
Slit2	Cofilin ↑	Piper et al. (2006)
	β-Actin ↔	
BDNF	β-Actin ↑	Yao et al. (2006)
Nerve growth factor (NGF)	cAMP-responsive element (CRE)-binding protein (CREB) ↑	Cox et al. (2008)
KCl depolarization	κ-Opioid receptor ↑	Bi et al. (2006)

targeted against the ATG site of β-actin mRNA disrupted the growth cone's steering in response to netrin-1. This indicates that β-actin translation is required for attractive turning to netrin-1. Interestingly, when growth cones are stimulated with a gradient of netrin-1, 4EBP phosphorylation and β-actin levels increase asymmetrically on the near-side of the growth cone, closest to the netrin-1 source. Vg1RBP was shown to interact directly with β-actin mRNA, raising the possibility that this interaction regulates asymmetrical translation of β-actin (Leung et al. 2006).

Similarly, a gradient of BDNF under attractive conditions elicits an asymmetric regulation of the β-actin level in growth cones of embryonic *Xenopus* spinal neurons (Yao et al. 2006). Fluorescent *in situ* hybridization (FISH) revealed that the intensity of the β-actin mRNA signal increases with BDNF stimulation. Furthermore, the co-localization of β-actin mRNA with ZBP1 increases with BDNF, suggesting that BDNF-induced β-actin mRNA localization occurs via ZBP1 binding. In a BDNF gradient, co-localization of β-actin mRNA and ZBP1, as well as β-actin protein expression increased rapidly (5 min) on the near-side of the growth cone closest to the BDNF source. A similar asymmetric rise was observed with the phosphorylated (active) form of src kinase. Interfering with the β-actin mRNA binding to ZBP1, using antisense oligonucleotides, abolished the growth cone's turning response to BDNF, signifying that the asymmetric β-actin translation is critical for BDNF-induced steering (Yao et al. 2006).

Together, these studies demonstrate that both netrin and BDNF-induced steering under attractive conditions is in part mediated by asymmetric translation of β-actin. Newly synthesized β-actin may mediate steering by stimulating actin polymerization that helps the growth cone to extend closer to the source of an attractive cue. Vg1RBP/ZBP1 may also be involved in this process by transporting β-actin mRNA and regulating its local translation (Lin and Holt 2008). Indeed, extracellular signals exert significant and selective effects on the transport of mRNAs in axons (Willis et al. 2007; Zhang et al. 1999).

6.2 Repulsive Cues

In contrast to attractive conditions, netrin-1 and BDNF under repulsive conditions do not elicit translation of β-actin (Leung et al. 2006; Yao et al. 2006). Furthermore, repulsive turning induced by BDNF was associated with an asymmetric decrease of β-actin on the near-side of the growth cone, closest to the BDNF source (Yao et al. 2006). Turning by netrin-1 under repulsive conditions is also not affected by morpholino-based blockade of β-actin translation (Leung et al. 2006). These findings indicate that attractive cues and repulsive cues elicit the translation of distinct subsets of mRNAs (Table 1), and that this helps to direct axon pathfinding.

Repulsive cues such as Sema3A and Slit-2 elicit steering away of growth cones from the source of the cue or collapse of growth cones, processes that require breakdown of the cytoskeleton. Indeed, the proteins identified as being synthesized in response to repulsive cues are mainly involved in cytoskeletal disassembly. For instance, Wu et al. (2005) showed that Sema3A stimulates the intra-axonal translation of RhoA mRNA, and that this local translation is necessary for growth cone collapse in embryonic rat DRG explant cultures (Wu et al. 2005). RhoA is a member of the Rho small guanosine triphosphatase (GTPase) protein known to regulate actin dynamics. In particular, RhoA-mediated signaling triggers growth cone collapse and neurite retraction by upregulating myosin II activity (for review, see Gallo and Letourneau 2004). To examine the dynamics of RhoA translation within axons and growth cones, Wu et al. expressed myristoylated-destabilized EGFP, regulated by RhoA 3′UTR. Myristoylation limits diffusion of the fluorescent protein, and destabilization of EGFP markedly reduces its half-life, facilitating visualization of *de novo* protein synthesis by reducing diffusion and persistence of existing proteins. When Sema3A was added, EGFP puncta significantly increased within 60 min, suggesting that Sema3A induces local translation of this RhoA reporter mRNA. Furthermore, intra-axonal knockdown of RhoA transcripts, using small interfering RNA (siRNA) directed against the 5′UTR of RhoA, significantly reduced Sema3A-mediated growth cone collapse (Hengst et al. 2006). Expression in these neurons of exogenous RhoA restored Sema3A-mediated growth cone collapse, indicating that RhoA translation is necessary for Sema3A's effect.

Other studies also identified proteins that are locally translated by repulsive cue stimulation. For instance, Slit elicits a rapid translation-dependent increase in actin-depolymerizing factor (ADF)/cofilin-1 (Piper et al. 2006). A selective knockdown of β-thymosin mRNA within axons was also shown to enhance neurite outgrowth in *Lymnea* neurons (van Kesteren et al. 2006). β-Thymosin sequesters monomeric actin, thereby preventing actin polymerization (For review, see Huff et al. 2001). This finding suggests that locally translated β-thymosin may inhibit neurite outgrowth by preventing actin polymerization. However, it remains to be determined whether local translation of ADF/cofilin-1 and β-thymosin is necessary for growth cone steering or collapse.

All considered, these findings suggest that attractants and repellents stimulate the synthesis of distinct subsets of proteins. Furthermore, guidance cue gradients appear to induce spatially regulated mRNA translation. It has yet to be elucidated whether a polarized repulsive cue elicits spatially asymmetric protein synthesis in growth cones like an attractive cue; however, the differential translation model predicts that near-side synthesis of proteins involved in cytoskeletal disassembly would induce directional turning away from the cue (Fig. 3).

7 Modulation of Chemotropic Response

7.1 Adaptation to Cue Stimulation

A growth cone's response to guidance cues like netrin-1 and BDNF involves cycles of desensitization and resensitization, sometimes manifested by a zigzag pattern of migration towards a cue gradient (Ming et al. 2002). Such adaptation to guidance cues is critical for long-term navigation, enabling growth cones to respond to a dynamic range of cue concentrations during their journey (Ming et al. 2002). Growth cones of *Xenopus* spinal neurons have reduced, or desensitized, responses to micropipette-ejected netrin-1 in the presence of bath-applied netrin-1 within

30 min. Interestingly, they eventually recover from desensitization, or are resensitized, to the cue after an additional 90 min of incubation. Experiments with BDNF also showed similar results. When either anisomycin or cycloheximide was added to the isolated growth cones, they no longer exhibited resensitization, indicating that local protein synthesis is critical for the recovery of cue response (Ming et al. 2002).

Desensitization occurs on a rapid timescale in *Xenopus* retinal growth cones, as demonstrated by the finding that brief exposure (2 min) to a sub-threshold level of Sema3A prior to stimulation with full-collapse-inducing levels of Sema3A significantly reduces the collapse response (Piper et al. 2005). When the sub-threshold preincubation time is extended from 2 to 5 min, growth cones resensitize and exhibit full collapse in response to Sema3A. Netrin-1 under repulsive conditions elicits similar responses. Translation inhibitors abolish the resensitization phase of the response to both Sema3A and netrin-1. Interestingly, the desensitization step is not affected by protein synthesis inhibitors, but is sensitive to inhibitors of endocytosis, indicating that protein synthesis and endocytosis together coordinately regulate cue-driven guidance (Piper et al. 2005). Of note, local translation is also involved in sensitization of fast-nociceptive A-fibers from the DRG. This implies that local protein synthesis may have a broader role in fine-tuning the neuron's response to external stimuli (Jimenez-Diaz et al. 2008).

7.2 Intermediate Targets

During development, growth cones encounter multiple intermediate targets before reaching their final destination. One of the best-characterized examples is the midline of CNS. Developing commissural axons of vertebrates, insects, and nematodes, for example, initially are attracted to the midline by netrins. After crossing, however, they are repelled from the midline and project to the contralateral side (for review, see Tessier-Lavigne and Goodman 1996). This process is mediated in part by receptor–receptor "silencing" interactions between the netrin receptor, deleted in colorectal carcinoma (DCC), and the slit receptor Roundabout (Robo) (for review, see Stein and Tessier-Lavigne 2001). However, it is not yet clear how axons are capable of responding to a new set of guidance cues after crossing the midline.

Findings that L1 receptors are expressed exclusively on the contralateral segment of spinal cord commissural axons (Dodd et al. 1988) raise the possibility that new proteins are synthesized after midline crossing. Brittis et al. (2002) examined this issue using chick commissural axons and found that ephrinA receptor, EphA2, is upregulated in the post-midline segment of spinal cord commissural axons. EphA2 transcripts are present in these axons, and the expression of a reporter driven by EphA2 3'UTR is upregulated in growth cones only after they cross the midline. The EphA2 3'UTR contains a CPE sequence, and when CPE mutations were introduced, the increase of the reporter upregulation was markedly reduced (Brittis et al. 2002). Spatial restriction of EphA2 to the post-midline segment of axons may thus be controlled by CPE-mediated polyadenylation and translation of localized mRNAs.

8 Intracellular Signaling Pathways Mediating Local Protein Synthesis

The transduction of extracellular stimuli requires highly regulated pathways involving multiple downstream effectors. Although the signaling processes that convert cue stimulation to local translation are not fully elucidated, evidence so far suggest that different cues are transduced via specific yet inter-related intracellular signaling pathways.

A common theme of the findings so far is that the cue-stimulated signal transduction utilizes pathways similar to those that regulate protein synthesis elsewhere. For instance, extracellular signals such as growth factors modulate protein synthesis via the phosphatidyl inositol-3 kinase (PI3K) and mammalian target of rapamycin (mTOR)-dependent pathway. Both PI3K and mTOR activation leads to phosphorylation of eIF-4EBP1, which in turn leads to the release of eIF-4E and translation initiation (for review, see Gingras et al. 1999). Protein synthesis induced by Sema3A and netrin-1 is inhibited by rapamycin, a pharmacological inhibitor mTOR. Interestingly, a PI3K inhibitor, wortmannin, inhibits netrin-1, but not Sema3A-mediated protein synthesis, indicating that the two guidance cues work via similar but distinct signaling pathways (Campbell and Holt 2003).

Mitogen-activated protein kinases (MAPKs) have also been shown to be involved in the cue-mediated protein synthesis. The MAPK family includes signal transducing enzymes that are involved in many aspects of cellular responses, including cell growth and differentiation (p42/p44 MAPK), as well as stress responses (p38 and c-Jun N-ternimal kinase/stress-activated protein kinase) (Chang and Karin 2001). MAPKs modulate protein translation by regulating MAP kinase-interacting protein kinase-1 (MNK1), which in turn phosphorylates eIF4E (for review, see Gingras et al. 1999). In embryonic *Xenopus* RGC axonal growth cones, netrin-1 and Sema3A activate p42/p44, and inhibition of p42/p44 abolishes cue-induced protein synthesis (Campbell and Holt 2003). The activation of translation initiation regulators (i.e., phosphorylation of eIF-4E, eIF-4EBP1, and Mnk-1) in growth cones is also dependent on p42/p44 and p38 MAPK. Interestingly, p38 is critical for netrin-1, but not for Sema3A-mediated protein synthesis, indicating that the two cues activate distinct pathways upstream of translation initiation (Campbell and Holt 2003). MAPK signaling is also critical for resensitization in *Xenopus* spinal axons (Ming et al. 2002).

It should be noted that these findings represent an incomplete view of the intracellular processes mediating local protein synthesis. Other second messenger molecules such as calcium (Ca^{2+}) are involved in this process, as Ca^{2+}-mediated bidirectional turning of *Xenopus laevis* growth cones is also dependent on protein synthesis (Yao et al. 2006). In addition, specific mRNAs are tightly regulated by various RNBPs as discussed previously. Therefore, the regulation of cue-induced local protein synthesis involves multiple targets and mechanisms, many of which remain to be characterized.

9 Conclusion and Perspectives

mRNA trafficking and local protein synthesis have emerged as a critical mechanism through which axons respond and navigate through a complex environment. Messenger RNAs are targeted to axons and growth cones, and different guidance cues elicit translation of a distinct set of proteins in a spatially restricted manner. The entire process is tightly regulated by multiple mediators that act on mRNA transport and translation, as well as on signal transduction of extracellular stimuli. In this way, a neuron is able to rapidly and faithfully respond to the extracellular environment. Findings so far, however, represent only a snapshot of a highly complex and dynamic process. For instance, there remain many *cis*- and *trans*-acting elements regulating mRNA transport and translation to be identified. Furthermore, more work is needed to determine exactly what proteins are synthesized at different stages of pathfinding, and to identify intracellular signals controlling this protein synthesis. Elucidating these mechanisms will not only help us understand the remarkable journey of an axon to its target, but also help us identify the potential therapeutic targets in diseased or injured conditions, when these carefully balanced processes go awry.

References

Alvarez J, Giuditta A, Koenig E (2000) Protein synthesis in axons and terminals: significance for maintenance, plasticity and regulation of phenotype. With a critique of slow transport theory. *Prog Neurobiol* 62:1–62

Ando R, Hama H, Yamamoto-Hino M, Mizuno H, Miyawaki A (2002) An optical marker based on the UV-induced green-to-red photoconversion of a fluorescent protein. *Proc Natl Acad Sci USA* 99:12651–12656

Antar LN, Li C, Zhang H, Carroll RC, Bassell GJ (2006) Local functions for FMRP in axon growth cone motility and activity-dependent regulation of filopodia and spine synapses. *Mol Cell Neurosci* 32:37–48

Aronov S, Marx R, Ginzburg I (1999) Identification of 3'UTR region implicated in tau mRNA stabilization in neuronal cells. *J Mol Neurosci* 12:131–145

Aschrafi A, Schwechter AD, Mameza MG, Natera-Naranjo O, Gioio AE, Kaplan BB (2008) MicroRNA-338 regulates local cytochrome c oxidase IV mRNA levels and oxidative phosphorylation in the axons of sympathetic neurons. *J Neurosci* 28:12581–12590

Baas PW, Deitch JS, Black MM, Banker GA (1988) Polarity orientation of microtubules in hippocampal neurons: uniformity in the axon and nonuniformity in the dendrite. *Proc Natl Acad Sci USA* 85:8335–8339

Bassell GJ, Kelic S (2004) Binding proteins for mRNA localization and local translation, and their dysfunction in genetic neurological disease. *Curr Opin Neurobiol* 14:574–581

Bassell GJ, Zhang H, Byrd AL, Femino AM, Singer RH, Taneja KL, Lifshitz LM, Herman IM, Kosik KS (1998) Sorting of beta-actin mRNA and protein to neurites and growth cones in culture. *J Neurosci* 18:251–265

Bi J, Tsai NP, Lin YP, Loh HH, Wei LN (2006) Axonal mRNA transport and localized translational regulation of kappa-opioid receptor in primary neurons of dorsal root ganglia. *Proc Natl Acad Sci USA* 103:19919–19924

Bodian D (1965) A suggestive relationship of nerve cell RNA with specific synaptic sites. *Proc Natl Acad Sci USA* 53:418–425

Bramham CR, Wells DG (2007) Dendritic mRNA: transport, translation and function. *Nat Rev Neurosci* 8:776–789

Brittis PA, Lu Q, Flanagan JG (2002) Axonal protein synthesis provides a mechanism for localized regulation at an intermediate target. *Cell* 110:223–235

Brunet I, Weinl C, Piper M, Trembleau A, Volovitch M, Harris W, Prochiantz A, Holt C (2005) The transcription factor Engrailed-2 guides retinal axons. *Nature* 438:94–98

Campbell DS, Holt CE (2001) Chemotropic responses of retinal growth cones mediated by rapid local protein synthesis and degradation. *Neuron* 32:1013–1026

Campbell DS, Holt CE (2003) Apoptotic pathway and MAPKs differentially regulate chemotropic responses of retinal growth cones. *Neuron* 37:939–952

Chang L, Karin M (2001) Mammalian MAP kinase signalling cascades. *Nature* 410:37–40

Cox LJ, Hengst U, Gurskaya NG, Lukyanov KA, Jaffrey SR (2008) Intra-axonal translation and retrograde trafficking of CREB promotes neuronal survival. Nat Cell Biol

Dickson BJ (2002) Molecular mechanisms of axon guidance. *Science* 298:1959–1964

Dodd J, Morton SB, Karagogeos D, Yamamoto M, Jessell TM (1988) Spatial regulation of axonal glycoprotein expression on subsets of embryonic spinal neurons. *Neuron* 1:105–116

Eng H, Lund K, Campenot RB (1999) Synthesis of beta-tubulin, actin, and other proteins in axons of sympathetic neurons in compartmented cultures. *J Neurosci* 19:1–9

Eom T, Antar LN, Singer RH, Bassell GJ (2003) Localization of a beta-actin messenger ribonucleoprotein complex with zipcode-binding protein modulates the density of dendritic filopodia and filopodial synapses. *J Neurosci* 23:10433–10444

Gallo G, Letourneau PC (2004) Regulation of growth cone actin filaments by guidance cues. *J Neurobiol* 58:92–102

Gingras AC, Raught B, Sonenberg N (1999) eIF4 initiation factors: effectors of mRNA recruitment to ribosomes and regulators of translation. *Annu Rev Biochem* 68:913–963

Giuditta A, Chun JT, Eyman M, Cefaliello C, Bruno AP, Crispino M (2008) Local gene expression in axons and nerve endings: the glia-neuron unit. *Physiol Rev* 88:515–555

Gu W, Pan F, Zhang H, Bassell GJ, Singer RH (2002) A predominantly nuclear protein affecting cytoplasmic localization of beta-actin mRNA in fibroblasts and neurons. *J Cell Biol* 156:41–51

Guirland C, Buck KB, Gibney JA, DiCicco-Bloom E, Zheng JQ (2003) Direct cAMP signaling through G-protein-coupled receptors mediates growth cone attraction induced by pituitary adenylate cyclase-activating polypeptide. *J Neurosci* 23:2274–2283

Hachet O, Ephrussi A (2004) Splicing of oskar RNA in the nucleus is coupled to its cytoplasmic localization. *Nature* 428:959–963

Harris WA, Holt CE, Bonhoeffer F (1987) Retinal axons with and without their somata, growing to and arborizing in the tectum of Xenopus embryos: a time-lapse video study of single fibres in vivo. *Development* 101:123–133

Hengst U, Cox LJ, Macosko EZ, Jaffrey SR (2006) Functional and selective RNA interference in developing axons and growth cones. *J Neurosci* 26:5727–5732

Hirokawa N (2006) mRNA transport in dendrites: RNA granules, motors, and tracks. *J Neurosci* 26:7139–7142

Hirokawa N, Takemura R (2005) Molecular motors and mechanisms of directional transport in neurons. *Nat Rev Neurosci* 6:201–214

Huang YS, Carson JH, Barbarese E, Richter JD (2003) Facilitation of dendritic mRNA transport by CPEB. *Genes Dev* 17:638–653

Huff T, Muller CS, Otto AM, Netzker R, Hannappel E (2001) beta-Thymosins, small acidic peptides with multiple functions. *Int J Biochem Cell Biol* 33:205–220

Huttelmaier S, Zenklusen D, Lederer M, Dictenberg J, Lorenz M, Meng X, Bassell GJ, Condeelis J, Singer RH (2005) Spatial regulation of beta-actin translation by Src-dependent phosphorylation of ZBP1. *Nature* 438:512–515

Jimenez-Diaz L, Geranton SM, Passmore GM, Leith JL, Fisher AS, Berliocchi L, Sivasubramaniam AK, Sheasby A, Lumb BM, Hunt SP (2008) Local translation in primary afferent fibers regulates nociception. *PLoS ONE* 3:e1961

Jin P, Zarnescu DC, Ceman S, Nakamoto M, Mowrey J, Jongens TA, Nelson DL, Moses K, Warren ST (2004) Biochemical and genetic interaction between the fragile X mental retardation protein and the microRNA pathway. *Nat Neurosci* 7:113–117

Jirikowski GF, Sanna PP, Bloom FE (1990) mRNA coding for oxytocin is present in axons of the hypothalamo-neurohypophysial tract. *Proc Natl Acad Sci USA* 87:7400–7404

Kanai Y, Dohmae N, Hirokawa N (2004) Kinesin transports RNA: isolation and characterization of an RNA-transporting granule. *Neuron* 43:513–525

Kim J, Krichevsky A, Grad Y, Hayes GD, Kosik KS, Church GM, Ruvkun G (2004) Identification of many microRNAs that copurify with polyribosomes in mammalian neurons. *Proc Natl Acad Sci USA* 101:360–365

Kindler S, Wang H, Richter D, Tiedge H (2005) RNA transport and local control of translation. *Annu Rev Cell Dev Biol* 21:223–245

Kislauskis EH, Zhu X, Singer RH (1994) Sequences responsible for intracellular localization of beta-actin messenger RNA also affect cell phenotype. *J Cell Biol* 127:441–451

Knowles RB, Sabry JH, Martone ME, Deerinck TJ, Ellisman MH, Bassell GJ, Kosik KS (1996) Translocation of RNA granules in living neurons. *J Neurosci* 16:7812–7820

Koenig E (1967) Synthetic mechanisms in the axon. IV. In vitro incorporation of [3H]precursors into axonal protein and RNA. *J Neurochem* 14:437–446

Koenig E, Adams P (1982) Local protein synthesizing activity in axonal fields regenerating in vitro. *J Neurochem* 39:386–400

Koenig E, Giuditta A (1999) Protein-synthesizing machinery in the axon compartment. *Neuroscience* 89:5–15

Koenig E, Koelle GB (1960) Acetylcholinesterase regeneration in peripheral nerve after irreversible inactivation. *Science* 132:1249–1250

Koenig E, Martin R (1996) Cortical plaque-like structures identify ribosome-containing domains in the Mauthner cell axon. *J Neurosci* 16:1400–1411

Koenig E, Martin R, Titmus M, Sotelo-Silveira JR (2000) Cryptic peripheral ribosomal domains distributed intermittently along mammalian myelinated axons. *J Neurosci* 20:8390–8400

Kress TL, Yoon YJ, Mowry KL (2004) Nuclear RNP complex assembly initiates cytoplasmic RNA localization. *J Cell Biol* 165:203–211

Krichevsky AM, Kosik KS (2001) Neuronal RNA granules: a link between RNA localization and stimulation-dependent translation. *Neuron* 32:683–696

Kye MJ, Liu T, Levy SF, Xu NL, Groves BB, Bonneau R, Lao K, Kosik KS (2007) Somatodendritic microRNAs identified by laser capture and multiplex RT-PCR. *Rna* 13:1224–1234

Lee SK, Hollenbeck PJ (2003) Organization and translation of mRNA in sympathetic axons. *J Cell Sci* 116:4467–4478

Leung KM, Holt CE (2008) Live visualization of protein synthesis in axonal growth cones by microinjection of photoconvertible Kaede into Xenopus embryos. *Nat Protoc* 3:1318–1327

Leung KM, van Horck FP, Lin AC, Allison R, Standart N, Holt CE (2006) Asymmetrical beta-actin mRNA translation in growth cones mediates attractive turning to netrin-1. *Nat Neurosci* 9:1247–1256

Lin AC, Holt CE (2008) Function and regulation of local axonal translation. *Curr Opin Neurobiol* 18:60–68

Litman P, Barg J, Rindzoonski L, Ginzburg I (1993) Subcellular localization of tau mRNA in differentiating neuronal cell culture: implications for neuronal polarity. *Neuron* 10:627–638

Mann F, Miranda E, Weinl C, Harmer E, Holt CE (2003) B-type Eph receptors and ephrins induce growth cone collapse through distinct intracellular pathways. *J. Neurobiology* 57:323–336.

Merianda TT, Lin A, Lam J, Vuppalanchi D, Willis DE, Karin N, Holt CE, Twiss JL (2009) A functional equivalent of endoplasmic reticulum and Golgi in axons for secretion of locally synthesized proteins. *Mol Cell Neurosci* 40:128–142

Min H, Turck CW, Nikolic JM, Black DL (1997) A new regulatory protein, KSRP, mediates exon inclusion through an intronic splicing enhancer. *Genes Dev* 11:1023–1036

Ming GL, Wong ST, Henley J, Yuan XB, Song HJ, Spitzer NC, Poo MM (2002) Adaptation in the chemotactic guidance of nerve growth cones. *Nature* 417:411–418

Olink-Coux M, Hollenbeck PJ (1996) Localization and active transport of mRNA in axons of sympathetic neurons in culture. *J Neurosci* 16:1346–1358

Palacios IM, St. Johnston D (2001) Getting the message across: the intracellular localization of mRNAs in higher eukaryotes. *Annu Rev Cell Dev Biol* 17:569–614

Piper M, Anderson R, Dwivedy A, Weinl C, van Horck F, Leung KM, Cogill E, Holt C (2006) Signaling mechanisms underlying Slit2-induced collapse of Xenopus retinal growth cones. *Neuron* 49:215–228

Piper M, Salih S, Weinl C, Holt CE, Harris WA (2005) Endocytosis-dependent desensitization and protein synthesis-dependent resensitization in retinal growth cone adaptation. *Nat Neurosci* 8:179–186

Rehbein M, Wege K, Buck F, Schweizer M, Richter D, Kindler S (2002) Molecular characterization of MARTA1, a protein interacting with the dendritic targeting element of MAP2 mRNAs. *J Neurochem* 82:1039–1046

Richter JD (1999) Cytoplasmic polyadenylation in development and beyond. *Microbiol Mol Biol Rev* 63:446–456

Rosoff WJ, Urbach JS, Esrick MA, McAllister RG, Richards LJ, Goodhill GJ (2004) A new chemotaxis assay shows the extreme sensitivity of axons to molecular gradients. *Nat Neurosci* 7:678–682

Ross AF, Oleynikov Y, Kislauskis EH, Taneja KL, Singer RH (1997) Characterization of a beta-actin mRNA zipcode-binding protein. *Mol Cell Biol* 17:2158–2165

Schratt GM, Tuebing F, Nigh EA, Kane CG, Sabatini ME, Kiebler M, Greenberg ME (2006) A brain-specific microRNA regulates dendritic spine development. *Nature* 439:283–289

Sossin WS, DesGroseillers L (2006) Intracellular trafficking of RNA in neurons. *Traffic* 7:1581–1589

Sotelo-Silveira JR, Calliari A, Kun A, Benech JC, Sanguinetti C, Chalar C, Sotelo JR (2000) Neurofilament mRNAs are present and translated in the normal and severed sciatic nerve. *J Neurosci Res* 62:65–74

Stein E, Tessier-Lavigne M (2001) Hierarchical organization of guidance receptors: silencing of netrin attraction by slit through a Robo/DCC receptor complex. *Science* 291:1928–1938

Steward O, Levy WB (1982) Preferential localization of polyribosomes under the base of dendritic spines in granule cells of the dentate gyrus. *J Neurosci* 2:284–291

Tennyson VM (1970) The fine structure of the axon and growth cone of the dorsal root neuroblast of the rabbit embryo. *J Cell Biol* 44:62–79

Tessier-Lavigne M, Goodman CS (1996) The molecular biology of axon guidance. *Science* 274:1123–1133

Tohda C, Sasaki M, Konemura T, Sasamura T, Itoh M, Kuraishi Y (2001) Axonal transport of VR1 capsaicin receptor mRNA in primary afferents and its participation in inflammation-induced increase in capsaicin sensitivity. *J Neurochem* 76:1628–1635

Trembleau A, Morales M, Bloom FE (1996) Differential compartmentalization of vasopressin messenger RNA and neuropeptide within the rat hypothalamo-neurohypophysial axonal tracts: light and electron microscopic evidence. *Neuroscience* 70:113–125 van Kesteren RE, Carter C, Dissel HM, van Minnen J, Gouwenberg Y, Syed NI, Spencer GE, Smit AB (2006) Local synthesis of actin-binding protein beta-thymosin regulates neurite outgrowth. *J Neurosci* 26:152–157

Weiner OD, Zorn AM, Krieg PA, Bittner GD (1996) Medium weight neurofilament mRNA in goldfish Mauthner axoplasm. *Neurosci Lett* 213:83–86

Wensley CH, Stone DM, Baker H, Kauer JS, Margolis FL, Chikaraishi DM (1995) Olfactory marker protein mRNA is found in axons of olfactory receptor neurons. *J Neurosci* 15:4827–4837

Willis D, Li KW, Zheng JQ, Chang JH, Smit A, Kelly T, Merianda TT, Sylvester J, van Minnen J, Twiss JL (2005) Differential transport and local translation of cytoskeletal, injury-response, and neurodegeneration protein mRNAs in axons. *J Neurosci* 25:778–791

Willis DE, van Niekerk EA, Sasaki Y, Mesngon M, Merianda TT, Williams GG, Kendall M, Smith DS, Bassell GJ, Twiss JL (2007) Extracellular stimuli specifically regulate localized levels of individual neuronal mRNAs. *J Cell Biol* 178:965–980

Wu KY, Hengst U, Cox LJ, Macosko EZ, Jeromin A, Urquhart ER, Jaffrey SR (2005) Local translation of RhoA regulates growth cone collapse. *Nature* 436:1020–1024

Yamada KM, Spooner BS, Wessells NK (1971) Ultrastructure and function of growth cones and axons of cultured nerve cells. *J Cell Biol* 49:614–635

Yao J, Sasaki Y, Wen Z, Bassell GJ, Zheng JQ (2006) An essential role for beta-actin mRNA localization and translation in Ca2+-dependent growth cone guidance. *Nat Neurosci* 9:1265–1273

Zalfa F, Achsel T, Bagni C (2006) mRNPs, polysomes or granules: FMRP in neuronal protein synthesis. *Curr Opin Neurobiol* 16:265–269

Zelenà J (1970) Ribosome-like particles in myelinated axons of the rat. *Brain Res* 24:359–363

Zhang HL, Eom T, Oleynikov Y, Shenoy SM, Liebelt DA, Dictenberg JB, Singer RH, Bassell GJ (2001) Neurotrophin-induced transport of a beta-actin mRNP complex increases beta-actin levels and stimulates growth cone motility. *Neuron* 31:261–275

Zhang HL, Singer RH, Bassell GJ (1999) Neurotrophin regulation of beta-actin mRNA and protein localization within growth cones. *J Cell Biol* 147:59–70

Spinal Muscular Atrophy and a Model for Survival of Motor Neuron Protein Function in Axonal Ribonucleoprotein Complexes

Wilfried Rossoll and Gary J. Bassell

Abstract Spinal muscular atrophy (SMA) is a neurodegenerative disease that results from loss of function of the *SMN1* gene, encoding the ubiquitously expressed survival of motor neuron (SMN) protein, a protein best known for its housekeeping role in the SMN–Gemin multiprotein complex involved in spliceosomal small nuclear ribonucleoprotein (snRNP) assembly. However, numerous studies reveal that SMN has many interaction partners, including mRNA binding proteins and actin regulators, suggesting its diverse role as a molecular chaperone involved in mRNA metabolism. This review focuses on studies suggesting an important role of SMN in regulating the assembly, localization, or stability of axonal messenger ribonucleoprotein (mRNP) complexes. Various animal models for SMA are discussed, and phenotypes described that indicate a predominant function for SMN in neuronal development and synapse formation. These models have begun to be used to test different therapeutic strategies that have the potential to restore SMN function. Further work to elucidate SMN mechanisms within motor neurons and other cell types involved in neuromuscular circuitry hold promise for the potential treatment of SMA.

1 Spinal Muscular Atrophy and Survival of Motor Neuron Protein

Spinal muscular atrophy (SMA) is an inherited autosomal recessive neurodegenerative disease, primarily affecting α-motor neurons of the lower spinal cord, in which proximal muscles are more severely affected then the distal muscles. SMA has an estimated incidence between 1-in-6,000 and 1-in-10,000 births, and about 1 in 35–40 people are genetic carriers (Wirth et al. 2006). SMA is classified clinically by age of onset and highest level of motor function achieved (Wirth et al. 2006; Lunn and

---

W. Rossoll and G.J. Bassell (✉)
Departments of Cell Biology and Neurology, Center for Neurodegenerative Disease,
Emory University School of Medicine, Atlanta, GA, 30322, USA
e-mail: wrossol@emory.edu, gbassel@emory.edu

Wang 2008; Oskoui and Kaufmann 2008). In its most severe form, known as SMA type I or Werdnig-Hoffman Disease (Online *Mendelian Inheritance in Man* (OMIM) #253300), SMA is the most common genetic cause of infant mortality and the second most common lethal, autosomal recessive disease after cystic fibrosis (Melki et al. 1994). Degeneration and death of the anterior horn motor neurons in the brain stem and spinal cord produce weakness in the limb muscles, as well as in muscles involved in swallowing and breathing (Nicole et al. 2002). Children with SMA present with generalized muscle weakness and hypotonia, and either do not acquire or progressively lose the ability to walk, stand, sit, and, eventually, move. Previously, SMA type I patients were predicted to die before the age of two, but in recent years, more proactive clinical care has improved survival (Oskoui et al. 2007; Oskoui and Kaufmann 2008). Intermediate SMA type II (OMIM #253550) is characterized by the onset after the age of 6 months. Patients acquire the ability to sit, but can never walk unaided. Patients with juvenile SMA type III (Kugelberg Welander disease; OMIM #253400) typically reach major milestones and can walk independently. The adult form, SMA type IV (OMIM #271150), is characterized by an age of onset beyond 30 years and relatively mild motor impairment.

SMA is caused by deletions or mutations of the survival of motor neuron gene (*SMN1*), which was originally cloned and characterized by Melki and colleagues (Lefebvre et al. 1995). *SMN1* is an essential gene in divergent organisms, in which null mutations are lethal during early development (Schmid and DiDonato 2007). The survival of motor neuron (SMN) protein is ubiquitously expressed in all cells and tissues, with high levels in the nervous system, especially in spinal cord (Battaglia et al. 1997). Unlike other species that have only one copy of the *SMN* gene (e.g., mice), the *SMN* gene is present on human chromosome 5q13 as a single copy of the telomeric *SMN1* gene, and a variable number of centromeric *SMN2* genes. *SMN2* appears to be unique to humans, since chimpanzees contain multiple copies of *SMN1*, but no *SMN2* (Rochette et al. 2001). The majority of mRNAs (90%) from *SMN1* encode for the full-length protein; whereas, the majority of mRNAs (90%) from *SMN2* encode for a truncated and unstable protein lacking the carboxy-terminal exon-7 due to a translationally silent mutation in an exonic splicing enhancer (Lorson et al. 1999). The full-length transcripts of *SMN1* and *SMN2* encode proteins with an identical sequence. The most commonly inherited forms of SMA are caused by large deletions that inactivate the *SMN1* gene and, while the unique presence of *SMN2* in humans can protect against lethality, a neurodegenerative process occurs, leading to SMA. *A major challenge is to understand how a reduction in total SMN levels results in neuronal dysfunction that leads to SMA.*

The function of SMN was addressed by Dreyfuss and colleagues, who were seeking to identify proteins binding to hnRNP-U, a member of a family of heterogenous ribonucleoproteins. SMN was identified from their screen and shown to localize to nuclear structures, termed gems, based on their proximity to coiled bodies/ Cajal bodies (i.e., Gemini of Cajal bodies) (Liu and Dreyfuss 1996). SMN was found tightly associated with a novel protein, SIP1, subsequently referred to as Gemin2, and together, they form a specific complex with several spliceosomal snRNP proteins (Liu et al. 1997). The Dreyfuss lab and others subsequently identified other

SMN interacting proteins (see Sect. 2.2), and extensively studied the critical role of the SMN–Gemin multiprotein complex in the assembly of spliceosomal snRNPs (Gubitz et al. 2004; Yong et al. 2004; Battle et al. 2006a; Eggert et al. 2006). The SMN–Gemin complex acts as a specificity factor to promote high-fidelity interactions between Sm core proteins and snRNAs, which prevent promiscuous interactions with other RNAs (Pellizzoni et al. 2002b). Recent data indicate that SMN deficiency alters stoichiometry of snRNAs and leads to splicing defects for numerous genes in all cells, including motor neurons (Zhang et al. 2008). In SMN-deficient mouse models of SMA, there is a preferential reduction in snRNP species that function in the minor spliceosome required for processing a rare class of introns (Gabanella et al. 2007). The function of the SMN–Gemin complex in snRNP assembly thus represents the most well understood function of SMN.

A major question needed to be addressed is how reduction of SMN in all tissues leads to a preferential impairment of motor neurons, although other neurons (e.g., sensory) and other tissues (e.g., muscle fibers) are also affected. Since the effects of SMN-deficiency on snRNP species and the stoichiometry of spliced introns appear to vary across tissues, one can also speculate that motor neurons may be more vulnerable to splicing alterations; however, to test this model, it would be necessary to demonstrate that altered splicing can have effects on protein expression and sequence composition that are deleterious to motor neurons. It is more than likely, however, that there are other functions of the SMN–Gemin complex, or even other types of SMN complexes, which contribute to the pathogenic mechanism in motor neurons. Inasmuch as SMN is localized to both the nucleus and cytoplasm (Young et al. 2000a), it becomes critical to understand whether the cytoplasmic pool of SMN is involved in activities other than that involved in the assembly of snRNPs.

2 SMN Domains and Interacting Proteins

2.1 SMN Domain Structure

The *SMN1* gene contains nine exons and eight introns in a genomic region of ca. 20 kb. The ubiquitously expressed *SMN1* transcript of ca. 1.7 kb encodes a highly conserved 294 amino acid protein of 38 kDa. SMN contains several functionally important regions for self assembly and protein interaction (see Fig. 1). The N-terminal part contains the binding sites that have been implicated in the interaction with Gemin2 from amino acids 13–44 (Liu et al. 1997) and 52–91 (Young et al. 2000b). The region encoded by exon2 has been shown to bind RNA directly in vitro (Lorson and Androphy 1998; Bertrandy et al. 1999). Exon 3 encodes a central Tudor domain, comprising amino acids 90–160. Tudor domains are conserved sequence motifs that were originally described for the *Drosophila* tudor protein. They are thought to mediate protein–protein interactions and are often found in RNA-associated proteins (Ponting 1997; Selenko et al. 2001). Tudor domain proteins have

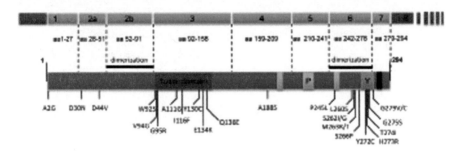

Fig. 1 SMN1 exon boundaries and protein domain structure. *Top*: SMN1 is encoded by 9 exons. Coding regions are indicated in *cyan* and untranslated regions in *purple*. *Bottom*: Domains required for oligomerization of SMN are indicated as *black bars*. Other regions of interest are the Tudor domain (*red*), poly-proline regions (*green*), the YG box (*orange*), and the cytoplasmic targeting motif (*black*). Point mutations identified in *SMN1* genes of SMA patients are indicated below (Wirth 2000; Alias et al. 2008)

been shown to interact with proteins that contain methylated arginine and lysine residues (Brahms et al. 2001; Sprangers et al. 2003; Cote and Richard 2005). The SMN Tudor domain mediates interaction with arginine and glycine-rich motifs in several proteins (Meister et al. 2002; Paushkin et al. 2002; Gubitz et al. 2004), including the Sm core proteins. Symmetrical dimethylation of specific arginine residues enhances their affinity for SMN (Friesen et al. 2001a, b; Boisvert et al. 2002; Hebert et al. 2002; Meister and Fischer 2002). Exons 4–6 contain three stretches of poly-proline sequences that mediate the interaction with the small actin-binding protein profilin (Giesemann et al. 1999b). A conserved tyrosine/glycine-rich motif (YG box) that is found in many RNA-binding proteins is localized in exon 6, from amino acid 258–279. Regions required for dimerization or oligomerization of SMN have been mapped to self-association domains in exon 2b (52–91) and exon 6 (242–279) (Lorson et al. 1998; Young et al. 2000b).

SMA is characterized by a reduced level of full length SMN in the presence of SMN that lacks exon 7 (SMNΔ7). Although the expression of the SMNΔ7 isoform is beneficial, it cannot fully compensate for loss of full length SMN (Le et al. 2005). Therefore, the function of the protein domain encoded by exon 7 from amino acid 280–294 is of special interest. SMNΔ7 has been reported to encode an unstable and rapidly degraded protein (Le et al. 2000; Lorson and Androphy 2000) that is deficient in oligomerization activity (Lorson et al. 1999; Young et al. 2000b), binding to Sm core proteins (Pellizzoni et al. 1999), and formation of gems (Frugier et al. 2000). The SMNΔ7 protein has a twofold shorter half-life than full length SMN in cells, despite similar turnover rates, mediated by the ubiquitin-proteasome system in an in vitro assay (Burnett et al. 2008). SMNΔ7 can be stabilized by the coexpression of full length SMN, suggesting that it is stabilized by recruitment into oligomeric SMN complexes (Le et al. 2005). It is noteworthy that exon7 has also been shown to contain a cytoplasmic targeting signal that is required for the active transport of SMN into neuronal processes (Zhang et al. 2003).

Although in the large majority of cases SMA is caused by homozygous deletion of the SMN1 gene, a small percentage of SMA patients bear one *SMN1* copy with small mutations. This is of considerable interest, since missense mutations may affect certain properties of SMN without causing a general loss of function. As depicted in Fig. 1, most SMA mutations have been found clustered in the tudor domain (W92S, V94G, G95R, A111G, I116F, Y130C, E134K, and Q136E) and in exon 6, within and near the Y/G box (L260S, S262G, S262I, M263R, M263T, S266P, Y272C, H273R, T274I, G275S, G279C, and G279V). This distribution of mutations suggests that the Tudor domain and the Y/G box with its flanking region are essential for SMN function. Studying these and synthetic SMN mutations may make it possible to uncouple functions required for snRNP assembly and proposed axonal functions in genetic rescue experiments (Beattie et al. 2007).

2.2 SMN Interacting Proteins

SMN is part of a well-characterized complex that facilitates assembly of Sm proteins on multiple U snRNAs to form the snRNP core (Meister et al. 2002; Paushkin et al. 2002; Gubitz et al. 2004; Kolb et al. 2007). To date, nine proteins have been identified as core components of this complex: SMN, Gemins 2–8 and unrip. SIP1 (Smn interacting protein) (Liu and Dreyfuss 1996), subsequently referred to as Gemin2, was identified as the first component of the SMN complex and shown to have a critical function in the assembly of spliceosomal snRNPs (Fischer et al. 1997; Liu et al. 1997), which has since been extensively studied by the Dreyfuss lab and others (Gubitz et al. 2004; Yong et al. 2004; Battle et al. 2006a; Eggert et al. 2006). Gemin2, Gemin3, and Gemin8 bind SMN directly (Otter et al. 2007). Gemin3 is a DEAD box RNA helicase (Charroux et al. 1999) and Gemin5 is the snRNA binding protein (Gubitz et al. 2002; Battle et al. 2006b). Gemin8 is needed for the structural organization of the SMN–Gemin complex (Carissimi et al. 2006). This SMN–Gemin core complex is found associated with spliceosomal Sm/LSm core proteins (SmB/B', D1–3, E, F, G, LSm10, 11). It catalyzes the assembly of ring structures consisting of seven Sm core proteins onto stem loop structures of uridine-rich snRNAs (Chari et al. 2008). The high molecular weight of the SMN complex suggests that SMN and other core components are present as multimers. Oligomerization of SMN is also a prerequisite for high-affinity binding of the SMN complex to spliceosomal snRNPs (Pellizzoni et al. 1999).

In addition to the SMN complex components, SMN interacts directly or is associated with a remarkably large number of other proteins (see Table 1). This suggests that SMN may have a pleiotropic function that goes beyond the well-characterized role in snRNP assembly.

The largest group of SMN-associated proteins comprises components of RNP complexes that play a role in some aspects of RNA metabolism. Many of them contain domains that are enriched in arginine and glycine residues that are required for interaction with SMN (Meister et al. 2002; Paushkin et al. 2002). Arginine residues in these

Table 1 SMN-associated proteins, functions, and motifs[a]

SMN associated protein	Function	Motifs[b]	References
SMN-Gemin complex			
Gemin1/SMN	snRNP assembly	Tudor, Y/G box	Liu et al. (1997)
Gemin2/SIP1	snRNP assembly	–	Liu et al. (1997)
Gemin3/DDX20/DP103	snRNP assembly	DEXDc, HELICc	Charroux et al. (1999); Meister et al. (2000)
Gemin4/GIP1	snRNP assembly	Leu zipper	Charroux et al. (2000); Meister et al. (2000)
Gemin5/p175	snRNP assembly	WD40 repeats	Gubitz et al. (2002)
Gemin6	snRNP assembly	–	Pellizzoni et al. (2002a)
Gemin7/SIP3	snRNP assembly	RG-rich	Baccon et al. (2002)
Gemin8	snRNP assembly	–	Carissimi et al. (2006)
unrip	snRNP assembly	WD40 repeats	Meister et al. (2001)
snRNP core proteins			
Sm proteins/Sm B/B′, D1–3,E,F,G	snRNP assembly	Sm domain, RG rich	Liu et al. (1997); Friesen and Dreyfuss (2000)
Sm-like proteins/Lsm 10,11	snRNP assembly	Sm domain, RG rich	Friesen and Dreyfuss (2000); Brahms et al. (2001)
snoRNP core proteins			
Fibrillarin/FBL	snoRNP assembly	RGG box, RNP-2	Liu and Dreyfuss (1996); Jones et al. (2001); Pellizzoni et al. (2001a)
GAR1/Nola1	snoRNP assembly	RGG box	Pellizzoni et al. (2001a)
snRNP import			
Snurportin and importin β	Nuclear import of snRNPs	HEAT repeat	Narayanan et al. (2002)
TGSI (trimethylguanosine synth. 1)	Nuclear import of snRNPs	–	Mouaikel et al. (2003)
Cajal body/coiled body			
Coilin/p80	Recruitment of SMN to Cajal bodies	–	Hebert et al. (2001)
Transcription			
mSin3A	Transcriptional regulation	PAH, HDAC interact	Zou et al. (2004)
EWS (Ewing Sarcoma)	Transcriptional regulation	RGG box	Young et al. (2003)
Viral proteins			
Papilloma virus E2	Transcriptional regulation	–	Strasswimmer et al. (1999)
Epstein-Barr virus nuclear antigen	Transcriptional regulation	–	Barth et al. (2003)
Minute virus NS1 and NS2	Viral replication and transcriptional regulation	–	Young et al. (2002b)

(continued)

Table 1 (continued)

SMN associated protein	Function	Motifs[b]	References
Actin metabolism			
Profilin	Control of actin dynamics	–	Giesemann et al. (1999b)
T-Plastin/PLS3	Actin bundling	EFh, CH	Oprea et al. (2008)
Apoptosis			
p53	Apoptosis	–	Young et al. (2002a)
Bcl-2	Antiapoptosis	BH4, BCL	Iwahashi et al. (1997)
RNA metabolism			
hnRNP U	RNA metabolism	SAP, SPRY, RGG	Liu and Dreyfuss (1996)
hnRNP Q and R	RNA metabolism	RRM, RGG	Mourelatos et al. (2001); Rossoll et al. (2002)
FUSE-binding protein/FBP	RNA metabolism	KH	Williams et al. (2000b)
KSRP/FBP2/ZBP2/MARTA1	RNA metabolism	KH	Tadesse et al. (2008)
FMRP	RNA metabolism	KH, Agenet	Piazzon et al. (2008)
U1A	pre-mRNA splicing	RRM	Liu et al. (1997)
Galectin 1 and 3/LGALS1	pre-mRNA splicing	GLECT	Park et al. (2001)
RNA helicase A	Transcription	DSRM, DEXDc, HELICc	Pellizzoni et al. (2001b)
RNA polymerase II	Transcription	–	Pellizzoni et al. (2001a)
Rpp20 (Ribonuclease P 20 kDa subunit)	tRNA and rRNA metabolism	–	Hua and Zhou (2004a)
Nucleolin and B23	rRNA metabolism	RRM, RGG-box	Lefebvre et al. (2002)
ISG20 (Interferon stimul. gene 20)	Degradation of ssRNA	EXOIII	Espert et al. (2006)
TIAR (TIA-1-related protein)	Stress granule formation	RRM	Hua and Zhou (2004b)
TDP-43 (TAR DNA-binding protein)	mRNA splicing and transcription	RRM	Wang et al. (2002)
NFAR-1/2/nuclear factor associated with dsRNA	RNA metabolism	DZF, DSRM	Saunders et al. (2001)
Others			
OSF (osteoclast-stimulating factor)	Src-related signaling	SH3, ANK	Kurihara et al. (2001)
USP9X (Ubiquitin-specific protease 9)	Deubiquitylating enzyme	–	Trinkle-Mulcahy et al. (2008)
PPP4 (protein phosphatase 4)	Ser/Thr protein phosphatase	PP2Ac	Carnegie et al. (2003)
FGF-2 (fibroblast growth factor 2)	Growth factor	FGF	Claus et al. (2003)
hsc70 (heat shock cognate prot. 70)	Protein folding/ trafficking	–	Meister et al. (2001)
ZPR1 (zinc-finger protein 1)	Protein translation?	Zpr1	Gangwani et al. (2001)
BAT3/HLA-B assoc. transcript 3[c]	Apoptosis/regulation of p53	UBQ domain	Stelzl et al. (2005)
UNC119/HRG4[c]	Photoreceptor synaptic protein	GMP-PDE delta	Stelzl et al. (2005)
RIF1 (receptor interact. factor 1)[c]	Transcriptional repressor	–	Stelzl et al. (2005)

(continued)

Table 1 (continued)

SMN associated protein	Function	Motifs[b]	References
GDF9 (growth diff. factor 9)[c]	Growth factor	TGFB	Stelzl et al. (2005)
COPS6 (COP9 signalosome subunit 6)[c]	Regulation of ubiquitin ligases	JAB/MPN	Stelzl et al. (2005)
Leukocyte receptor cluster LRC member 8[c]	–	SAC3/GANP	Rual et al. (2005)
FLJ10204/C8orf32[c]	–	–	Rual et al. (2005)

[a] Meister et al. (2002); Peri et al. (2003); Stark et al. (2006); Wirth et al. (2006)
[b] Based on Schultz et al. (1998)
[c] Interaction partners identified in high throughput screens may have not been confirmed by independent methods

RG-rich domain are often methylated, which can enhance the interaction with SMN (Paushkin et al. 2002). The other interaction partners are very heterogenous and include viral and cellular transcription factors, regulators of apoptosis, and growth factors.

2.3 SMN Interacting Proteins Associated with Actin Metabolism

Interestingly, several interacting proteins have been implicated in either β-actin mRNA transport or in the regulation of actin dynamics. Examples of interacting partners that could potentially affect actin-based functions are briefly outlined as follows and further discussed below.

1. *SMN interactions with regulators of actin dynamics.* SMN was found to interact with the small actin-binding proteins profilin I and II in yeast-two-hybrid assays and to colocalize in motor neurons (Giesemann et al. 1999a). Profilin II and SMN were shown to colocalize in neurites and in growth cones of differentiating rat PC12 cells (Sharma et al. 2005). Antisense knockdown of profilin I and II isoforms inhibited neurite outgrowth of PC12 cells and caused accumulation of SMN and its associated proteins in cytoplasmic aggregates. SMN can modulate actin polymerization in vitro by reducing the inhibitory effect of profilin IIa. Therefore, reduced SMN-levels in SMA could disturb the normal regulation of microfilament growth by profilins and cause defects in axonal outgrowth. In a related study, SMN knockdown in PC12 cells altered the expression pattern of profilin II, leading to an increased formation of ROCK/profilin IIa complexes (Bowerman et al. 2007). Inappropriate activation of the RhoA/ROCK actin-remodeling pathway may result in altered cytoskeletal integrity and impaired neurite outgrowth.
2. *Reduced actin expression in Smn-deficient distal axons.* The distal distribution of β-actin mRNA and protein is defective in SMN-deficient motor neurons (see Sect. 3.1) (Rossoll et al. 2003). Treatment of primary motor neurons from *Smn−/−; SMN2* embryos with a cell-permeable cAMP analog (8-CPT-cAMP) not only leads to an increased β-actin mRNA and protein level in the growth cone but also normalized the axon length and growth cone size defects to control

levels (Jablonka et al. 2007). Thus, pharmacological intervention can increase distal β-actin mRNA and protein levels, and at the same time correct morphological axonal defects observed in SMA.
3. *SMN and transport/regulation of β-actin mRNA and other transcripts.* SMN interacts with several proteins thought to be involved in the transport and post-transcriptional regulation of β-actin mRNA and other transcripts: KSRP/ZBP2/MARTA1 (Tadesse et al. 2008), FBP (Williams et al. 2000b), hnRNP R, and hnRNP Q (Mourelatos et al. 2001; Rossoll et al. 2002, 2003). This is discussed in more detail in Sects. 3.1 and 3.2.
4. *Protection from SMA by plastin 3 expression.* Plastin 3/T-plastin (PLS3) has been identified as a protective modifier shown to physically interact with SMN complexes in a study of siblings discordant for SMA (Oprea et al. 2008). SMA-unaffected *SMN1*-deleted females exhibit significantly higher expression of PLS3 than their SMA-affected siblings. The actin-bundling protein PLS3 is important for axonogenesis through the increase in the F-actin level. Proteins of the fimbrin/plastin family share the unique property of cross-linking actin filaments into tight bundles that assist in stabilizing and rearranging the organization of the actin cytoskeleton in response to external stimuli, as well as perhaps by actin filament stabilization and anti-depolymerization activities (Oprea et al. 2008). Overexpression of PLS3 in SMN-deficient motor neurons rescued the axon length and outgrowth defects associated with SMN down-regulation.
5. *SMN complexes with ZPR1 and eEF1A.* SMN assembles into complexes with the zinc finger protein ZPR1 and eukaryotic translation elongation factor 1A (eEF1A) (Mishra et al. 2007). eEF1A binds to actin and has a noncanonical function in actin bundling (Gross and Kinzy 2005).

All considered, these data emphasize a potential role of SMN in the regulation of the axonal actin cytoskeleton, in part, by influencing β-actin mRNA transport and local translation and/or actin filament formation. Failure of locally modulated β-actin synthesis or F-actin localization in axon terminals may have a severe impact on axon growth, synapse differentiation and maintenance, the scaffolding of regulatory molecules such as synapsin, and the trafficking and release of synaptic vesicles (Shupliakov et al. 2002).

3 Localization of SMN and Links to Axonal mRNA Regulation

Previous immunocytochemical studies have localized SMN in dendrites and axons of spinal cord motor neurons in vivo (Bechade et al. 1999; Pagliardini et al. 2000). These immuno-EM analyses also depicted SMN on cytoskeletal filaments and associated with polyribosomes. Several immunofluorescence (IF) studies have detected Smn in neurites of cultured P19 cells (Fan and Simard 2002) and axons of cultured motor neurons (Rossoll et al. 2002). High-resolution imaging has revealed the presence of SMN in granules that localize to axons and growth cones of cultured neurons and align along microtubules (Zhang et al. 2003). As shown in Fig. 3, SMN-containing granules are not only abundant in the cell body and dendrites, but also along the axons and growth cones of motor neurons. SMN granules were shown by IF double labeling and fluorescence in situ hybridization (FISH) to colocalize with mRNA and ribosomes. Fluorescence imaging methods applied to living neurons showed that SMN granules are actively transported into neuronal processes and growth cones at rates over 1 µm s^{-1} (Zhang et al. 2003), consistent with fast axonal transport. Depolymerization of microtubules impaired the long-range transport of SMN granules, whereas disruption of F-actin impaired short-range trafficking (Zhang et al. 2003).

Double label IF studies have suggested possible colocalization between SMN and Gemin proteins in neurites of PC12 cells (Sharma et al. 2005). However, significant levels of SMN appear not to colocalize with Gemin2 in axons of primary motor neurons (Jablonka et al. 2001). A later study, using high-resolution IF, digital imaging analysis, and 3D reconstruction of growth cones, showed that a population of SMN granules contained Gemin2 and Gemin3 (30–40%); however, the majority of SMN lacked Gemin proteins (Zhang et al. 2006). FRET analysis of fluorescently tagged SMN and Gemin proteins demonstrated interactions within individual granules, which were also observed to move as a complex in live neurons (Zhang et al. 2006). The QNQKE sequence from exon-7 was necessary for the sorting of the SMN–Gemin complex into the cytoplasm (Zhang et al. 2007). Of interest, the spliceosomal Sm proteins, necessary for snRNP assembly, were confined to the cell body and exhibited little colocalization with Smn–Gemin complex in neuronal processes (Zhang et al. 2006). Collectively, these studies indicate the presence of distinct SMN ribonucleoprotein complexes in neuronal processes that may play a role in mRNA regulation.

There appears to be some relationship between the neuritogenic effects of SMN and its cytoplasmic localization. The SMNΔ7 form of SMN, which is the predominant form in SMA, was enriched within the nucleus, with much lower levels in the cytoplasm (Zhang et al. 2003). A QNQKE sequence with the carboxy-terminus of SMN was found to be necessary for cytoplasmic localization. Overexpression of SMN was neuritogenic in comparison to SMNΔ7. This defect could be rescued by fusion of the membrane targeting sequence from GAP-43. SMNΔ7 fused to this GAP-43 sequence now was targeted to axons and stimulated neurite growth.

A shorter splicing isoform of SMN called axonal SMN (a-Smn) has been reported to have a neuritogenic effect in NSC34 cells (Setola et al. 2007). The identified mRNA retains intron 3, which contains an in-frame stop codon. a-SMN is encoded by exons 1–3 and a small region of intron 3 that is not conserved between species. Future work is needed to address how prevalent this form of SMN is in axons. However, there has been some controversy as to whether this form of SMN could play a role in the pathogenesis of SMA (Burghes 2008). Most missense mutations found in SMA patients lie downstream of exon 3 and are clustered in exon 6. Also, there is strong evidence that SMA is caused by lack of exon7, and we know from experiments with SMA mouse models that expression of full-length murine Smn cDNA and human SMN2 can rescue the mutant phenotype. It will be interesting to assess the effects of a-SMN on rescue of axon phenotypes in animal models (discussed below).

3.1 Mechanism of β-Actin mRNA Localization in Neurons and Possible Connection to SMN and SMA

RNA localization is an essential and highly conserved biological mechanism for protein sorting that plays critical roles in neuronal polarity, axon guidance, and synaptic plasticity (Kindler et al. 2005; Bramham and Wells 2007; Lin and Holt 2008). Several mRNAs have been shown to be targeted to dendrites and/or axons, both in vivo and in cultured neurons. The molecular mechanism of mRNA localization involves recognition of *cis*-acting sequences within the 3'-untranslated region (UTR) by specific mRNA binding proteins and accessory factors, which are assembled into large mRNP complexes, often termed "*granules*" (Kiebler and Bassell 2006). mRNA granules are actively transported along microtubules and contain mRNA binding proteins, translation factors, ribosomal subunits, and accessory factors. mRNAs present within transport granules are often translationally repressed, which become derepressed in response to receptor signaling (Besse and Ephrussi 2008).

β-actin mRNA represents the most extensively studied localized mRNA, and hence provides a logical starting point to assess a possible role for SMN in the mechanism of mRNA granule localization. FISH analysis revealed that β-actin mRNAs are localized to developing axons and growth cones of primary cortical neurons in culture (Bassell et al. 1998). EM analysis showed the presence of polyribosomes within axon growth cones. A neurotrophin (NT-3) signaling pathway increased localization of β-actin mRNA from the cell body into processes and growth cones, while γ-actin mRNA remained in the cell body (Bassell et al. 1998; Zhang et al. 1999). The neurotrophin-induced localization of β-actin mRNA resulted in a protein-synthesis dependent increase in β-actin protein levels in growth cones (Zhang et al. 2001). The molecular mechanism of β-actin mRNA localization in chick forebrain neurons was shown to involve a highly conserved zipcode (54 nt) sequence, only present in the 3'UTR of β-actin, which is recognized by the mRNA binding protein, zipcode binding protein ZBP1 (Ross et al. 1997). ZBP1 has four KH-domains, which are necessary for binding to the zipcode, RNA granule formation, and association with the cytoskeleton (Farina et al. 2001).

ZBP1 also has NLS and NES and has been shown to shuttle into the nucleus where it first associates with nascent β-actin mRNA (Oleynikov and Singer 2003). Once in the cytoplasm, ZBP1 is hypothesized to function as an adapter for microfilament- and/or microtubule-dependent motors to facilitate the transport of β-actin mRNA in fibroblasts and neurons, respectively. In neurons, antisense-mediated disruption of ZBP1 binding to the zipcode impaired NT-3-induced localization of β-actin mRNA into neurites, resulting in reduced enrichment of β-actin protein and impaired growth cone dynamics (Zhang et al. 2001). In *Xenopus* neurons, binding of ZBP1 (VgRBP) to the β-actin zipcode was required for local β-actin synthesis and axon guidance in vitro in response to BDNF or netrin (Leung et al. 2006; Yao et al. 2006). Recent evidence indicates that ZBP1 acts to repress β-actin mRNA translation, and that in response to Src activation, ZBP1 is phosphorylated, resulting in its release from the mRNA, allowing for translation (Huttelmaier et al. 2005).

Since β-actin appears to be the preferred actin isoform to induce rearrangements of the actin cytoskeleton beneath the plasma membrane (Bassell et al. 1998), it is hypothesized that local β-actin mRNA translation is necessary to promote enrichment of β-actin protein at the membrane and influence growth cone motility and guidance. ZBP1 is thus a critical factor for both β-actin mRNA transport and local protein synthesis; however, there may be other proteins within the complex that contribute to these processes, and Smn is attractive candidate based on its trafficking in the form of RNA granules.

In a study providing the first link between Smn and β-actin mRNA, Smn-deficient motor neurons, cultured from a transgenic mouse model of SMA, were shown to have reduced localization of β-actin mRNA in axons and reduced levels of β-actin protein in axonal growth cones (Rossoll et al. 2003). Of interest, Smn-deficient motor neurons displayed only defects in axonal morphology, that is, shorter axons and smaller growth cones, whereas dendrites appeared otherwise normal in morphology. These results suggest some role of SMN in the molecular mechanism of β-actin mRNA localization to axons. These results also imply that axonal defects in SMA may be due, in part, to altered localization of β-actin mRNA. Future work will be necessary to explore a possible interaction between SMN and ZBP1. However, it also possible that Smn may interact with ZBP2 to facilitate β-actin mRNA localization. ZBP2 is a KH-domain-containing protein that is predominantly localized to the nucleus, where it facilitates the hand-off of β-actin mRNA to ZBP1 that exports the mRNA into the cytoplasm (Gu et al. 2002; Pan et al. 2007). Of interest, SMN has been shown to interact with ZBP2 homologs, viz., FBP and KSRP. By yeast two-hybrid and co-immunoprecipitation (IP) analysis, SMN was shown to bind FUSE binding protein (FBP) (Williams et al. 2000a), a KH domain-containing RNA binding protein that regulates the stability of GAP-43 mRNA (Irwin et al. 1997), which is localized to axons and growth cones (Smith et al. 2004). More recent work has shown that SMN binds KSRP (Tadesse et al. 2008), a neuron-specific splicing factor that is also involved in targeting AU-rich mRNAs for degradation (Min et al. 1997). KSRP was shown to be arginine-methylated and interact with the Tudor domain of SMN. Smn and KSRP colocalized in granules within neurites and missense mutations in the Tudor domain that cause SMA abolished the colocalization with KSRP. In SMA transgenic mice, a KSRP target mRNA, p21 (cip/waf1), was increased in spinal cord, suggesting possible impairment of mRNA stability. How this relates to altered mRNA localization is unknown, but there are known links between mRNA localization and stability mechanisms; thus, it will be important for future work to assess a possible alternative or related role of SMN in regulation of mRNA stability in axons. Other studies have demonstrated a role for MARTA1, the rat KSRP ortholog, to bind the dendritic targeting element of MAP2 mRNA and facilitate its targeting in dendrites (Rehbein et al. 2002). Since Smn is localized to dendrites and axons (Bechade et al. 1999; Zhang et al. 2006), it will be important to assess a possible role of Smn in regulation of dendritic mRNA localization, which may affect postsynaptic function in spinal cord motor neurons.

3.2 SMN Interactions with hnRNP R and hnRNP Q: Another Possible Link to β-Actin mRNP Complexes

Yeast two-hybrid (Y2H) screens using SMN as a bait have identified two highly related proteins, hnRNP R (heterogenous nuclear ribonucleoprotein R) and hnRNP Q, previously discovered as glycine–arginine–tyrosine-rich RNA-binding protein (gry-rbp). hnRNP R and Q interact with wild-type SMN, but not with mutated versions of the protein found in SMA patients (Rossoll et al. 2002). In a parallel Y2H study, Mourelatos et al. have identified three human splicing isoforms of hnRNP Q as SMN interacting proteins (Mourelatos et al. 2001). hnRNP Q3/gry-rbp encodes a protein of 623 aa that is composed of an N-terminal acidic domain, followed by three consecutive RNA recognition motifs (RRM1–3) and an arginine- and glycine-rich domain (RGG box). hnRNP Q3 and hnRNP R share 82% identity and 90% similarity on the protein level. They also identified two smaller isoforms that are derived by alternative splicing of hnRNP Q3. hnRNP Q2 lacks 34 amino acids from RRM2, and in hnRNP Q1, the last 74 amino acids from the C-terminus of hnRNP Q3 are replaced by the sequence VKGVEAGPDLLQ. All yeast two-hybrid clones identified in both studies contain the C-termini, including most of the RGG box, suggesting that the RGG box of hnRNP Q is necessary and sufficient for binding to SMN. This was confirmed by GST-pull down assays that identified an RG dipeptide-rich sequence from amino acids 518–549 as the minimal SMN-binding domain.

Although hnRNP R is mainly a nuclear protein, it was also found to colocalize with Smn in axons of motor neurons (Rossoll et al. 2002). hnRNP R has been shown to associate with the 3′ UTR of β-actin mRNA and to modulate localization of actin mRNA in neurites (Rossoll et al. 2003). While hnRNP Q3 has been isolated as a predominantly nuclear protein associated with the spliceosome complex (Neubauer et al. 1998), hnRNP Q1 has been identified in several independent studies on various aspects of mRNA metabolism in the cytoplasm. Taken together, they suggest a role of hnRNP Q1 in mRNA transport, regulation of mRNA stability, and/or translation. Previously, hnRNP Q1 has been described as NSAP1, a protein interacting with the NS1 protein of minute virus of mice that is essential for viral replication in the cytoplasm (Harris et al. 1999). Regulation of mRNA stability by hnRNP Q1/NSAP1 was shown in a study on the major protein-coding-region determinant of instability (mCRD) of the c-fos proto-oncogene mRNA. hnRNP Q1/NSAP1 binds the mCRD and modulates the translationally coupled process of mRNA turnover. HnRNP Q1/NSAP overexpression stabilizes mCRD-containing mRNA by inhibiting deadenylation (Chen and Shyu 2003). In addition, hnRNP Q1/NSAP overexpression resulted in a dramatic stabilization of AU-rich element-containing mRNAs (Grosset et al. 2000). Work from several groups has demonstrated that hnRNP Q1/NSAP1 modulates cap-independent translation. It enhances IRES-dependent translation through interaction downstream of the hepatitis C virus polyprotein initiation codon (Kim et al. 2004) and it has also been shown to modulate IRES-dependent translation of cellular BiP mRNA through an RNA–protein interaction under heat stress conditions (Cho et al. 2007). Another substrate of hnRNP Q1 is mRNA related to the circadian rhythm. hnRNP Q1 has been shown to bind serotonin N-acetyltransferase (AANAT) both at its 3' untranslated region to mediate mRNA degradation (Kim et al. 2005) and its 5' region to regulate cap-independent translation (Kim et al. 2007). Other studies have connected hnRNP Q1 to vesicle transport and RNP transport granules. hnRNP Q1 has been described as SYNCRIP, a protein that interacts with the C2B domains of ubiquitous synaptotagmin isoforms; thereby, suggesting a potential link between vesicles and mRNA transport (Mizutani et al. 2000). hnRNP Q1/SYNCRIP has also been identified as a component of inositol 1,4,5-triphosphate receptor type I mRNA transport granules in rat hippocampal dendrites (Bannai et al. 2004). In another proteomic study, hnRNP Q1/SYNCRIP was isolated as a component of kinesin (KIF5)-containing RNA transport granules (Kanai et al. 2004). Therefore, it will be important to investigate whether SMN associates with hnRNP R/Q in neuronal processes, and whether this interaction is necessary for hnRNP R/Q mediated regulation of mRNA granules in neuronal processes.

4 A Role for SMN in Stress Granule Formation

Stress granules (SGs) are an additional type of RNA granule within neurons whose formation within the cell body and processes can be induced in response to low dose arsenate treatment (Vessey et al. 2006). In contrast to RNA transport granules, SGs have limited motility, are larger morphologically, and may assemble de-novo, or perhaps be converted from RNA transport granules (Kiebler and Bassell 2006; Anderson and Kedersha 2008). SGs serve as a depot to recruit mRNAs and translational factors (however, lacking some ribosomal components), where mRNAs are poised in translational arrest and also protected from degradation (Kedersha et al. 2005). A number of mRNA binding proteins are recruited to SGs in response to various stressors (Anderson and Kedersha 2006, 2008). The RGG box has been shown to be necessary for recruitment of some mRNA binding proteins to SGs (Yang et al. 2006). SMN was shown to target to SGs upon arsenate exposure and SMN-deficient fibroblasts from SMA patients showed impaired formation of SGs and recruitment of mRNAs (Hua and Zhou 2004b). The SG hypothesis to explain the neurodegeneration in SMA posits that impaired formation of SGs in axons, in response to various insults (i.e., oxidative stress), may lead to increased degradation of axonal mRNAs (Hua and Zhou 2004b). In that, many of the SMN-associated mRNA binding proteins are also localized to stress granules, where they play a role in mRNA stability regulation (Stohr et al. 2006), it will be important to assess how these interactions may affect the trafficking of mRNAs between various types of granules involved in mRNA transport, stability, and degradation, including stress granules and P-bodies (Kiebler and Bassell 2006). Collectively, these dynamics may have a great impact on mRNA translation in axons.

4.1 The SMN-mRNA Granule Hypothesis

As discussed earlier, a critical goal will be to continue to identify interactions between SMN and specific mRNA binding proteins that are localized to neuronal processes, and assess the function of these interactions in mRNA regulation. Although it has been shown that SMN can bind RNA in vitro, there is no evidence so far that it acts as an RNA-binding protein in vivo (Lorson and Androphy 1998; Bertrandy et al. 1999). SMN may impart binding specificity to RNA binding proteins through its role in the formation of the ribonucleoprotein complex. The past work on the SMN–Gemin complex and its role to facilitate Sm core assembly onto stem loops of snRNA may suggest a mechanism that could be used for mRNP assembly. Whether SMN interactions with mRNA binding proteins depend on Gemins will need to be addressed. Nonetheless, SMN interactions with mRNA binding proteins may modulate the binding of mRNA binding proteins to *cis*-acting elements and assembly of RNA transport granules. A model for the proposed role of SMN complexes in axonal RNA localization is shown in Fig. 2.

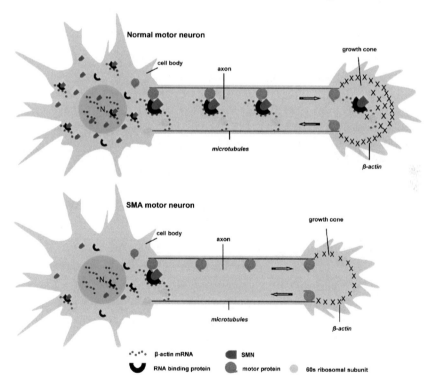

Fig. 2 Model of a putative function of SMN complexes in axonal RNA localization. Data from different labs suggest a model whereby SMN-associated RNP complexes contribute to the localization of β-actin mRNA and probably other transcripts. SMN may facilitate the assembly of RNA cargo molecules with different RNA-binding proteins, adaptor proteins, molecular motor proteins, translational components, and auxiliary factors required for efficient transport and/or local translation (Based on a model by Lei Xing.)

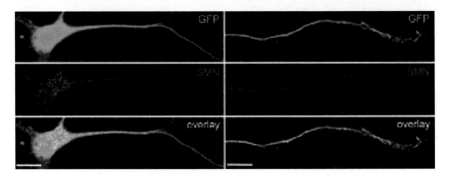

Fig. 3 SMN-containing granules are transported along motor axons. SMN is localized in granules that are actively transported into neuronal processes and growth cones (Zhang et al. 2006). Shown above are murine primary embryonic motor neurons expressing GFP from the motor neuron-specific HB9 promoter. SMN-containing granules stained with specific antibodies are present in the cell body, dendrites, and axon, including the growth cone (*red*). Size bar = 10 μm (photomicrograph provided by Lei Xing)

5 Animal Models of SMA

SMN is present in all metazoan cells with an evolutionary conserved role in small nuclear ribonucleoprotein assembly and pre-mRNA splicing. SMN orthologs have been disrupted in different genetic model organisms, ranging from yeast to mouse, in an effort to model the disease phenotype (Schmid and DiDonato 2007). These SMA models serve as powerful research tools to elucidate the pathogenic mechanism, and also to aid in the development of therapies for the treatment of spinal muscular atrophy. Using a variety of model organisms makes it possible to exploit their respective strengths and to validate results gained from different species. Analyzing the phenotype of these model organisms suggests a unique requirement for SMN orthologs in the development and maintenance of motor neurons (Monani et al. 2000a).

5.1 Caenorhabditis elegans

The nematode *Caenorhabditis elegans* has been used as a valuable research tool in many fields of biological research. Establishment of a cell lineage map, sequencing of the entire genome, and a well-characterized wiring diagram of its neurons allow an in-depth investigation of the development of its nervous system (Girard et al. 2007). CeSMN is an essential protein that is required for embryonic viability. Targeting *smn-1* with RNAi in all tissues (including maternal transcripts) and overexpression of the gene gives rise to locomotive defects and embryonic lethality (Miguel-Aliaga et al. 1999). Surviving hypomorphs demonstrate pleiotropic developmental defects in neuronal, muscular, and reproductive tissue that lead to locomotive defects, lack of muscle tone, vulval abnormalities, and sterility caused by a defect in germ cell maturation (Miguel-Aliaga et al. 1999).

Recently, a deletion mutant of *smn-1 (ok355)* has been characterized, which removes most of the gene including the translation start codon (Briese et al. 2009). Unlike the embryonic lethality caused by targeting *smn-1* with RNAi, maternally contributed CeSMN in the mutant allows for normal early larval development of *smn-1(ok355)* mutant animals, but causes late larval arrest, sterility, decreased lifespan, as well as impaired locomotion and pharyngeal pumping activity. The *smn-1(ok355)* phenotype can be partially rescued by neuronal, but not muscle-directed, expression of *smn-1*. A *C. elegans* ortholog of the human *Gemin2* gene and several RNA-binding proteins have also been identified as interaction partners of CeSMN (Burt et al. 2006). Knockdown of *smn-1* by RNAi, as well as deletion mutants have the potential to be utilized in screens searching for phenotypic modifiers as a further step toward potential SMA therapies (Briese et al. 2009).

5.2 Drosophila melanogaster

The fruit fly *Drosophila melanogaster* has proven to be an excellent invertebrate model organism for studying human neurodegenerative diseases. Mutational analyses and the existence of a broad array of screening techniques allow direct observation of the effects of the loss of gene function, making it an excellent in vivo system for the identification of genetic modifiers and the testing of therapeutic compounds (Lu and Vogel 2008).

The *Drosophila* genome bears a single copy of the *Smn* gene, which encodes a highly conserved homologue of SMN (DmSMN) (Miguel-Aliaga et al. 2000). *Drosophila* contains a remarkably simple SMN complex that lacks most Gemin proteins (Kroiss et al. 2008). This minimal complex mediates the assembly of spliceosomal snRNPs in a manner very similar to its vertebrate counterpart, by also preventing misassembly onto nontarget RNAs (Kroiss et al. 2008). Larval-lethal *Smn*-null mutations do not show detectable differences in spliceosomal snRNA levels, suggesting that these animals do not die from global snRNP depletion (Rajendra et al. 2007). Ectopic expression of human SMN exerts a dominant-negative effect that leads to developmental defects and pupal lethality, presumably, by forming a nonfunctional complex with endogenous DmSMN (Miguel-Aliaga et al. 2000). *Drosophila* mutant strains that contain point mutations in *Smn* similar to those found in SMA patients show abnormal motor behavior (Chan et al. 2003). Because of maternal contribution of the smn transcript, smn zygotic mutant embryos survive early development. Mutant larvae show severe motor abnormalities, defects in late developmental stages, and late larval lethality. Defects at the neuromuscular junction (NMJ) include disorganized synaptic motor neuron boutons, reduced clustering of the neurotransmitter receptor subunit GluR IIA, and reduced excitatory post-synaptic currents. Rescue experiments show that *Smn* gene activity is required in both neurons and muscle to alleviate this phenotype. Taken together, these studies point to a functional role of DmSMN at the NMJ (Chan et al. 2003).

Hypomorphic mutations in *Smn* with a tissue specific reduction of DmSMN protein levels in the adult thorax cause flightlessness and acute muscular atrophy (Rajendra et al. 2007). Mutant flight muscle motor neurons display motor axon routing and arborization defects coupled with a loss of the flight muscle-specific actin isoform Act88F expression. The authors also observed colocalization of DmSMN with sarcomeric actin, association between SMN and the thin filament crosslinker α-actinin, and disruption of α-actinin localization in the thoracic muscles of *Smn* mutants. These observations suggest a specific function for DmSMN in muscle and would underline the importance of this tissue in modulating the disease phenotype (Rajendra et al. 2007). However, it is not clear whether the defects observed in the *Smn* hypomorphs are caused by reduced DmSMN expression in the motor neurons, the thoracic muscles, or a combination of both.

Recently, a screen for genetic modifiers of an *Smn* phenotype, using the Exelixis collection of transposon-induced mutations, identified 17 enhancers and 10 suppressors (Chang et al. 2008). Among these were several genes not previously known to be associated with the genetic circuitry of *Smn*. These include several members of the bone morphogenic protein (BMP) signaling pathway. Interestingly, the observed NMJ defects in the SMA model can be ameliorated by increased BMP signaling, suggesting that drugs that stimulate BMP activity may provide potential SMA therapeutics (Chang et al. 2008).

5.3 Danio rerio

The zebrafish *Danio rerio* has been developed as an excellent vertebrate model organism for studying developmental biology and genetics (Beattie et al. 2007). A very stereotyped and well characterized neuromuscular system and external embryonic development allow for easy analysis of early nervous system development. Recent advances, such as large genetic screens for mutations that disrupt development, the establishment of an effective antisense approach, and sequencing of the zebrafish genome, make them an ideal model system to study developing motor neurons and their axonal projections. Although targeted gene knockout is not yet established in the zebrafish model, most loss-of-function studies use a morpholino (MO) knockdown approach.

Beattie and colleagues have developed a zebrafish model for SMA, which demonstrate an axonal function for SMN. Using antisense morpholinos injected into zebrafish embryos at the 1–2 cell stage to reduce Smn levels lead to motor axon-specific pathfinding defects (McWhorter et al. 2003). While the number of motor neurons present at 24 h was not affected, motor axons were short or excessively branched. Importantly, specific reduction of Smn in individual motor neurons revealed that Smn is acting cell-autonomously. This study indicates an essential function of Smn in motor axon development in vivo, and suggests that these early developmental defects may contribute to later defects in synapse development that may precede motor neuron loss. To further characterize Smn functions necessary for motor axon outgrowth, *smn* MO were coinjected with various human *SMN* RNAs and assayed for their effect on motor axons (Carrel et al. 2006). Motor axon defects caused by Smn reduction were rescued by wild-type human SMN but not by SMN lacking exon-7, the predominant form in SMA, or other mutations in human SMN that also cause SMA. The identification of mutations that failed to rescue motor axon defects, but retained properties needed for snRNP assembly such as oligomerization and Sm protein binding, would indicate a critical function of SMN in motor axons that is relevant to SMA and is independent of snRNP biosynthesis (Carrel et al. 2006). SMNΔ7, SMN G279V, and SMN Y272C were defective in these properties and failed to rescue axonal defects. The human SMA mutant A111G was also unable to rescue motor axon defects while retaining Sm binding and oligomerization properties. In contrast, SMNΔ7 fused to an exon-7 sequence, containing the minimal cytoplasmic localization motif (QNQKE) (Zhang et al. 2007), was able to rescue motor axon defects, although it was deficient in oligomerization and Sm proteins binding in vitro (Carrel et al. 2006; Beattie et al. 2007). A synthetic point mutant of an exon-7 sequence (SMN Q282A) important for the cytoplasmic localization of SMN did not rescue the axon guidance defects, although this mutant retained snRNP function. Further work will be needed to show that this point mutant is fully competent for snRNP function, and it will be interesting if interactions with proteins other than Gemins are impaired. Such efforts to identify SMN domains that compromise SMN function in axon guidance, without affecting snRNP function, represent a useful approach toward identification of alternate mechanisms underlying SMA.

Another study used the morpholino knock-down approach to show an effect of SMN reduction on U snRNP metabolism (Winkler et al. 2005). Knockdown of SMN and the U snRNP assembly factors Gemin2 and pICln led to motor axon degeneration. This developmental defect was rescued by injection of purified U snRNPs into SMN- or Gemin2-deficient zebrafish embryos. These findings would imply that motor neuron degeneration in SMA patients may be a direct consequence of impaired production of U snRNPs (Winkler et al. 2005). However, these findings are contradicted by a study that did not find any aberrant motor axons caused by Gemin2 knockdown in morphologically normal embryos (McWhorter et al. 2008). Specific knockdown of Gemin2 in motor neurons by either injection into motor neurons or generation of genetic mosaics did not lead to motor axon defects. Since Gemin2 and SMN both function in snRNP biogenesis, yet, only SMN knockdown causes motor axon defects, these findings would support an axonal role for SMN independent of snRNP assembly. The authors have suggested that motor axon defects observed by Winkler et al. (2005) upon Gemin knockdown may be secondary to the overall morphological defects – including widespread defects in the nervous system – observed in these embryos.

5.4 Mouse Models of SMA

With the advent of gene targeting technology that is now routinely used to manipulate the genome, mouse models of human diseases have become the most widely used vertebrate animal models in biomedical research. Like all model organisms studied so far, mice have only a single copy of the SMN gene corresponding to human SMN1. As will be evident in the discussion below, mouse models of SMA have provided new insight into SMN function in synapse development. A list of commonly available SMA mouse models is shown in Table 2.

Mice that lack Smn due to homozygous insertion of a lacZ reporter gene after exon2a display massive cell death during early embryonic development and die during the morula stage prior to transition to the blastocyst stage and uterine implantation (Schrank et al. 1997). A follow-up study on Smn+/− heterozygous mice found a 46% reduction of Smn protein levels in the spinal cord and motor neuron degeneration resembling human type III SMA with normal lifespan and no reported overt phenotype (Jablonka et al. 2000). Between birth and six months of age, 40% of spinal cord motor neurons were lost relative to wild-type mice, but no further statistically significant loss occurred later. This order of pathophysiological events corresponds to an early phase of deterioration, followed by a phase of stabilization observed in SMA III patients (Crawford and Pardo 1996).

Two groups have chosen a very similar approach to model the SMA disease phenotype. These most commonly used SMA mouse models faithfully replicate the situation in human SMA patients by expressing human *SMN2* transgenes in an *Smn* knockout background. The Burghes lab used the knockout mouse from the Sendtner lab (Schrank et al. 1997) and inserted a 35.5 kb fragment containing only the *SMN2* transgene and its promoter region (*Tg(SMN2)89Ahmb*) (Monani et al. 2000b). The Li laboratory pursued a similar strategy by disrupting *Smn* exon 7 and

Table 2 SMA mouse models available from The Jackson Laboratories

Stock	Strain genotype	Average survival	References
006214	FVB.Cg-Smn1$^{tm1Msd/J}$	90–100 h	Schrank et al. (1997)
005024	FVB.Cg-Tg(SMN2)89Ahmb Smn1$^{tm1Msd/J}$	0–5 days[a]	Monani et al. (2000b)
006570	STOCK Smn1^{tm1Msd} Tg(Hlxb9-GFP)1Tmj Tg(SMN2)89Ahmb/J	0–5 days[a]	McGovern et al. (2008)
005025	FVB.Cg-Tg(SMN2*delta7)4299Ahmb Tg(SMN2)89Ahmb Smn1$^{tm1Msd/J}$	ca. 13 days	Le et al. (2005)
005026	FVB.Cg-Tg(SMN2)89Ahmb Tg(SMN1*A2G)2023Ahmb Smn1$^{tm1Msd/J}$	<1 year[a]	Monani et al. (2003)
006146	B6.129-Smn1$^{tm1Jme/J}$	Normal[b]	Frugier et al. (2000)
006138	FVB.129(B6)-Smn1$^{tm1Jme/J}$	Normal[b]	Frugier et al. (2000)
005058	FVB.Cg-Tg(SMN2)2Hung Smn1$^{tm1Hung/J}$	<P10–normal[c]	Hsieh-Li et al. (2000)

[a] Difference between the survival times reported in the original publication and the survival of the strains available from The Jackson Laboratory may be due to inbreeding (http://jaxmice.jax.org/list/ra1733.html). It is the experience at The Jackson Laboratory that mice hemizygous for the *SMN2* transgene do not survive

[b] When crossed to a strain expressing Cre recombinase in a conditional manner, this mutant mouse strain will develop SMN-deficiency and associated defects in the affected tissue

[c] Average time of survival depends on the transgene copy number

using a larger 115 kb fragment of the human *SMN2* region that also contained flanking genes (*Tg(SMN2)2Hung*) (Hsieh-Li et al. 2000). *Smn* heterozygous mice, containing the transgene, were bred to give progeny lacking endogenous *Smn*, but carrying the respective transgene. Aside from minor differences between these transgenic models, both had strong SMA-like characteristics. Furthermore, in both models, the human *SMN2* transgene was able to complement the embryonic lethality of *Smn*–/– mice. The phenotype was modulated by the expression levels of *SMN2* and even rescued by sufficiently high copy number of the inserted *SMN2* transgene. The severely affected mice (*Smn*–/–, *SMN2*) have only one or two copies of the *SMN2* transgene on an *Smn* null background and frequently died perinatally. Surviving mice initially appeared normal but became progressively smaller and weaker than their normal littermates and presented with decreased mobility and suckling, labored breathing, and finally, limb tremor followed by death within days after birth. These mice produce low levels of SMN protein and suffered a 35–40% loss of spinal cord and lower-brainstem motor neurons by day five, despite normal numbers at birth. Primary embryonic motor neurons cultured from these mice showed defective axon outgrowth, but normal survival (Rossoll et al. 2003). Isolated sensory neurons were less severely affected (Jablonka et al. 2006). This is in agreement with a study on SMA patients, who underwent sural nerve biopsy that showed significant sensory nerve pathology in severely affected SMA type I patients, while no sensory nerve alterations were detected in SMA type II or III patients (Rudnik-Schoneborn et al. 2003). Taken together, these results suggest that motor neuron loss is a late event, occurring subsequent to axonal denervation (discussed below), and the timing of motor neuron loss is critical for disease severity. In both mouse models, expression from several copies of the *SMN2* transgene completely rescued the disease phenotype, with the exception of short, thick tails that became necrotic after about 2 weeks of age. The cause of this phenotype is currently unknown. A study with cultured primary motor neurons isolated from the *Smn*–/–, *SMN2* mice found defective accumulation and clustering of $Ca_v2.2$ Ca^{2+} channels in the axonal growth cones that was correlated with a reduction of local Ca^{2+} transients in distal axons and growth cones (Jablonka et al. 2007). The observed functional abnormalities of Ca^{2+} channels could contribute to maturation defects found in SMA.

Using the severe SMA models as a basis, several groups have inserted additional transgenes to modify the disease severity and answer specific questions about SMN function. Studies in *Drosophila* suggested that expression of *SMN* in both muscle and nervous tissue was required for complete rescue of an SMA phenotype (see above), implying that high levels of *SMN* expression in muscle may impact the SMA phenotype. To investigate whether this is also the case in severe mouse models of SMA, *SMN* was expressed in skeletal muscle fibers under the human skeletal actin (HSA) promoter and in neurons under the prion promoter (PrP) (Gavrilina et al. 2008). The results suggest that only neuronal *SMN* expression corrected the severe SMA phenotype in mice, whereas high *SMN* levels restricted to skeletal muscle fibers alone had no impact on the SMA phenotype or survival. It should be noted, however, that the transgene expressed from the PrP promoter showed "leaky" expression in skeletal muscle. Although it is possible that the timing and protein levels expressed from the HSA promoter was not sufficient to have an effect on the SMA genotype, an HSA-SMN strain that showed low levels of expression in the spinal cord and brain had significantly increased survival. Mice homozygous for PrP-SMN in an *Smn−/−; SMN2* background survived for an average of 210 days, and motor neuron counts in these mice were normal. This demonstrates that increased levels of full-length *SMN* in neurons, but not muscle, were required to correct the SMA phenotype.

Expression of a mutant SMN transgene harboring the A2G SMA mutation in a severe SMA background (*Smn−/−; SMN2*) delayed the onset of motor neuron loss, resulting in mice with symptoms and neuropathology similar to patients afflicted with SMA type III. These mild SMA mice presented with motor neuron degeneration, muscle atrophy, and abnormal electromyographs, and lived to about 8 months of age (Monani et al. 2003). In the absence of the SMN2 gene, the mutant SMN A2G transgene did not rescue the embryonic lethality, suggesting that low levels of wild-type SMN are required for the formation of functional SMN complexes.

To investigate motor axon development and innervation in severe SMA mice, the HB9:GFP transgene was introduced into the *Smn−/−; SMN2 +/+* strain by cross breeding to specifically label motor neurons with GFP (McGovern et al. 2008). Motor neurons innervating various muscles were examined from embryonic day 10.5 to postnatal day 2 for abnormalities in axon formation and outgrowth, but no defects were detected at any stage of development. However, a significant increase in unoccupied AChR clusters in intercostal muscles and the presence of axonal varicosities in motor axons in embryonic SMA mice indicated denervation during embryogenesis as one of the earliest detectable morphological defects in the SMA mice.

The question of whether SMNΔ7, the major splicing isoforms expressed from *SMN2* locus lacking exon 7, has a beneficial or detrimental effect has been answered unequivocally by several independent studies. Transgenic mice expressing *SMNΔ7* cDNA under the control of the SMN promoter were crossed onto a severe SMA background (*Smn−/−; SMN2*). Added expression of the *SMNΔ7* transgene (*Tg(SMN2*delta7) 4299Ahmb*) turned out to be beneficial by extending mean survival of SMA mice homozygous for the targeted mutant *Smn* allele and homozygous for the two transgenic alleles from 5 to 13 days (Le et al. 2005). It is thought that expression of SMNΔ7 can rescue the SMA phenotype if a low level full-length SMN is present, probably by oligomerizing with full-length SMN and thus raising the effective concentration of SMN complexes (Le et al. 2005). In a detailed study of mice homozygous or hemizygous for the *SMNΔ7* transgene, the initial effects of reduced SMN on the neuromuscular system were described as a NMJ synaptopathy (Kariya et al. 2008). The earliest structural defects appear distally prior to overt symptoms and involve the NMJ. Reduced levels of the SMN protein impair the normal maturation of acetylcholine receptor (AChR) clusters and lead to presynaptic defects, including poor terminal arborization and intermediate filament aggregation causing neurotransmission defects.

In a related study, the synaptic pathology at the neuromuscular junction in two different mouse models of SMA, the *Smn–/–;SMN2* mouse model of type I SMA and the *Smn–/–;SMN2;Δ7* mouse model, was analyzed in detail (Murray et al. 2008). NMJs in various muscles were disrupted in both mouse models, especially in postural muscles in the trunk. Loss of presynaptic inputs and nerve terminals with abnormal accumulations of neurofilament protein occurred even in early/mid-symptomatic animals in the most severely affected muscle groups. Skeletal muscles can be assigned to one of the two distinct classes of muscles, termed "fast synapsing" (FaSyn) and "delayed synapsing" (DeSyn) muscles, which differ in their pattern of neuromuscular synaptogenesis during embryonic development and maintenance of NMJs in adult mice (Pun et al. 2002). This study suggests that motor neurons with FaSyn-like, nonplastic characteristics may be more vulnerable to SMA-induced synapse loss than those with a DeSyn phenotype. Treatments that challenge synaptic stability result in nerve sprouting, NMJ remodeling, and ectopic synaptogenesis selectively on DeSyn muscles. Possibly, NMJs on FaSyn muscles lacking the synaptic plasticity are lost selectively and very early in SMA mice, suggesting that disease-associated motor neuron dysfunction may selectively fail to initiate maintenance processes at nonplastic FaSyn NMJs (Santos and Caroni 2003).

Electrophysiology and ultra-structure was the focus of a recent study on the function and structure of NMJ synapses in the *Smn–/–;SMN2;Δ7* mouse model of severe SMA (Kong et al. 2009). While NMJ synapses remained well connected late in the disease course, decreased density of synaptic vesicles and reduced probability of vesicle release at SMA NMJs was found to be associated with impaired morphological and biochemical maturation of postsynaptic NMJ terminals and myofibers. These findings suggest that NMJ synaptic dysfunction precedes axonal degeneration in severe SMA mouse models. Similar defects were described in an earlier investigation of the NMJ ultrastructure in SMA type 1 that showed shallow subsynaptic clefts and axon terminals devoid of transmitter vesicles (Diers et al. 2005).

A different approach for the generation of SMA mouse models was chosen in the Melki lab (Frugier et al. 2000). Mice that harbor a conditional allele of *Smn*, in which LoxP sites flank *Smn* exon 7 (SmnF7), allow conditional excision of Smn exon 7 when crossed with mice expressing Cre recombinase in a tissue-specific manner. Mice carrying a homozygous deletion of *Smn* exon 7 directed to neurons display progressive loss of motor axons, defects in NMJs, and massive accumulation of neurofilaments in terminal axons of the remaining neuromuscular junctions (Cifuentes-Diaz et al. 2002). Selective deletion of exon7 in either skeletal muscle (Cifuentes-Diaz et al. 2001) or liver (Vitte et al. 2004) also leads to severe defects in these tissues, confirming that full length *Smn* is essential for all tissues and cell types. The relevance of these findings for the disease phenotype of SMA is currently not clear. The observed defects differ from those seen in human SMA patients or the other mouse models that both retain at least some expression of full length SMN. It is likely that selective reduction of SMN levels under a certain threshold level in any tissue will lead to defects due to the essential function of SMN. Future studies are needed to investigate to what extent SMN deficiency in other tissues contributes to the SMA disease phenotype.

SMA mouse models have been crossbred with other mutant mice to study the effect of neuroprotective and anti-apoptotic mutations on the SMA disease phenotype. Slow Wallerian degeneration (Wld^s) mice express a fusion of nicotinamide mononucleotide adenlyl transferase 1 (Nmnat-1) and ubiquitination factor e4b (Ube4b) that protects axons from Wallerian degeneration in genetic mutants such as the pmn mouse (Ferri et al. 2003). Wld^s in a SMA mouse model has no protective effect to alter the SMA phenotype, indicating that Wallerian degeneration does not directly contribute to the pathogenesis of SMA development (Rose et al. 2008; Kariya et al. 2009). Deletion of Bax, the major proapoptotic member of the Bcl-2 family, and overexpression of Bcl-xL, an anti-apoptotic of the Bcl-2 family, prevent neuronal cell death. Crossbreeding of Bax-deficient or Bcl-xL transgenic mice with SMA mice showed milder disease severity and longer life spans of the progeny as compared to their SMA littermates (Tsai et al. 2008). These results suggest that suppression of neuronal apoptosis has a potential to ameliorate the disease phenotype of SMA and may be a therapeutic target for future drug development.

Taken together, several recent studies on severe SMA mouse models emphasize early defects in NMJ synapses that include nerve terminal loss, neurofilament accumulation in presynaptic terminals, immature postsynaptic terminals, and functional abnormalities (Cifuentes-Diaz et al. 2002; Kariya et al. 2008; McGovern et al. 2008; Murray et al. 2008; Kong et al. 2009).

It will be important to compare the above phenotypes in animal models with observations from human samples. Likewise, we need to compare observations in human patients with the findings in model organisms. For example, in a recent study, SMA-I and SMA-II spinal cords showed a significant number of motor neurons that had migrated aberrantly toward the ventral spinal roots (Simic 2008; Simic et al. 2008). These heterotopic motor neurons (HMN) were located mostly in the ventral white matter and had no axon or dendrites. The authors propose that abnormal migration, differentiation, and lack of axonal outgrowth may induce cell death in displaced HMNs and contribute to the pathogenesis of SMA. It will be interesting to see whether similar phenotypes can be observed in animal models of SMA.

6 Therapeutic Strategies

SMA is the leading genetic killer of infants and toddlers with an incidence of 1/10,000 and a carrier frequency of one in 35 (Wirth et al. 2006). Currently, there is no cure for SMA or treatment to stop its progression. Treatment is symptomatic and supportive and includes treating pneumonia, curvature of the spine, and respiratory infections, if present (*http://www.ninds.nih.gov/disorders/sma/sma.htm*). The human toll of this disease, combined with its unique genetic profile, has focused attention from the NIH, industry, academia, and advocacy organizations on developing a treatment. SMA was selected in 2003 by the National Institute of Neurological Disorders and Stroke (NINDS) for a model translational research program aimed at accelerating drug discovery efforts against neurodegenerative diseases.

To uncover therapeutic options for SMA patients, research groups typically focus on one of the following strategies (Lunn and Stockwell 2005; Sumner 2006): (1) Elevation of endogenous full length SMN levels from the SMN2 gene by upregulating the SMN2 promoter, correcting alternative SMN2 pre-mRNA splicing, increasing read-through of SMNΔ7 mRNA, stabilization of SMN protein, and modulating SMN2 translation; (2) Neuroprotection; (3) Delivery of SMN with viral vectors (gene therapy); and (4) Implantation of stem cell derived precursors into the spinal cord (stem cell therapy). Beyond the scope of this review, many drugs, including histone deacetylase inhibitors, such as valproic acid (VPA) or 4-phenyl-butyrate (PBA), have been shown to increase full length SMN2-derived RNA and protein levels in cell culture. None of these drugs is believed to act specifically on expression of the SMN gene, but is expected to change the expression level of hundreds of genes. Clinical trials are underway to investigate the effect of VPA, PBA, and other drugs on motor function in SMA patients (Lunn and Wang 2008; Oskoui and Kaufmann 2008). Also, beyond the scope of this review, many different approaches have been used successfully to specifically restore inclusion of exon7 in *SMN2* mRNA in vitro by using small antisense RNA molecules that modulate exon7 inclusion by directly binding to *SMN2* transcripts (Madocsai et al. 2005; Coady et al. 2007; Marquis et al. 2007; Singh 2007; Hua et al. 2008). While this approach holds a lot of promise, it remains to be seen whether such therapeutic oligonucleotides can be successfully delivered to spinal motor neurons.

6.1 Gene Therapy

Multiple single injections of a lentiviral vector construct that was pseudotyped with rabies virus glycoprotein in various muscles of SMA mice restored SMN expression to motor neurons, reduced motor neuron death, and increased the life expectancy by an average of 5 days, compared with untreated animals (Azzouz et al. 2004). Recently, adeno-associated virus (AAV) 9 injected intravenously has been shown to bypass the blood–brain barrier and efficiently target cells of the central nervous system (Foust et al. 2009). Intravenous injection of AAV9-GFP into neonatal mice through the facial vein resulted in extensive transduction of dorsal root ganglia and motor neurons throughout the spinal cord. These promising results suggest that lentiviral or AAV based vectors may enable the future development of gene therapies for SMA.

6.2 Stem Cells

Stem cells can be utilized as a biological tool to examine the abnormal cellular and molecular processes in these cells to gain a better understanding of the SMA pathophysiology. In a recent study, neurosphere-derived neural stem cells from the severe

SMA mice *(Smn−/−;SMN2)* showed differences in proliferation and differentiation as compared to wild-type controls (Shafey et al. 2008). Creating a cell-culture model of human disease can also be a useful tool in developing screens for potential therapeutic agents. Human ES cell-derived neuroprogenitor cells have been generated as a new primary cell source for the purpose of developing assays for new SMA therapeutics (Wilson et al. 2007). Endogenous stem cells present in the mammalian nervous system may be manipulated to induce neural tissue repair. Alternatively, neuronal progenitor cells derived from stem cells that are committed to the motor neuron lineage can be transplanted into the injured nervous system as a therapeutic strategy. Transplanted cells may contribute to the repair of the damaged neuronal circuits, not only by directly replacing degenerated cells, but also by providing support in many other ways (Nayak et al. 2006).

In a study on paralyzed rats, the transplantation of ES cells that were treated to induce differentiation into motor neurons has shown promising results (Deshpande et al. 2006). Multipotent neural stem cells isolated from the spinal cord of GFP-expressing mice that were also primed for differentiation into motor neurons were transplanted intrathecally into SMA mice *(Smn−/−; SMN2+/+; SMNΔ7+/+)* (Corti et al. 2008). Treated mice showed improved motor function and extended survival from 13 to 18 days. Transplanted cells even generated a small number of motor neurons that were properly localized to the ventral horn of the spinal cord, but secretion of trophic factors from the transplanted cells may also be responsible for the observed effect (Corti et al. 2008). Stem cell transplantation, combined with other molecular and pharmacologic approaches, may be a promising strategy in the development of SMA therapies.

An especially promising resource to develop an in vitro SMA disease model and develop novel therapies is based on two recent developments: (1) the generation of induced pluripotent stem (iPS) cells from patient dermal fibroblasts (reviewed by (Lowry and Plath 2008), and (2) the differentiation of stem cells into motor neurons (Wichterle et al. 2002). A combination of these approaches was used to generate ALS patient-specific motor neurons (Dimos et al. 2008). Recently, these two combined approaches were used in a study to generate iPS cells from skin fibroblast samples taken from a child with SMA (Ebert et al. 2008). These cells were differentiated into motor neurons that showed selective deficits compared to those derived from the child's unaffected mother. Initially, iPS-SMA cells produced similar numbers of motor neurons, but showed a decline at later time points. Diffuse synapsin staining suggested specific presynaptic maturation deficits in the iPS-SMA cultures. These results validate motor neurons derived from iPS cells as a promising new in vitro model for studying the SMA disease mechanism and drug discovery.

7 Open Questions and Perspectives

Despite recent progress in research of SMN function and the pathophysiology of SMA, many open questions remain that have important implications for SMA therapy, and are considered below.

Does a defect in the canonical housekeeping function of SMN in snRNP assembly lead to SMA? While it is known that extremely low SMN levels affect splicing, it is not clear yet, whether a more moderate reduction, as seen in patients, cause splicing defects in motor neurons that may explain the disease phenotype. Identifying aberrantly spliced transcripts with arrays and other methods will be required to answer this question.

Does SMN play a role in the assembly of neuronal RNP complexes? While there is evidence that SMN affects local β-actin mRNA levels in axons, it will be interesting to see whether other mRNAs are targets of SMN complexes. While SMN is transported in axons and co-localizes with interacting proteins, the composition of these putative RNA granules is not known. Proteomic studies of RNA granules to date have not identified SMN as a component. It will be important to further characterize a putative function of SMN in the assembly of axonal RNP complexes and to assess its role in the pathogenesis of SMA.

Are motor neurons the only cells affected by SMN deficiency? The question whether SMA is caused by SMN deficiency in neurons, muscle fibers, or both appears still controversial. Most current studies suggest that low levels of SMN in motor neurons lead to SMA. It is important to clarify whether upregulation of SMN function in motor neurons is necessary and sufficient to prevent SMA. Presently, it cannot be excluded that other cells in the circuitry of the spinal cord or muscle tissue contribute to the pathophysiology of SMA. Indeed, more broadly restored SMN expression throughout the neuromuscular system, including neurons and muscle, may be necessary.

Which other genetic modifiers affect the disease phenotype of SMA? A study in SMA discordant families with affected and asymptomatic siblings carrying the same *SMN1* mutations has identified PLS3 as a modifier of SMA severity (Oprea et al. 2008). It would be interesting to see whether expression of a PLS3 transgene in spinal cord or other tissues can indeed rescue the SMA phenotype in mouse models of SMA. Comparisons across several animal models will help to further characterize these targets and assess epistatic relationships of modifiers, especially, to conduct genetic modifier screens in invertebrate animals, such as *Drosophila* (Chang et al. 2008) and *C. elegans*, which may lead to the identification of pathways that modulate the SMA phenotype and the discovery of novel therapeutic targets.

When do first neuromuscular defects develop, and can further progression be stopped? Studies in several animal models of SMA point to early developmental defects of motor axons and NMJs and a dysfunction of motor neurons preceding cell death. The interaction of motor axons with muscle fibers in neuromuscular synapses is an important distinctive feature that distinguishes the motor neuron from all other neurons, which may underlie its specific vulnerability. It will be important to investigate how early developmental defects influence the establishment and maturation of muscle end-plate innervation patterns and may set the stage for subsequent denervation and eventual motor neuron cell death (Hausmanowa-Petrusewicz and Vrbova 2005). A question that is highly relevant for therapeutic intervention is the time window during which treatment, such as raising SMN levels, will benefit patients. The generation of conditional mouse models that allow for the induced expression of SMN at various time points (e.g., via tetracycline or tamoxifen) or conditional alleles that are engineered to revert to fully functional SMN via Cre-mediated recombination will be required to answer that question. Studies in several severe SMA models suggests that low SMN levels may lead to defects very early in development, at least in severe SMA type I, which has important implications for the development of a therapy for SMA. If an effective SMA drug treatment becomes available, intervention at a presymptomatic stage will require the establishment of neonatal screening to identify children at risk to develop SMA.

Results derived from addressing these questions will ultimately aid in the development of therapies for the treatment of SMA. Since axonal transport defects are common to many neurological disorders, including Alzheimer disease (AD), Huntington disease (HD), amyotrophic lateral sclerosis (ALS), and other forms of motor neurons disease, elucidating the cause of axonal defects in SMA may have widespread applicability to other types of neurological diseases (Duncan and Goldstein 2006).

Acknowledgment The authors thank Claudia Fallini, Lei Xing, and Kristy Welshhans for helpful comments. Special thanks to Lei Xing for contributing to Figs. 2 and 3. The authors gratefully acknowledge funding from the Spinal Muscular Atrophy Foundation and NIH (HD055835) to GJB and Families of SMA and NIH (HD056130) to WR.

References

Alias L, Bernal S, Fuentes-Prior P, Barcelo MJ, Also E, Martinez-Hernandez R, Rodriguez-Alvarez FJ, Martin Y, Aller E, Grau E, Pecina A, Antinolo G, Galan E, Rosa AL, Fernandez-Burriel M, Borrego S, Millan JM, Hernandez-Chico C, Baiget M, Tizzano EF (2009) Mutation update of spinal muscular atrophy in Spain: molecular characterization of 745 unrelated patients and identification of four novel mutations in the SMN1 gene. Hum Genet 125(1):29–39
Anderson P, Kedersha N (2006) RNA granules. J Cell Biol 172:803–808
Anderson P, Kedersha N (2008) Stress granules: the Tao of RNA triage. Trends Biochem Sci 33:141–150
Azzouz M, Le T, Ralph GS, Walmsley L, Monani UR, Lee DC, Wilkes F, Mitrophanous KA, Kingsman SM, Burghes AH, Mazarakis ND (2004) Lentivector-mediated SMN replacement in a mouse model of spinal muscular atrophy. J Clin Invest 114:1726–1731
Baccon J, Pellizzoni L, Rappsilber J, Mann M, Dreyfuss G (2002) Identification and characterization of Gemin7, a novel component of the survival of motor neuron complex. J Biol Chem 277:31957–31962
Bannai H, Fukatsu K, Mizutani A, Natsume T, Iemura S, Ikegami T, Inoue T, Mikoshiba K (2004) An RNA-interacting protein, SYNCRIP (heterogeneous nuclear ribonuclear protein Q1/NSAP1) is a component of mRNA granule transported with inositol 1,4,5-trisphosphate receptor type 1 mRNA in neuronal dendrites. J Biol Chem 279:53427–53434
Barth S, Liss M, Voss MD, Dobner T, Fischer U, Meister G, Grasser FA (2003) Epstein-Barr virus nuclear antigen 2 binds via its methylated arginine-glycine repeat to the survival motor neuron protein. J Virol 77:5008–5013
Bassell GJ, Zhang H, Byrd AL, Femino AM, Singer RH, Taneja KL, Lifshitz LM, Herman IM, Kosik KS (1998) Sorting of beta-actin mRNA and protein to neurites and growth cones in culture. J Neurosci 18:251–265
Battaglia G, Princivalle A, Forti F, Lizier C, Zeviani M (1997) Expression of the SMN gene, the spinal muscular atrophy determining gene, in the mammalian central nervous system. Hum Mol Genet 6:1961–1971

Battle DJ, Kasim M, Yong J, Lotti F, Lau CK, Mouaikel J, Zhang Z, Han K, Wan L, Dreyfuss G (2006a) The SMN complex: an assembly machine for RNPs. Cold Spring Harb Symp Quant Biol 71:313–320

Battle DJ, Lau CK, Wan L, Deng H, Lotti F, Dreyfuss G (2006b) The Gemin5 protein of the SMN complex identifies snRNAs. Mol Cell 23:273–279

Beattie CE, Carrel TL, McWhorter ML (2007) Fishing for a mechanism: using zebrafish to understand spinal muscular atrophy. J Child Neurol 22:995–1003

Bechade C, Rostaing P, Cisterni C, Kalisch R, La Bella V, Pettmann B, Triller A (1999) Subcellular distribution of survival motor neuron (SMN) protein: possible involvement in nucleocytoplasmic and dendritic transport. Eur J Neurosci 11:293–304

Bertrandy S, Burlet P, Clermont O, Huber C, Fondrat C, Thierry-Mieg D, Munnich A, Lefebvre S (1999) The RNA-binding properties of SMN: deletion analysis of the zebrafish orthologue defines domains conserved in evolution. Hum Mol Genet 8:775–782

Besse F, Ephrussi A (2008) Translational control of localized mRNAs: restricting protein synthesis in space and time. Nat Rev Mol Cell Biol 9:971–980

Boisvert FM, Cote J, Boulanger MC, Cleroux P, Bachand F, Autexier C, Richard S (2002) Symmetrical dimethylarginine methylation is required for the localization of SMN in Cajal bodies and pre-mRNA splicing. J Cell Biol 159:957–969

Bowerman M, Shafey D, Kothary R (2007) Smn Depletion Alters Profilin II Expression and Leads to Upregulation of the RhoA/ROCK Pathway and Defects in Neuronal Integrity. J Mol Neurosci 32:120–131

Brahms H, Meheus L, de Brabandere V, Fischer U, Luhrmann R (2001) Symmetrical dimethylation of arginine residues in spliceosomal Sm protein B/B' and the Sm-like protein LSm4, and their interaction with the SMN protein. RNA 7:1531–1542

Bramham CR, Wells DG (2007) Dendritic mRNA: transport, translation and function. Nat Rev Neurosci 8:776–789

Briese M, Esmaeili B, Fraboulet S, Burt EC, Christodoulou S, Towers PR, Davies KE, Sattelle DB (2009) Deletion of smn-1, the *Caenorhabditis elegans* ortholog of the spinal muscular atrophy gene, results in locomotor dysfunction and reduced lifespan. Hum Mol Genet 18:97–104

Burghes HM (2008) Other forms of survival motor neuron protein and spinal muscular atrophy: an opinion. Neuromuscul Disord 18:82–83

Burnett BG, Munoz E, Tandon A, Kwon DY, Sumner CJ, Fischbeck KH (2008) Regulation of SMN protein stability. Mol Cell Biol 29:1107–1115

Burt EC, Towers PR, Sattelle DB (2006) *Caenorhabditis elegans* in the study of SMN-interacting proteins: a role for SMI-1, an orthologue of human Gemin2 and the identification of novel components of the SMN complex. Invert Neurosci 6:145–159

Carissimi C, Saieva L, Baccon J, Chiarella P, Maiolica A, Sawyer A, Rappsilber J, Pellizzoni L (2006) Gemin8 is a novel component of the survival motor neuron complex and functions in small nuclear ribonucleoprotein assembly. J Biol Chem 281:8126–8134

Carnegie GK, Sleeman JE, Morrice N, Hastie CJ, Peggie MW, Philp A, Lamond AI, Cohen PT (2003) Protein phosphatase 4 interacts with the Survival of Motor Neurons complex and enhances the temporal localisation of snRNPs. J Cell Sci 116:1905–1913

Carrel TL, McWhorter ML, Workman E, Zhang H, Wolstencroft EC, Lorson C, Bassell GJ, Burghes AH, Beattie CE (2006) Survival motor neuron function in motor axons is independent of functions required for small nuclear ribonucleoprotein biogenesis. J Neurosci 26:11014–11022

Chan YB, Miguel-Aliaga I, Franks C, Thomas N, Trulzsch B, Sattelle DB, Davies KE, van den Heuvel M (2003) Neuromuscular defects in a Drosophila survival motor neuron gene mutant. Hum Mol Genet 12:1367–1376

Chang HC, Dimlich DN, Yokokura T, Mukherjee A, Kankel MW, Sen A, Sridhar V, Fulga TA, Hart AC, Van Vactor D, Artavanis-Tsakonas S (2008) Modeling spinal muscular atrophy in Drosophila. PLoS ONE 3:e3209

Chari A, Golas MM, Klingenhager M, Neuenkirchen N, Sander B, Englbrecht C, Sickmann A, Stark H, Fischer U (2008) An assembly chaperone collaborates with the SMN complex to generate spliceosomal SnRNPs. Cell 135:497–509

Charroux B, Pellizzoni L, Perkinson RA, Shevchenko A, Mann M, Dreyfuss G (1999) Gemin3: A novel DEAD box protein that interacts with SMN, the spinal muscular atrophy gene product, and is a component of gems. J Cell Biol 147:1181–1194

Charroux B, Pellizzoni L, Perkinson RA, Yong J, Shevchenko A, Mann M, Dreyfuss G (2000) Gemin4. A novel component of the SMN complex that is found in both gems and nucleoli. J Cell Biol 148:1177–1186

Chen CY, Shyu AB (2003) Rapid deadenylation triggered by a nonsense codon precedes decay of the RNA body in a mammalian cytoplasmic nonsense-mediated decay pathway. Mol Cell Biol 23:4805–4813

Cho S, Park SM, Kim TD, Kim JH, Kim KT, Jang SK (2007) BiP internal ribosomal entry site activity is controlled by heat-induced interaction of NSAP1. Mol Cell Biol 27:368–383

Cifuentes-Diaz C, Frugier T, Tiziano FD, Lacene E, Roblot N, Joshi V, Moreau MH, Melki J (2001) Deletion of murine SMN exon 7 directed to skeletal muscle leads to severe muscular dystrophy. J Cell Biol 152:1107–1114

Cifuentes-Diaz C, Nicole S, Velasco ME, Borra-Cebrian C, Panozzo C, Frugier T, Millet G, Roblot N, Joshi V, Melki J (2002) Neurofilament accumulation at the motor endplate and lack of axonal sprouting in a spinal muscular atrophy mouse model. Hum Mol Genet 11:1439–1447

Claus P, Doring F, Gringel S, Muller-Ostermeyer F, Fuhlrott J, Kraft T, Grothe C (2003) Differential intranuclear localization of fibroblast growth factor-2 isoforms and specific interaction with the survival of motoneuron protein. J Biol Chem 278:479–485

Coady TH, Shababi M, Tullis GE, Lorson CL (2007) Restoration of SMN function: delivery of a trans-splicing RNA re-directs SMN2 pre-mRNA splicing. Mol Ther 15:1471–1478

Corti S, Nizzardo M, Nardini M, Donadoni C, Salani S, Ronchi D, Saladino F, Bordoni A, Fortunato F, Del Bo R, Papadimitriou D, Locatelli F, Menozzi G, Strazzer S, Bresolin N, Comi GP (2008) Neural stem cell transplantation can ameliorate the phenotype of a mouse model of spinal muscular atrophy. J Clin Invest 118:3316–3330

Cote J, Richard S (2005) Tudor domains bind symmetrical dimethylated arginines. J Biol Chem 280:28476–28483

Crawford TO, Pardo CA (1996) The neurobiology of childhood spinal muscular atrophy. Neurobiol Dis 3:97–110

Deshpande DM, Kim YS, Martinez T, Carmen J, Dike S, Shats I, Rubin LL, Drummond J, Krishnan C, Hoke A, Maragakis N, Shefner J, Rothstein JD, Kerr DA (2006) Recovery from paralysis in adult rats using embryonic stem cells. Ann Neurol 60:32–44

Diers A, Kaczinski M, Grohmann K, Hubner C, Stoltenburg-Didinger G (2005) The ultrastructure of peripheral nerve, motor end-plate and skeletal muscle in patients suffering from spinal muscular atrophy with respiratory distress type 1 (SMARD1). Acta Neuropathol 110:289–297

Dimos JT, Rodolfa KT, Niakan KK, Weisenthal LM, Mitsumoto H, Chung W, Croft GF, Saphier G, Leibel R, Goland R, Wichterle H, Henderson CE, Eggan K (2008) Induced pluripotent stem cells generated from patients with ALS can be differentiated into motor neurons. Science 321:1218–1221

Duncan JE, Goldstein LS (2006) The genetics of axonal transport and axonal transport disorders. PLoS Genet 2:e124

Ebert AD, Yu J, Rose FF Jr, Mattis VB, Lorson CL, Thomson JA, Svendsen CN (2009) Induced pluripotent stem cells from a spinal muscular atrophy patient. Nature 457:277–280

Eggert C, Chari A, Laggerbauer B, Fischer U (2006) Spinal muscular atrophy: the RNP connection. Trends Mol Med 12:113–121

Espert L, Eldin P, Gongora C, Bayard B, Harper F, Chelbi-Alix MK, Bertrand E, Degols G, Mechti N (2006) The exonuclease ISG20 mainly localizes in the nucleolus and the Cajal (Coiled) bodies and is associated with nuclear SMN protein-containing complexes. J Cell Biochem 98:1320–1333

Fan L, Simard LR (2002) Survival motor neuron (SMN) protein: role in neurite outgrowth and neuromuscular maturation during neuronal differentiation and development. Hum Mol Genet 11:1605–1614

Farina K, Oleynikov Y, Singer RH (2001) Interaction of ZBP1 KH Domains with the b-actin zipcode and other RNA sequences. Molecular Biology Cell 12:1996

Ferri A, Sanes JR, Coleman MP, Cunningham JM, Kato AC (2003) Inhibiting axon degeneration and synapse loss attenuates apoptosis and disease progression in a mouse model of motoneuron disease. Curr Biol 13:669–673

Fischer U, Liu Q, Dreyfuss G (1997) The SMN-SIP1 complex has an essential role in spliceosomal snRNP biogenesis. Cell 90:1023–1029

Foust KD, Nurre E, Montgomery CL, Hernandez A, Chan CM, Kaspar BK (2009) Intravascular AAV9 preferentially targets neonatal neurons and adult astrocytes. Nat Biotechnol 27:59–65

Friesen WJ, Dreyfuss G (2000) Specific sequences of the Sm and Sm-like (Lsm) proteins mediate their interaction with the spinal muscular atrophy disease gene product (SMN). J Biol Chem 275:26370–26375

Friesen WJ, Massenet S, Paushkin S, Wyce A, Dreyfuss G (2001a) SMN, the product of the spinal muscular atrophy gene, binds preferentially to dimethylarginine-containing protein targets. Mol Cell 7:1111–1117

Friesen WJ, Paushkin S, Wyce A, Massenet S, Pesiridis GS, Van Duyne G, Rappsilber J, Mann M, Dreyfuss G (2001b) The methylosome, a 20S complex containing JBP1 and pICln, produces dimethylarginine-modified Sm proteins. Mol Cell Biol 21:8289–8300

Frugier T, Tiziano FD, Cifuentes-Diaz C, Miniou P, Roblot N, Dierich A, Le Meur M, Melki J (2000) Nuclear targeting defect of SMN lacking the C-terminus in a mouse model of spinal muscular atrophy. Hum Mol Genet 9:849–858

Gabanella F, Butchbach ME, Saieva L, Carissimi C, Burghes AH, Pellizzoni L (2007) Ribonucleoprotein assembly defects correlate with spinal muscular atrophy severity and preferentially affect a subset of spliceosomal snRNPs. PLoS ONE 2:e921

Gangwani L, Mikrut M, Theroux S, Sharma M, Davis RJ (2001) Spinal muscular atrophy disrupts the interaction of ZPR1 with the SMN protein. Nat Cell Biol 3:376–383

Gavrilina TO, McGovern VL, Workman E, Crawford TO, Gogliotti RG, DiDonato CJ, Monani UR, Morris GE, Burghes AH (2008) Neuronal SMN expression corrects spinal muscular atrophy in severe SMA mice while muscle-specific SMN expression has no phenotypic effect. Hum Mol Genet 17:1063–1075

Giesemann T, Rathke-Hartlieb S, Rothkegel M, Bartsch JW, Buchmeier S, Jockusch BM, Jockusch H (1999a) A role for polyproline motifs in the spinal muscular atrophy protein SMN. Profilins bind to and colocalize with smn in nuclear gems. J Biol Chem 274:37908–37914

Giesemann T, Rathke-Hartlieb S, Rothkegel M, Bartsch JW, Buchmeier S, Jockusch BM, Jockusch H (1999b) A role for polyproline motifs in the spinal muscular atrophy protein SMN. Profilins bind to and colocalize with smn in nuclear gems. J Biol Chem 274:37908–37914

Girard LR, Fiedler TJ, Harris TW, Carvalho F, Antoshechkin I, Han M, Sternberg PW, Stein LD, Chalfie M (2007) WormBook: the online review of *Caenorhabditis elegans* biology. Nucleic Acids Res 35:D472–D475

Gross SR, Kinzy TG (2005) Translation elongation factor 1A is essential for regulation of the actin cytoskeleton and cell morphology. Nat Struct Mol Biol 12:772–778

Grosset C, Chen CY, Xu N, Sonenberg N, Jacquemin-Sablon H, Shyu AB (2000) A mechanism for translationally coupled mRNA turnover: interaction between the poly(A) tail and a c-fos RNA coding determinant via a protein complex. Cell 103:29–40

Gu W, Pan F, Zhang H, Bassell GJ, Singer RH (2002) A predominantly nuclear protein affecting cytoplasmic localization of beta-actin mRNA in fibroblasts and neurons. J Cell Biol 156:41–51

Gubitz AK, Mourelatos Z, Abel L, Rappsilber J, Mann M, Dreyfuss G (2002) Gemin5, a novel WD repeat protein component of the SMN complex that binds Sm proteins. J Biol Chem 277:5631–5636

Gubitz AK, Feng W, Dreyfuss G (2004) The SMN complex. Exp Cell Res 296:51–56

Harris CE, Boden RA, Astell CR (1999) A novel heterogeneous nuclear ribonucleoprotein-like protein interacts with NS1 of the minute virus of mice. J Virol 73:72–80

Hausmanowa-Petrusewicz I, Vrbova G (2005) Spinal muscular atrophy: a delayed development hypothesis. Neuroreport 16:657–661

Hebert MD, Szymczyk PW, Shpargel KB, Matera AG (2001) Coilin forms the bridge between Cajal bodies and SMN, the spinal muscular atrophy protein. Genes Dev 15:2720–2729

Hebert MD, Shpargel KB, Ospina JK, Tucker KE, Matera AG (2002) Coilin methylation regulates nuclear body formation. Dev Cell 3:329–337

Hsieh-Li HM, Chang JG, Jong YJ, Wu MH, Wang NM, Tsai CH, Li H (2000) A mouse model for spinal muscular atrophy. Nat Genet 24:66–70

Hua Y, Zhou J (2004a) Rpp20 interacts with SMN and is re-distributed into SMN granules in response to stress. Biochem Biophys Res Commun 314:268–276

Hua Y, Zhou J (2004b) Survival motor neuron protein facilitates assembly of stress granules. FEBS Lett 572:69–74

Hua Y, Vickers TA, Okunola HL, Bennett CF, Krainer AR (2008) Antisense masking of an hnRNP A1/A2 intronic splicing silencer corrects SMN2 splicing in transgenic mice. Am J Hum Genet 82:834–848

Huttelmaier S, Zenklusen D, Lederer M, Dictenberg J, Lorenz M, Meng X, Bassell GJ, Condeelis J, Singer RH (2005) Spatial regulation of β-actin translation by Src- dependent phosphorylation of ZBP1. Nature 438:512–515

Irwin N, Baekelandt V, Goritchenko L, Benowitz LI (1997) Identification of two proteins that bind to a pyrimidine-rich sequence in the 3′-untranslated region of GAP-43 mRNA. Nucleic Acids Res 25:1281–1288

Iwahashi H, Eguchi Y, Yasuhara N, Hanafusa T, Matsuzawa Y, Tsujimoto Y (1997) Synergistic anti-apoptotic activity between Bcl-2 and SMN implicated in spinal muscular atrophy. Nature 390:413–417

Jablonka S, Rossoll W, Schrank B, Sendtner M (2000) The role of SMN in spinal muscular atrophy. J Neurol 247:I37–42

Jablonka S, Bandilla M, Wiese S, Buhler D, Wirth B, Sendtner M, Fischer U (2001) Co-regulation of survival of motor neuron (SMN) protein and its interactor SIP1 during development and in spinal muscular atrophy. Hum Mol Genet 10:497–505

Jablonka S, Karle K, Sandner B, Andreassi C, von Au K, Sendtner M (2006) Distinct and overlapping alterations in motor and sensory neurons in a mouse model of spinal muscular atrophy. Hum Mol Genet 15:511–518

Jablonka S, Beck M, Lechner BD, Mayer C, Sendtner M (2007) Defective Ca2+ channel clustering in axon terminals disturbs excitability in motoneurons in spinal muscular atrophy. J Cell Biol 179:139–149

Jones KW, Gorzynski K, Hales CM, Fischer U, Badbanchi F, Terns RM, Terns MP (2001) Direct interaction of the spinal muscular atrophy disease protein SMN with the small nucleolar RNA-associated protein fibrillarin. J Biol Chem 276:38645–38651

Kanai Y, Dohmae N, Hirokawa N (2004) Kinesin transports RNA: isolation and characterization of an RNA-transporting granule. Neuron 43:513–525

Kariya S, Park GH, Maeno-Hikichi Y, Leykekhman O, Lutz C, Arkovitz MS, Landmesser LT, Monani UR (2008) Reduced SMN protein impairs maturation of the neuromuscular junctions in mouse models of spinal muscular atrophy. Hum Mol Genet 17:2552–2569

Kariya S, Mauricio R, Dai Y, Monani UR (2009) The neuroprotective factor Wld(s) fails to mitigate distal axonal and neuromuscular junction (NMJ) defects in mouse models of spinal muscular atrophy. Neurosci Lett 449:246–251

Kedersha N, Stoecklin G, Ayodele M, Yacono P, Lykke-Andersen J, Fitzler MJ, Scheuner D, Kaufman RJ, Golan DE, Anderson P (2005) Stress granules and processing bodies are dynamically linked sites of mRNP remodeling. J Cell Biol 169:871–884

Kiebler MA, Bassell GJ (2006) Neuronal RNA granules: movers and makers. Neuron 51:685–690

Kim JH, Paek KY, Ha SH, Cho S, Choi K, Kim CS, Ryu SH, Jang SK (2004) A cellular RNA-binding protein enhances internal ribosomal entry site-dependent translation through an interaction downstream of the hepatitis C virus polyprotein initiation codon. Mol Cell Biol 24:7878–7890

Kim TD, Kim JS, Kim JH, Myung J, Chae HD, Woo KC, Jang SK, Koh DS, Kim KT (2005) Rhythmic serotonin N-acetyltransferase mRNA degradation is essential for the maintenance of its circadian oscillation. Mol Cell Biol 25:3232–3246

Kim TD, Woo KC, Cho S, Ha DC, Jang SK, Kim KT (2007) Rhythmic control of AANAT translation by hnRNP Q in circadian melatonin production. Genes Dev 21:797–810

Kindler S, Wang H, Richter D, Tiedge H (2005) RNA transport and local control of translation. Annu Rev Cell Dev Biol 21:223–245

Kolb SJ, Battle DJ, Dreyfuss G (2007) Molecular functions of the SMN complex. J Child Neurol 22:990–994

Kong L, Wang X, Choe DW, Polley M, Burnett BG, Bosch-Marce M, Griffin JW, Rich MM, Sumner CJ (2009) Impaired synaptic vesicle release and immaturity of neuromuscular junctions in spinal muscular atrophy mice. J Neurosci 29:842–851

Kroiss M, Schultz J, Wiesner J, Chari A, Sickmann A, Fischer U (2008) Evolution of an RNP assembly system: a minimal SMN complex facilitates formation of UsnRNPs in Drosophila melanogaster. Proc Natl Acad Sci USA 105:10045–10050

Kurihara N, Menaa C, Maeda H, Haile DJ, Reddy SV (2001) Osteoclast-stimulating factor interacts with the spinal muscular atrophy gene product to stimulate osteoclast formation. J Biol Chem 276:41035–41039

Le TT, Coovert DD, Monani UR, Morris GE, Burghes AH (2000) The survival motor neuron (SMN) protein: effect of exon loss and mutation on protein localization. Neurogenetics 3:7–16

Le TT, Pham LT, Butchbach ME, Zhang HL, Monani UR, Coovert DD, Gavrilina TO, Xing L, Bassell GJ, Burghes AH (2005) SMNDelta7, the major product of the centromeric survival motor neuron (SMN2) gene, extends survival in mice with spinal muscular atrophy and associates with full-length SMN. Hum Mol Genet 14:845–857

Lefebvre S, Burglen L, Reboullet S, Clermont O, Burlet P, Viollet L, Benichou B, Cruaud C, Millasseau P, Zeviani M, Le Paslier D, Frézal J, Cohen D, Weissenbach J, Munnich A, Melki J (1995) Identification and characterization of a spinal muscular atrophy-determining gene. Cell 80:155–165

Lefebvre S, Burlet P, Viollet L, Bertrandy S, Huber C, Belser C, Munnich A (2002) A novel association of the SMN protein with two major non-ribosomal nucleolar proteins and its implication in spinal muscular atrophy. Hum Mol Genet 11:1017–1027

Leung KM, van Horck FP, Lin AC, Allison R, Standart N, Holt CE (2006) Asymmetrical β-actin mRNA translation in growth cones mediates attractive turning to netrin-1. Nat Neurosci 9:1247–1256

Lin AC, Holt CE (2008) Function and regulation of local axonal translation. Curr Opin Neurobiol 18:60–68

Liu Q, Dreyfuss G (1996) A novel nuclear structure containing the survival of motor neurons protein. EMBO J 15:3555–3565

Liu Q, Fischer U, Wang F, Dreyfuss G (1997) The spinal muscular atrophy disease gene product, SMN, and its associated protein SIP1 are in a complex with spliceosomal snRNP proteins. Cell 90:1013–1021

Lorson CL, Androphy EJ (1998) The domain encoded by exon 2 of the survival motor neuron protein mediates nucleic acid binding. Hum Mol Genet 7:1269–1275

Lorson CL, Strasswimmer J, Yao JM, Baleja JD, Hahnen E, Wirth B, Le T, Burghes AH, Androphy EJ (1998) SMN oligomerization defect correlates with spinal muscular atrophy severity. Nat Genet 19:63–66

Lorson CL, Hahnen E, Androphy EJ, Wirth B (1999) A single nucleotide in the SMN gene regulates splicing and is responsible for spinal muscular atrophy. Proc Natl Acad Sci USA 96:6307–6311

Lorson CL, Androphy EJ (2000) An exonic enhancer is required for inclusion of an essential exon in the SMA-determining gene SMN. Hum Mol Genet 9:259–265

Lowry WE, Plath K (2008) The many ways to make an iPS cell. Nat Biotechnol 26:1246–1248

Lu B, Vogel H (2008) Drosophila Models of Neurodegenerative Diseases. Annu Rev Pathol doi:10.1146/annurev.pathol.3.121806.151529

Lunn MR, Stockwell BR (2005) Chemical genetics and orphan genetic diseases. Chem Biol 12:1063–1073

Lunn MR, Wang CH (2008) Spinal muscular atrophy. Lancet 371:2120–2133

Madocsai C, Lim SR, Geib T, Lam BJ, Hertel KJ (2005) Correction of SMN2 Pre- mRNA splicing by antisense U7 small nuclear RNAs. Mol Ther 12:1013–1022

Marquis J, Meyer K, Angehrn L, Kampfer SS, Rothen-Rutishauser B, Schumperli D (2007) Spinal muscular atrophy: SMN2 pre-mRNA splicing corrected by a U7 snRNA derivative carrying a splicing enhancer sequence. Mol Ther 15:1479–1486

McGovern VL, Gavrilina TO, Beattie CE, Burghes AH (2008) Embryonic motor axon development in the severe SMA mouse. Hum Mol Genet 17:2900–2909

McWhorter ML, Monani UR, Burghes AH, Beattie CE (2003) Knockdown of the survival motor neuron (Smn) protein in zebrafish causes defects in motor axon outgrowth and pathfinding. J Cell Biol 162:919–931

McWhorter ML, Boon KL, Horan ES, Burghes AH, Beattie CE (2008) The SMN binding protein Gemin2 is not involved in motor axon outgrowth. Dev Neurobiol 68:182–194

Meister G, Buhler D, Laggerbauer B, Zobawa M, Lottspeich F, Fischer U (2000) Characterization of a nuclear 20S complex containing the survival of motor neurons (SMN) protein and a specific subset of spliceosomal Sm proteins. Hum Mol Genet 9:1977–1986

Meister G, Buhler D, Pillai R, Lottspeich F, Fischer U (2001) A multiprotein complex mediates the ATP-dependent assembly of spliceosomal U snRNPs. Nat Cell Biol 3:945–949

Meister G, Eggert C, Fischer U (2002) SMN-mediated assembly of RNPs: a complex story. Trends Cell Biol 12:472–478

Meister G, Fischer U (2002) Assisted RNP assembly: SMN and PRMT5 complexes cooperate in the formation of spliceosomal UsnRNPs. EMBO J 21:5853–5863

Melki J, Lefebvre S, Burglen L, Burlet P, Clermont O, Millasseau P, Reboullet S, Benichou B, Zeviani M, Le Paslier D, et al. (1994) De novo and inherited deletions of the 5q13 region in spinal muscular atrophies. Science 264:1474–1477

Miguel-Aliaga I, Culetto E, Walker DS, Baylis HA, Sattelle DB, Davies KE (1999) The *Caenorhabditis elegans* orthologue of the human gene responsible for spinal muscular atrophy is a maternal product critical for germline maturation and embryonic viability. Hum Mol Genet 8:2133–2143

Miguel-Aliaga I, Chan YB, Davies KE, van den Heuvel M (2000) Disruption of SMN function by ectopic expression of the human SMN gene in Drosophila. FEBS Lett 486:99–102

Min H, Turck CW, Nikolic JM, Black DL (1997) A new regulatory protein, KSRP, mediates exon inclusion through an intronic splicing enhancer. Genes Dev 11:1023–1036

Mishra AK, Gangwani L, Davis RJ, Lambright DG (2007) Structural insights into the interaction of the evolutionarily conserved ZPR1 domain tandem with eukaryotic EF1A, receptors, and SMN complexes. Proc Natl Acad Sci USA 104:13930–13935

Mizutani A, Fukuda M, Ibata K, Shiraishi Y, Mikoshiba K (2000) SYNCRIP, a cytoplasmic counterpart of heterogeneous nuclear ribonucleoprotein R, interacts with ubiquitous synaptotagmin isoforms. J Biol Chem 275:9823–9831

Monani UR, Coovert DD, Burghes AH (2000a) Animal models of spinal muscular atrophy. Hum Mol Genet 9:2451–2457

Monani UR, Sendtner M, Coovert DD, Parsons DW, Andreassi C, Le TT, Jablonka S, Schrank B, Rossol W, Prior TW, Morris GE, Burghes AH (2000b) The human centromeric survival motor neuron gene (SMN2) rescues embryonic lethality in Smn(−/−) mice and results in a mouse with spinal muscular atrophy. Hum Mol Genet 9:333–339

Monani UR, Pastore MT, Gavrilina TO, Jablonka S, Le TT, Andreassi C, DiCocco JM, Lorson C, Androphy EJ, Sendtner M, Podell M, Burghes AH (2003) A transgene carrying an A2G missense mutation in the SMN gene modulates phenotypic severity in mice with severe (type I) spinal muscular atrophy. J Cell Biol 160:41–52

Mouaikel J, Narayanan U, Verheggen C, Matera AG, Bertrand E, Tazi J, Bordonne R (2003) Interaction between the small-nuclear-RNA cap hypermethylase and the spinal muscular atrophy protein, survival of motor neuron. EMBO Rep 4:616–622

Mourelatos Z, Abel L, Yong J, Kataoka N, Dreyfuss G (2001) SMN interacts with a novel family of hnRNP and spliceosomal proteins. EMBO J 20:5443–5452

Murray LM, Comley LH, Thomson D, Parkinson N, Talbot K, Gillingwater TH (2008) Selective vulnerability of motor neurons and dissociation of pre- and post-synaptic pathology at the neuromuscular junction in mouse models of spinal muscular atrophy. Hum Mol Genet 17:949–962

Narayanan U, Ospina JK, Frey MR, Hebert MD, Matera AG (2002) SMN, the spinal muscular atrophy protein, forms a pre-import snRNP complex with snurportin1 and importin beta. Hum Mol Genet 11:1785–1795

Nayak MS, Kim YS, Goldman M, Keirstead HS, Kerr DA (2006) Cellular therapies in motor neuron diseases. Biochim Biophys Acta 1762:1128–1138

Neubauer G, King A, Rappsilber J, Calvio C, Watson M, Ajuh P, Sleeman J, Lamond A, Mann M (1998) Mass spectrometry and EST-database searching allows characterization of the multi-protein spliceosome complex. Nat Genet 20:46–50

Nicole S, Diaz CC, Frugier T, Melki J (2002) Spinal muscular atrophy: Recent advances and future prospects. Muscle Nerve 26:4–13

Oleynikov Y, Singer RH (2003) Real-time visualization of ZBP1 association with β-actin mRNA during transcription and localization. Curr Biol 13:199–207

Oprea GE, Krober S, McWhorter ML, Rossoll W, Muller S, Krawczak M, Bassell GJ, Beattie CE, Wirth B (2008) Plastin 3 is a protective modifier of autosomal recessive spinal muscular atrophy. Science 320:524–527

Oskoui M, Levy G, Garland CJ, Gray JM, O'Hagen J, De Vivo DC, Kaufmann P (2007) The changing natural history of spinal muscular atrophy type 1. Neurology 69:1931–1936

Oskoui M, Kaufmann P (2008) Spinal muscular atrophy. Neurotherapeutics 5:499–506

Otter S, Grimmler M, Neuenkirchen N, Chari A, Sickmann A, Fischer U (2007) A comprehensive interaction map of the human survival of motor neuron (SMN) complex. J Biol Chem 282:5825–5833

Pagliardini S, Giavazzi A, Setola V, Lizier C, Di Luca M, DeBiasi S, Battaglia G (2000) Subcellular localization and axonal transport of the survival motor neuron (SMN) protein in the developing rat spinal cord. Hum Mol Genet 9:47–56

Pan F, Huttelmaier S, Singer RH, Gu W (2007) ZBP2 facilitates binding of ZBP1 to β-actin mRNA during transcription. Mol Cell Biol 27:8340–8351

Park JW, Voss PG, Grabski S, Wang JL, Patterson RJ (2001) Association of galectin-1 and galectin-3 with Gemin4 in complexes containing the SMN protein. Nucleic Acids Res 29:3595–3602

Paushkin S, Gubitz AK, Massenet S, Dreyfuss G (2002) The SMN complex, an assemblyosome of ribonucleoproteins. Curr Opin Cell Biol 14:305–312

Pellizzoni L, Charroux B, Dreyfuss G (1999) SMN mutants of spinal muscular atrophy patients are defective in binding to snRNP proteins. Proc Natl Acad Sci USA 96:11167–11172

Pellizzoni L, Baccon J, Charroux B, Dreyfuss G (2001a) The survival of motor neurons (SMN) protein interacts with the snoRNP proteins fibrillarin and GAR1. Curr Biol 11:1079–1088

Pellizzoni L, Charroux B, Rappsilber J, Mann M, Dreyfuss G (2001b) A functional interaction between the survival motor neuron complex and RNA polymerase II. J Cell Biol 152:75–85

Pellizzoni L, Baccon J, Rappsilber J, Mann M, Dreyfuss G (2002a) Purification of native survival of motor neurons complexes and identification of Gemin6 as a novel component. J Biol Chem 277:7540–7545

Pellizzoni L, Yong J, Dreyfuss G (2002b) Essential role for the SMN complex in the specificity of snRNP assembly. Science 298:1775–1779

Peri S et al. (2003) Development of human protein reference database as an initial platform for approaching systems biology in humans. Genome Res 13:2363–2371

Piazzon N, Rage F, Schlotter F, Moine H, Branlant C, Massenet S (2008) In vitro and in cellulo evidences for association of the survival of motor neuron complex with the fragile X mental retardation protein. J Biol Chem 283:5598–5610

Ponting CP (1997) Tudor domains in proteins that interact with RNA. Trends Biochem Sci 22:51–52

Pun S, Sigrist M, Santos AF, Ruegg MA, Sanes JR, Jessell TM, Arber S, Caroni P (2002) An intrinsic distinction in neuromuscular junction assembly and maintenance in different skeletal muscles. Neuron 34:357–370

Rajendra TK, Gonsalvez GB, Walker MP, Shpargel KB, Salz HK, Matera AG (2007) A *Drosophila melanogaster* model of spinal muscular atrophy reveals a function for SMN in striated muscle. J Cell Biol 176:831–841

Rehbein M, Wege K, Buck F, Schweizer M, Richter D, Kindler S (2002) Molecular characterization of MARTA1, a protein interacting with the dendritic targeting element of MAP2 mRNAs. J Neurochem 82:1039–1046

Rochette CF, Gilbert N, Simard LR (2001) SMN gene duplication and the emergence of the SMN2 gene occurred in distinct hominids: SMN2 is unique to Homo sapiens. Hum Genet 108:255–266

Rose FF Jr, Meehan PW, Coady TH, Garcia VB, Garcia ML, Lorson CL (2008) The Wallerian degeneration slow (Wld(s)) gene does not attenuate disease in a mouse model of spinal muscular atrophy. Biochem Biophys Res Commun 375:119–123

Ross AF, Oleynikov Y, Kislauskis EH, Taneja KL, Singer RH (1997) Characterization of a β-actin mRNA zipcode-binding protein. Mol Cell Biol 17:2158–2165

Rossoll W, Kroning AK, Ohndorf UM, Steegborn C, Jablonka S, Sendtner M (2002) Specific interaction of Smn, the spinal muscular atrophy determining gene product, with hnRNP-R and gry-rbp/hnRNP-Q: a role for Smn in RNA processing in motor axons? Hum Mol Genet 11:93–105

Rossoll W, Jablonka S, Andreassi C, Kroning AK, Karle K, Monani UR, Sendtner M (2003) Smn, the spinal muscular atrophy-determining gene product, modulates axon growth and localization of beta-actin mRNA in growth cones of motoneurons. J Cell Biol 163:801–812

Rual JF et al. (2005) Towards a proteome-scale map of the human protein-protein interaction network. Nature 437:1173–1178

Rudnik-Schoneborn S, Goebel HH, Schlote W, Molaian S, Omran H, Ketelsen U, Korinthenberg R, Wenzel D, Lauffer H, Kreiss-Nachtsheim M, Wirth B, Zerres K (2003) Classical infantile spinal muscular atrophy with SMN deficiency causes sensory neuronopathy. Neurology 60:983–987

Santos AF, Caroni P (2003) Assembly, plasticity and selective vulnerability to disease of mouse neuromuscular junctions. J Neurocytol 32:849–862

Saunders LR, Perkins DJ, Balachandran S, Michaels R, Ford R, Mayeda A, Barber GN (2001) Characterization of two evolutionarily conserved, alternatively spliced nuclear phosphoproteins, NFAR-1 and -2, that function in mRNA processing and interact with the double-stranded RNA-dependent protein kinase, PKR. J Biol Chem 276:32300–32312

Schmid A, DiDonato CJ (2007) Animal models of spinal muscular atrophy. J Child Neurol 22:1004–1012

Schrank B, Gotz R, Gunnersen JM, Ure JM, Toyka KV, Smith AG, Sendtner M (1997) Inactivation of the survival motor neuron gene, a candidate gene for human spinal muscular atrophy, leads to massive cell death in early mouse embryos. Proc Natl Acad Sci USA 94:9920–9925

Schultz J, Milpetz F, Bork P, Ponting CP (1998) SMART, a simple modular architecture research tool: identification of signaling domains. Proc Natl Acad Sci U S A 95:5857–5864

Selenko P, Sprangers R, Stier G, Buhler D, Fischer U, Sattler M (2001) SMN tudor domain structure and its interaction with the Sm proteins. Nat Struct Biol 8:27–31

Setola V, Terao M, Locatelli D, Bassanini S, Garattini E, Battaglia G (2007) Axonal- SMN (a-SMN), a protein isoform of the survival motor neuron gene, is specifically involved in axonogenesis. Proc Natl Acad Sci USA 104:1959–1964

Shafey D, MacKenzie AE, Kothary R (2008) Neurodevelopmental abnormalities in neurosphere-derived neural stem cells from SMN-depleted mice. J Neurosci Res 86:2839–2847

Sharma A, Lambrechts A, Hao le T, Le TT, Sewry CA, Ampe C, Burghes AH, Morris GE (2005) A role for complexes of survival of motor neurons (SMN) protein with gemins and profilin in neurite-like cytoplasmic extensions of cultured nerve cells. Exp Cell Res 309:185–197

Shupliakov O, Bloom O, Gustafsson JS, Kjaerulff O, Low P, Tomilin N, Pieribone VA, Greengard P, Brodin L (2002) Impaired recycling of synaptic vesicles after acute perturbation of the presynaptic actin cytoskeleton. Proc Natl Acad Sci USA 99:14476–14481

Simic G (2008) Pathogenesis of proximal autosomal recessive spinal muscular atrophy. Acta Neuropathol 116:223–234

Simic G, Mladinov M, Seso Simic D, Jovanov Milosevic N, Islam A, Pajtak A, Barisic N, Sertic J, Lucassen PJ, Hof PR, Kruslin B (2008) Abnormal motoneuron migration, differentiation, and axon outgrowth in spinal muscular atrophy. Acta Neuropathol 115:313–326

Singh RN (2007) Evolving concepts on human SMN pre-mRNA splicing. RNA Biol 4:7–10

Smith CL, Afroz R, Bassell GJ, Furneaux HM, Perrone-Bizzozero NI, Burry RW (2004) GAP-43 mRNA in growth cones is associated with HuD and ribosomes. J Neurobiol 61:222–235

Sprangers R, Groves MR, Sinning I, Sattler M (2003) High-resolution X-ray and NMR structures of the SMN Tudor domain: conformational variation in the binding site for symmetrically dimethylated arginine residues. J Mol Biol 327:507–520

Stark C, Breitkreutz BJ, Reguly T, Boucher L, Breitkreutz A, Tyers M (2006) BioGRID: a general repository for interaction datasets. Nucleic Acids Res 34:D535–539

Stelzl U et al. (2005) A human protein–protein interaction network: a resource for annotating the proteome. Cell 122:957–968

Stohr N, Lederer M, Reinke C, Meyer S, Hatzfeld M, Singer RH, Huttelmaier S (2006) ZBP1 regulates mRNA stability during cellular stress. J Cell Biol 175:527–534

Strasswimmer J, Lorson CL, Breiding DE, Chen JJ, Le T, Burghes AH, Androphy EJ (1999) Identification of survival motor neuron as a transcriptional activator-binding protein. Hum Mol Genet 8:1219–1226

Sumner CJ (2006) Therapeutics development for spinal muscular atrophy. NeuroRx 3:235–245

Tadesse H, Deschenes-Furry J, Boisvenue S, Cote J (2008) KH-type splicing regulatory protein interacts with survival motor neuron protein and is misregulated in spinal muscular atrophy. Hum Mol Genet 17:506–524

Trinkle-Mulcahy L, Boulon S, Lam YW, Urcia R, Boisvert FM, Vandermoere F, Morrice NA, Swift S, Rothbauer U, Leonhardt H, Lamond A (2008) Identifying specific protein interaction partners using quantitative mass spectrometry and bead proteomes. J Cell Biol 183:223–239

Tsai LK, Tsai MS, Ting CH, Wang SH, Li H (2008) Restoring Bcl-x(L) levels benefits a mouse model of spinal muscular atrophy. Neurobiol Dis 31:361–367

Vessey JP, Vaccani A, Xie Y, Dahm R, Karra D, Kiebler MA, Macchi P (2006) Dendritic localization of the translational repressor Pumilio 2 and its contribution to dendritic stress granules. J Neurosci 26:6496–6508

Vitte JM, Davoult B, Roblot N, Mayer M, Joshi V, Courageot S, Tronche F, Vadrot J, Moreau MH, Kemeny F, Melki J (2004) Deletion of murine Smn exon 7 directed to liver leads to severe defect of liver development associated with iron overload. Am J Pathol 165:1731–1741

Wang IF, Reddy NM, Shen CK (2002) Higher order arrangement of the eukaryotic nuclear bodies. Proc Natl Acad Sci USA 99:13583–13588

Wichterle H, Lieberam I, Porter JA, Jessell TM (2002) Directed differentiation of embryonic stem cells into motor neurons. Cell 110:385–397

Williams BY, Hamilton SL, Sarkar HK (2000a) The SMN protein interacts with the tranactivator FUSE Binding protein from human fetal brain. FEBBS 470:207–210

Williams BY, Hamilton SL, Sarkar HK (2000b) The survival motor neuron protein interacts with the transactivator FUSE binding protein from human fetal brain. FEBS Lett 470:207–210

Wilson PG, Cherry JJ, Schwamberger S, Adams AM, Zhou J, Shin S, Stice SL (2007) An SMA project report: neural cell-based assays derived from human embryonic stem cells. Stem Cells Dev 16:1027–1041

Winkler C, Eggert C, Gradl D, Meister G, Giegerich M, Wedlich D, Laggerbauer B, Fischer U (2005) Reduced U snRNP assembly causes motor axon degeneration in an animal model for spinal muscular atrophy. Genes Dev 19:2320–2330

Wirth B (2000) An update of the mutation spectrum of the survival motor neuron gene (SMN1) in autosomal recessive spinal muscular atrophy (SMA). Hum Mutat 15:228–237

Wirth B, Brichta L, Hahnen E (2006) Spinal muscular atrophy: from gene to therapy. Semin Pediatr Neurol 13:121–131

Yang WH, Yu JH, Gulick T, Bloch KD, Bloch DB (2006) RNA-associated protein 55 (RAP55) localizes to mRNA processing bodies and stress granules. RNA 12:547–554

Yao J, Sasaki Y, Wen Z, Bassell GJ, Zheng JQ (2006) An essential role for beta-actin mRNA localization and translation in Ca2+-dependent growth cone guidance. Nat Neurosci 9:1265–1273

Yong J, Wan L, Dreyfuss G (2004) Why do cells need an assembly machine for RNA- protein complexes? Trends Cell Biol 14:226–232

Young PJ, Le TT, thi Man N, Burghes AH, Morris GE (2000a) The relationship between SMN, the spinal muscular atrophy protein, and nuclear coiled bodies in differentiated tissues and cultured cells. Exp Cell Res 256:365–374

Young PJ, Man NT, Lorson CL, Le TT, Androphy EJ, Burghes AH, Morris GE (2000b) The exon 2b region of the spinal muscular atrophy protein, SMN, is involved in self- association and SIP1 binding. Hum Mol Genet 9:2869–2877

Young PJ, Day PM, Zhou J, Androphy EJ, Morris GE, Lorson CL (2002a) A direct interaction between the survival motor neuron protein and p53 and its relationship to spinal muscular atrophy. J Biol Chem 277:2852–2859

Young PJ, Jensen KT, Burger LR, Pintel DJ, Lorson CL (2002b) Minute virus of mice NS1 interacts with the SMN protein, and they colocalize in novel nuclear bodies induced by parvovirus infection. J Virol 76:3892–3904

Young PJ, Francis JW, Lince D, Coon K, Androphy EJ, Lorson CL (2003) The Ewing's sarcoma protein interacts with the Tudor domain of the survival motor neuron protein. Brain Res Mol Brain Res 119:37–49

Zhang H, Xing L, Rossoll W, Wichterle H, Singer RH, Bassell GJ (2006) Multiprotein complexes of the survival of motor neuron protein SMN with Gemins traffic to neuronal processes and growth cones of motor neurons. J Neurosci 26:8622–8632

Zhang H, Xing L, Singer RH, Bassell GJ (2007) QNQKE targeting motif for the SMN- Gemin multiprotein complex in neurons. J Neurosci Res 85:2657–2667

Zhang HL, Singer RH, Bassell GJ (1999) Neurotrophin regulation of beta-actin mRNA and protein localization within growth cones. J Cell Biol 147:59–70

Zhang HL, Eom T, Oleynikov Y, Shenoy SM, Liebelt DA, Dictenberg JB, Singer RH, Bassell GJ (2001) Neurotrophin-induced transport of a β-actin mRNP complex increases beta-actin levels and stimulates growth cone motility. Neuron 31:261–275

Zhang HL, Pan F, Hong D, Shenoy SM, Singer RH, Bassell GJ (2003) Active transport of the survival motor neuron protein and the role of exon-7 in cytoplasmic localization. J Neurosci 23:6627–6637

Zhang Z, Lotti F, Dittmar K, Younis I, Wan L, Kasim M, Dreyfuss G (2008) SMN deficiency causes tissue-specific perturbations in the repertoire of snRNAs and widespread defects in splicing. Cell 133:585–600

Zou J, Barahmand-pour F, Blackburn ML, Matsui Y, Chansky HA, Yang L (2004) Survival motor neuron (SMN) protein interacts with transcription corepressor mSin3A. J Biol Chem 279:14922–14928

Retrograde Injury Signaling in Lesioned Axons

Keren Ben-Yaakov and Mike Fainzilber

Abstract The cell body of a lesioned neuron must receive accurate and timely information on the site and extent of axonal damage, in order to mount an appropriate response. Specific mechanisms must therefore exist to transmit such information along the length of the axon from the lesion site to the cell body. Three distinct types of signals have been postulated to underlie this process, starting with injury-induced discharge of axon potentials, and continuing with two distinct types of retrogradely transported macromolecular signals. The latter includes, on the one hand, an interruption of the normal supply of retrogradely transported trophic factors from the target, and, on the other hand, activated proteins originating from the injury site. This chapter reviews the progress on understanding the different mechanistic aspects of the axonal response to injury, and how the information is conveyed from the injury site to the cell body to initiate regeneration.

1 Introduction

Axonal injury to peripheral neurons induces a cell body reaction, which involves profound changes in transcription, translation, and posttranslational modifications in the neuronal cell body to achieve successful regeneration (Abe and Cavalli 2008; Hanz and Fainzilber 2006; Rossi et al. 2007). Studies in peripheral sensory neurons have provided compelling evidence for the importance of retrogradely transported injury signals for initiation of a regeneration response. Sensory neurons of the dorsal root ganglion (DRG) possess two axonal branches – a peripheral branch that regenerates when injured, and a central branch that enters the spinal cord and does not regenerate following injury. The failure of the latter is partially attributable to an inhibitory environment in the damaged central nervous system (CNS), where myelin-associated inhibitors and glial scar tissues are considered the main obstacles, providing inhibitory signals and physical barriers (Gervasi et al. 2008; Kim and Snider 2008;

K. Ben-Yaakov and M. Fainzilber (✉)
Department of Biological Chemistry, Weizmann Institute of Science, 76100 Rehovot, Israel
e-mail: mike.fainzilber@weizmann.ac.il

Yamashita et al. 2005). However, a conditioning lesion (Smith and Skene 1997) of the peripheral branch 1–2 weeks before a central tract lesion will enhance regeneration of the centrally projecting neurites (Neumann and Woolf 1999; Richardson and Issa 1984). These results indicate that retrograde injury signals travel from the peripheral injury site back to the cell body to increase the intrinsic growth capacity of the neuron, helping to overcome inhibitory cues in the central tract.

At least three distinct types of temporally graded signals have been postulated to act complementarily to signal injury from axonal lesion sites to the corresponding neuronal cell bodies (Perlson et al. 2004). Immediately after axotomy there is a rapid depolarization at the injury site, followed by a burst of action potentials termed an "injury discharge" (Berdan et al. 1993). After this first wave of antidromic electrical signals, two distinct types of retrogradely transported macromolecular signals are thought to impinge on the cell body. These include, on the one hand, an interruption of the normal supply of retrogradely transported trophic factors from the target and, on the other hand, "activated" proteins originating from the injury site (Ambron and Walters 1996). The first category of proteins was termed "negative injury signals" by some investigators, because the signal is conveyed by an interruption of their normal arrival in the cell body, while the activated/modified protein ensembles were conversely termed "positive injury signals" (Ambron et al. 1996; Zhang and Ambron 2000). In this chapter we describe the current state of knowledge of these distinct signaling mechanisms and their contribution to the enhancement of intrinsic growth capacity in peripheral neurons following injury.

2 Rapid Signaling: The Electrophysiological Response

Rapid ion fluxes emanating from the lesion site are likely to be the first indication of a breach of axonal integrity. Axotomy of *Aplysia* neurons in culture elevates intra-axonal calcium concentrations to levels above 1 mM near the tip of the cut in the axon, and to hundreds-of-micromolar along the axon (Ziv and Spira 1995). A wave of increased calcium then propagates at a rate of approximately 1 mm min^{-1} from the point of transection towards the intact portions of the cell. Calcium recovery in these invertebrate neurons is fairly rapid (within minutes) once the cut ends are resealed, and calcium levels recover as a retreating front, traveling back towards the lesion site (Ziv and Spira 1993). In mammalian systems, in vitro axotomy of neurons cultured from the rodent cortex caused an increase of axonal calcium, propagated to the soma via a mechanism dependent on voltage-dependent sodium channels (Mandolesi et al. 2004). Injury was followed by vigorous spiking activity that caused a sodium load and an activation of transient calcium currents by each action potential. The decrease of the electrochemical gradient of sodium caused inversion of the sodium–calcium exchange pump; thus, providing an additional and prolonged means of entry for calcium (Mandolesi et al. 2004). Stretch-induced injuries in neuron-like cell lines, or in embryonic cortical neurons in vitro also caused a continued increase in axonal calcium via activation of tetrodotoxin-sensitive sodium channels and inversion of sodium–calcium exchange (Iwata et al. 2004; Wolf et al. 2001). This response was facilitated by calcium-dependent proteolysis of an intra-axonal domain of tetrodotoxin-sensitive sodium channels, providing an additional mechanism for propagation of the injury signal (Iwata et al. 2004). In vivo, resealing of lesioned mammalian axons can take hours, and is dependent on axon diameter and on calcium in the extracellular environment (Howard et al. 1999); thus, the rapid changes in axonal calcium levels following injury may be sustained over a period of time to a degree that is proportional to severity of the injury, and the resealing capacity of the axon. Intriguingly, differences in the resealing capacity of the central versus peripheral axons may contribute to differences in regeneration capacity (Ahmed et al. 2001).

The electrophysiological response was shown to be necessary for neurite outgrowth in vitro, since axotomy of cortical cultured neurons in the presence of tetrodotoxin reduced the regenerative response (Mandolesi et al. 2004), while electrical stimulation of an intact sciatic nerve increased neurite outgowth (Udina et al. 2008). Prevention of calcium influx also reduced regeneration response in vitro (Chierzi et al. 2005); however, this effect might be due to the role of calcium in cytoskeletal rearrangement and growth cone formation (Spira et al. 2003) in addition to its contribution to retrograde signaling. In vivo studies showed that electrical stimulation accelerates both motor and sensory axon outgrowth, and increases intracellular cAMP levels in DRG neurons as effectively as the conditioning lesion (Al-Majed et al. 2004; Udina et al. 2008). However, the electrical stimulation did not mimic the conditioning lesion effect in promoting axonal regeneration within the CNS lesion site, suggesting that additional signals are needed for complete regenerative response (Udina et al. 2008). In contrast to axons of the peripheral nervous system (PNS), electrical stimulation of CNS axons does not promote regeneration, even in the permissive growth environment of a peripheral nerve graft (Harvey et al. 2005). Thus, electrical activity may play an important role as an early priming injury signal in the PNS, but might be insufficient to initiate regeneration of CNS neurons.

3 Negative Injury Signals

Some time after arrival of the ion fluxes described above, signals dependent on motor-driven transport systems start to affect the cell body. This phase may include an interruption of the normal supply of retrogradely transported molecules from the nerve terminal to the cell body. Once a neuron is connected to its target, target-derived retrograde signals must repress the intrinsic neuronal growth activity to allow for proper synaptic development. This repression has to be relieved to allow regeneration to occur. Neurotrophins represent the ideal candidate for such signals, since their continuous supply from axon terminals to the cell body is important for neuronal survival during development, and in the maintenance of the phenotype and modulation of plasticity in diverse neuronal populations in the adult (Bronfman et al. 2007; Ibanez 2007). Following sciatic nerve axotomy, there is a tenfold decrease in the levels of retrogradely transported nerve growth factor (NGF) (Raivich et al. 1991) (see Campenot 2009). Artificial interruption of NGF supply induced axotomy-like alterations in gene expression in injured sensory or sympathetic neurons (Shadiack et al. 2001), while intrathecal infusion of NGF delayed GAP-43 induction (Hirata et al. 2002) and nerve regeneration (Gold 1997; Hirata et al. 2002). These observations suggest that reductions in the levels of retrogradely transported trophic factors might be decoded as an injury signal by the cell body. On the other hand, retrograde transport of other types of neurotrophins may actually be increased after injury and may serve as positive injury signal (see below).

Apart from neurotrophins, another recently identified modulator of regeneration is the PTEN/mTOR pathway. In wild-type adult mice, the mammalian target of rapamycin (mTOR) activity, which promotes translational activity, was suppressed, and new protein synthesis was thereby impaired in axotomized retinal ganglion cells (RGCs), which may contribute to regeneration failure. Reactivating this pathway by conditional knockout of phosphatase and tensin homolog (PTEN), a negative regulator of the mTOR pathway, leads to axon regeneration after optic nerve injury (Park et al. 2008). Since mature RGCs have very restricted axon growth capacity and the mature optic nerve has a strong inhibitory environment, axon regeneration of RGCs after PTEN elimination is surprising. Additional studies are required to determine how applicable this paradigm might be to other injury models.

4 Positive Injury Signals: The Importance of Importins

A second and prominent aspect of the change in retrogradely transported macromolecular signals after injury is the de novo appearance of activated, or modified proteins originating from the injury site (Hanz and Fainzilber 2006; Perlson et al. 2004). Early work in *Aplysia* indicated that some injury signaling proteins are dependent on nuclear localization signals (NLS) for access to the retrograde transport system (Zhang and Ambron 2000). The evidence that NLS sequences might target injury-signaling proteins to the retrograde transport system suggested that NLS-binding proteins should be found in axons. Classical NLS-dependent nuclear import is dependent on transport factors termed importins or karyopherins. Importin α binds NLS directly, and its affinity for NLS is increased by interaction with importin β, which conveys transport of the complex through the nuclear pore (Weis 2003). In rodent sciatic nerve, a number of importin α's were found in axons of both control and injured nerve in constitutive association with dynein, while importin β1 protein was present only after injury (Hanz et al. 2003). Interestingly, mRNA for importin β1 was found in axons, and the up-regulation of importin β1 protein after injury was attributed to local translation in the axon (Hanz et al. 2003). This leads to the formation of importin α/β heterodimers bound to the retrograde motor dynein; thereby, potentially enabling transport of signaling cargos that bind to the importins. Retrograde transport of fluorescently labeled NLS peptides was indeed observed in lesioned sciatic nerve, and the introduction of excess NLS peptides into lesioned DRG axons inhibited, or delayed both in vitro regenerative, and in vivo conditioning lesion responses (Hanz et al. 2003). These data suggest that in parallel with local axonal synthesis of importin β1, local activation of NLS-bearing signaling proteins creates a signaling cargo that binds to the α/β high affinity NLS binding site; thus, accessing the retrograde transport pathway. This mechanism appears to play a major role in signaling the regeneration response in peripheral sensory neurons (Hanz and Fainzilber 2004; Hanz et al. 2003).

A recent study revealed an additional level of regulation for the formation of importin signaling complexes in injured axons, carried out by Ran GTPase, and its associated effectors RanBP1 and RanGAP (Yudin et al. 2008). A gradient of nuclear RanGTP versus cytoplasmic RanGDP regulates nuclear import, and is thought to be fundamental for the organization of eukaryotic cells. Surprisingly, Yudin and colleagues found RanGTP in sciatic nerve axoplasm, distant from neuronal cell bodies and nuclei, and in association with dynein and importin α. Following an injury, localized translation of RanBP1 stimulates RanGTP dissociation from importins and subsequent hydrolysis, thereby allowing binding of newly synthesized importin β to importin α and dynein. Perturbation of RanGTP hydrolysis, or RanBP1 blockade at axonal injury sites reduces the neuronal conditioning lesion response (Yudin and Fainzilber 2009; Yudin et al. 2008). Thus, transmitting an injury signal requires not only the appearance of a signal, but also the local synthesis of multiple components of the system in the axon (Fig. 1). Clearly, RNA localization in axons is critical for this response system (for a detailed discussion of RNA localization and local translation mechanisms in axons, see Vuppalanchi et al. 2009).

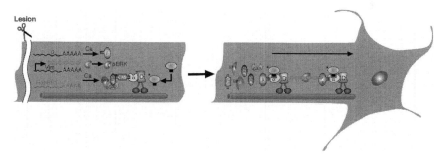

Fig. 1 Importin-mediated retrograde injury signaling in axons. Importin α protein is constitutively associated with the retrograde motor dynein (D) in axons; whereas, importin β1, Vimentin, and RanBP1 are normally present in transcript form. Upon injury, local translation of these mRNAs leads to expression and up-regulation of the corresponding proteins. Newly synthesized RanBP1 stimulates dissociation of RanGTP and RanGAP synergized hydrolysis; thereby, allowing formation of a cargo-binding complex of importin α with de novo synthesized importin β1. Vimentin is transported in the retrograde injury-signaling complex via a direct interaction with importin β1

4.1 Signaling Components of the Importin-Mediated Retrograde Injury Complex

The description of an importin mediated mechanism for retrograde injury signaling directed attention to the identity of the potential cargos. Although NLS-containing proteins such as transcription factors are obvious candidates for this role, paradoxically, the first validated signals found to be trafficked in an importin-dependent manner in axons do not contain NLS. A type III intermediate filament vimentin was found to be locally translated in axoplasm after injury, and cleaved to soluble fragments by the calcium-activated protease calpain (Perlson et al. 2005). These vimentin fragments bind directly to both phosphorylated Erks and importin β; thereby, linking activated Erks to importin-mediated retrograde transport (Perlson et al. 2005). Strikingly, the vimentin–Erk complex protects Erk from dephosphorylation, and since the interaction is calcium dependent, the signal generated may provide information both on the injury and on the degree of damage, as reflected by sustained calcium elevation (Perlson et al. 2006). Thus, even though Erk1 and Erk2 do not contain classical NLS sequences, they are dependent on importins for retrograde transport after nerve injury via a series of protein–protein interactions between Erk, vimentin, importin β, importin α, and dynein (Fig. 1).

Several transcription factors have been identified in peripheral axons that may be implicated in retrograde signaling. Peripheral nerve axotomy causes up-regulation of the gp130 cytokines interleukin-6 (IL-6), leukemia inhibitory factor (LIF), and ciliary neurotrophic factor (CNTF), which activate the JAK–STAT3 signaling pathway (Banner and Patterson 1994; Bolin et al. 1995; Sendtner et al. 1992). Phosphorylation of STAT3 was demonstrated in axons at the site of sciatic nerve lesion, and this response extended backwards toward the neuronal nuclei over a period of time, suggesting a retrograde transport of axonal phospho-STAT3 (Lee et al. 2004). Interestingly, STAT3 activation occurs in DRG neuronal cell bodies after peripheral, but not central, lesion (Qiu et al. 2005; Schwaiger et al. 2000). Recent data from the laboratory of Martin Kanje indicate that activated transcription factor 2 (ATF2) and ATF3 are also transported retrogradely following sciatic nerve injury (Lindwall and Kanje, 2005). It remains to be determined whether the retrograde transport of STAT3, the ATF's and other transcription factors is mediated by axonal importins.

5 Other Modes of Linkage to the Retrograde Injury-Signaling Complex

Several studies have implicated kinases in retrograde injury signaling. RISK-1, an *Aplysia* homolog of Erk, and *Aplysia* protein kinase G (apPKG) are activated at lesion sites in *Aplysia* nerves, and subsequently undergo retrograde transport (Sung et al. 2001, 2004). A number of kinases, including Erk1/2, p38 MAPK and Jnk, are activated in lesioned sciatic nerve (Lindwall and Kanje 2005; Zrouri et al. 2004). As noted above, phosphorylated Erk is retrogradely transported following injury (Perlson et al. 2005; Reynolds et al. 2001). Phosphorylated Jnk is also retrogradely transported in injured sciatic nerve, apparently together with the scaffold Jnk-interacting protein (JIP), and other upstream kinases (Lindwall and Kanje 2005).

Most of the candidate signaling cargos listed above lack an obvious NLS, and might not directly associate with importins. Although linker molecules such as vimentin provide a solution for importin-mediated transport of activated Erks, additional mechanisms for linking signaling molecules to retrograde transport may be important. For example, the Jnk scaffold protein Sunday driver (Syd) may link activated Jnk to injury signaling (Cavalli et al. 2005). Syd and Jnk3 are present on vesicular structures in the axons and are transported in both anterograde and retrograde directions. Nerve injury induces axonal activation of Jnk3, after which the activated Jnk is transported together with Syd, predominantly in the retrograde direction, most likely due to an enhanced interaction between Syd and dynactin, a dynein motor regulator, after injury. Cavalli and colleagues proposed that the Jnk–Syd complex acts as a damage surveillance system, and that the direction switch after injury provides a rapid response mechanism for propagation of retrograde injury signals (Abe and Cavalli 2008; Cavalli et al. 2005).

The studies described above suggest that different compositions of retrograde injury signaling complexes are possible, with at least three nonexclusive potential binding sites for signaling molecules identified to date; namely, dynactin-Syd for Jnk and associated molecules, the classical NLS binding site on importin α, and importin β-vimentin for activated Erk (Hanz and Fainzilber 2006). Variability in subunit composition within the dynein complex might also allow for differential cargo binding (Pfister et al. 2006), leading to different combinations of signals being transported in different cell types.

6 Neurotrophins as Positive Injury Signals

Apart from their role in neuronal survival during development, the function of neurotrophins has been recently extended to other aspects of neuronal function, including regeneration (Cui 2006). Nerve injury induces up-regulation of the glial-derived neurotrophic factor GDNF, and one of its receptors GFRα1, suggesting that GDNF provides neurotrophic support for injured DRG neurons (Hoke et al. 2000). In support of this notion, delivery of GDNF directly to DRG cell bodies facilitates the conditioning injury-induced growth promoting effect (Mills et al. 2007). GDNF and GFRα1 are retrogradely transported in peripheral axons (Coulpier and Ibanez 2004), but a role of GDNF in injury signaling has not yet been examined. Fibroblast growth factor-2 (FGF-2) is another growth factor that seems to contribute to nerve regeneration (Grothe and Nikkhah 2001). FGF-2 is up-regulated following injury both at the lesion site, and in the cell bodies of peripheral nerves; moreover, transgenic mice overexpressing FGF-2 show a greater increase in the number of regenerating axons after sciatic nerve transaction (Jungnickel et al. 2006). Neurotrophins are thought to be retrogradely transported by dynein motors via signaling endosomes, containing endocytosed ligand–receptor complexes and downstream effectors (Ibanez 2007)). The effects of neurotrophin signaling in injured nerves emphasize the need for further study of the crosstalk between neuronal survival and axon regeneration pathways.

7 The Cell Body Response to Axonal Injury

The signals arriving from the axonal injury site must activate transcription and translational programs to elicit regrowth of the injured axon. Our mechanistic understanding of these processes is very limited, and there is no clear understanding to date what a neuron must do in order to grow. Several kinases are activated in cell bodies in response to an injury event. Among these are the MAP kinases Erk1 and Erk2 (Obata et al. 2004), and Jnk (Kenney and Kocsis 1998). Interestingly, Jnk activation in DRG neurons was shown to be dependent on the distance of the site of axotomy (Kenney and Kocsis 1998). Axotomy may also induce elevated transcription and translation of kinases in cell bodies, as exemplified in the case of Janus kinase (JAK) (Rajan et al. 1995; Yao et al. 1997). Another well characterized cell body response to peripheral nerve injury is elevation of cAMP levels in the cell body, which contributes to the initiation of axonal regrowth (for a recent review see (Hannila and Filbin 2008)). cAMP not only increases the growth capacity of injured neurons, but also partly relieves CNS growth inhibition, since artificial elevation of cAMP by microinjection of analogs in lumbar DRG markedly increases the regeneration of injured central sensory branches (Neumann et al. 2002; Qiu et al. 2002). The effects of cAMP are transcription-dependent (Smith and Skene 1997), and involve activation of PKA (Qiu et al. 2002), and the transcription factor cAMP response element binding protein (CREB) (Gao et al. 2004).

Downstream events influenced by axotomy-activated kinases include up-regulation, or activation of transcription factors. For example lesion-induced Jnk induces up-regulation and phosphorylation of the transcription factors c-Jun, JunD and Fos, and also translocation of ATF3 into the nucleus, leading to formation of complexes with DNA binding activity (Kenney and Kocsis 1998; Lindwall et al. 2004; Lindwall and Kanje 2005; Seijffers et al. 2006). Activated retrogradely transported Erk induces phosphorylation of the ETS domain transcription factor Elk-1 in DRG neuronal cell bodies (Perlson et al. 2005). Other transcription factors, including STAT3, P311, Sox11 and C/EBPb, were also found to be up-regulated and activated following injury (Nadeau et al. 2005; Schwaiger et al. 2000), while NF-kB in contrast is decreased (Povelones et al. 1997). These alterations in the transcription factor activity result in changes of gene expression in the injured neuron, with up-regulation of a growing list of "regeneration-associated genes." Transgenic and overexpression approaches with a few regeneration-associated genes, primarily GAP-43/CAP-43, or c-jun, have shown modest improvements in regeneration in peripheral neurons, but these results have unfortunately not been recapitulated in the outgrowth refractory central neurons (Rossi et al. 2007). Clearly, much more additional work will be required to define the critical elements of a neuronal regeneration program.

8 Summary

A successful axonal response to injury requires retrograde signaling to induce changes in the cell body response, culminating in successful regeneration. A single signaling pathway is unlikely to fully mediate nerve regeneration. Several classes of injury signals may coexist to ensure precise information on the nature of the damage and its distance from cell body. Recent work has highlighted the importance of local synthesis, and localized axonal reactions to elicit and coordinate the retrograde signaling response. Much work is still needed to fully understand the different mechanistic aspects of the axonal response to injury, and how it contributes to neuronal regeneration. Given the many debilitating CNS disorders caused by axonal damage, such as spinal cord injury, stroke and neurodegeneration, there are urgent reasons to move forward as quickly and as scientifically as possible in this field.

Acknowledgements The authors' research on these topics was supported by the Adelson Medical Research Foundation, the Christopher Reeve Foundation and the International Institute for Research in Paraplegia. M.F. is the incumbent of the Chaya Professorial Chair in Molecular Neuroscience at the Weizmann Institute of Science.

References

Abe N, Cavalli V (2008) Nerve injury signaling. Curr Opin Neurobiol 18:276–283
Ahmed FA, Ingoglia NA, Sharma SC (2001) Axon resealing following transection takes longer in central axons than in peripheral axons: implications for axonal regeneration. Exp Neurol 167:451–455
Al-Majed AA, Tam SL, Gordon T (2004) Electrical stimulation accelerates and enhances expression of regeneration-associated genes in regenerating rat femoral motoneurons. Cell Mol Neurobiol 24:379–402
Ambron RT, Walters ET (1996) Priming events and retrograde injury signals. A new perspective on the cellular and molecular biology of nerve regeneration. Mol Neurobiol 13:61–79
Ambron RT, Zhang XP, Gunstream JD, Povelones M, Walters ET (1996) Intrinsic injury signals enhance growth, survival, and excitability of Aplysia neurons. J Neurosci 16:7469–7477
Banner LR, Patterson PH (1994) Major changes in the expression of the mRNAs for cholinergic differentiation factor/leukemia inhibitory factor and its receptor after injury to adult peripheral nerves and ganglia. Proc Natl Acad Sci U S A 91:7109–7113
Berdan RC, Easaw JC, Wang R (1993) Alterations in membrane potential after axotomy at different distances from the soma of an identified neuron and the effect of depolarization on neurite outgrowth and calcium channel expression. J Neurophysiol 69:151–164
Bolin LM, Verity AN, Silver JE, Shooter EM, Abrams JS (1995) Interleukin-6 production by Schwann cells and induction in sciatic nerve injury. J Neurochem 64:850–858

Bronfman FC, Escudero CA, Weis J, Kruttgen A (2007) Endosomal transport of neurotrophins: roles in signaling and neurodegenerative diseases. Dev Neurobiol 67:1183–1203

Campenot RB (2009) NGF uptake and retrograde signaling mechanisms in sympathetic neurons in compartmented cultures. Results Probl Cell Differ. doi: 10.1007/400_2009_7

Cavalli V, Kujala P, Klumperman J, Goldstein LS (2005) Sunday Driver links axonal transport to damage signaling. J Cell Biol 168:775–787

Chierzi S, Ratto GM, Verma P, Fawcett JW (2005) The ability of axons to regenerate their growth cones depends on axonal type and age, and is regulated by calcium, cAMP and ERK. Eur J Neurosci 21:2051–2062

Coulpier M, Ibanez CF (2004) Retrograde propagation of GDNF-mediated signals in sympathetic neurons. Mol Cell Neurosci 27:132–139

Cui Q (2006) Actions of neurotrophic factors and their signaling pathways in neuronal survival and axonal regeneration. Mol Neurobiol 33:155–179

Gao Y, Deng K, Hou J, Bryson JB, Barco A, Nikulina E, Spencer T, Mellado W, Kandel ER, Filbin MT (2004) Activated CREB is sufficient to overcome inhibitors in myelin and promote spinal axon regeneration in vivo. Neuron 44:609–621

Gervasi NM, Kwok JC, Fawcett JW (2008) Role of extracellular factors in axon regeneration in the CNS: implications for therapy. Regen Med 3:907–923

Gold BG (1997) Axonal regeneration of sensory nerves is delayed by continuous intrathecal infusion of nerve growth factor. Neuroscience 76:1153–1158

Grothe C, Nikkhah G (2001) The role of basic fibroblast growth factor in peripheral nerve regeneration. Anat Embryol 204:171–177

Hannila SS, Filbin MT (2008) The role of cyclic AMP signaling in promoting axonal regeneration after spinal cord injury. Exp Neurol 209:321–332

Hanz S, Fainzilber M (2004) Integration of retrograde axonal and nuclear transport mechanisms in neurons: implications for therapeutics. Neuroscientist 10:404–408

Hanz S, Fainzilber M (2006) Retrograde signaling in injured nerve--the axon reaction revisited. J Neurochem 99:13–19

Hanz S, Perlson E, Willis D, Zheng JQ, Massarwa R, Huerta JJ, Koltzenburg M, Kohler M, van-Minnen J, Twiss JL, Fainzilber M (2003) Axoplasmic importins enable retrograde injury signaling in lesioned nerve. Neuron 40:1095–1104

Harvey PJ, Grochmal J, Tetzlaff W, Gordon T, Bennett DJ (2005) An investigation into the potential for activity-dependent regeneration of the rubrospinal tract after spinal cord injury. Eur J Neurosci 22:3025–3035

Hirata A, Masaki T, Motoyoshi K, Kamakura K (2002) Intrathecal administration of nerve growth factor delays GAP 43 expression and early phase regeneration of adult rat peripheral nerve. Brain Res 944:146–156

Hoke A, Cheng C, Zochodne DW (2000) Expression of glial cell line-derived neurotrophic factor family of growth factors in peripheral nerve injury in rats. Neuroreport 11:1651–1654

Howard MJ, David G, Barrett JN (1999) Resealing of transected myelinated mammalian axons in vivo: evidence for involvement of calpain. Neuroscience 93:807–815

Ibanez CF (2007) Message in a bottle: long-range retrograde signaling in the nervous system. Trends Cell Biol 17:519–528

Iwata A, Stys, PK, Wolf JA, Chen XH, Taylor AG, Meaney DF, Smith DH (2004) Traumatic axonal injury induces proteolytic cleavage of the voltage-gated sodium channels modulated by tetrodotoxin and protease inhibitors. J Neurosci 24:4605–4613

Jungnickel J, Haase K, Konitzer J, Timmer M, Grothe C (2006) Faster nerve regeneration after sciatic nerve injury in mice over-expressing basic fibroblast growth factor. J Neurobiol 66:940–948

Kenney AM, Kocsis JD (1998) Peripheral axotomy induces long-term c-Jun amino-terminal kinase-1 activation and activator protein-1 binding activity by c-Jun and junD in adult rat dorsal root ganglia in vivo. J Neurosci 18:1318–1328

Kim WY, Snider WD (2008) Neuroscience. Overcoming inhibitions. Science 322:869–872

Lee N, Neitzel KL, Devlin BK, MacLennan AJ (2004) STAT3 phosphorylation in injured axons before sensory and motor neuron nuclei: potential role for STAT3 as a retrograde signaling transcription factor. J Comp Neurol 474:535–545

Lindwall C, Kanje M (2005) Retrograde axonal transport of JNK signaling molecules influence injury induced nuclear changes in p-c-Jun and ATF3 in adult rat sensory neurons. Mol Cell Neurosci 29:269–282

Lindwall C, Dahlin L, Lundborg G, Kanje M (2004) Inhibition of c-Jun phosphorylation reduces axonal outgrowth of adult rat nodose ganglia and dorsal root ganglia sensory neurons. Mol Cell Neurosci 27:267–279

Mandolesi G, Madeddu F, Bozzi Y, Maffei L, Ratto GM (2004) Acute physiological response of mammalian central neurons to axotomy: ionic regulation and electrical activity. FASEB J 18:1934–1936

Mills CD, Allchorne AJ, Griffin RS, Woolf CJ, Costigan M (2007) GDNF selectively promotes regeneration of injury-primed sensory neurons in the lesioned spinal cord. Mol Cell Neurosci 36:185–194

Nadeau S, Hein P, Fernandes KJ, Peterson AC, Miller FD (2005) A transcriptional role for C/EBP beta in the neuronal response to axonal injury. Mol Cell Neurosci 29:525–535

Neumann S, Woolf CJ (1999) Regeneration of dorsal column fibers into and beyond the lesion site following adult spinal cord injury. Neuron 23:83–91

Neumann S, Bradke F, Tessier-Lavigne M, Basbaum AI (2002) Regeneration of sensory axons within the injured spinal cord induced by intraganglionic cAMP elevation. Neuron 34:885–893

Obata K, Yamanaka H, Dai Y, Mizushima T, Fukuoka T, Tokunaga A, Noguchi K (2004) Differential activation of MAPK in injured and uninjured DRG neurons following chronic constriction injury of the sciatic nerve in rats. Eur J Neurosci 20:2881–2895

Park KK, Liu K, Hu Y, Smith PD, Wang C, Cai B, Xu B, Connolly L, Kramvis I, Sahin M, He Z (2008) Promoting axon regeneration in the adult CNS by modulation of the PTEN/mTOR pathway. Science 322:963–966

Perlson E, Hanz S, Medzihradszky KF, Burlingame AL, Fainzilber M (2004) From snails to sciatic nerve: retrograde injury signaling from axon to soma in lesioned neurons. J Neurobiol 58:287–294

Perlson E, Hanz S, Ben-Yaakov K, Segal-Ruder Y, Seger R, Fainzilber M (2005) Vimentin-dependent spatial translocation of an activated MAP kinase in injured nerve. Neuron 45:715–726

Perlson E, Michaelevski I, Kowalsman N, Ben-Yaakov K, Shaked M, Seger R, Eisenstein M, Fainzilber M (2006) Vimentin binding to phosphorylated Erk sterically hinders enzymatic dephosphorylation of the kinase. J Mol Biol 364:938–944

Pfister KK, Shah PR, Hummerich H, Russ A, Cotton J, Annuar AA, King SM, Fisher EM (2006) Genetic analysis of the cytoplasmic dynein subunit families. PLoS Genet 2:e1

Povelones M, Tran K, Thanos D, Ambron RT (1997) An NF-kappaB-like transcription factor in axoplasm is rapidly inactivated after nerve injury in Aplysia. J Neurosci 17:4915–4920

Qiu J, Cai D, Dai H, McAtee M, Hoffman PN, Bregman BS, Filbin MT (2002) Spinal axon regeneration induced by elevation of cyclic AMP. Neuron 34:895–903

Qiu J, Cafferty WB, McMahon SB, Thompson SW (2005) Conditioning injury-induced spinal axon regeneration requires signal transducer and activator of transcription 3 activation. J Neurosci 25:1645–1653

Raivich G, Hellweg R, Kreutzberg GW (1991) NGF receptor-mediated reduction in axonal NGF uptake and retrograde transport following sciatic nerve injury and during regeneration. Neuron 7:151–164

Rajan P, Stewart CL, Fink JS (1995) LIF-mediated activation of STAT proteins after neuronal injury in vivo. Neuroreport 6:2240–2244

Reynolds AJ, Hendry IA, Bartlett SE (2001) Anterograde and retrograde transport of active extracellular signal-related kinase 1 (ERK1) in the ligated rat sciatic nerve. Neuroscience 105:761–771

Richardson PM, Issa VM (1984) Peripheral injury enhances central regeneration of primary sensory neurones. Nature 309:791–793

Rossi F, Gianola S, Corvetti L (2007) Regulation of intrinsic neuronal properties for axon growth and regeneration. Prog Neurobiol 81:1–28

Schwaiger FW, Hager G, Schmitt AB, Horvat A, Streif R, Spitzer C, Gamal S, Breuer S, Brook GA, Nacimiento W, Kreutzberg GW (2000) Peripheral but not central axotomy induces changes in Janus kinases (JAK) and signal transducers and activators of transcription (STAT). Eur J Neurosci 12:1165–1176

Seijffers R, Allchorne AJ, Woolf CJ (2006) The transcription factor ATF-3 promotes neurite outgrowth. Mol Cell Neurosci 32:143–154

Sendtner M, Stockli KA, Thoenen H (1992) Synthesis and localization of ciliary neurotrophic factor in the sciatic nerve of the adult rat after lesion and during regeneration. J Cell Biol 118:139–148

Shadiack AM, Sun Y, Zigmond RE (2001) Nerve growth factor antiserum induces axotomy-like changes in neuropeptide expression in intact sympathetic and sensory neurons. J Neurosci 21:363–371

Smith DS, Skene JH (1997) A transcription-dependent switch controls competence of adult neurons for distinct modes of axon growth. J Neurosci 17:646–658

Spira ME, Oren R, Dormann A, Gitler D (2003) Critical calpain dependent ultrastructural alterations underlie the transformation of an axonal segment into a growth cone after axotomy of cultured Aplysia neurons. J Comp Neurol 457:293–312

Sung YJ, Povelones M, Ambron RT (2001) RISK-1: a novel MAPK homologue in axoplasm that is activated and retrogradely transported after nerve injury. J Neurobiol 47:67–79

Sung YJ, Walters ET, Ambron, RT (2004) A neuronal isoform of protein kinase G couples mitogen-activated protein kinase nuclear import to axotomy-induced long-term hyperexcitability in Aplysia sensory neurons. J Neurosci 24:7583–7595

Udina E, Furey M, Busch S, Silver J, Gordon T, Fouad K (2008) Electrical stimulation of intact peripheral sensory axons in rats promotes outgrowth of their central projections. Exp Neurol 210:238–247

Vuppalanchi D, Willis DE, Twiss JL (2009) Regulation of mRNA transport and translation in axons. Results Probl Cell Differ. doi: 10.1007/400_2009_16

Weis K (2003) Regulating access to the genome: nucleocytoplasmic transport throughout the cell cycle. Cell 112:441–451

Wolf JA, Stys PK, Lusardi T, Meaney D, Smith DH (2001) Traumatic axonal injury induces calcium influx modulated by tetrodotoxin-sensitive sodium channels. J Neurosci 21;1923–1930

Yamashita T, Fujitani M, Yamagishi S, Hata K, Mimura F (2005) Multiple signals regulate axon regeneration through the Nogo receptor complex. Mol Neurobiol 32:105–111

Yao GL, Kato H, Khalil M, Kiryu S, Kiyama H (1997) Selective upregulation of cytokine receptor subchain and their intracellular signalling molecules after peripheral nerve injury. Eur J Neurosci 9:1047–1054

Yudin D, Fainzilber M (2009) Ran on tracks - cytoplasmic roles for a nuclear regulator. J Cell Sci 122:587–593

Yudin D, Hanz S, Yoo S, Iavnilovitch E, Willis D, Gradus T, Vuppalanchi D, Segal-Ruder Y, Ben-Yaakov K, Hieda M, Yoneda Y, Twiss JL, Fainzilber M (2008) Localized regulation of axonal RanGTPase controls retrograde injury signaling in peripheral nerve. Neuron 59:241–252

Zhang XP, Ambron RT (2000) Positive injury signals induce growth and prolong survival in Aplysia neurons. J Neurobiol 45:84–94

Ziv NE, Spira ME (1993) Spatiotemporal distribution of Ca^{2+} following axotomy and throughout the recovery process of cultured Aplysia neurons. Eur J Neurosci 5:657–668

Ziv NE, Spira ME (1995) Axotomy induces a transient and localized elevation of the free intracellular calcium concentration to the millimolar range. J Neurophysiol 74:2625–2637

Zrouri H, Le Goascogne C, Li WW, Pierre M, Courtin F (2004) The role of MAP kinases in rapid gene induction after lesioning of the rat sciatic nerve. Eur J Neurosci 20:1811–1818

Axon Regeneration in the Peripheral and Central Nervous Systems

Eric A. Huebner and Stephen M. Strittmatter

Abstract Axon regeneration in the mature mammalian central nervous system (CNS) is extremely limited after injury. Consequently, functional deficits persist after spinal cord injury (SCI), traumatic brain injury, stroke, and related conditions that involve axonal disconnection. This situation differs from that in the mammalian peripheral nervous system (PNS), where long-distance axon regeneration and substantial functional recovery can occur in the adult. Both extracellular molecules and the intrinsic growth capacity of the neuron influence regenerative success. This chapter discusses determinants of axon regeneration in the PNS and CNS.

1 Introduction

Central nervous system (CNS) axons do not spontaneously regenerate after injury in adult mammals. In contrast, peripheral nervous system (PNS) axons readily regenerate, allowing recovery of function after peripheral nerve damage. Aguayo and colleagues demonstrated that at least some mature CNS neurons retain the capacity to regenerate when provided with a permissive peripheral nerve graft (Richardson et al. 1980, 1984; David and Aguayo, 1981; Benfey and Aguayo, 1982). This work suggested that the PNS environment is stimulatory and/or that the CNS environment is inhibitory for axon growth. Subsequent studies identified both growth-promoting factors in the PNS and growth-inhibiting factors in the CNS. Inhibitors of regeneration include specific proteins in CNS myelin and molecules associated with the astroglial scar. In addition, slower debris clearance in the CNS relative to the PNS may impede axonal re-growth. The failure of axotomized CNS neurons to induce growth-promoting genes also limits brain and spinal cord repair.

E.A. Huebner and S.M. Strittmatter (✉)
Program in Cellular Neuroscience, Neurodegeneration and Repair,
Yale University School of Medicine, New Haven, CT, USA
e-mail: stephen.strittmatter@yale.edu

An understanding of factors which influence axon growth is critical for the development of therapeutics to promote CNS regeneration.

2 Axon Regeneration in the Peripheral Nervous System

2.1 Overview of Peripheral Nervous System Regeneration

After peripheral nerve injury, axons readily regenerate. The distal portion of the axon, which is disconnected from the cell body, undergoes Wallerian degeneration. This active process results in fragmentation and disintegration of the axon. Debris is removed by glial cells, predominantly macrophages. Proximal axons can then regenerate and re-innervate their targets, allowing recovery of function.

2.2 Regeneration-Associated Genes

Following axotomy, PNS neurons upregulate numerous regeneration-associated genes (RAGs). Some of these genes have a direct role in axon regeneration, while others do not. A number of RAGs have been shown to be important for neurite outgrowth and/or regeneration. These include c-Jun (Raivich et al. 2004), activating transcription factor-3 (ATF-3) (Seijffers et al. 2006), SRY-box containing gene 11 (Sox11) (Jankowski et al. 2009), small proline-repeat protein 1A (SPRR1A) (Bonilla et al. 2002), growth-associated protein-43 (GAP-43) and CAP-23 (Bomze et al. 2001).

One strategy to identify RAGs involves injuring a peripheral nerve, and then observing gene expression changes in the corresponding cell bodies (Bonilla et al. 2002; Tanabe et al. 2003; Costigan et al. 2002). A number of such studies have used gene profiling technology to examine gene expression changes in sensory neurons following axotomy. For example, Bonilla et al. (2002) demonstrated that SPRR1A is highly induced in dorsal root ganglion (DRG) neurons one week after sciatic nerve transection (protein increased more than 60-fold from whole DRGs). Immunohistochemistry demonstrated expression of SPRR1A in DRG neuronal cell bodies and regenerating peripheral axons. SPRR1A expression is also increased after sciatic nerve injury in the ventral horn motor neuron cell bodies and sensory fibers within the spinal cord (Fig. 1). Herpes simplex virus-mediated overexpression of SPRR1A in embryonic chick DRG neurons promotes neurite outgrowth. The association of SPRR1A expression with regeneration and its ability to promote neurite outgrowth suggest that it may have a role in axon regeneration.

ATF-3 is a transcriptional factor induced in sensory neurons after injury (Tanabe et al. 2003; Boeshore et al. 2004). Over expression of ATF-3 promotes neurite outgrowth (Seijffers et al. 2006). Sox11 and c-Jun are injury-induced transcription

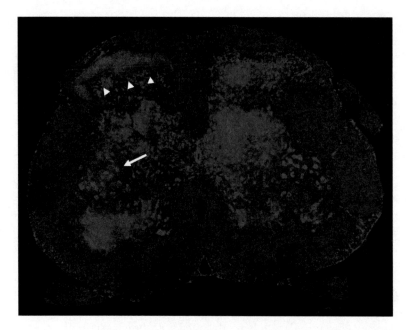

Fig. 1 SPRR1A upregulation in the central process of primary afferent sensory neurons and in motoneurons after sciatic nerve injury. The sciatic nerve was crushed at the mid-thigh on one side of an adult mouse. Seven days later, the animal was sacrificed, and L5 spinal cord transverse sections were processed for anti-SPRR1A immunohistology (*red*) and for Nissl Stain (*blue*). Note the intense SPRR1A protein upregulation in afferent terminals in the dorsal horn (*arrowheads*) and in ventral horn motoneurons (*arrow*). Upregulation is confined to the injured side (*left*) with very low levels of SPRR1A on the intact side (*right*). Dorsal is *up*. Methods as in Bonilla et al. (2002). Image courtesy of Dr. William B. Cafferty

factors and are required for efficient nerve regeneration (Jankowski et al. 2009; Raivich et al. 2004). Transcriptional factors associated with regeneration, such as c-Jun, appear to induce the expression of other RAGs and thereby may promote a growth state (Raivich et al. 2004).

3 Axon Regeneration in the Central Nervous System

3.1 Overview of Central Nervous System Regeneration

Pioneering work by Aguayo and colleagues demonstrated that adult mammalian CNS neurons, which normally do not regenerate, are able to grow for long distances into the permissive environment of a peripheral nerve graft (Richardson et al. 1980, 1984; David and Aguayo 1981; Benfey and Aguayo 1982). These studies demonstrated that the environment is a critical determinant of axon regeneration.

Subsequently, numerous molecules were identified within the CNS that limit regeneration.

The two major classes of CNS regeneration inhibitors are the myelin-associated inhibitors (MAIs) and the chondroitin sulfate proteoglycans (CSPGs). These molecules limit axon regeneration, and, by interfering with their function, some degree of growth in the adult CNS is achieved.

Cell-autonomous factors are also important determinants of CNS regeneration failure. CNS neurons do not upregulate growth-associated genes to the same extent as do PNS neurons. Consequently, their ability to regenerate is limited even in the absence of inhibitors. Increasing the intrinsic growth capacity of neurons allows modest axon regeneration within the CNS (Bomze et al. 2001; Neumann and Woolf 1999).

Axon regeneration is one of many factors influencing recovery after CNS damage. Sprouting of uninjured axons can also contribute dramatically to functional improvements. Additionally, plasticity at the synaptic level may underlie a certain degree of recovery seen even in the absence of treatments (i.e., learning to use spared neuronal circuitry in new ways). CNS regeneration studies do not always distinguish between these different mechanisms, and, for the purpose of this discussion, they will be considered together. Replacement of lost neuronal cell bodies, a prominent component of many CNS disorders, is beyond the scope of this chapter.

3.2 Myelin-Associated Inhibitors

MAIs are proteins expressed by oligodendrocytes as components of CNS myelin. MAIs impair neurite outgrowth in vitro and are thought to limit axon growth in vivo after CNS damage. MAIs include Nogo-A (Chen et al. 2000; GrandPre et al., 2000), myelin-associated glycoprotein (MAG) (McKerracher et al., 1994), oligodendrocyte myelin glycoprotein (OMgp) (Kottis et al. 2002), ephrin-B3 (Benson et al. 2005) and Semaphorin 4D (Sema4D) (Moreau-Fauvarque et al. 2003). Three of these (Nogo-A, MAG and OMgp) interact with a neuronal Nogo-66 receptor 1 (NgR1) to limit axon growth. These three structurally unrelated ligands also show affinity for a second axon growth-inhibiting receptor, paired immunoglobulin-like receptor B (PirB) (Atwal et al. 2008). For the most part, MAIs are not found in PNS myelin, which is produced by Schwann cells rather than oligodendrocytes. An exception is MAG, which is present in PNS myelin but is cleared much more rapidly by glial cells in the periphery than in the brain and spinal cord.

One of the most well- characterized MAIs is Nogo-A. Genetic deletion of Nogo-A promotes corticospinal (Fig. 2) and raphespinal tract growth and enhances functional recovery after SCI, although this phenotype is modulated by strain background, age and axonal tract (Kim et al. 2003b; Simonen et al. 2003; Zheng et al. 2003; Dimou et al., 2006). Even after controlling for these factors, certain targeted mutations create a greater axon growth response than others (Cafferty et al. 2007b). However, even the least growth-promoting mutation of the Nogo gene has an

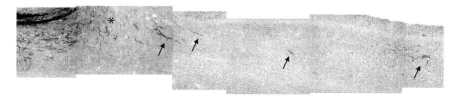

Fig. 2 Corticospinal tract (CST) axonal tracing in mice lacking Nogo-A/B after mid-thoracic spinal cord dorsal hemisection. A parasagittal section of thoracic spinal cord from a mouse with a mutation in the Nogo gene that prevents Nogo-A and Nogo-B expression. The CST is traced from a cortical biotin dextran amine (BDA) injection after dorsal hemisection. Rostral is *left*; dorsal is *up*. The lesion is indicated by the *asterisk*. Note the evidence of CST fiber growth caudal to the lesion site (*arrows*). Significantly less BDA tracing is present caudal to the lesion in control animals (not shown). This photographic montage is a different image from the same mice described in Kim et al. (2003b)

enhanced growth phenotype after pyramidotomy (Cafferty and Strittmatter 2006). In addition, antibodies that target Nogo-A promote axonal growth and functional recovery after CNS injury (Z'Graggen et al., 2000; Wiessner et al. 2003; Seymour et al. 2005). An anti-Nogo-A antibody has advanced to clinical trials for SCI.

Two inhibitory portions of Nogo-A have been identified. The first, termed Nogo-66, is a 66 amino acid fragment which interacts with NgR1 on the neuronal membrane (Fournier et al. 2001). An adjacent 24 amino acid sequence, although not inhibitory in itself, facilitates picomolar-affinity binding of the Nogo-66 to NgR1 (Hu et al. 2005). The Nogo-66 domain can also interact directly with a secondary receptor, PirB, on the surface of neurons (Atwal et al. 2008). The other inhibitory portion of Nogo-A Amino-Nogo, acts via an independent mechanism to disrupt neuronal integrin function (Hu and Strittmatter 2008).

Two other Nogo isoforms exist (Nogo-B and Nogo-C), which contain the inhibitory Nogo-66 loop found in Nogo-A but lack the Amino-Nogo sequence. These isoforms are not found naturally in myelin, but transgenic overexpression of Nogo-C in Schwann cells, which normally do not express any Nogo isoform, delays peripheral nerve regeneration. This demonstrates the ability of Nogo-66 to limit axon regeneration in vivo (Kim et al. 2003a).

MAG is another inhibitory protein present in CNS myelin (McKerracher et al., 1994). MAG interacts with several neuronal receptors to limit neurite outgrowth in vitro, including NgR1 (Liu et al. 2002; Domeniconi et al. 2002), gangliosides (Vyas et al. 2002; Mehta et al. 2007), Nogo receptor 2 (NgR2) (Venkatesh et al. 2005) and PirB (Atwal et al. 2008). The relative importance of each receptor varies with neuronal type. For example, in postnatal DRG neurons, NgR1 mediates the majority of inhibition by MAG, whereas in postnatal cerebellar granule neurons, GT1b appears to be more important (Mehta et al. 2007). Although MAG inhibits neurite outgrowth in vitro, no evidence of enhanced corticospinal tract (CST) regeneration or sprouting after SCI or optic nerve injury is observed in MAG knockout mice (Bartsch et al. 1995). Thus, MAG appears to have a less important role in limiting CNS axon growth than other inhibitors such as Nogo-A.

OMgp, another MAI that interacts with NgR1 and PirB, also limits neurite outgrowth in vitro (Kottis et al. 2002). CNS regeneration studies with OMgp knockout mice have not been described.

NgR1 is a glycosylphosphatidylinositol (GPI)-linked membrane receptor, which lacks a transmembrane or cytoplasmic domain. It interacts with coreceptors LINGO-1 (Mi et al. 2004) and p75 (Wang et al. 2002) or TAJ/TROY (Park et al. 2005; Shao et al. 2005), depending on neuronal type, to limit axon growth. Enhanced rubrospinal and raphespinal, but not corticospinal, axon regeneration is observed in NgR1 knockout mice after SCI (Kim et al. 2004). The enhanced axonal growth is correlated with improved functional recovery. Although CST regeneration is not observed after SCI in mice lacking NgR1, corticofugal axons do show enhanced growth in these mice after a stroke (Li et al. 2004) or pyramidotomy (Cafferty and Strittmatter 2006). Thus, NgR1 limits axonal growth and functional recovery after CNS damage.

NgR1 function can be blocked by a soluble form of extracellular NgR1 fused to human Fc (NgR(310)ecto-Fc). NgR(310)ecto-Fc promotes corticospinal and raphespinal growth and functional recovery after SCI in rats (Li et al. 2004). In addition, transgenic mice which secrete NgR(310)ecto under control of the GFAP promoter (causing reactive astrocytes to secrete high levels of the protein after CNS injury) show enhanced functional recovery after SCI (Li et al. 2005).

A competitive NgR1 antagonist, Nogo-extracellular peptide, residues 1–40 (NEP1-40), binds to, but does not activate, NgR1. NEP1-40 attenuates inhibition of neurite outgrowth by Nogo-A and CNS myelin. After SCI, NEP1-40 promotes corticospinal and raphespinal regeneration and functional recovery, even when the initiation of treatment is delayed for one week (GrandPre et al. 2002; Li and Strittmatter, 2003). Because NEP1-40 blocks Nogo-66, but not other NgR1 ligands, it is less effective than NgR(310)ecto-Fc.

The studies described above confirm the importance of NgR1 and its ligands in limiting CNS regeneration. In vivo functional studies of other MAI receptors have not yet been reported.

3.3 Chondroitin Sulfate Proteoglycans

The astroglial scar, which forms after CNS injury, is a physical barrier to regeneration, and also contains inhibitory molecules that impede axon growth. CSPGs are the main inhibitory molecules found in the glial scar (reviewed in Morgenstern et al. 2002). CSPGs are upregulated by reactive astrocytes after CNS damage and are both membrane bound and secreted into the extracellular space. Members of this class of inhibitors include neurocan (Asher et al. 2000), versican (Schmalfeldt et al. 2000), brevican (Yamada et al. 1997), phosphacan (Inatani et al. 2001), aggrecan (Chan et al. 2008) and NG2 (Dou and Levine, 1994). A receptor for CSPGs has not been identified.

Interfering with CSPG function promotes axon regeneration in the CNS. CPSGs contain core proteins with attached glycosaminoglycan (GAG) side chains, which can be cleaved by the bacterial enzyme Chondroitinase ABC. This enzyme reduces the inhibitory activity of CSPGs in vitro (McKeon et al. 1995). Moreover, when Chondroitinase ABC is administered after spinal contusion in rats, regeneration of both descending CST and ascending sensory fibers can be detected (Bradbury et al. 2002). This axonal growth is accompanied by enhanced recovery of associated locomotor and proprioceptive functions. Several other studies have confirmed that Chondrotinase ABC promotes axonal growth after CNS injury (Barritt et al. 2006; Massey et al. 2006; Cafferty et al. 2007a).

3.4 Other Axon Regeneration Inhibitors

Axon regeneration inhibitors (ARIs) found in the CNS that are not present in myelin or the glial scar include repulsive guidance molecule (RGM) and semaphorin 3A (Sema3A). Evidence that these molecules limit CNS regeneration include studies demonstrating that an anti-RGMa antibody (Hata et al. 2006) or a small molecule inhibitor of Sema3A (Kaneko et al. 2007) promote functional recovery after SCI in rats.

3.5 Inhibitory Signaling Pathways

Multiple ARIs have been shown to activate the small GTPase ras homolog gene family, member A (RhoA) (Niederost et al. 2002; Fournier et al. 2003; Shao et al. 2005). Activated RhoA, in turn, activates Rho-associated coiled-coil containing protein kinase 2 (ROCK2), a kinase that regulates actin cytoskeletal dynamics (reviewed in Schmandke et al. 2007). Activation of ROCK2 results in cessation of neurite growth. Interfering with RhoA or ROCK2 activity promotes CNS axon regeneration and functional recovery.

Ibuprofen, which inhibits RhoA, promotes corticospinal and raphespinal sprouting after spinal contusion (Fu et al. 2007; Wang et al. 2009), as well as long- distance raphespinal axon regeneration after a complete spinal cord transection (Wang et al. 2009). Tissue sparing at the lesion site is also enhanced by ibuprofen and thus may contribute to functional recovery (Wang et al. 2009).

The ROCK2 inhibitor Y27632 promotes CST sprouting and locomotor recovery after dorsal hemi section spinal injury in rats (Fournier et al. 2003). In addition, ROCK2 knockout mice show enhanced functional recovery after SCI (Duffy PJ, Schmandke A, and Strittmatter SM, unpublished observations). Thus, ROCK2 is an important mediator of CNS regeneration failure.

Some evidence suggests that epidermal growth factor receptor (EGFR) contributes to CNS regeneration failure. One study demonstrated enhanced optic nerve regeneration after treatment with the irreversible EGFR inhibitor PD168393 (Koprivica et al. 2005). This study provides evidence that *trans-activation* of EGFR mediates inhibition of neurite outgrowth by MAIs and CSPGs. Another study observed that PD168393 enhances sparing, and/or regeneration of 5-hydroxytryptophan-immunoreactive (serotonergic) fibers caudal to a spinal cord lesion (Erschbamer et al. 2007). Thus, EGFR activation appears to limit recovery after CNS trauma.

Other molecules that have been implicated in ARI- signaling include protein kinase C, (Sivasankaran et al. 2004), LIM kinase, Slingshot phosphatase and cofilin (Hsieh et al. 2006).

3.6 Intrinsic Growth State of the Neuron

In contrast to the PNS, the upregulation of peripheral RAGs (see Sect. 2.2) is relatively modest in the CNS after injury (Fernandes et al. 1999; Marklund et al. 2006). This paucity of RAG expression appears to be partially responsible for the limited ability of CNS neurons to regenerate. Increasing RAG expression in CNS neurons improves their regenerative ability. For example, Bomze et al. (2001) demonstrated that overexpressing GAP-43 and CAP-23 together promotes sensory axon regeneration after SCI.

DRG neurons have a peripheral process and a central process. Injury to the peripheral process results in robust upregulation of RAGs, as described above. However, injury to the central process by dorsal rhizotomy or spinal cord dorsal hemi section does not induce nearly as robust of a regenerative response, and central processes fail to regenerate in the CNS. Injury of peripheral axons one week prior to central injury (termed a conditioning lesion) allows some degree of sensory fiber regeneration within the spinal cord (Neumann and Woolf 1999). The conditioning lesion appears to enhance the growth state of the neuron such that its central process is able to regenerate in the CNS environment.

Cyclic adenosine monophosphate (cAMP) is a second messenger molecule which influences the growth state of the neuron. cAMP levels are increased by a peripheral conditioning lesion (Qiu et al. 2002). Elevation of cAMP levels by intra-ganglionic injection of a membrane-permeable cAMP analog, dibutyryl cAMP (db-cAMP), mimics the growth-promoting effects of a conditioning lesion, promoting regeneration of sensory axons within the spinal cord (Neumann et al. 2002; Qiu et al. 2002). In vivo injection of db-cAMP prior to DRG removal also improves the ability of dissociated DRG cultures to grow on MAG or CNS myelin in vitro, indicating that cAMP elevation can promote growth in the presence of MAIs (Qiu et al. 2002).

Rolipram, a phosphodiesterase 4 inhibitor, increases cAMP by interfering with its hydrolysis. When delivered 2 weeks after spinal cord hemisection, rolipram increases serotonergic axon regeneration into embryonic spinal tissue grafts implanted at the lesion site at the time of injury (Nikulina et al. 2004). Reactive gliosis is reduced by rolipram, and this might contribute to the functional recovery observed. Additionally, enhanced axonal sparing and myelination are induced by cAMP elevation in combination with Schwann cell grafts after SCI, compared to Schwann cell grafts alone (Pearse et al. 2004), suggesting additional mechanisms by which cAMP elevation could lead to functional improvements. Nonetheless, the demonstration of serotonergic axon growth into grafts at the lesion site in both of these studies indicates that cAMP elevation can induce CNS axon regeneration.

cAMP elevation activates protein kinase A (PKA) and induces CREB-mediated transcription of various growth-associated genes, including IL-6 and arginase I. Subsequent synthesis of polyamines by arginase I has been proposed as a possible mechanism by which cAMP increases neurite growth (Cai et al. 2002).

4 Conclusion

Regeneration of the injured mammalian CNS was once thought to be an unachievable goal. Recent advances in our understanding of factors which limit CNS regeneration and those which facilitate PNS regeneration have lead to therapies which allow some degree of recovery from brain and SCI in animal models. These findings open the possibility of promoting regeneration of the damaged human CNS.

References

Asher RA, Morgenstern DA, Fidler PS, Adcock KH, Oohira A, Braistead JE, Levine JM, Margolis RU, Rogers JH, Fawcett JW (2000) Neurocan is upregulated in injured brain and in cytokine-treated astrocytes. J Neurosci 20:2427–2438

Atwal JK, Pinkston-Gosse J, Syken J, Stawicki S, Wu Y, Shatz C, Tessier-Lavigne M (2008) PirB is a functional receptor for myelin inhibitors of axonal regeneration. Science 322:967–970

Barritt AW, Davies M, Marchand F, Hartley R, Grist J, Yip P, McMahon SB, Bradbury EJ (2006) Chondroitinase ABC promotes sprouting of intact and injured spinal systems after spinal cord injury. J Neurosci 26:10856–10867

Bartsch U, Bandtlow CE, Schnell L, Bartsch S, Spillmann AA, Rubin BP, Hillenbrand R, Montag D, Schwab ME, Schachner M (1995) Lack of evidence that myelin-associated glycoprotein is a major inhibitor of axonal regeneration in the CNS. Neuron 15:1375–1381

Benfey M, Aguayo AJ (1982) Extensive elongation of axons from rat brain into peripheral nerve grafts. Nature 296:150–152

Benson MD, Romero MI, Lush ME, Lu QR, Henkemeyer M, Parada LF (2005) Ephrin-B3 is a myelin-based inhibitor of neurite outgrowth. Proc Nat Acad Sci USA 102:10694–10699

Boeshore KL, Schreiber RC, Vaccariello SA, Sachs HH, Salazar R, Lee J, Ratan RR, Leahy P, Zigmond RE (2004) Novel changes in gene expression following axotomy of a sympathetic ganglion: a microarray analysis. J Neurobiol 59:216–235

Bomze HM, Bulsara KR, Iskandar BJ, Caroni P, Pate Skene JH (2001) Spinal axon regeneration evoked by replacing two growth cone proteins in adult neurons. Nat Neurosci 4:38–43

Bonilla IE, Tanabe K, Strittmatter SM (2002) Small proline-rich repeat protein 1A Is expressed by axotomized neurons and promotes axonal outgrowth. J Neurosci 22:1303–1315

Bradbury EJ, Moon LDF, Popat RJ, King VR, Bennett GS, Patel PN, Fawcett JW, McMahon SB (2002) Chondroitinase ABC promotes functional recovery after spinal cord injury. Nature 416:636–640

Cafferty WB, Strittmatter SM (2006) The Nogo-Nogo receptor pathway limits a spectrum of adult cns axonal growth. J Neurosci 26:12242–12250

Cafferty WB, Yang S-H, Duffy PJ, Li S, Strittmatter SM (2007a) Functional axonal regeneration through astrocytic scar genetically modified to digest chondroitin sulfate proteoglycans. J Neurosci 27:2176–2185

Cafferty WB, Kim J-E, Lee J-K, Strittmatter SM (2007b) Response to correspondence: Kim et al. "Axon regeneration in young adult mice lacking Nogo-A/B." Neuron 38, 187–199. Neuron 54:195–199

Cai D, Deng K, Mellado W, Lee J, Ratan RR, Filbin MT (2002) Arginase I and polyamines act downstream from cyclic AMP in overcoming inhibition of axonal growth MAG and myelin in vitro. Neuron 35:711–719

Chan CC, Roberts CR, Steeves JD, Tetzlaff W (2008) Aggrecan components differentially modulate nerve growth factor-responsive and neurotrophin-3-responsive dorsal root ganglion neurite growth. J Neurosci Res 86:581–592

Chen MS, Huber AB, van der Haar ME, Frank M, Schnell L, Spillmann AA, Christ F, Schwab ME (2000) Nogo-A is a myelin-associated neurite outgrowth inhibitor and an antigen for monoclonal antibody IN-1. Nature 403:434–439

Costigan M, Befort K, Karchewski L, Griffin R, D'Urso D, Allchorne A, Sitarski J, Mannion J, Pratt R, Woolf C (2002) Replicate high-density rat genome oligonucleotide microarrays reveal hundreds of regulated genes in the dorsal root ganglion after peripheral nerve injury. BMC Neurosci 3:16

David S, Aguayo AJ (1981) Axonal elongation into peripheral nervous system "bridges" after central nervous system injury in adult rats. Science 214:931–933

Dimou L, Schnell L, Montani L, Duncan C, Simonen M, Schneider R, Liebscher T, Gullo M, Schwab ME (2006) Nogo-A-deficient mice reveal strain-dependent differences in axonal regeneration. J Neurosci 26:5591–5603

Domeniconi M, Cao Z, Spencer T, Sivasankaran R, Wang KC, Nikulina E, Kimura N, Cai H, Deng K, Gao Y, He Z, Filbin MT (2002) Myelin-associated glycoprotein interacts with the Nogo66 receptor to inhibit neurite outgrowth. Neuron 35:283–290

Dou CL, Levine JM (1994) Inhibition of neurite growth by the NG2 chondroitin sulfate proteoglycan. J Neurosci 14:7616–7628

Erschbamer M, Pernold K, Olson L (2007) Inhibiting epidermal growth factor receptor improves structural, locomotor, sensory, and bladder recovery from experimental spinal cord injury. J Neurosci 27:6428–6435

Fernandes KJ, Fan DP, Tsui BJ, Cassar SL, Tetzlaff W (1999) Influence of the axotomy to cell body distance in rat rubrospinal and spinal motoneurons: differential regulation of GAP-43, tubulins, and neurofilament-M. J Comp Neurol 414:495–510

Fournier AE, GrandPre T, Strittmatter SM (2001) Identification of a receptor mediating Nogo-66 inhibition of axonal regeneration. Nature 409:341–346

Fournier AE, Takizawa BT, Strittmatter SM (2003) Rho kinase inhibition enhances axonal regeneration in the injured CNS. J Neurosci 23:1416–1423

Fu Q, Hue J, Li S (2007) Nonsteroidal anti-inflammatory drugs promote axon regeneration via RhoA inhibition. J Neurosci 27:4154–4164

GrandPre T, Nakamura F, Vartanian T, Strittmatter SM (2000) Identification of the Nogo inhibitor of axon regeneration as a Reticulon protein. Nature 403:439–444

GrandPre T, Li S, Strittmatter SM (2002) Nogo-66 receptor antagonist peptide promotes axonal regeneration. Nature 417:547–551

Hata K, Fujitani M, Yasuda Y, Doya H, Saito T, Yamagishi S, Mueller BK, Yamashita T (2006) RGMa inhibition promotes axonal growth and recovery after spinal cord injury. J Cell Biol 173:47–58

Hsieh SH, Ferraro GB, Fournier AE (2006) Myelin-associated inhibitors regulate cofilin phosphorylation and neuronal inhibition through LIM kinase and slingshot phosphatase. J Neurosci 26:1006–1015

Hu F, Strittmatter SM (2008) The N-terminal domain of Nogo-A inhibits cell adhesion and axonal outgrowth by an integrin-specific mechanism. J Neurosci 28:1262–1269

Hu F, Liu BP, Budel S, Liao J, Chin J, Fournier A, Strittmatter SM (2005) Nogo-A interacts with the Nogo-66 receptor through multiple sites to create an isoform-selective subnanomolar agonist. J Neurosci 25:5298–5304

Inatani M, Honjo M, Otori Y, Oohira A, Kido N, Tano Y, Honda Y, Tanihara H (2001) Inhibitory effects of neurocan and phosphacan on neurite outgrowth from retinal ganglion cells in culture. Invest Ophthalmol Vis Sci 42:1930–1938

Jankowski MP, McIlwrath SL, Jing X, Cornuet PK, Salerno KM, Koerber HR, Albers KM (2009) Sox11 transcription factor modulates peripheral nerve regeneration in adult mice. Brain Research 1256:43–54

Kaneko S, Iwanami A, Nakamura M, Kishino A, Kikuchi K, Shibata S, Okano HJ, Ikegami T, Moriya A, Konishi O, Nakayama C, Kumagai K, Kimura T, Sato Y, Goshima Y, Taniguchi M, Ito M, He Z, Toyama Y, Okano H (2007) A selective Sema3A inhibitor enhances regenerative responses and functional recovery of the injured spinal cord. Nat Med 12:1380–1389

Kim J-E, Bonilla IE, Qiu D, Strittmatter SM (2003a) Nogo-C is sufficient to delay nerve regeneration. Mol Cell Neurosci 23:451–459

Kim J-E, Li S, GrandPré T, Qiu D, Strittmatter SM (2003b) Axon regeneration in young adult mice lacking Nogo-A/B. Neuron 38:187–199

Kim J-E, Liu BP, Park JH, Strittmatter SM (2004) Nogo-66 receptor prevents raphespinal and rubrospinal axon regeneration and limits functional recovery from spinal cord injury. Neuron 44:439–451

Koprivica V, Cho K-S, Park JB, Yiu G, Atwal J, Gore B, Kim JA, Lin E, Tessier-Lavigne M, Chen DF, He Z (2005) EGFR activation mediates inhibition of axon regeneration by myelin and chondroitin sulfate proteoglycans. Science 310:106–110

Kottis V, Thibault P, Mikol D, Xiao ZC, Zhang R, Dergham P, Braun PE (2002) Oligodendrocyte-myelin glycoprotein (OMgp) is an inhibitor of neurite outgrowth. J Neurochem 82:1566–1569

Li S, Strittmatter SM (2003) Delayed systemic Nogo-66 receptor antagonist promotes recovery from spinal cord injury. J Neurosci 23:4219–4227

Li S, Liu BP, Budel S, Li M, Ji B, Walus L, Li W, Jirik A, Rabacchi S, Choi E, Worley D, Sah DWY, Pepinsky B, Lee D, Relton J, Strittmatter SM (2004) Blockade of Nogo-66, myelin-associated glycoprotein, and oligodendrocyte myelin glycoprotein by soluble Nogo-66 receptor promotes axonal sprouting and recovery after spinal injury. J Neurosci 24:10511–10520

Li S, Kim J-E, Budel S, Hampton TG, Strittmatter SM (2005) Transgenic inhibition of Nogo-66 receptor function allows axonal sprouting and improved locomotion after spinal injury. Mol Cell Neurosci 29:26–39

Liu BP, Fournier A, GrandPre T, Strittmatter SM (2002) Myelin-associated glycoprotein as a functional ligand for the Nogo-66 receptor. Science 297:1190–1193

Marklund N, Fulp CT, Shimizu S, Puri R, McMillan A, Strittmatter SM, McIntosh TK (2006) Selective temporal and regional alterations of Nogo-A and small proline-rich repeat protein 1A (SPRR1A) but not Nogo-66 receptor (NgR) occur following traumatic brain injury in the rat. Exp Neurol 197:70–83

Massey JM, Hubscher CH, Wagoner MR, Decker JA, Amps J, Silver J, Onifer SM (2006) Chondroitinase ABC digestion of the perineuronal net promotes functional collateral sprouting in the Cuneate nucleus after cervical spinal cord injury. J Neurosci 26:4406–4414

McKeon RJ, Höke A, Silver J (1995) Injury-induced proteoglycans inhibit the potential for laminin-mediated axon growth on astrocytic scars. Experimental Neurology 136:32–43

McKerracher L, David S, Jackson DL, Kottis V, Dunn RJ, Braun PE (1994) Identification of myelin-associated glycoprotein as a major myelin-derived inhibitor of neurite growth. Neuron 13:805–811

Mehta NR, Lopez PH, Vyas AA, Schnaar RL (2007) Gangliosides and Nogo Receptors Independently Mediate Myelin-associated Glycoprotein Inhibition of Neurite Outgrowth in Different Nerve Cells. J Biol Chem 282:27875–27886

Mi S, Lee X, Shao Z, Thill G, Ji B, Relton J, Levesque M, Allaire N, Perrin S, Sands B, Crowell T, Cate RL, McCoy JM, Pepinsky RB (2004) LINGO-1 is a component of the Nogo-66 receptor/p75 signaling complex. Nat Neurosci 7:221–228

Moreau-Fauvarque C, Kumanogoh A, Camand E, Jaillard C, Barbin G, Boquet I, Love C, Jones EY, Kikutani H, Lubetzki C, Dusart I, Chedotal A (2003) The Transmembrane Semaphorin Sema4D/CD100, an inhibitor of axonal growth, is expressed on oligodendrocytes and upregulated after CNS lesion. J Neurosci 23:9229–9239

Morgenstern DA, Asher RA, Fawcett JW (2002) Chondroitin sulphate proteoglycans in the CNS injury response. Prog Brain Res 137:313–332

Neumann S, Woolf CJ (1999) Regeneration of dorsal column fibers into and beyond the lesion site following adult spinal cord injury. Neuron 23:83–91

Neumann S, Bradke F, Tessier-Lavigne M, Basbaum AI (2002) Regeneration of sensory axons within the injured spinal cord induced by intraganglionic cAMP elevation. Neuron 34:885–893

Niederost B, Oertle T, Fritsche J, McKinney RA, Bandtlow CE (2002) Nogo-A and myelin-associated glycoprotein mediate neurite growth inhibition by antagonistic regulation of RhoA and Rac1. J Neurosci 22:10368–10376

Nikulina E, Tidwell JL, Dai HN, Bregman BS, Filbin MT (2004) The phosphodiesterase inhibitor rolipram delivered after a spinal cord lesion promotes axonal regeneration and functional recovery. Proc Natl Acad Sci USA 101:8786–8790

Park JB, Yiu G, Kaneko S, Wang J, Chang J, He Z (2005) A TNF receptor family member, TROY, is a coreceptor with Nogo receptor in mediating the inhibitory activity of myelin inhibitors. Neuron 45:345–351

Pearse DD, Pereira FC, Marcillo AE, Bates ML, Berrocal YA, Filbin MT, Bunge MB (2004) cAMP and Schwann cells promote axonal growth and functional recovery after spinal cord injury. Nat Med 10:610–616

Qiu J, Cai D, Dai H, McAtee M, Hoffman PN, Bregman BS, Filbin MT (2002) Spinal axon regeneration induced by elevation of cyclic AMP. Neuron 34:895–903

Raivich G, Bohatschek M, Da Costa C, Iwata O, Galiano M, Hristova M, Nateri AS, Makwana M, Ls R-S, Wolfer DP, Lipp H-P, Aguzzi A, Wagner EF, Behrens A (2004) The AP-1 transcription factor c-Jun is required for efficient axonal regeneration. Neuron 43:57–67

Richardson PM, McGuinness UM, Aguayo AJ (1980) Axons from CNS neurons regenerate into PNS grafts. Nature 284:264–265

Richardson PM, Issa VM, Aguayo AJ (1984) Regeneration of long spinal axons in the rat. J Neurocytol 13:165–182

Schmalfeldt M, Bandtlow CE, Dours-Zimmermann MT, Winterhalter KH, Zimmermann DR (2000) Brain derived versican V2 is a potent inhibitor of axonal growth. J Cell Sci 113:807–816

Schmandke A, Schmandke A, Strittmatter SM (2007) ROCK and Rho: biochemistry and neuronal functions of Rho-associated protein kinases. Neuroscientist 13:454–469

Seijffers R, Allchorne AJ, Woolf CJ (2006) The transcription factor ATF-3 promotes neurite outgrowth. Mol Cell Neurosci 32:143–154

Seymour AB, Andrews EM, Tsai S-Y, Markus TM, Bollnow MR, Brenneman MM, O'Brien TE, Castro AJ, Schwab ME, Kartje GL (2005) Delayed treatment with monoclonal antibody IN-1 1 week after stroke results in recovery of function and corticorubral plasticity in adult rats. J Cereb Blood Flow Metab 25:1366–1375

Shao Z, Browning JL, Lee X, Scott ML, Shulga-Morskaya S, Allaire N, Thill G, Levesque M, Sah D, McCoy JM, Murray B, Jung V, Pepinsky RB, Mi S (2005) TAJ/TROY, an orphan TNF receptor family member, binds Nogo-66 receptor 1 and regulates axonal regeneration. Neuron 45:353–359

Simonen M, Pedersen V, Weinmann O, Schnell L, Buss A, Ledermann B, Christ F, Sansig G, van der Putten H, Schwab ME (2003) Systemic deletion of the myelin-associated outgrowth inhibitor

Nogo-A improves regenerative and plastic responses after spinal cord injury. Neuron 38: 201–211

Sivasankaran R, Pei J, Wang KC, Zhang YP, Shields CB, Xu X-M, He Z (2004) PKC mediates inhibitory effects of myelin and chondroitin sulfate proteoglycans on axonal regeneration. Nat Neurosci 7:261–268

Tanabe K, Bonilla I, Winkles JA, Strittmatter SM (2003) Fibroblast growth factor-inducible-14 is induced in axotomized neurons and promotes neurite outgrowth. J Neurosci 23:9675–9686

Venkatesh K, Chivatakarn O, Lee H, Joshi PS, Kantor DB, Newman BA, Mage R, Rader C, Giger RJ (2005) The Nogo-66 receptor homolog NgR2 is a sialic acid-dependent receptor selective for myelin-associated glycoprotein. J Neurosci 25:808–822

Vyas AA, Patel HV, Fromholt SE, Heffer-Lauc M, Vyas KA, Dang J, Schachner M, Schnaar RL (2002) Gangliosides are functional nerve cell ligands for myelin-associated glycoprotein (MAG), an inhibitor of nerve regeneration. Proc Natl Acad Sci USA 99:8412–8417

Wang KC, Kim JA, Sivasankaran R, Segal R, He Z (2002) p75 interacts with the Nogo receptor as a co-receptor for Nogo, MAG and OMgp. Nature 420:74–78

Wang X, Budel S, Baughman K, Gould G, Song KH, Strittmatter SM (2009) Ibuprofen enhances recovery from spinal cord injury by limiting tissue loss and stimulating axonal growth. J Neurotrauma 26:81–95

Wiessner C, Bareyre FM, Allegrini PR, Mir AK, Frentzel S, Zurini M, Schnell L, Oertle T, Schwab ME (2003) Anti-Nogo-A antibody infusion 24 hours after experimental stroke improved behavioral outcome and corticospinal plasticity in normotensive and spontaneously hypertensive rats. J Cereb Blood Flow Metab 23:154–165

Yamada H, Fredette B, Shitara K, Hagihara K, Miura R, Ransht B, Stallcup WB, Yamaguchi Y (1997) The brain chondroitin sulfate proteoglycan brevican associates with astrocytes ensheathing cerebellar glomeruli and inhibits neurite outgrowth from granule neurons. J Neurosci 17:7784–7795

Z'Graggen WJ, Fouad K, Raineteau O, Metz GAS, Schwab ME, Kartje GL (2000) Compensatory sprouting and impulse rerouting after unilateral pyramidal tract lesion in neonatal rats. J Neurosci 20:6561–6569

Index

A

Actin, 65–86, 110, 113, 116, 117, 126, 129–132
Actin-binding proteins (ABP), 65–70, 72, 74, 77–82, 84–86, 292, 296
β-Actin mRNA localization, 299–300
Actin related protein 1 (Arp1), 116, 123, 124
β-Actin synthesis, 280
  effects of repulsive cues, 280–281
ADF/cofilin, 67, 74, 79–81
Adhesive contacts, 74, 76, 79, 81
Akt, 153, 155
Alamar Blue (AB), 234
ALS. *See* Amyotrophic lateral sclerosis.
Alzheimers disease, 85
Amyotrophic lateral sclerosis (ALS), 39, 58–59
  spheroids, disorganized NFs, 39
  spheroids, motor entrapment, 40
  spheroids, phospho-epitopes, 39
  superoxide dismutase (SOD1), 58–59
  transport perturbation, 59, 60
Ankyrin, 71
Ankyrin G, 5
Aplysia neurons, 247, 249, 259
  axonal mRNAs, 249, 259
  expression in terminals, 247
Apoptosis, 142, 148, 152, 154, 155
  neuronal culling, 142
ARIs. *See* Axon regeneration inhibitors
Arp2/3 complex, 67, 69, 74
Array analyses, 198
Astroglial scar. *See* Glial scar
ATP, 230–232, 234, 236–239
ATP synthase, 228–230
Autophagy, 110
Axo-glial junctions, 2, 4, 7–9, 11–13, 17–20
  nodal stability, 18
Axon, 226–240, 269–284

acetylcholinesterase (AChE) resynthesis, 272
β-actin mRNA, 274–277, 279, 280
β-actin mRNA zipcode, 274, 275
cis-acting mRNA targeting, 274–275
local protein synthesis, 270, 271, 277–279, 282–284
microtubules, 273, 274, 276, 277
miR-338, 276
mRNAs, 270–281, 283, 284
pathfinding, 269–284
signal transduction for protein synthesis, 283
tau mRNA 3′UTR, 274, 275
translational machinery, 271, 273, 274, 276
ZBP1-β-mRNA transport, 275, 276, 280
Axonal branching, 82–83
Axonal domains formation, 12, 16–20
Axonal outgrowth, 329
  electrical stimulation effect, 329
Axonal protein synthesis, 226–240
Axonal RNAs, 251, 254–259
  origin, 251, 254–256, 258, 259
Axonal transport, 91–96, 99, 100
  actin-based system, 98
  dynein, retrograde motor, 92–94
  kinesin, anterograde motor, 92
  long-range *versus* short-range, 93
  microtubule-based system, 93, 94, 98
  myosin Va and kinesin interactions, 98
  neurodegenerative diseases, 92
  overview, 92
Axon guidance cues, 269, 270, 277, 278, 281–284
Axon initial segment, 71
Axon injury. *See* Negative injury signals; Positive injury signals
  activation of kinases, 334
  calcium influx, 329
  disto-proximal Ca^{2+} wave, 328
  downstream events, 334

353

Axon injury (cont.)
  GDNF, GFRα1 upregulation, 334
  injury discharge, 328
  local vimentin synthesis, 332
  transcription factors upregulated, 334
  upstream events, 334
Axon pathfinding, 269–284
Axon regeneration, 73, 85
Axon regeneration inhibitors (ARIs), 345, 346
Axoplasmic filter, 70
Axoplasmic whole-mounts, 177–180, 182–187
  preparation, properties, 177

B

Barbed end, 66–67, 73, 74, 80, 81
Bax, 312
Bcl-xL, 312
BC1 RNA transport, 185, 186, 251
  Mauthner axon, 187
  targeting to Mauthner axon, 251–252
BDNF. *See* Brain-derived neurotrophic factor
Brain-derived neurotrophic factor (BDNF), 201, 202, 211, 213, 278, 280–282
  β-actin synthesis, 280
  induced steering, 280

C

Caenorhabditis elegans, 304–305
Calcitonin gene-related peptide (CGRP), 211, 213, 214
  injured axon secretion, 211
Calyx of Held, 109, 126
cAMP. *See* Cyclic adenosine monophosphate
cAMP response element binding (CREB), 153, 155, 202, 203, 209
Campenot chambers, 229, 232, 234, 237
Caspr, 4, 9–14, 16, 18–20
Caspr2, 11, 14, 15, 20
Central nervous system (CNS), 339–347
  environment, 339, 346
Charcot-Marie-Tooth disease, 19, 108, 115
Chloramphenicol, 228, 230
Chondroitinase, 345
Chondroitin sulfate proteoglycans (CSPGs), 342, 344–346
Cis-acting elements, 200–203, 206
  β-actin mRNA zip-code, 201
  EphA2 mRNA, 203
  κ-opioid receptor mRNA 3′, 5′ UTR, 202
  RanBP1 mRNAs, 203
  secondary structures, 203
  3′ untranslated region (3′ UTR), 200

CNS. *See* Central nervous system
Cofilin, 273, 276, 281
Cofilin translation, 209
Compartmented culture, 197
Compartmented neuron culture, 141–157
Contactin, 9–14, 16, 19
Corticospinal tract (CST), 343–345
CPSGs. *See* Chondroitin sulfate proteoglycans
CREB. *See* cAMP response element binding
CSPGs. *See* Chondroitin sulfate proteoglycans
CST. *See* Corticospinal tract
Cyclic adenosine monophosphate (cAMP), 346, 347
Cyclin-dependent kinase 5 (Cdk5), 132
Cycloheximide, 230–232
Cytochrome c oxidase IV (COXIV), 228, 233
Cytoskeleton, 29–30, 38, 41
  alpha-internexin, 30
  organization and dynamics, 29–30, 38, 41

D

Death signal, 143, 154–157
  neurodegeneration, 157
  neurotrauma, 157
Dendrites, 194, 195, 197–202, 204, 208, 209, 214
  localized synthesis, 194
Desiprimine, 234
Dicer, 233
Differential mRNA display, 227
DNA polymerase γ, 228–230, 238
Dorsal root ganglion (DRG), 197, 199, 210, 214, 327, 329, 331–334
  conditioning lesion effects, 329
  peripheral *versus* central regeneration, 327
Dorsal root ganglion cells, 273
Downstream signaling pathways, 210
  effects on mRNA transport, 210
DRG. *See* Dorsal root ganglion
Drosophila glial cells, 8
Drosophila melanogaster, 305–306
Dynactin, 116, 121–125
Dynamitin, 116, 123, 124
Dynein, 33, 34, 37–41, 111, 113–116, 119–125, 207, 209
  cargo based NF transport, 37
  La protein retrograde transport, 207
  regulator of anterograde NF transport, 38
  retrograde NF transport, 33, 36–38
Dystroglycan, 6, 7, 17, 18

E

EARP domains. *See* Endoaxoplasmic ribosomal plaques
Electron spectroscopic imaging, 176–178, 180–185, 187, 188
    principles, 177
    rabbit axons, 185
    ribosome P signal, 177
Electrophoretic RNA profiles, 174, 175
    Mauthner and squid axons, 174
Emetine, 230–232
Ena, 67, 74, 77, 78, 82
Endoaxoplasmic ribosomal plaques domains, 187–189
    squid axon, 189
EphA2, 272, 282, 283
EphB, 273, 278
EphB2, 273
EphrinA, 282
EphrinB, 278
ER chaperone proteins, 213
    in axons, 213
Ezrin/Radixin/Moesin (ERM), 67, 74, 80

F

F-actin, 66–79, 81–85, 297, 298
Filamin, 67, 69
Fluorescent dyes, 231, 232
    Alexa Fluor 488, 232
    JC-1, 231, 232
    TMRE, 231, 232
Fluorescent reporter proteins, 203
    photobleaching, 203
    photoconversion, 203
Fragile X mental retardation, 199, 204
Fragile X related proteins, 208

G

G-actin, 66–67, 74, 81
GAP43, 78, 79
Gemins, 290, 291, 293, 298, 303, 305, 307
Gene ontology analyses, 198, 214
    mRNA levels, 214
Gene therapy, 313
Glial cells, 245, 246, 252, 253, 258, 260
    source of neuronal RNA, 251–258
Glial scar, 344, 345
Glia-neuron transfer, 252–254, 258
    Deiters cell RNA, 252–254
    mechanism, 258
    superior cervical ganglion cells, 253
Glia-neuron unit concept, 243–260

Gliomedin, 6, 7, 17, 18
Gö6976, 152, 153
Goldberg-Shprintzen syndrome, 117
GRIF1. *See* Milton
Growth cone, 68, 73, 75–82, 84, 85, 108, 113, 126, 129, 130, 196, 197, 200, 201, 207, 210, 211, 213–215, 269, 270, 272, 274, 275, 277–284
    local protein synthesis, 270, 277–279, 282–284
    retinal ganglion cells (RGCs), 270, 272, 277, 278, 283
Growth cone formation, 213
    translation and proteolysis, 213
Guidance cues, 269, 270, 277, 278, 281–284
    adaptation, 281–282
    differential protein synthesis, 278
    local protein synthesis, 270, 277–278, 282, 284
    sensitization, desensitization cycles, 281, 282
Gurken, 211, 212

H

Hereditary spastic paraplegia (HSP), 53–56, 59
    gain-of-function hypothesis, 55, 59
    spastin mutation, 54, 55
    strategy of spastin-based HSP, 55, 56
Hippocampal neuron, 74–75, 79, 85
hnRNP Q, 297, 301–302
hnRNP-R, 297, 301–302
Huntington's disease (HD), 59
    disrupted transport, 59
    huntingtin mutations, 59

I

IMP1. *See* Zipcode binding protiein-1 (ZBP1)
Injury-conditioned neurons, 212
    importin β1, 213
    RanBP1, 213
In situ hybridization, 202, 204, 209, 214
In situ hybridization histochemistry, 228, 229
Internode, 1–3, 14–16, 18
    nectin-like proteins, 15

J

Juxtaparanode, 1–4, 7, 9, 11, 14, 15

K

K252a, 150, 152, 153, 155
KIF5, 114, 115, 117–119, 122, 124, 125, 127, 128

KIF1B, 114, 115, 117
Kinesin, 31–38, 40, 41, 111, 113–115,
    117–125, 127, 128, 195, 200, 208
  interaction with NFs, 30
KSRP, 297, 300
Kuhn, T.S., 152, 153

L
LB diseases. *See* Lewy body diseases
LBs. *See* Lewy bodies
Leukemia inhibitory factor (LIF), 156
Lewy bodies (LBs), 159, 160, 162, 164
Lewy body diseases, 159–162, 164, 165
Lewy neurites, 162, 164
LNs. *See* Lewy neurites
Lobster stretch receptor, 253
  stretch-induced RNA changes, 253
Lymnea stagnalis axons, 249, 251
  membrane receptor synthesis, 251
  reporter gene translation, 250–251

M
MAIs. *See* Myelin-associated inhibitors
MAP2, 198, 200
Mauthner axon, 248, 252, 254
  RNA changes after transection, 248
  RNAs, 248
Mauthner axon collaterals, 180
Membrane potential, 110, 111
Membrane proteins, 197, 211
  axon mRNAs, 211
Metabolic labeling, 195, 197, 230
Microfilament-based transport, 207
  RNPs, 207
Microfluidic device, 197, 213
  isolation of axon growth, 213
MicroRNAs, 233, 239, 240
Microtubule-associated proteins (MAPs),
    49, 56–58
Microtubule-based transport, 196, 200, 207, 215
Microtubules (MTs), 32, 35, 37–39, 110,
    113, 129
  crosslinking by motor proteins, 38–39
Microtubule-severing, 53–58
  in microtubule dynamics, 56
  proteins, 49, 51, 52, 57
    katanin, 49, 53, 54, 57
    katanin, spastin compared, 55
    katanin, spastin expression, 54
    spastin isoforms, 54
Microtubules in axons, 1–53
  arrangement,

dynamics, 49, 56
functions, 49, 52, 54
interaction with actin, 52
microtubule transport, 49, 52
+tip proteins, 51–52
Milton, 114, 118–119, 124–125, 127, 128
miRNAs, 204, 209, 257, 259, 273, 275, 276
  miR-338 and CoxIV expression, 204
Miro, 114, 118–119, 124–129, 131
Mitochondria, 72, 73, 227, 228, 230–232, 234,
    236, 238–240
  nuclear-encoded mitochondrial mRNAs,
    226, 228, 229, 231, 233, 239
  nuclear-encoded mitochondrial proteins,
    226, 228, 238
Mitochondrial activity, 230, 234, 237–239
Mitochondrial membrane potential,
    230–232, 239
Molecular chaperones, 227, 228, 230, 231, 238
  Hsp70, 227, 228, 230, 231
  Hsp90, 227, 228, 230–232
Motor coordination, 120–124
Motor proteins, 91–100
  processive properties, 97
Mouse models, 291, 298, 300, 307–312, 315
mRNA transport, 193–216
MTs. *See* Microtubules
Multiple sclerosis, 19–21
Myelin, 339, 342, 343, 345, 346
Myelin-associated glycoprotein, 209–211
  growth inhibiting, 209
Myelin associated glycoprotein (MAG), 342,
    343, 346
Myelin-associated inhibitors (MAIs),
    342–344, 346
Myelination, 1–21
  oligodenrocytes, 1–3, 8, 12, 13, 16–18, 20
  Schwann cells, 1–3, 7, 12, 16–18
Myelin lipids, 13
  during myelination, 3, 6, 12, 13
Myosin, 113, 114, 116–117, 200, 207
Myosin II, 72, 74, 76, 78, 81, 84, 85
Myosin classes, 95–99
  in neurons, 94
Myosin I isoforms, 95
  role in endocytosis, 95
Myosin II (nonmuscle), 95–96
  isoforms, 95
  roles in neurons, 95–96
Myosin V, 73, 96–98
  dilute lethal, myosin Va null, 97, 98
  expression in nervous system, 96
  motor properties, 97
  myosin Va passive transport, 97

myosin Va properties, 96
myosin Vb, 96–98
processivity, 98
role in dendrites, 97
short-range transport, 97, 98
subunits and properties, 96, 97
Myosin VI, 99
 hair cell function and deafness, 99
 properties, nonhair cells, 99
Myosin X, 99
 properties and roles, 99

N

Navigation, 73, 80–82
Negative injury signals, 328–330
 interrupted NGF transport, 330
Nerve, 339–341, 343, 345
 optic, 343, 345
 sciatic, 340, 341
Nerve growth factor (NGF), 126, 129–131, 141–157
 linked to polystyrene beads, 148
 quantum dot, 151–152
 transport in compartmented culture, 146, 147
Netrin, 269, 277–283
 β-actin synthesis, 278–280
 induced steering, 280
Netrin-1, 211
Neurites, 74, 76, 77, 79, 80
Neurofascin, 18, 20
 NF155, 4, 6, 9, 10, 12–14, 18–20
 NF186, 4–7, 12, 16, 17, 19–20
Neurofilaments (NFs), 29–41, 113
 complexity of phsophorylation, 30
 divalent ion-induced bundllling, 37
 mitochondrial binding, 40
 NF-NF associations, 33–35, 37–41
 phosphorylation dependent interactions, 30
 RT97 phospho-epitope, 31, 41
 subunit composition, 30–32, 36
Neurohypophyseal axons, 196
 mRNAs, 196
Neurohypophyseal tract,
Neuromuscular junction (NMJ), 305, 306, 310–312, 315
Neuronal polarity, 196
 invertebrate neurons, 196
Neurotrophins, 209–211
 effects on axonal mRNAs, 210
 NT3, RNP movements, 209
NFs. *See* Neurofilaments
NF transport, 30–40
 Cdk5 kinase modulation, 33, 39

 counterbalancing motor forces, 37
 extensively-phosphorylated NFs, 32, 34
 MAP kinase inhibition, 33, 36, 39
 monitoring, 33
 NF-H deleion, 31, 32, 39, 41
 NF-H phosphorylation, 33–39
 p24/44 MAPK modulation, 33, 36
  regulation, 30–33
 site-directed NF-H mutagenesis, 33
 tail-less NF-H, 31
NF transport modeling, 34, 35, 40
 phospho-dependent dynamics, 36, 40
 phospho NF dynamics, 30, 31
 stationary phase, 32, 35
NGF. *See* Nerve growth factor
NGF withdrawal, 148, 153, 156
 c-jun nuclear accumulation, 148, 153–156
NMDA receptor, 131, 132
NMJ. *See* Neuromuscular junction
Nodal assembly, 16, 18
 CNS intrinsic pattern, 17, 18
 gliomedin, 17, 18
 perinodal astrocytes, 18
 PNS extrinsic pattern, 16–18
Nodes of Ranvier, 2–7, 71, 182, 183
 RNA labeling, 182
 voltage-gated sodium channels, 4
Nogo, 342, 343
 Amino-Nogo, 343
 Nogo-66, 342–344
 Nogo-A, 342–344
 Nogo-B, 343
 Nogo-C, 343
Nogo-66 receptor, 342
Norepinephrine (NE), 234, 236–238
NrCAM, 5–7, 16, 17
Nuclear import, 330, 331
 RanGTP/RanGDP regulation, 331

O

OIP106. *See* Milton
Olfactory axons, 196
 odorant receptor mRNAs, 196
OMgp, 342, 344
Oxidative phosphorylation, 233

P

Par-3, 15–16
 Drosophila homolog, 15
Paranode, 1, 3–14, 16–20
 paranodal loops, 7, 10–13

Parkinson's disease (PD), 60
  parkin, 60
  parkin mutants, 60
PARP domains, 178–189, 200
  β-actin mRNA, 200
  F-actin distribution, 180
  mammalian axons, 182–185
  Mauthner axon, EM level, 180–182
  Mauthner axon, LM level, 178–180
  sizes in rabbit axons, 182
  YOYO-1 RNA fluorescence, 178
PARP markers, 179, 183, 188
  β-actin mRNA, 187
  kinesin II, 185
  myosin Va, 185
  ZBP-1, 188
PARP matrix, 182
  matrix hypothesis, 189
  rabbit axons, 184
  ribosome binding, Mauthner, 180
  ribosome binding, rabbit, 185
PARPs. *See* Periaxoplasmic ribosomal plaques
P-bodies, 206, 208–209
  RNA degradation, 208
PC12 cell line, 150
Periaxoplasmic ribosomal plaques. *See* PARP domains
Periaxoplasmic ribosomal plaques (PARPs), 67–68, 71
Peripheral nervous system (PNS), 339–347
  environment, 339
p150Glued, 116, 123, 124
Plastins, 297
Pleckstrin, 130
p75 neurotrophin receptor, 155
PNS. *See* Peripheral nervous system
Pointed end, 66, 67, 74
Polymerase chain reaction (PCR), 196, 209
Polysomes, 227, 240
Positive injury signals, 328, 330–332
  activated proteins with NLS signals, 330
  importins α/β transport, 330–331
  local importin β1 synthesis, 330
  local RanBP1 synthesis, 331
  signaling molecules, 333
Presynaptic nerve terminal, 226, 227, 234, 238
Presynaptic RNA, 247, 256–259
  local TH mRNA synthesis, 256, 259
Presynaptic terminal, 68, 72
Profilin, 292, 296
Properties, 66–67, 80, 85
Protein synthesis, 174, 176, 180, 189
  dependence on F-actin, 180

R

RAGs. *See* Regeneration-associated genes
Regeneration-associated genes, 340, 341, 346
Regeneration modulation, 330
  mTOR stimulation of translation, 330
RER and Golgi, 211, 212
  axonal equivalents, 211
Retrograde apoptotic signal, 153–156
Retrograde flow, 74, 76, 78
Retrograde injury complex, 332
  importin-dependent vimentin-Erk transport, 332
  Jnk-Syn-dynactin, 333
Retrograde signaling, 141–157
  endosome hypothesis, 146, 152
  significance, 147, 152
Retrograde signaling complex, 212, 215
Retrograde transport, 143, 146–154, 156
  [125II]NGF rat sympathetic neurons, 146, 147
  phosphorylated TrkA, 146, 147, 152–153
RhoA, 202, 204, 209, 273, 275, 280, 281
Rho GTPases, 72, 78
Ribonucleoprotein particles (mRNPs), 239, 240
Ribosomal RNA, 174–176
  Mauthner and squid axons, 175
Ribosomes, 173, 174, 176–178, 180–187, 189
  EM detection, 176
  Schwann cell-axon transcytosis, 185–186
RNA, 174–180, 182–189
  disproportional 4S, 175, 189
  early detection, 174
  Mauthner myelin sheath, 175
  tagging, 202, 203
RNA binding proteins (RNBPs), 201, 202, 206, 215, 273, 275–277, 284
  Cop1b, 206, 211, 212
  CPEB, 208
  Elav protein, 206
  functional roles, 275
  growth cones, 275, 277
  nuclear function, 276
  RanBP1, 202
  RNP transport particles, 273, 274
  transport complexes, 273, 274, 276
  ZBP1, 206, 208
  ZBP2, 206
  ZBP1 ortholog, 201
RNAi, 204, 208, 209
RNA-induced silencing complex (RISC), 233
RNA interference, 238. *See* miRNA
RNA localization, 194, 199, 201–203, 207, 271, 273, 276
  neurons, dendrites, 271–275

Index   359

RNBPs. *See* RNA binding proteins
RNP, 200, 201, 206–209, 211–213, 215, 216
RNP transport, 206–209, 211, 215, 216
    associated components, 207, 211
    live cell imaging, 207

S

Schwann cell microvilli, 6, 11, 18
SCI. *See* Spinal cord injury
Semaphorin3A (Sema3A), 277, 278, 280–283
Semaphorins, 209, 210
    semaphorin 3A growth inhibiting, 209, 210
Sensory neurons, 197, 198, 202, 207, 209, 210, 212, 213
Septate junctions, 7–10, 12
    axo-glial junctions, 9
    Drosophila, 8–10, 12, 15
    epithelial cell, 8, 9, 15
Shaker-like potassium channels, 11, 14
Short inhibitory RNAs (siRNA), 237, 238
Signal recognition particle, 176, 183–184
    axon compartment, 184
    SRP54, rabbit axons, 184
Slit, 269, 278–282
Slow component-b (Scb), 68, 72
SMA. *See* Spinal muscular atrophy
Small nuclear ribonucleoprotein (snRNP), 290, 291, 293, 298, 304, 305, 307, 315
Sm/LSm core proteins, 293
SMN. *See* Survival of motor neuron
Spectrin, 67, 69, 71, 74, 78, 84
    αII Spectrin, 5–6
    βIV Spectrin, 5–6, 16
Spinal cord injury, 339, 342–347
Spinal cord transection, Mauthner fiber RNA, 174
Spinal muscular atrophy, 199
Spinal muscular atrophy (SMA), 289–316
    animal models, 298, 304–312, 315
    classification, 289
    Danio rerio, 306–307
SPRR1A, 340, 341
Squid
    giant axon, 226, 228, 230, 240
    photoreceptor neurons, 226, 227
Squid axon, 247–249, 255, 257
    internal perfusion, 255, 257
    lmRNA sequence complexity, 248
    local RNA synthesis, 255, 257, 259
    mRNAs, 248
    pharmacologic modulation of glial RNA transfer, 253
    protein synthesis, 247–249, 257

Squid giant axon, 194, 195
    β-actin mRNA, 195
    β-tubulin mRNA, 195
    tRNAs, 204
Squid photoreceptor terminals, 249–251
    mRNAs, 249–251
    nuclear encoded mitochondrial proteins, 249, 250
    RNA synthesis, 256
SRP. *See* Signal recognition particle
Stem cells, 313–314
Stress granules (SGs), 206, 208, 209, 302
Subaxolemmal space, 68–69
Superior cervical ganglia (SCG), 228–237
Survival of motor neuron (SMN), 289–316
    axonal SMN (a-Smn), 298
    domains, 291–297, 300–302, 307
    interacting proteins, 291–297, 301, 315
    knockout, 308
    localization in axons, 297–303
    mutations, 290, 292, 293, 298, 300, 305–307, 315
    oligomerization, 292, 293, 307
    SMNΔ7 transgene, 310
    SMN2 transgene, 308–310
Sympathetic neurons, 196, 197, 204, 228, 233, 236–238
Synaptic plasticity, 72, 78, 194, 204
    localized protein synthesis, 194
Synaptogenesis, 214
Synaptosomes, 227, 228, 230, 245–247, 249, 251, 256, 257, 259
    expression in optic lobe, 246, 256
    mitochondrial proteins, 246, 247, 249
    mitochondrial RNA synthesis, 257
    poly(A)⁺RNA, 257, 259
    protein synthesis, 245–247, 257
    squid optic lobe terminals, 246, 256
Syntabulin, 114, 118, 129
Syntaphilin, 114, 128–129
α-Synuclein, 159–169
    accumulation, 160, 164
    biochemical characteristics, 162
    biogenesis, 165–169
    in disease pathogenesis, 160
    fibrillar structures, 159
    LB and LN fibrils, 162, 164
    mislocalization/accummulation, 160, 164–165
    mutant forms, 161
    targeting mechanisms, 168–169
    transport of mutants, 167
Synucleinopathies, 159, 160
    definition, 159

T

TAG-1, 14, 15, 20
Tau, 49, 51, 52, 56–58, 200, 202, 206
  expression and isoform ratios, 57
  functions, 57
  MAP protection hypothesis, 58
  molecular alterations, 56
  mutation-based diseases, 56
Tauopathies, 56–58
  abnormal microtubule-severing, 57, 58
  microtubule-severing, 57, 58
Therapy, 313, 314, 316
TOM70 receptor, 230–232
Trailer hitch, 211–212
TRAK2. *See* Milton
Trans-acting mRNA binding proteins. *See* RNA binding proteins (RNBPs)
Transcript profiling, 197, 198
Transport RNP, 206–209, 211, 215, 216
Transport RNP granules, 252
TrkA receptor, 130, 142, 143, 147–150
Trophic capacity, 244, 258, 259
  conditions for peripheral independence, 258
  neuron soma, 244, 258, 259
Tudor domain, 291–293, 300
Tug-of-war, 120–123
Tyrosine hydroxylase (TH), 245, 246, 259
  expression in terminals, 246, 259

U

3' Untranslated region (3'UTR), 233

V

Vertebrate axons, 251, 259
  local RNA synthesis, 259
  Schwann cell-axon transfer of ribosomes, 255, 256
Vg1RBP, 277, 279, 280
  actin dependent transport, 277, 280
  growth cone, 277, 279, 280
  netrin-1 stimulated transport, 277, 280

W

Wallerian degeneration (Wld^s), 312
WAVE1, 117, 131, 132
Wld^s mouse, 185

Z

ZBP1. *See* Zipcode binding protiein-1
ZBP2. *See* KSRP; Zipcode binding protiein-2
Zebrafish. *See Danio rerio*
Zipcode-binding protein, 67, 68
Zipcode binding protein-2 (ZBP2), 276
  nuclear β-actin mRNA binding, 276
Zipcode binding protiein-1 (ZBP1), 275–277, 280, 299, 300
  β-actin mRNA binding, 275, 276, 280